Nanotheranostics in Metabolic Syndrome

Vivek K. Chaturvedi • Anurag Kumar Singh
Jay Singh • Dawesh P. Yadav
Editors

Nanotheranostics in Metabolic Syndrome

Transforming Diagnosis and Therapeutics

Springer

Editors

Vivek K. Chaturvedi
Department of Gastroenterology
Banaras Hindu University
Varanasi, Uttar Pradesh, India

Anurag Kumar Singh
Department of Pharmaceutical Engineering and Technology
Indian Institute of Technology (Banaras Hindu University)
Varanasi, Uttar Pradesh, India

Jay Singh
Department of Chemistry
Banaras Hindu University
Varanasi, Uttar Pradesh, India

Dawesh P. Yadav
Department of Gastroenterology
Banaras Hindu University
Varanasi, Uttar Pradesh, India

ISBN 978-981-95-3745-7 ISBN 978-981-95-3746-4 (eBook)
https://doi.org/10.1007/978-981-95-3746-4

This Springer imprint is published by the registered company Springer Nature Singapore Pte Ltd.
The registered company address is: 152 Beach Road, #21-01/04 Gateway East, Singapore 189721, Singapore

This book, Nanotheranostics in Metabolic Syndrome, *is respectfully dedicated to Bharat Ratna* ***Pt. Madan Mohan Malaviya Ji****, the visionary founder of Banaras Hindu University. His steadfast dedication to education, innovation, and the quest for knowledge has inspired countless scholars and researchers. It is his pioneering vision that established a world-class institution, nurturing excellence in science, technology, and the humanities. May this work, born from the very university he created, serve as a tribute to his enduring legacy and his commitment to societal progress through education and research.*
With profound respect and gratitude.

Vivek K. Chaturvedi
Anurag Kumar Singh
Jay Singh
Dawesh P. Yadav

Preface

One of the primary advantages of nanotechnology in the context of metabolic syndrome is its capacity to enhance drug delivery systems. The nanoscale formulations can navigate biological barriers more effectively, reaching their designated sites with precision. This targeted drug delivery not only improves the bioavailability of therapeutic agents but also minimizes off-target effects, thereby enhancing the overall effectiveness of treatment. Moreover, nanotechnology is pivotal in early disease detection and diagnosis through innovative nano-imaging techniques. These cutting-edge approaches enable the visualization of molecular and cellular events at unprecedented resolutions, facilitating the identification of metabolic syndrome markers and associated disorders at their incipient stages. Early detection is paramount in initiating timely interventions and preventing the progression of these conditions to more advanced and challenging states. The versatility of nanotechnology extends to addressing the inflammatory component of metabolic syndrome. Nanoparticles can be engineered to carry anti-inflammatory agents, providing targeted relief at sites of inflammation. This targeted approach mitigates inflammation and reduces the systemic side effects associated with conventional anti-inflammatory treatments. Furthermore, nanotechnology in managing metabolic syndrome holds promise for disorders like diabetes and obesity. By leveraging nanoscaled therapeutics, it becomes possible to modulate metabolic processes with enhanced precision. This opens avenues for developing personalized treatment regimens tailored to the specific needs of individuals, potentially revolutionizing the approach to metabolic disorders. In summary, nanotechnology is a transformative tool in managing metabolic syndrome and its associated disorders. Through targeted drug delivery, early disease detection, and precision therapeutics, nanotechnology addresses the limitations of current therapies. By increasing bioavailability, permeability, and treatment effectiveness while reducing adverse effects, nanotechnology paves the way for a new era in healthcare—one where the challenges posed by metabolic syndrome can be met with innovative and tailored solutions. As research in this field progresses, the integration of nanotechnology holds the potential to redefine the landscape of metabolic syndrome management, offering hope for improved patient outcomes and a healthier global population. This book, *Nanotheranostics in Metabolic Syndrome: Transforming Diagnosis and Therapeutic*, aims to attract scientists and researchers in metabolic syndrome, drug delivery, biosensors, and nanomaterials, covering the fundamental aspects of the syndrome, including insulin resistance, hypertension, and obesity. It compiles up-to-date literature on its mechanisms and therapies, emphasizing the potential of nanotechnology as a transformative biomedical tool. Additionally, it serves as a valuable resource for UG/PG students, clinicians, and professionals in advanced materials science, flexible electronics, and biomedical sciences, highlighting innovative solutions for improved patient outcomes and a healthier population.

Varanasi, Uttar Pradesh, India

Vivek K. Chaturvedi
Anurag Kumar Singh
Jay Singh
Dawesh P. Yadav

Acknowledgments

Vivek K. Chaturvedi sincerely acknowledges the Department of Health Research (DHR), Ministry of Health and Family Welfare (MoHFW), Government of India, for their support through the Young Scientist Grant (R.12014/56/2022-HR). Anurag Kumar Singh extends his gratitude to the Indian Council of Medical Research (ICMR), Government of India, for support through the Research Associate (ICMR-RA) Award (No: 3/1/2/110Neuro/2019-NCD-I). Jay Singh expresses his appreciation to the Institutes of Eminence (IoE)-BHU Grant, Ministry of Education, India, for their unwavering support. Dawesh P. Yadav gratefully acknowledges funding from the IOE-BHU (SEED grant & Credit Research Incentive Grant), and Indian Council of Medical Research (ICMR), New Delhi, for their Multicentric Grants (EMDR/CARE/12/2023-0000572; DDR/IntermediateGrant/MCH/HRPIG-2023-0000769). We extend our heartfelt gratitude to all contributors whose quality material has enriched this book. We are deeply thankful to our beloved family members, friends, and colleagues for their joyful support, encouragement, and assistance throughout the writing process. Finally, we acknowledge the Springer publishing team for their professionalism and expertise in design and production, which have been crucial in bringing this book, *Nanotheranostics in Metabolic Syndrome: Transforming Diagnosis and Therapeutics*, to completion.

Vivek K. Chaturvedi
Anurag Kumar Singh
Jay Singh
Dawesh P. Yadav

Contents

Editors and Contributors

About the Editors

Vivek K. Chaturvedi is a DHR-Young Scientist (Department of Health Research, Ministry of Health and Family Welfare) at Institute of Medical Sciences, Banaras Hindu University, Varanasi, India. Prior to this role, he worked as a postdoctoral research associate in the Department of Gastroenterology at IMS-BHU, Varanasi. He earned his PhD in Biotechnology from the University of Allahabad (a central university), Prayagraj, India. He also obtained his BSc in Zoology and Chemistry, as well as his MSc in Biochemistry, from Veer Bahadur Singh Purvanchal University, Jaunpur, India. His research focuses on Nanobiotechnology in Cancer Biology and Inflammatory Bowel Diseases. He has voluntarily served as a reviewer and editorial board member, and has published numerous original articles in peer-reviewed journals, along with several international books.

Anurag Kumar Singh is currently working as a Research Associate (ICMR-RA) in the Department of Pharmaceutical Engineering and Technology at Indian Institute of Technology (Banaras Hindu University), Varanasi, India. Previously, he served as a postdoctoral research assistant in Cancer Biology at the Department of Biological Sciences, Alabama State University, USA, and at the Institute of Nanotechnology & Advanced Materials, Department of Chemistry, Bar-Ilan University, Israel. He brings over 2.5 years of postdoctoral research experience to his current role. He earned his doctoral degree in pharmaceutical chemistry from the

Centre of Experimental Medicine and Surgery, Institute of Medical Sciences, Banaras Hindu University, Varanasi, and his MPharm degree from the School of Chemical Sciences and Pharmacy, Central University of Rajasthan, Ajmer, India. His research focuses on the development of innovative nano-particulate systems for applications in chronic pain management, neurodegenerative disorders, and brain-targeted drug delivery. This includes expertise in dendrimers, nanoporous silica/silicon materials, and polymeric nanoparticles for controlled drug delivery, diagnostics, and nanotheranostics. His work aims to integrate engineering, pharmaceutical, and medical sciences to advance nanomedicine for healthcare and biomedical applications. He has an impressive academic record with over 50 research articles, reviews, and journal editorials published in esteemed national and international journals. He has edited or authored more than 5 books and contributed over 15 book chapters published by internationally renowned presses.

Jay Singh is an Assistant Professor in the Department of Chemistry at the Institute of Sciences, Banaras Hindu University, Varanasi, Uttar Pradesh, a position he has held since 2017. He earned his PhD in Polymer Science from Motilal Nehru National Institute of Technology in 2010, and he completed both his MSc and BSc at Allahabad University, Uttar Pradesh, India. His postdoctoral research took him to the National Physical Laboratory in New Delhi, Chonbuk National University in South Korea, and Delhi Technological University. He has been awarded several prestigious fellowships, including the CSIR Research Associate, DST Young Scientist Fellowship, and the DST-INSPIRE Faculty Award. His research focuses on the development of sustainable nanomaterials, including nanobiocomposites, conducting polymers, and self-assembled monolayers for clinically significant biosensors. These biosensors are designed for the estimation of bioanalytes such as cholesterol, xanthine, glucose, pathogens, and pesticides/toxins using DNA and antibodies. He has published over 172 international research papers, which have garnered more than 6200 citations and an h-index of 44. He is also involved in multiple research projects funded by various agencies and has edited more than 21 books and authored over 65 book chapters with renowned publishers such as Elsevier, Springer Nature, IOP, Wiley, and CRC. His work includes the fabrication of metal oxide-based biosensors for clinical diagnostics, food packaging, drug delivery, and tissue engineering, significantly advancing our understanding of interfacial charge transfer processes and the sensing capabilities of metal nanoparticles.

Dawesh P. Yadav is currently an Associate Professor and Head in the Department of Gastroenterology at Banaras Hindu University in Varanasi, India. He earned his MBBS and MD in General Medicine from the Institute of Medical Sciences (IMS) at Banaras Hindu University. He is a member of both the Indian National Association for the Study of the Liver (INASL) and the Indian Society of Gastroenterology. He has received numerous prestigious awards for his contributions to the field. Before his current role, he worked as a Senior Resident at the All-India Institute of Medical Sciences (AIIMS) in New Delhi from 2013 to 2017. He joined IMS, BHU, as an Assistant Professor in 2017 and was promoted to Associate Professor in the Department of Gastroenterology in 2019. His research focuses on Inflammatory Bowel Disease, Liver Cirrhosis, and Nanobiotechnology. He has published over 50 articles in peer-reviewed, high-impact journals and contributed several book chapters, in addition to editing multiple international books. He also serves on the editorial boards of several reputable journals.

Contributors

Amitabh Aharwar Government College Harrai, Chhindwara, Madhya Pradesh, India

Ankur Phytomedicine Research Laboratory, Department of Pharmaceutical Engineering & Technology, IIT (BHU), Varanasi, India

Rachayeeta Bera NSHM College of Pharmaceutical Technology, NSHM Knowledge Campus, Kolkata-Group of Institutions, Kolkata, West Bengal, India

Sudarsan Biswal Directorate of Drugs Control Administration, Health and Family Welfare Department, Bhubaneswar, Government of Odisha, Bhubaneswar, Odisha, India

Vivek K. Chaturvedi Department of Gastroenterology, Banaras Hindu University, Varanasi, Uttar Pradesh, India

Gyanada Sankar Dash KIIT School of Pharmacy, KIMS, KIIT Deemed to University, Patia, Bhubaneswar, Odisha, India

Pooja Dubey Department of Anatomy, Institute of Medical Science, Banaras Hindu University, Varanasi, Uttar Pradesh, India

Sunil Dutt Maharaja Agrasen School of Pharmacy, Maharaja Agrasen University, Solan, Himachal Pradesh, India

Muhammad Delwar Hussain Department of Pharmaceutical Sciences, School of Pharmacy and Health Professions, University of Maryland Eastern Shore, Princess Anne, MD, USA

Komal Kumar Ingle Algology Laboratory, CSIR-National Botanical Research Institute, Lucknow, Uttar Pradesh, India

Srishti Jain Department of Anatomy, All India Institute of Medical Sciences, New Delhi, India

Gautam Kumar Jena Department of Pharmaceutics, Roland Institute of Pharmaceutical Sciences, Berhampur, Odisha, India

Pankaj Kalia Maharaja Agrasen School of Pharmacy, Maharaja Agrasen University, Solan, Himachal Pradesh, India

Pooja Kathait Phytomedicine Research Laboratory, Department of Pharmaceutical Engineering & Technology, IIT (BHU), Varanasi, India

Nisha Kharka Algology Laboratory, CSIR-National Botanical Research Institute, Lucknow, Uttar Pradesh, India
Academy of Scientific and Innovative Research, Ghaziabad, Uttar Pradesh, India

Ashutosh Kumar Department of Ophthalmology, University of California Los Angeles, Los Angeles, CA, USA

Brijesh Kumar Department of Pharmacology, Institute of Medical Sciences, Banaras Hindu University, Varanasi, India

Devinder Kumar Maharaja Agrasen School of Pharmacy, Maharaja Agrasen University, Solan, Himachal Pradesh, India

Amar Shaheed Baba Ajit Singh Jujhar Singh Memorial College of Pharmacy (An Autonomous College), Bela, Ropar, Punjab, India

Rajesh Kumar Maharaja Agrasen School of Pharmacy, Maharaja Agrasen University, Solan, Himachal Pradesh, India

Vijay Kumar Algology Laboratory, CSIR-National Botanical Research Institute, Lucknow, Uttar Pradesh, India

Sutapa Biswas Majee NSHM College of Pharmaceutical Technology, NSHM Knowledge Campus, Kolkata-Group of Institutions, Kolkata, West Bengal, India

Soki Daeme Malang Phytomedicine Research Laboratory, Department of Pharmaceutical Engineering & Technology, IIT (BHU), Varanasi, India

Debadatta Mohapatra Phytomedicine Research Laboratory, Department of Pharmaceutical Engineering & Technology, IIT (BHU), Varanasi, India

Leslie Dan Faculty of Pharmacy, University of Toronto, Toronto, ON, Canada

Renu Negi Department of Pharmacology and Experimental Therapeutics, Boston University School of Medicine, Boston, MA, USA

Sunday O. Otimenyin Department of Pharmacology, Faculty of Pharmaceutical Sciences, University of Jos, Jos, Plateau, Nigeria

Niranjan Panda School of Pharmacy, The Neotia University, Sarisa, West Bengal, India

D. K. Parihar Department of Biotechnology, Guru Ghasidas University, Bilaspur, Chhattisgarh, India

Pradeep Kumar Patel Phytomedicine Research Laboratory, Department of Pharmaceutical Engineering & Technology, IIT (BHU), Varanasi, India

Parameswar Patra School of Pharmacy, The Neotia University, Sarisa, West Bengal, India

Vipin Rai Rutgers University, Newark, NJ, USA

Sachitra Kumar Ratha Algology Laboratory, CSIR-National Botanical Research Institute, Lucknow, Uttar Pradesh, India

Academy of Scientific and Innovative Research, Ghaziabad, Uttar Pradesh, India

Hassan Raza NSHM College of Pharmaceutical Technology, NSHM Knowledge Campus, Kolkata-Group of Institutions, Kolkata, West Bengal, India

Rishika Munjal AIIMS, New Delhi, India

Smriti Sachan Department of Gastroenterology, Banaras Hindu University, Varanasi, India

Rakesh Kumar Sahoo Department of Ophthalmology and Visual Sciences, University of Illinois, Chicago, IL, USA

Alakh N. Sahu Phytomedicine Research Laboratory, Department of Pharmaceutical Engineering & Technology, IIT (BHU), Varanasi, India

Gitali Sen Phytomedicine Research Laboratory, Department of Pharmaceutical Engineering & Technology, IIT (BHU), Varanasi, India

Prakash Ch. Senapati School of Pharmaceutical Sciences-II, Phulnakhara Campus, Siksha 'O' Anusandhan (Deemed to be University), Bhubaneswar, Odisha, India

Abhinav Sharma Algology Laboratory, CSIR-National Botanical Research Institute, Lucknow, Uttar Pradesh, India

Bhaskar Sharma Department of Anatomy, All India Institute of Medical Sciences, New Delhi, India

Padma Sharma Department of Environment Studies, Panjab University, Chandigarh, India

Singh Shreya SBS College of Pharmacy, Fatehpur, India

Phytomedicine Research Laboratory, Department of Pharmaceutical Engineering & Technology, IIT (BHU), Varanasi, India

Anurag Kumar Singh Department of Pharmaceutical Engineering and Technology, Indian Institute of Technology (Banaras Hindu University), Varanasi, Uttar Pradesh, India

Aprajita Singh Department of Microbiology, Institute of Medical Sciences (Banaras Hindu University), Varanasi, Uttar Pradesh, India

Jaikee Kumar Singh Department of Bioinformatics, Faculty of Engineering and Technology, Marwadi University, Rajkot, Gujarat, India

Jay Singh Department of Chemistry, Banaras Hindu University, Varanasi, India

Kshitij R. B. Singh Graduate School of Life Science and Systems Engineering, Kyushu Institute of Technology, Wakamatsu, Kitakyushu, Japan

Nitish Kumar Singh Multidisciplinary Research Unit, ICMR-DHR, Institute of Medical Sciences, Banaras Hindu University, Varanasi, Uttar Pradesh, India

Department of Biosciences Manipal University, Jaipur, Rajasthan, India

Royana Singh Department of Anatomy, Institute of Medical Science, Banaras Hindu University, Varanasi, Uttar Pradesh, India

Surbhi Singh Department of Anatomy, Institute of Medical Science, Banaras Hindu University, Varanasi, Uttar Pradesh, India

Pooja Thakur Maharaja Agrasen School of Pharmacy, Maharaja Agrasen University, Solan, Himachal Pradesh, India

Kiran Toppo Algology Laboratory, CSIR-National Botanical Research Institute, Lucknow, Uttar Pradesh, India

Vivek Chandra Verma Department of Biochemistry, Punjab University, Chandigarh, India

Arshad Warsi Gomal Center of Biochemistry and Biotechnology (GCBB), Gomal University, Dera Ismail Khan, Pakistan

Anjali Yadav Department of Anatomy, All India Institute of Medical Sciences, New Delhi, India

Dawesh P. Yadav Department of Gastroenterology, Banaras Hindu University, Varanasi, India

Navneet Yadav Department of Mechanical Engineering, Faculty of Science and Engineering, Swansea University, Sketty, Swansea, UK

Ravi Kumar Yadav Department of Botany, Kashi Naresh Government Post Graduate College, Gyanpur, Bhadohi, Uttar Pradesh, India

Introduction to Nanotheranostics and Metabolic Syndrome

1

Smriti Sachan, Vivek K. Chaturvedi, Anurag Kumar Singh, Kshitij R. B. Singh, Jay Singh, and Dawesh P. Yadav

Abstract

Central obesity, insulin resistance, hypertension, dyslipidemia, and hyperglycemia are among the metabolic abnormalities that make up the multifactorial health condition known as metabolic syndrome (MetS). It presents serious cardiovascular and neurodegenerative risks, contributing to an increasing global burden. Because of the syndrome's complex pathophysiology and diverse etiology, current diagnostic and treatment methods frequently fall short. A new, integrated approach to real-time disease monitoring and focused intervention is provided by nanotheranostics, a combination of diagnostics and therapeutics employing nanoscale platforms. A thorough review of MetS, including its molecular foundations, diagnostic standards, and clinical stages, is given in this chapter. It explores the rapidly developing field of nanotheranostics, highlighting its therapeutic (drug delivery, gene therapy) and diagnostic (biosensors, imaging) uses. Advanced nanomaterials, present issues (biocompatibility, scalability, regulations), and new future directions (personalized nanomedicine, AI integration) are also covered. There are encouraging opportunities for precision healthcare in the treatment of MetS due to the convergence of metabolic science and nanotechnology.

S. Sachan · V. K. Chaturvedi · D. P. Yadav (✉)
Department of Gastroenterology, Banaras Hindu University, Varanasi, India

A. K. Singh
Department of Pharmaceutical Engineering and Technology, Indian Institute of Technology (Banaras Hindu University), Varanasi, Uttar Pradesh, India

K. R. B. Singh
Graduate School of Life Science and Systems Engineering, Kyushu Institute of Technology, Wakamatsu, Kitakyushu, Japan

J. Singh
Department of Chemistry, Banaras Hindu University, Varanasi, India

V. K. Chaturvedi et al. (eds.), *Nanotheranostics in Metabolic Syndrome*,
https://doi.org/10.1007/978-981-95-3746-4_1

Keywords

Metabolic syndrome · Nanotheranostics · Insulin resistance · Targeted drug delivery · Biosensors · Oxidative stress · Epigenetics · Lipid nanoparticles · Gene therapy · Personalized medicine · Artificial intelligence · Inflammation · Obesity · Dyslipidemia

1 Introduction

Metabolic syndrome (MetS) is a multifaceted clinical syndrome with a cluster of interrelated metabolic disturbances such as abdominal fatness, hypertension, impaired glucose metabolism, insulin resistance, and dyslipidemia. It is also linked with an increased risk of cardiovascular disease, chronic kidney disease, hepatic steatosis, and chronic low-grade inflammation [3]. At the cellular level, MetS originates from dysregulated metabolic pathways such as glycolysis, the tricarboxylic acid (TCA) cycle, and lipid metabolism, which play a significant role in the regulation of energy balance and cellular homeostasis. Evidence to date suggests that over two-thirds of adults in Western populations are overweight or obese with nearly one-quarter of the world's adult population having MetS. The syndrome is also rising very fast among children and adolescents in urban populations primarily due to physical inactivity and diet [5].

Despite being common and with serious consequences on health, MetS is difficult to diagnose and treat due to its multifactorial etiology, overlapping diagnostic criteria, and multifactorial pathophysiology. Several international guidelines (e.g., ATP III, IDF, and AHA/NHLBI) yield disparate prevalence rates, complicating global surveillance and treatment [1]. Moreover, conventional therapeutic measures like lifestyle modification, diet, and pharmacotherapy are usually insufficient to reverse the underlying metabolic derangements. The chronic state of low-grade inflammation, adipose tissue dysfunction, and new mechanisms like ferroptosis perpetuate the disease cascade but are yet to be investigated and applied in the clinical setting. These limitations highlight the urgent need for new diagnostic modalities and tailored treatment strategies [11].

The last decade has seen the advent of nanotechnology and digital health platforms driving the revolution in disease management. One of the most promising of these is nanotheranostics—hybrid technology that binds therapeutic and diagnostic functionality in a single nanoscale platform. Nanotheranostic platforms have the potential for real-time monitoring, targeted drug delivery, and minimally invasive diagnosis and therefore hold particular interest in the fight against heterogeneity and systemic nature of MetS. Alongside, technological innovation is revolutionizing disease monitoring and management [2, 14]. The advent of electronic health tools, wearable sensors, and mobile health platforms is facilitating a more proactive and personalized approach to patient care. Theranostic approaches—blending diagnostic and therapeutic potential—hold new hope for continuous monitoring and timely intervention of metabolic disorders [10].

This book chapter attempts to provide a state-of-the-art overview of the molecular and clinical pathophysiology of MetS, with particular emphasis on the new role of natural bioactive compounds and integrative treatments. By describing some of the existing knowledge gaps and through a critical overview of existing and future therapeutic strategies, we hope to orient future research and lead the way toward more effective, more integrated therapies for the prevention and treatment of MetS.

2 Metabolic Syndrome: An Overview

2.1 Definition and Components

The term metabolic syndrome (MetS) describes a group of connected metabolic disorders that dramatically increase the risk of type 2 diabetes mellitus (T2DM), cardiovascular disease (CVD), and other complications [1, 7, 14, 15]. Among the essential elements are:

1. *Obesity (especially central/visceral)*: In addition to serving as an energy store, excess adipose tissue secretes adipokines like resistin and leptin, which act as an active endocrine organ that affects insulin resistance and systemic inflammation.
2. *Diabetes and Hyperglycemia*: The primary diagnostic components of MetS include insulin resistance, impaired fasting glucose, or full-blown type 2 diabetes. These are caused by β-cell dysfunction and compromised insulin signaling.
3. *Hypertension*: High blood pressure is a recurring feature in the diagnostic criteria for MetS and is frequently associated with obesity and endothelial dysfunction.
4. *Dyslipidemia*: Usually consists of low HDL cholesterol and elevated triglyceride levels, which raise the risk of CVD and atherosclerosis.
5. *Inflammation*. Inflammation: A key factor in the pathophysiology of MetS is chronic low-grade inflammation, which involves cytokines such as TNF-α, IL-6, and CRP.
6. *Vascular and Neurological Complications*: Through oxidative and inflammatory processes, MetS is linked to endothelial dysfunction and is becoming more and more connected to neurodegenerative diseases.

2.2 Causative Factors

Several interlinked lifestyles, genetic, and environmental elements contribute to MetS:

2.2.1 Lifestyle Factors

One of the main causes of metabolic syndrome is lifestyle factors. The increasing incidence of MetS worldwide is caused by a shift toward sedentary behavior, poor eating habits, and physical inactivity [1, 14].

Table 1.1 Lifestyle factors contributing to metabolic syndrome

Factor	Mechanism/Impact
Sedentary lifestyle	Reduces glucose uptake, promotes fat accumulation and insulin resistance
High-calorie diet	Especially high in sugar and fat; increases visceral fat and lipotoxicity
Poor sleep	Affects cortisol and insulin regulation, increasing appetite and glucose levels
Alcohol consumption	Leads to hepatic steatosis, oxidative stress, and impaired metabolism
Smoking	Induces inflammation and vascular dysfunction

1. *Sedentary Lifestyle*: Lack of exercise causes weight gain, decreased insulin sensitivity, and changes in lipid metabolism. It promotes insulin resistance and adiposity by reducing skeletal muscle's capacity to oxidize fatty acids and absorb glucose.
2. *High-Calorie Diets*: diets high in refined carbohydrates, especially fructose and sucrose, saturated fats, and beverages with added sugar encourage.
3. *Circadian Disruption and Sleep Disturbance*: Inconsistent sleep-wake cycles or poor sleep quality affect hormonal regulation (cortisol, insulin, and leptin), and they are associated with increased appetite, obesity, and glucose intolerance.
4. *Smoking and Alcohol Consumption*: Smoking causes inflammation and oxidative stress, which exacerbates vascular dysfunction and insulin resistance, while excessive alcohol consumption encourages the buildup of hepatic fat (Table 1.1).

2.2.2 Environmental Factors

Metabolic risk is greatly influenced by environmental exposures, such as pollution, urbanization, and socioeconomic status:

1. *Urbanization and the Environment of Processed Food*: Increased consumption of processed, high-calorie, low-nutrient foods has resulted from the global shift to urban living, which has raised the prevalence of obesity and MetS.
2. *Endocrine-Disrupting Chemicals (EDCs)*: Exposure to pollutants such as phthalates, bisphenol A (BPA), and persistent organic pollutants (POPs) disrupts hormone pathways, causing insulin signaling to be impaired and adipogenesis to be promoted.
3. *Socioeconomic Stressors*: Prolonged cortisol elevation brought on by chronic psychosocial stress encourages central adiposity, hyperglycemia, and hypertension—all of which are important elements of MetS (Table 1.2).

2.2.3 Genetic Factors

A person's susceptibility to MetS is greatly influenced by their genetic predisposition, even though lifestyle and environment play a crucial role:

Table 1.2 Environmental contributors

Environmental factor	Effects on metabolism
Urbanization and processed food	Increases availability of unhealthy food, decreases physical activity
Pollutants/EDCs (e.g., BPA)	Disrupt hormone signaling; promote adipogenesis and insulin resistance
Socioeconomic stress	Elevates cortisol chronically, increases central adiposity and glucose dysregulation

Table 1.3 Genetic factors in metabolic syndrome

Gene pathway affected	Associated effects
Insulin receptor signaling	Reduced insulin sensitivity and glucose uptake
Adipokine secretion (e.g., leptin, adiponectin)	Alters inflammatory state, energy homeostasis
Appetite and metabolism genes	Increases risk of obesity and energy imbalance
FTO, TCF7L2, PPARG	Associated with T2DM and obesity susceptibility

1. *Genes Related to Obesity*: Variants in genes that control insulin signaling, lipid metabolism, appetite, and adipogenesis (such as FTO, TCF7L2, and PPARG) are linked to an increased risk of type 2 diabetes and obesity.
2. *Insulin Signaling Pathways*: Individuals are at risk for insulin resistance and diabetes due to impaired glucose homeostasis caused by mutations or polymorphisms in genes related to insulin receptor function or glucose transport.
3. *Adipokine Regulation*: Insulin sensitivity and inflammatory tone are influenced by genes that regulate the release of adipokines, such as leptin and adiponectin. The pathophysiology of MetS is influenced by altered expression of these cytokines.

Crucially, although genetics are thought to be responsible for 10–15% of the risk, they interact strongly with environmental and lifestyle factors to modify the expression of phenotypes [1, 14] (Table 1.3).

2.2.4 Epigenetic Factors

Heritable variations in gene expression that are impacted by environmental factors rather than modifications in DNA sequence are referred to as epigenetics:

1. Histone modification and DNA methylation: Epigenetic alterations brought on by early exposure to stress, malnutrition, or overnutrition can make people more susceptible to obesity, insulin resistance, and dyslipidemia in later life. Genes related to glucose metabolism, lipid handling, and energy balance may be affected by these modifications.
2. Maternal Obesity and Fetal Programming: Exposure to maternal obesity or hyperglycemia during pregnancy can epigenetically rewire the fetus's metabolic

Table 1.4 Epigenetic influences

Epigenetic factor	Effect on metabolic programming
DNA methylation (early-life exposure)	Alters gene expression linked to insulin resistance and lipid metabolism
Fetal programming (maternal obesity)	Increases long-term risk of obesity and metabolic disorders
Transgenerational epigenetic inheritance	Sustains risk factors across generations despite stable environment

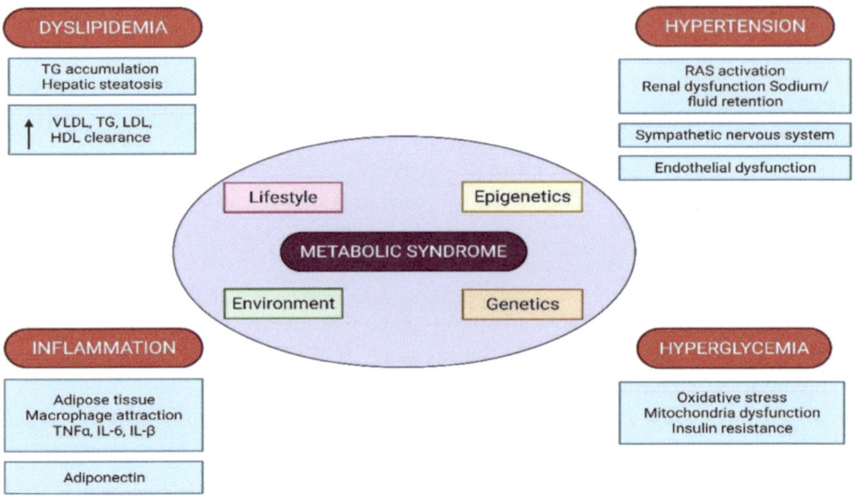

Fig. 1.1 Causative factors of metabolic syndrome

pathways, raising the adult risk of MetS. The "Developmental Origins of Health and Disease" (DOHaD) theory is supported by this.

3. Transgenerational Transmission: Epigenetic changes may endure from one generation to the next, providing an explanation for the familial accumulation of metabolic traits that goes beyond genetics (Table 1.4 and Fig. 1.1).

2.3 Pathogenesis and Pathophysiology of Metabolic Syndrome

Instead of being a single illness, metabolic syndrome is a collection of linked metabolic disorders with similar pathophysiological causes, including central obesity, insulin resistance, dyslipidemia, hypertension, and hyperglycemia. Insulin resistance, oxidative stress, chronic low-grade inflammation, dysfunctional adipose tissue, and endothelial dysfunction are some of these causes [1].

Insulin resistance (IR), or the reduced capacity of insulin-sensitive tissues, mainly muscle, liver, and adipose tissue, to react to insulin, is one of the main mechanisms. Increased levels of proinflammatory cytokines (such as TNF-α and IL-6), free fatty acids (FFAs), and dangerous lipid intermediates like ceramides and diacylglycerol are the main causes of this resistance [14]. By encouraging the phosphorylation of insulin receptor substrate-1 (IRS-1) at serine residues rather than tyrosine residues, these molecules disrupt insulin signaling and impede the downstream PI3K/Akt pathway. As a result, the liver produces more glucose (gluconeogenesis), muscle cells absorb less glucose, and glycogen synthesis is suppressed. This results in hyperglycemia, or persistently elevated blood sugar, which triggers hyperinsulinemia, or increased insulin production by the pancreas [15]. This compensatory mechanism eventually breaks down, resulting in β-cell dysfunction and exhaustion.

Adipose tissue inflammation and dysregulation, especially in visceral fat, are additional important factors. Adipocytes that undergo hypertrophy become dysfunctional and release chemokines that draw immune cells like macrophages. These immune cells induce a shift toward a proinflammatory state within adipose tissue. The balance of adipokines, which are bioactive substances released by fat cells, is upset by this chronic inflammation, leading to higher levels of leptin and resistin and lower levels of adiponectin, which typically improves insulin sensitivity [1, 7].

Despite an increase in leptin levels, the body becomes less susceptible to its effects, especially those related to energy expenditure and appetite suppression. Additionally, more FFAs are released into the bloodstream by increased lipolysis in inflammatory adipose tissue, exacerbating insulin resistance and promoting fat deposition in nonadipose tissues. The development of MetS is also significantly influenced by oxidative stress. When the body produces more reactive oxygen species (ROS) than it can fend off with antioxidant defenses, this condition develops. Excessive ROS generation is caused by chronic hyperglycemia, overnutrition, and mitochondrial overload from excess substrates (particularly glucose and FFAs). Cellular dysfunction can result from ROS's damage to lipids, proteins, and DNA. They also stimulate the expression of proinflammatory genes by activating transcription factors like NF-κB. Further hindering insulin secretion, oxidative stress also encourages the pancreatic β-cells that produce insulin to undergo apoptosis, or programmed cell death. Additionally, it results in the production of advanced glycation end-products (AGEs), which attach to cells' RAGE receptors to prolong inflammation and harm blood vessels [7, 15].

Another important mechanism is ectopic fat deposition and lipotoxicity. Excess lipids start to build up in organs like the liver, skeletal muscle, and pancreas that aren't made to store fat when adipose tissue can no longer hold them. By producing harmful lipid metabolites, this ectopic fat impairs insulin signaling and damages local tissue, disrupting cellular function. This leads to the development of nonalcoholic fatty liver disease (NAFLD) and hepatic insulin resistance in the liver. Lipid overload causes β-cell apoptosis in the pancreas, which impairs the production of insulin [1, 7, 15].

Vascular effects of MetS include endothelial dysfunction and hypertension. Nitric oxide (NO), a substance necessary for vasodilation and preserving vascular health, is normally produced by endothelial cells. But in MetS, oxidative stress and inflammation lower NO production, which impairs vasodilation. Furthermore, the MAPK signaling pathway is triggered by elevated insulin levels, which encourages the release of vasoconstrictive agents. Additionally, insulin stimulates the renin–angiotensin–aldosterone system (RAAS), increases sympathetic nervous system activity, and improves renal sodium reabsorption—all of which lead to hypertension, or elevated blood pressure [1, 15].

These problems are made worse by mitochondrial dysfunction. The mitochondria in healthy cells effectively oxidize fatty acids to produce energy. This process becomes ineffective in MetS, particularly in tissues that respond to insulin, such as the liver, skeletal muscle, and adipose tissues [15].

The burden of oxidative stress is increased as a result of mitochondria producing more ROS. Furthermore, AMP-activated protein kinase (AMPK) activity is inhibited by mitochondrial dysfunction. A crucial cellular energy sensor that encourages the oxidation of fatty acids and the uptake of glucose is inhibited by AMPK. Lipid accumulation deteriorates and energy metabolism is further compromised when AMPK is inhibited [1, 7].

Last but not least, a hallmark of MetS is dyslipidemia, which is defined by high triglycerides, low HDL levels, and a profusion of dense, small LDL particles. Triglyceride-carrying very-low-density lipoprotein (VLDL) particles are produced in greater quantities by the liver in response to insulin resistance. LDL particles get denser and smaller, which makes them more atherogenic (likely to encourage the formation of arterial plaques), while HDL levels decrease as a result of both increased breakdown and decreased synthesis. The risk of cardiovascular diseases is greatly increased by this lipid profile [1, 7, 15] (Fig. 1.2).

Table 1.5 explains the origin and contributing factors of MetS.

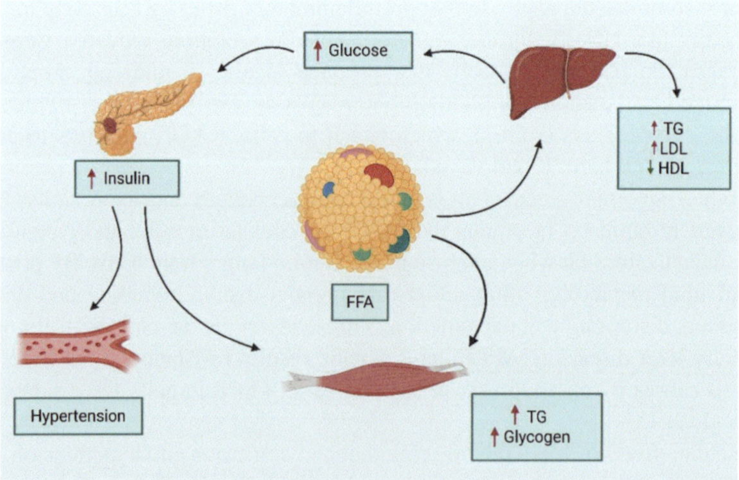

Fig. 1.2 Pathophysiology of metabolic syndrome. FFA free fatty acid, TG triglyceride, LDL low-density lipoprotein, HDL high-density lipoprotein

Table 1.5 Pathogenesis of metabolic syndrome

Category	Key factors	Mechanisms involved	References
Genetic factors	Gene variants (*TCF7L2*, *PPARγ*, *IRS1*, *FTO*)	Influence insulin signaling, adipogenesis, and glucose/lipid metabolism	[5, 13]
Epigenetic modifications	DNA methylation, histone modifications, miRNA regulation	Alter gene expression patterns in utero or postnatally, leading to lifelong metabolic disruption (DOHaD theory)	[5, 13]
Dietary triggers	High-fat, high-sugar diets (especially fructose); processed food consumption	Leads to visceral fat accumulation, dysbiosis, and increased free fatty acid release, promoting lipotoxicity and inflammation	[3, 5]
Physical inactivity	Sedentary lifestyle	Reduces insulin sensitivity, increases fat storage, suppresses mitochondrial fatty acid oxidation	[3, 13]
Environmental toxins	Endocrine-disrupting chemicals (e.g., BPA, phthalates), air pollution	Disrupt hormonal signaling, induce oxidative stress and epigenetic alterations	[5]
Gut microbiota dysbiosis	Altered composition (↑ Firmicutes, ↓ Bacteroidetes), ↑ LPS-producing bacteria	Triggers systemic low-grade inflammation via increased intestinal permeability and endotoxemia (↑ LPS→TLR4 activation)	[3, 5]
Psychosocial stress & sleep deprivation	Chronic stress (↑ cortisol), poor sleep hygiene	Disrupts circadian rhythm, increases appetite, alters glucose–insulin axis and lipid metabolism	[8, 9]

Table 1.6 illustrates the biological processes and dysfunctions driving the progression of MetS.

Summary Table: Pathogenesis vs. pathophysiology

Aspect	Pathogenesis	Pathophysiology
Triggers	Genetics, epigenetics, diet, physical inactivity, pollution, gut dysbiosis	Insulin resistance, inflammation, oxidative stress, mitochondrial dysfunction
Mechanism	Adipose dysfunction → Inflammatory cytokines → IR → Endocrine and metabolic imbalance	Impaired glucose uptake, increased gluconeogenesis, β-cell dysfunction, endothelial injury, and hypertension
Mediators	FFAs, TNF-α, IL-6, LPS, cortisol	ROS, AGEs, DAG, ceramides, NO reduction
Target organs	Adipose tissue, liver, muscle, pancreas, brain	Blood vessels, heart, pancreas, kidneys
Outcomes	Onset of central obesity, glucose intolerance, dyslipidemia	Hypertension, T2DM, atherosclerosis, NAFLD, neurodegeneration

Table 1.6 Pathophysiology of metabolic syndrome

Pathophysiological factor	Primary disturbances	Consequences/disease outcomes	References
Insulin resistance (IR)	Impaired insulin receptor signaling (via IRS-1 serine phosphorylation), ↓ GLUT4 translocation	Hyperglycemia, compensatory hyperinsulinemia, β-cell stress, T2DM progression	[3, 5, 13]
Adipose tissue inflammation	Hypertrophied adipocytes, macrophage infiltration, ↑ TNF-α, IL-6, ↓ adiponectin	Systemic low-grade inflammation, worsening of IR, atherogenesis	[3, 5]
Oxidative stress	↑ ROS (from NOX activation, mitochondrial overload), ↑ AGEs	Endothelial dysfunction, mitochondrial damage, DNA/protein/lipid oxidation, progression of CVD and diabetes	[6, 13]
Lipotoxicity	FFA accumulation in liver, muscle, pancreas; ↑ DAG, ceramides	β-cell apoptosis, NAFLD, IR, ER stress	[5, 13]
Dyslipidemia	↑ VLDL, ↑ triglycerides, ↓ HDL, formation of small dense LDL	Atherosclerosis, cardiovascular complications	[3]
Endothelial dysfunction	↓ NO bioavailability, ↑ vasoconstrictors (endothelin-1), RAAS and sympathetic overactivation	Hypertension, vascular stiffness, impaired glucose disposal	[3, 13]
Mitochondrial dysfunction	Impaired oxidative phosphorylation, ↓ ATP, ↑ ROS	Insulin resistance, metabolic inflexibility, lipid accumulation	[13]
AMPK inhibition	Inflammation and energy surplus suppress AMPK activity	↓ Fatty acid oxidation, ↑ lipogenesis, ↓ glucose uptake, loss of metabolic control	[5, 13]
Neuroendocrine disruption	Leptin resistance, ghrelin alterations, HPA axis overdrive	Appetite dysregulation, weight gain, sympathetic overactivity contributing to MetS manifestations	[11, 12]

2.4 Clinical Phases of Metabolic Syndrome

The three clinical phases of metabolic syndrome, each marked by unique physiological and biochemical alterations, are not abrupt; rather, they develop over time. Let's examine them in detail [1]:

Phase I: Early (Subclinical) Phase
Description: The body begins reacting negatively to risk factors like obesity, a sedentary lifestyle, and genetic predispositions during this first, frequently silent phase when there are no visible symptoms.

Underlying Pathogenesis:

- Insulin resistance starts to develop.
- Mild alterations in lipid profiles, inflammatory markers, and adipokine secretion may begin.
- There may be early indicators of mitochondrial dysfunction, oxidative stress, and ER stress.
- SIRT1 and AMPK activity begins to decrease, which affects energy regulation.

Clinical Clues:

- Weight gain, particularly central obesity, may be observed; routine checkups may reveal slight increases in blood pressure, triglycerides, or fasting glucose.
- There isn't a definitive diagnosis yet, but it's crucial to start lifestyle changes early.

Phase II: Intermediate (Preclinical/Borderline) Phase
Description: Although the metabolic abnormalities are now easier to see, patients may still be asymptomatic or exhibit vague symptoms (e.g., fatigue, weight problems).

Pathogenesis:

- There may be compensatory hyperinsulinemia as insulin resistance deteriorates.
- Lipid metabolism is further dysregulated.
- Chronic mitochondrial dysfunction, ectopic fat deposition, and low-grade inflammation.
- Some of these changes may start to be locked in by epigenetic modifications.

Clinical Findings:

- Laboratory findings may show:
  - Impaired fasting glucose or impaired glucose tolerance.
  - Borderline dyslipidemia (elevated TG, low HDL).
  - Mild hypertension.
  - Early signs of NAFLD on imaging.
- Diagnosis of "metabolic syndrome" may now be established if multiple criteria are met (e.g., waist circumference, BP, HDL, TG, glucose).

Phase III: Late (Clinical Disease) Phase

Description: The patient develops one or more chronic diseases during this full expression phase.

Pathogenesis:

- Type 2 diabetes mellitus may result from pancreatic beta-cell failure; persistent stress on metabolic pathways causes irreversible organ damage; Cardiovascular disease and atherosclerosis are facilitated by lipid abnormalities and chronic inflammation.
- Fibrosis and nonalcoholic steatohepatitis (NASH) can result from persistent fat buildup in the liver.

Clinical Outcomes:

- Established chronic conditions, including:
  - Type 2 diabetes
  - Hypertension
  - Dyslipidemia
  - Atherosclerotic cardiovascular disease (ASCVD)
  - NAFLD/NASH
  - Certain cancers (e.g., colon, breast)
  - Neurodegenerative diseases like Alzheimer's disease

Management Focus:

- Shift from prevention to control and management.
- Requires pharmacological intervention, aggressive lifestyle changes, and monitoring for complications.

Clinical Insight:

- *Early Phase (Phase I)*: Best window for preventive intervention (diet, exercise).
- *Intermediate Phase (Phase II)*: Screening is key in most cases can be halted or reversed here.
- *Late Phase (Phase III)*: Requires chronic disease management and often multidisciplinary care (Fig. 1.3).

Phase	Name	Key features	Pathogenesis	Pathophysiology	Clinical signs/ findings
Phase I	Early (subclinical)	No obvious symptoms	Triggered by obesity, genetic factors, inactivity initial insulin resistance low-level	Decreased AMPK and SIRT1 activity mild mitochondrial dysfunction Subtle lipid and glucose	Central/visceral fat increase slight rise in fasting glucose or triglycerides no overt disease yet

(continued)

Phase	Name	Key features	Pathogenesis	Pathophysiology	Clinical signs/ findings
			inflammation, oxidative stress	metabolism disruption	
Phase II	Intermediate (preclinical)	Metabolic alterations, often asymptomatic	Progression of insulin resistance ↑ Inflammatory cytokines and ectopic fat accumulation epigenetic changes start appearing	Dysregulation in lipid metabolism Fatty liver (NAFLD) starts developing further mitochondrial damage	Prediabetes (IFG/IGT), dyslipidemia elevated BP diagnosis of metabolic syndrome often made
Phase III	Late (clinical disease)	Full-blown metabolic and cardiovascular diseases	Persistent metabolic dysfunction advanced mitochondrial dysfunction chronic inflammation and oxidative stress cell signaling disruptions	β-cell dysfunction → T2DM vascular endothelial dysfunction → ASCVD hepatic steatosis → NASH Neurodegeneration	Type 2 diabetes mellitus hypertension atherosclerosis (ASCVD) NAFLD/NASH cancer, Alzheimer's disease

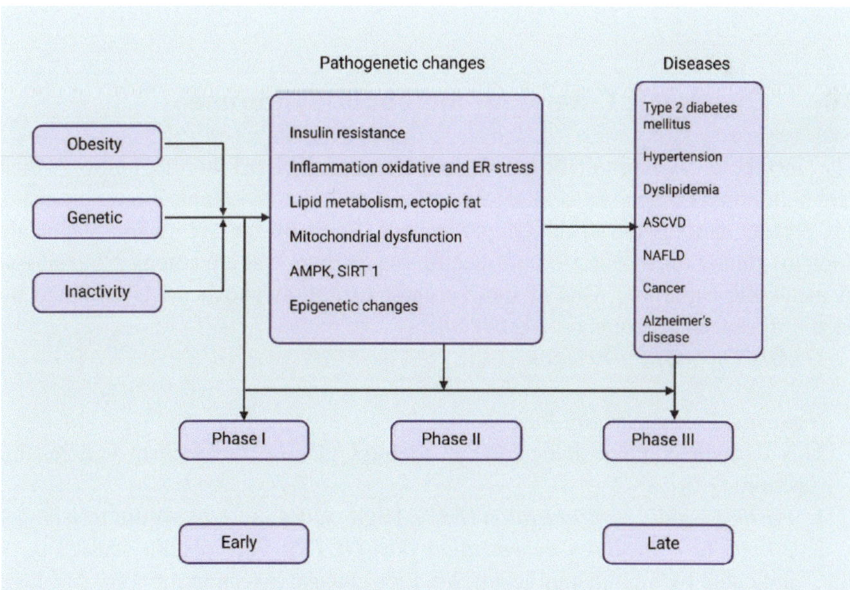

Fig. 1.3 Clinical phases of metabolic syndrome

2.5 Global Prevalence and Challenges in Management

Due in large part to urbanization, sedentary lifestyles, and rising obesity rates, MetS has spread throughout the world [3, 4, 6]:

- *Global Burden*: About 20–25% of adults worldwide suffer from MetS, with rates being higher in developed nations and urban areas. Over the past 20 years, the prevalence among American adults has increased from 25.3% to 34.2%.
- *Gender and Ethnic Variations*: Global estimates are impacted by the fact that prevalence is typically higher in women and that diagnostic thresholds differ by ethnicity.
- *Management Difficulties*:
  - I. *Polypharmacy*: Managing distinct conditions (such as hypertension, dyslipidemia, and type 2 diabetes) frequently necessitates taking several drugs, which lowers patient adherence.
  - II. *Lack of Unified Criteria*: Screening and epidemiological comparisons are made more difficult by the disparate definitions provided by the WHO, IDF, and ATP III.
  - III. *Underdiagnosis and Delayed Intervention*: A large number of patients do not receive a diagnosis until complications such as diabetes or cardiovascular disease appear.
  - IV. *Adherence to Lifestyle Modification*: Although fundamental, lifestyle modifications are challenging to maintain, particularly in the absence of behavioral or social support.

3 Diagnostic Criteria for Metabolic Syndrome

The metabolic syndrome diagnostic criteria are depicted in the figure, which highlights that a diagnosis is made when three or more biochemical measurements are present among the specified parameters. Each of the six major diagnostic domains stands for a collection of quantifiable markers that are commonly changed in metabolic syndrome. One or more criteria for the diagnosis are contributed by each of these six domains [1, 14, 15].

The Six Diagnostic Domains:

1. *Hyperinsulinemia/Insulin Resistance*

   This domain contains indicators of aberrant glucose metabolism and insulin function:

   1. *Fasting Insulin Concentration (FIC)*: High values indicate insulin resistance.
   2. *2-h IC (2-h Insulin Concentration post-OGTT)*: Poor insulin sensitivity is indicated by high insulin levels 2 h after glucose ingestion.
   3. *Insulin Resistance (IR)*: IR is a general term that is frequently determined using HOMA-IR or comparable indices.

4. *Fasting Blood Glucose (FBG)*: If the fasting glucose level is greater than 100 mg/dL, it indicates a problem with glucose regulation.
5. *Glycated Hemoglobin (or HbA1)*: It measures the average blood glucose levels over 2–3 months; a value of ≥5.7% is frequently regarded as prediabetic.
6. *2-h BG (2-h Blood Glucose)*: Blood glucose levels 2 h following OGTT; a value of ≥140 mg/dL indicates impaired glucose tolerance.
   Why it matters: Insulin resistance occurs before type 2 diabetes and is a key mechanism in metabolic syndrome.

2. *Hyperuricemia*
   - *Uric Acid Levels*: Hypertension, insulin resistance, and renal dysfunction are all linked to elevated serum uric acid levels.
     Why it matters: Hyperuricemia is frequently observed in metabolic syndrome and is associated with endothelial dysfunction and oxidative stress, though it is not a component of all diagnostic criteria.
3. *High Blood Pressure*
   1. Blood pressure, or BP, is a general term for high artery pressure.
   2. Systolic blood pressure, or SBP, is the pressure during a heartbeat; a reading of ≥130 mmHg is regarded as elevated.
   3. Diastolic blood pressure, or DBP, is the pressure at which the heart relaxes; a value of ≥85 mmHg is regarded as elevated.
      Why it matters: Blood vessel damage from high blood pressure raises the risk of heart disease and stroke.
4. *Kidney Function*
   - *Microalbuminuria*: When there is minimal albumin in the urine, it is a sign of early kidney damage.
     Why it matters: The significance of it is a predictor of renal and cardiovascular problems and is indicative of endothelial dysfunction.
5. *Dyslipidemia*
   - *Triglycerides (TG)*: A common finding is elevated TG (≥150 mg/dL).
   - *HDL (High-Density Lipoprotein)*: The metabolic syndrome is characterized by low HDL (<40 mg/dL in men and <50 mg/dL in women).
   - *LDL (Low-Density Lipoprotein)*: Usually small and dense (more atherogenic), but may be normal or elevated.
   - *Total cholesterol, or TC*, may or may not be elevated.
     Why it matters: Dyslipidemia raises the risk of cardiovascular disease and contributes to atherosclerosis.
6. *Obesity*
   Several physical and biochemical markers of excessive fat accumulation are included in this domain:
   - Body height (BH) and body weight (BW) are used to determine BMI.
   - Body fat (BF): Direct measurement using DEXA, bioimpedance, etc.
   - WHR (Waist–hip ratio): One important risk factor is increased abdominal fat.
   - Obesity is indicated by a BMI (body mass index) of ≥25 kg/m^2 for Asians or ≥30 kg/m^2 for Western populations.

Why it matters: Insulin resistance and inflammation are strongly influenced by central (abdominal) obesity.

Diagnostic Rule: 3 or More Criteria

The figure highlights that at least three of these changed parameters (from any domain) must be present for metabolic syndrome to be diagnosed. The idea is the same, even though different professional guidelines (such as NCEP ATP III, IDF, and WHO) may have slightly different thresholds or requirements [7, 15].

Clinical Insight:

Clinicians can detect at-risk individuals early and start lifestyle or medication changes to stop the development of cardiovascular disease, diabetes, or kidney failure by assessing these domains (Fig. 1.4).

Summary Table

Domain	Key markers
Hyperinsulinemia	FIC, 2-h IC, IR, FBG, HbA1c, 2-h BG
Hyperuricemia	Uric acid
Hypertension	SBP ≥130 mmHg, DBP ≥85 mmHg
Kidney function	Microalbuminuria
Dyslipidemia	TG ≥150 mg/dL, HDL <40/50 mg/dL, LDL, TC
Obesity	BMI ≥25/30, WHR ↑, BF ↑, BW ↑

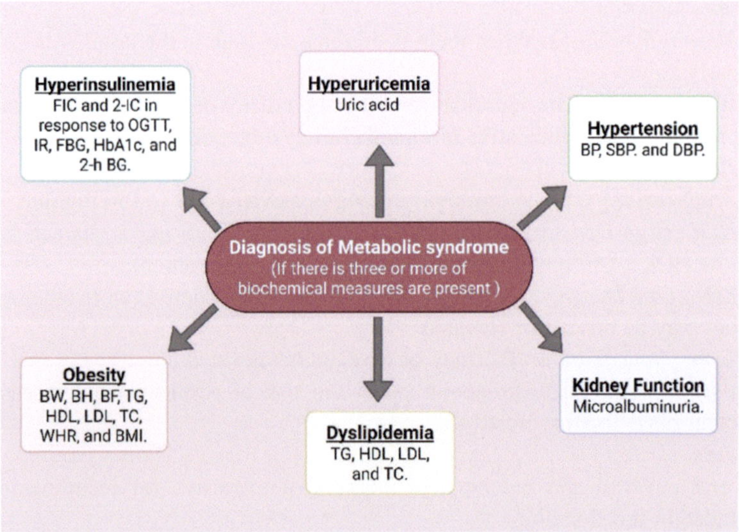

Fig. 1.4 Diagnosis of metabolic syndrome. BW Body weight, BH Body height, BF Body fat, BP Blood pressure, HbA1c Hemoglobin A1C, HDL High density lipoprotein, LDL Low density lipoprotein, TC Total cholesterol, TG Triglycerides, WHR Waist: hip ratio, BMI Body mass index, SBP Systolic blood pressure, DBP Diastolic blood pressure, FIC Fasting insulin concentration, 2-h IC 2 h insulin concentration after OGTT, OGTT Oral glucose tolerance test

4 Diagnostic Clinical Criteria of Metabolic Syndrome By Different Health Organizations

A number of clinical and biochemical anomalies are used to diagnose metabolic syndrome (MetS). To standardize its identification, a number of international health organizations have created diagnostic criteria. The presence of the fundamental metabolic risk factors such as central obesity, hyperglycemia, dyslipidemia, and elevated blood pressure, emphasized in all organizations; however, the thresholds and the number of criteria needed vary slightly [1].

4.1 World Health Organization (WHO, 1999)

The WHO was the first to provide formal criteria for MetS. Their definition places central emphasis on insulin resistance, requiring its presence for diagnosis, along with two or more of the following:

Component	Criteria
Insulin resistance	Type 2 diabetes mellitus, impaired fasting glucose, or impaired glucose tolerance
Dyslipidemia	Triglycerides ≥150 mg/dL and/or HDL cholesterol <35 mg/dL (men), <39 mg/dL (women)
Central obesity	Waist-to-hip ratio >0.90 (men), >0.85 (women) or BMI >30 kg/m^2
Hypertension	Blood pressure ≥140/90 mmHg
Microalbuminuria	Urinary albumin excretion rate ≥20 μg/min or albumin-to-creatinine ratio ≥30 mg/g

4.2 National Cholesterol Education Program—Adult Treatment Panel III (NCEP: ATP III, 2001)

This definition is widely used due to its simplicity and clinical practicality. It requires the presence of any three out of the five criteria:

Component	Criteria
Abdominal obesity	Waist circumference >102 cm (men), >88 cm (women)
Triglycerides	≥150 mg/dL
HDL cholesterol	<40 mg/dL (men), <50 mg/dL (women)
Blood pressure	≥130/85 mmHg or on antihypertensive treatment
Fasting glucose	≥110 mg/dL (revised later to ≥100 mg/dL)

4.3 International Diabetes Federation (IDF, 2005)

The IDF criteria emphasize central obesity as a mandatory criterion, using ethnicity-specific waist circumference cutoffs. Diagnosis requires central obesity plus any two of the remaining four factors.

Component	Criteria
Central obesity	Waist circumference (ethnicity-specific values)[a]
Triglycerides	≥150 mg/dL or on treatment
HDL cholesterol	<40 mg/dL (men), <50 mg/dL (women), or on treatment
Blood pressure	≥130/85 mmHg or on treatment
Fasting glucose	≥ 00 mg/dL or previously diagnosed type 2 diabetes

[a]For example, South Asians: >90 cm (men), >80 cm (women)

4.4 Harmonized Definition (IDF, AHA/NHLBI, 2009)

A collaborative effort among global organizations (including IDF and AHA/NHLBI) led to a harmonized set of criteria, where no single factor is mandatory. Three or more of the following five criteria are required:

Component	Criteria
Waist circumference	Population- and country-specific thresholds
Triglycerides	≥150 mg/dL or on lipid-lowering medication
HDL cholesterol	<40 mg/dL (men), <50 mg/dL (women), or on treatment
Blood pressure	≥130/85 mmHg or on antihypertensive medication
Fasting glucose	≥100 mg/dL or diagnosed diabetes

Comparison Summary Table

Criteria	WHO	NCEP: ATP III	IDF	Harmonized (2009)
Insulin resistance	Mandatory	Not required	Not required	Not required
Central obesity	Optional	One of five	Mandatory	One of five (ethnicity-specific)
Number of criteria	IR + 2 others	Any 3 of 5	Obesity +2 others	Any 3 of 5
Blood pressure cut-off	≥140/90 mmHg	≥130/85 mmHg	≥130/85 mmHg	≥130/85 mmHg
Fasting glucose cut-off	IGT/IFG/T2DM	≥110 mg/dL (later ≥100)	≥100 mg/dL	≥100 mg/dL

The fundamental characteristics of central obesity, insulin resistance or hyperglycemia, hypertension, and dyslipidemia are all the same, despite minor variations in definitions. Although certain populations may need modifications depending on ethnicity and regional health trends, the harmonized definition offers a universally recognized standard and streamlines comparative studies [1].

5 Nanotheranostics: Bridging Diagnostics and Therapeutics

5.1 Definition and Key Concepts

The term "nanotheranostics" describes the combination of therapeutics and diagnostics based on nanotechnology into a single, multipurpose platform. In a single system, these intelligent nanostructures can deliver treatment, track the course of the disease, and identify disease biomarkers. Precision and personalized medicine, which enables targeted interventions and real-time monitoring of therapeutic results, is the ultimate goal of nanotheranostics [2, 14].

5.2 Diagnostic Capabilities

High-sensitivity diagnostic techniques used in nanotheranostics allow for early disease detection, staging, and monitoring.

Modality	Description
MRI (magnetic resonance imaging)	Superparamagnetic nanoparticles enhance soft tissue contrast
PET/CT imaging	Radiolabeled nanoparticles for 3D molecular imaging
Fluorescence/NIR imaging	Optical imaging using dye-conjugated nanoparticles for in vivo visualization
Photoacoustic imaging	Combines optical and ultrasound for deeper tissue imaging
Biosensors	Detect circulating biomarkers like miRNA, cytokines, or tumor DNA

Example: Iron oxide nanoparticles functionalized with antibodies can be used for both MRI imaging and antibody-mediated tumor targeting

5.3 Therapeutic Capabilities

Beyond imaging, nanotheranostics provides individualized and tailored treatment options:

Therapeutic function	Mechanism/Example
Targeted drug delivery	Site-specific release using ligands, pH/temperature responsiveness (e.g., liposomes)
Gene therapy	Delivery of CRISPR-Cas9, siRNA, or miRNA to regulate gene expression

(continued)

Therapeutic function	Mechanism/Example
Photothermal therapy (PTT)	Gold nanoshells absorb light and convert it to heat to ablate cancer cells
Photodynamic therapy (PDT)	Light-activated agents produce reactive oxygen species (ROS) to kill tumor cells
Stimuli-responsive systems	Release drugs in response to pH, enzymes, redox changes at the disease site

Example: Polymeric micelles loaded with paclitaxel and conjugated with anti-EGFR antibodies for colorectal cancer therapy

5.4 Advantages Over Conventional Approaches

When compared to traditional stand-alone diagnostics and treatments, nanotheranostics provides several clinical benefits:

Parameter	Conventional approaches	Nanotheranostics
Diagnosis and therapy	Separate procedures	Integrated in one platform
Specificity	Often lacks target selectivity	Ligand-mediated targeting to disease sites
Side effects	High systemic toxicity	Site-specific action minimizes toxicity
Monitoring	Post-treatment evaluations	Real-time monitoring of drug response
Personalization	Limited	Highly tunable to patient-specific molecular profiles
Multimodal function	Unimodal (imaging or therapy only)	Combines imaging, sensing, and treatment in one system

6 Applications of Nanotheranostics in Metabolic Syndrome

6.1 Diagnostics

6.1.1 Biosensors for Glucose, Lipid Levels, and Inflammatory Markers

For the real-time detection and tracking of metabolic markers like glucose, cholesterol, and inflammatory cytokines, nanotheranostic tools are being utilized more and more. Nanoparticle-based smart biosensors provide quick, noninvasive, and extremely sensitive detection techniques. For instance, wearable biosensors composed of conductive polymers such as PEDOT: PSS can detect metabolic stress and insulin resistance early by measuring cortisol and other metabolic biomarkers from perspiration. Similar to this, biosensors incorporated into wristbands or skin patches can continuously measure variations in lipid or glucose levels, improving diabetes control. Highly sensitive and specific biosensors that can identify important metabolic syndrome biomarkers have been made possible by nanotechnology. These consist of stress hormones (like cortisol), inflammatory cytokines (like IL-6 and TNF-α), glucose, triglycerides, and cholesterol [2, 10].

Biosensor type	Target biomarker	Nanomaterial used	Application
Sweat-based electrochemical patch	Cortisol	PEDOT: PSS	Detects stress-related hormone levels
Wearable glucose monitor	Glucose	Carbon nanotubes, gold nanoparticles	Noninvasive glucose monitoring in diabetes
Point-of-care lipid detection sensor	Cholesterol, triglycerides	ZnO nanoparticles, AuNPs	Real-time lipid profile measurement
Cytokine biosensors	IL-6, TNF-α	Quantum dots, graphene oxide	Tracks systemic inflammation

6.1.2 Imaging Technologies Enhanced by Nanoparticles

Numerous imaging techniques, including MRI, CT, and PET scans, are made more sensitive and accurate by nanoparticles. In order to visualize inflammatory tissues, identify fatty tissue abnormalities, and detect vascular dysfunction, gold nanoparticles, iron oxide, and silica-based agents are being used more and more as contrast enhancers. For example, gold-based nanoprobes functionalized with gadolinium have enhanced the resolution of MRI scans to evaluate adipose tissue and liver fat accumulation, while 64Cu-labeled nanoparticles have been used in PET scans to track glucose metabolism. Furthermore, nanomaterial-supported photoacoustic and fluorescence imaging technologies enable deeper tissue visualization with fewer adverse effects. By serving as contrast agents, enhancing resolution, focusing on diseased tissues, and reducing side effects, nanoparticles enhance diagnostic imaging [14].

Imaging technique	Nanoparticle used	Target	Advantages
MRI	Gadolinium-functionalized gold nanoparticles	Adipose and liver tissue	Enhanced imaging with reduced toxicity
PET	64Cu-labeled liposomal nanoparticles	Tumor, metabolic tissues	High-resolution imaging of glucose metabolism
CT scan	PEG-bilirubin coated gold nanoparticles	Liver	Differentiates diseased vs. healthy liver tissue
Photoacoustic imaging	ICG-loaded liposomes, carbon nanotubes	Vascular inflammation	Deep-tissue imaging, noninvasive monitoring

6.2 Therapeutics

6.2.1 Nanocarriers for Delivering Antidiabetic, Antihypertensive, and Lipid-Lowering Drugs

To enhance the delivery of common medications like metformin, statins, or thiazolidinediones, nanocarriers such as liposomes, polymeric nanoparticles (like PLGA), and lipid-based nanogels are employed. By increasing a drug's bioavailability, preventing degradation, and enabling targeted delivery, these nanocarriers lessen adverse effects. In animal models, for instance, it was demonstrated that thymoquinone and metformin-loaded nanocapsules had more potent antidiabetic effects than either drug alone. Similarly, in obese mice, *Salacia chinensis* gold nanoparticles (SC-AuNPs) improved lipid profiles and decreased obesity-related markers. Drug solubility, bioavailability, and targeted delivery are all improved by nanocarriers, which also lessen systemic side effects and increase therapeutic effects [10].

Drug/compound	Nanocarrier used	Disease target	Outcome
Metformin + Thymoquinone	Polymeric nanocapsules	Type 2 diabetes	Enhanced glucose control in diabetic mice
Salacia chinensis extract	Gold nanoparticles (SC-AuNPs)	Obesity	Reduced fat mass, increased HDL
Curcumin	Nanocurcumin, PLGA nanoparticles	Dyslipidemia	Improved lipid profile, fewer side effects
Ursolic acid	Self-assembled nanoliposomes	Obesity	Decreased adipogenesis, enhanced fat metabolism

6.2.2 Gene Editing and RNA-Based Therapies

By employing nucleic acid nanoparticles like plasmid DNA (pDNA), siRNA, miRNA, and antisense oligonucleotides, nanotheranostics is also advancing gene-based therapies. Nonviral nanocarriers that improve cell uptake and shield nucleic acids from deterioration, such as lipoplexes, polyplexes, and exosomes, are used to deliver these. For example, miR-21-loaded nanoparticles have demonstrated encouraging outcomes in the treatment of metabolic diseases related to bone, such as osteoporosis. Additionally, liver fibrosis, a common metabolic complication, has been effectively targeted by antifibrotic siRNA-loaded nanogels. The precise targeting of disease-causing genes and enhanced therapeutic results are provided by the nanoscale delivery of nucleic acids such as siRNA, miRNA, and antisense oligonucleotides [5, 13].

Therapeutic type	Nanocarrier system	Target	Result
miR-21	Salicylic acid-based nanoparticles	Osteoporosis	Improved bone regeneration, low toxicity
siRNA (antifibrotic)	Ketal cross-linked nanohydrogel	Liver fibrosis	Reduced collagen deposition, improved liver function

(continued)

Therapeutic type	Nanocarrier system	Target	Result
miRNA-29b & 122	SPIO-loaded micelles	Hepatic stellate cells	Downregulated fibrosis-related genes
Antisense miR-375	Ginger-derived lipid nanoparticles	T2DM, IBD	Enhanced glucose control, anti-inflammatory effect

6.3 Theranostic Platforms: Combining Diagnostics and Therapy

Theranostic systems use a single nanoplatform to deliver therapeutic drugs and perform diagnostic imaging. This makes it possible to monitor treatment response in real time and modify therapy as necessary [14]. For instance:

- Because of their inherent fluorescence and stability in the GI tract, ginger-derived nanoparticles (GDNPs) have been used as imaging agents and to deliver microRNAs to treat diabetes and inflammation.
- Melanin-loaded liposomes have been created to deliver chemotherapy and photoacoustic imaging in a light-triggered fashion, particularly for cancers and tissue inflammation associated with obesity.
- Noninvasive organ function monitoring is possible with bio-adhesive ultrasound patches that are integrated with nanoparticle sensors. These patches can react with therapeutic measures like drug release or sonodynamic stimulation.

By combining targeted medication action, minimal invasiveness, and ongoing disease monitoring, these platforms offer a personalized medicine approach.

Platform	Functionality	Application
GDNP-loaded nanoparticles	miRNA delivery + colon-targeted bioimaging	T2DM, liver inflammation
ICG-melanin liposomes	Photoacoustic imaging + drug release	Obesity-linked inflammation
Smart bioadhesive ultrasound patches	Real-time monitoring + sonodynamic therapy	Glucose regulation, liver inflammation
Smartphone-controlled gene release systems	Remote-controlled therapeutic activation	Diabetes, metabolic stress management

7 Advanced Nanomaterials for Nanotheranostics in Metabolic Syndrome (MetS)

With the creation of specialized nanomaterials intended for both therapeutic and diagnostic applications, nanotheranostics in MetS has made significant strides. Better stability, biocompatibility, and the capacity to target particular tissues or biomarkers implicated in metabolic disorders such as diabetes, obesity, dyslipidemia, and chronic inflammation are all provided by these engineered nanomaterials [2, 10].

7.1 Nucleic Acid-Based Nanoparticles

In order to modify gene expression in metabolic disorders, nucleic acid-based nanoparticles (NANPs) use DNA, siRNA, miRNA, antisense oligonucleotides, and mRNA that are delivered via nanocarriers. High molecular precision and specificity are provided by these treatments [2].

Mechanism: Usually enclosed in liposomes, polyplexes, or exosomes, NANPs are shielded from enzymatic breakdown and improve cellular uptake.

- *Applications*:
  - *miR-21@PSA-NP* for bone metabolic disorders such as osteoporosis showed enhanced bone regeneration and long blood circulation time.
  - *Anti-col1a1 siRNA-Loaded Nanohydrogel* significantly reduced fibrosis in liver cells by targeting collagen expression.
  - *miRNA-122 and miRNA-29b Delivered via SPIO Nanomicelles* effectively reduced liver inflammation and fibrosis in hepatic stellate cells.

NANP type	Target	Outcome
miRNA-loaded NPs	Osteoporosis, liver fibrosis	Gene modulation and reduced inflammation
Antisense miRNA-375	T2DM, IBD	Glucose control, anti-inflammatory effect

7.2 Lipid-Based Nanoparticles: Drug Delivery for Diabetes and Cholesterol Management

In order to deliver lipid-lowering and antidiabetic medications, lipid-based nanoparticles such as liposomes, solid lipid nanoparticles (SLNs), and nanostructured lipid carriers (NLCs) are now essential [2, 10].

- *Advantages*: High biocompatibility, excellent drug loading capacity, and minimal toxicity.
- *Applications*:
  - *Metformin and thymoquinone* coencapsulated in lipid carriers showed superior glycemic control in T2DM models.
  - *Curcumin-loaded lipid nanoparticles* improved cholesterol levels with fewer side effects.

  – *Ginger-derived lipid NPs (GDNPs)* were used for miRNA delivery and shown to improve gut health and reduce inflammation in high-fat diet-induced diabetes.

Lipid NP example	Drug	Disease target	Benefit
Liposomal NPs	Metformin, Thymoquinone	T2DM	Enhanced glucose regulation
GDNPs	miR-375	T2DM, colitis	Gut protection, inflammation reduction
Curcumin-loaded liposomes	Curcumin	Hyperlipidemia	Improved lipid profiles, fewer side effects

7.3 Polymeric Nanoparticles: Sustained Drug Release Systems

Particularly derived from PLGA, PEG, or chitosan, polymeric nanoparticles (PNPs) provide controlled and sustained release, lower dosage frequency, and enhance drug targeting [2, 10].

- *Advantages*: Biodegradable, customizable surface, and long circulation time.
- *Applications*:
  – PLGA nanoparticles were used for prolonged release of *antidiabetic drugs*.
  – PEGylated NPs enhance systemic circulation and avoid immune clearance.
  – *Polymeric thymoquinone-metformin NPs* provided superior glycemic control in diabetic mice.

Polymeric NP type	Function	Outcome
PLGA	Slow drug release	Prolonged glucose control, reduced dosing
PEGylated NPs	Stealth delivery system	Enhanced blood circulation, immune evasion
PSA-NP (salicylic acid NP)	microRNA delivery (miR-21)	Improved bone health in metabolic bone disorders

7.4 Carbon-Based Nanomaterials: Noninvasive Glucose Monitoring and Lipid Profiling

Nanomaterials based on carbon, such as graphene oxide (GO), quantum dots, and carbon nanotubes (CNTs), are very biocompatible and conductive [2, 10]. They are extensively utilized in metabolic diagnostic biosensors.

- *Applications*:
  - Noninvasive *wearable glucose biosensors* built with carbon nanotubes allow real-time blood sugar monitoring via sweat.
  - Graphene oxide-based sensors are sensitive enough to detect minor changes in cholesterol and triglyceride levels.
  - Quantum dots functionalized with enzymes are used to measure free radicals and lipid markers.

Material	Sensor target	Use
Carbon nanotubes	Glucose, cortisol	Sweat-based diabetes management
Graphene oxide	Cholesterol, TGs	Lipid profiling and inflammation
Carbon quantum dots	ROS, insulin	Oxidative stress detection

7.5 Metallic Nanoparticles (Gold, Silver): Biosensors for Inflammatory Markers

Due to their large surface area and simple functionalization, metallic nanoparticles like gold (AuNPs) and silver (AgNPs) make great biosensor platforms [2, 10].

- Applications:
  1. Antibody-functionalized gold nanoparticles are employed to identify inflammatory cytokines such as TNF-α and IL-6.
  2. By lowering local inflammation, silver nanoparticles embedded in hydrogels (such as QCT-AgNPs) have been used to accelerate wound healing in diabetic ulcers.
  3. Targeting oxidative environments, AuNPs that have been modified for CT imaging can also reveal tissue inflammation (Fig. 1.5).

Metal NP type	Target marker	Diagnostic use
Gold nanoparticles	IL-6, TNF-α	Biosensing for chronic inflammation
Silver nanoparticles	Diabetic wound cytokines	Anti-inflammatory wound healing agent
Gold–Gd hybrid NPs	Inflamed tissues (MRI)	Imaging + inflammation detection

8 Challenges in Nanotheranostics for Metabolic Syndrome (MetS)

Despite the revolutionary potential of nanotheranostics for the diagnosis and treatment of metabolic syndrome, a number of significant obstacles still limit its clinical application. To guarantee the safe, efficient, and scalable use of nanotechnologies in actual healthcare settings, these obstacles need to be overcome.

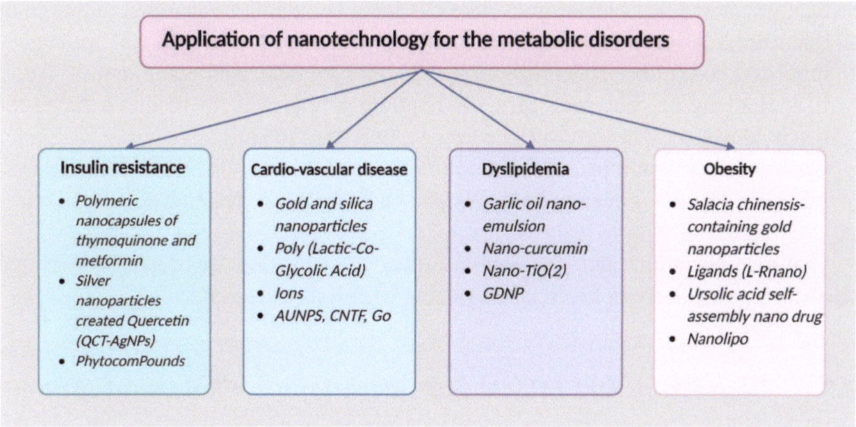

Fig. 1.5 Applications of nanotechnology for the metabolic disorder

8.1 Biocompatibility and Potential Toxicity of Nanomaterials

The biocompatibility of nanomaterials is one of the main issues in nanotheranostic applications. Preclinical research has shown that many nanoparticles (NPs) have promising therapeutic effects; however, their long-term toxicity and safety profiles are still unknown, particularly when used systemically or repeatedly [2, 10].

- Certain metallic nanoparticles (NPs), such as those based on silver or cadmium, have the potential to produce reactive oxygen species (ROS), which can cause oxidative stress, inflammation, or cytotoxicity in healthy tissues.
- Nondegradable NP buildup in organs like the kidneys, spleen, and liver may affect how well those organs function.
- If not appropriately optimized, even biodegradable NPs like PLGA or lipid-based carriers may trigger immunological reactions or alter cellular signaling pathways.

To prevent unforeseen side effects, it is essential to fully comprehend the biological fate of nanomaterials, including their distribution, metabolism, and excretion [2, 10].

8.2 High Production Costs and Scalability Issues

The expense and difficulty of large-scale manufacturing represent yet another significant barrier. The majority of nanotherapeutics are made in lab settings, which makes them challenging to duplicate in industrial settings [2, 10, 14].

- Scaling up synthesis techniques (like chemical conjugation, nanoprecipitation, or microfluidics) economically is difficult because they frequently call for highly regulated conditions, specialized reagents, and sophisticated equipment.
- It is challenging, but necessary for regulatory approval, to maintain batch-to-batch consistency in size, surface charge, and drug loading.
- Complex functionalities, such as dual-targeting or stimulus-responsive features, make nanoparticles even more costly and challenging to replicate on a large scale.

Commercialization and accessibility may be impeded by these cost-related obstacles, particularly in healthcare systems with limited resources.

8.3 Regulatory and Ethical Considerations in Clinical Adoption

The transition of nanotheranostic technologies from bench to bedside is fraught with ethical and regulatory issues, notwithstanding scientific advancements [2, 10]:

- Standardized international regulatory frameworks designed especially for nanotheranostic products do not exist. Regulation confusion results from existing guidelines that differ by region and frequently treat these tools as either drugs or devices.
- The majority of nanomaterials used in MetS lack long-term human safety data, which makes obtaining clinical trial approval challenging.

- *Ethical concerns* include:
  - I. Consent to be exposed to nanomaterials, particularly when miRNA delivery or gene editing are being used.
  - II. Uncertainties regarding possible off-target effects and the biodistribution of nanoparticles.
  - III. Data security and privacy issues when theranostic platforms use real-time biosensing or wearable monitoring.

Coordination between researchers, clinicians, regulatory agencies, and industry stakeholders will be necessary to address these problems.

Summary Table: Key challenges in nanotheranostics for MetS

Challenge area	Key issues	Implications
Biocompatibility and toxicity	ROS generation, immune response, long-term accumulation	Limits clinical safety and long-term usability
Cost and scalability	Complex fabrication, reproducibility, high raw material costs	Barriers to mass production and affordability
Regulatory and ethical considerations	Lack of unified guidelines, off-target risks, informed consent	Slows clinical approval and public acceptance

9 Future Directions in Nanotheranostics for Metabolic Syndrome (MetS)

The field of nanotheranostics is one that is developing quickly, and its use in the treatment of metabolic syndrome (MetS) has exciting potential to revolutionize healthcare. Future research is concentrating on innovation in material design, automation, and personalization to get around existing constraints and make these technologies more useful and accessible [2, 10, 14].

9.1 Personalized Nanotheranostics for Individualized Treatment Strategies

Individuals with MetS may exhibit a wide range of symptoms and severity levels, from mild dyslipidemia to advanced insulin resistance and organ complications. This variation emphasizes the necessity of individualized care as opposed to a "one-size-fits-all" strategy.

- By combining biosensor data (such as glucose, lipid levels, and inflammation markers), nanodiagnostic systems can be customized to a patient's specific metabolic profile in order to identify the best drug dosage, timing, and delivery method.
- For example, implantable nanodevices and smart wearable sensors can release therapeutic agents (like insulin and anti-inflammatory medications) only when necessary and continuously monitor physiological conditions.
- Target-specific ligands or antibodies can be used to customize platforms so that nanocarriers only attach to disease-affected tissues like vascular, hepatic, or adipose cells, minimizing off-target effects and increasing efficacy.

This vision is in line with precision medicine's overarching objectives, which include improving treatment efficacy, safety, and patient-centeredness.

9.2 Integration of Artificial Intelligence (AI) for Predictive Diagnostics and Therapy Optimization

By introducing intelligence and automation into the diagnosis and treatment process, artificial intelligence (AI) and machine learning have the potential to completely transform nanotheranostics.

- To forecast the course of a disease or a patient's reaction to medication, artificial intelligence (AI) algorithms can examine sizable datasets gathered from wearables, biosensors, and electronic medical records.

- Based on a patient's real-time biomarker data, predictive modeling with AI can help clinicians choose the best kind of nanocarrier, medication formulation, and dosage schedule.
- AI can also optimize the functionalization and design of nanoparticles, minimizing laboratory trial-and-error and speeding up clinical translation.

AI and nanotheranostics working together could result in closed-loop systems that seamlessly integrate therapy, diagnosis, monitoring, and decision-making.

9.3 Development of Biodegradable and Multifunctional Nanomaterials

Future nanotheranostic platforms will prioritize eco-friendly, nontoxic, and biodegradable materials to guarantee long-term safety and regulatory compliance.

- Natural polymer-based nanoparticles (such as chitosan, alginate, and gelatin) and lipid nanocarriers are examples of advanced nanomaterials that are being developed to break down into innocuous byproducts after delivering their payload.
- Targeting ligands, imaging agents, and therapeutic molecules will all be combined into a single platform by multifunctional nanoparticles, eliminating the need for multiple interventions.
- On-demand drug release is being developed for responsive systems that respond to particular stimuli, such as pH, temperature, enzymes, or glucose concentration, enhancing treatment control.

By including these characteristics, the systems will be more palatable in clinical settings and improve therapeutic efficacy while reducing side effects.

Summary Table: Future trends in nanotheranostics for MetS

Direction	Description	Impact
Personalized nanotheranostics	Tailored drug delivery based on individual biomarker profiles	Better outcomes, fewer side effects
AI integration	Predictive modeling, real-time monitoring, therapy optimization	Automated and intelligent decision-making
Biodegradable & multifunctional materials	Safe, eco-friendly carriers with combined imaging and therapy capabilities	Safer systems, improved patient compliance

10 Conclusion

The complex pathophysiology, multifactorial origin, and strong correlation with life-threatening conditions like cardiovascular disease and type 2 diabetes make metabolic syndrome (MetS) a growing global health concern. Because of this syndrome's progressive nature and heterogeneity, traditional treatment approaches frequently fail to meet its needs. Nanotheranostics presents a promising way to close this gap by combining targeted therapies and accurate diagnostics into a single platform. Nanotheranostic tools allow for highly effective, minimally invasive, and personalized management of MetS through sophisticated biosensors, real-time imaging, and intelligent drug delivery systems. Numerous nanomaterials have shown great promise in monitoring metabolic biomarkers, delivering medications to particular tissues, and reducing systemic side effects. These include lipid-based carriers, polymeric systems, carbon-based structures, and metallic nanoparticles.

Clinical application is still significantly hampered by issues with biocompatibility, large-scale manufacturing, and regulatory approval. Future developments in MetS management are anticipated to be reshaped by the incorporation of artificial intelligence, the creation of biodegradable and multifunctional nanomaterials, and a move toward personalized nanomedicine. A new era of precision healthcare may soon be ushered in by nanotheranostics, which could revolutionize the way we identify, track, and treat metabolic disorders with sustained interdisciplinary research and strategic clinical translation.

References

1. Ananthy, V., Priyadharsini, R. P., & Subramanian, U. (2021). Pathogenesis, diagnosis, and management of metabolic syndrome: A comprehensive review. *SBV Journal of Basic, Clinical and Applied Health Science, 4*(2), 39–45. https://doi.org/10.5005/jp-journals-10082-03111
2. Chu, D. T., Thi, H. V., Nguyen, T. T., et al. (2023). Nanotechnology and nucleic acid nanoparticles for treatment of metabolic disorders. *OpenNano, 13*, 100181. https://doi.org/10.1016/j.onano.2023.100181
3. DiBaise, J. K., Zhang, H., Crowell, M. D., & Rittmann, B. E. (2008). Gut microbiota and its possible relationship with obesity. *Mayo Clinic Proceedings, 83*(4), 460–469. https://doi.org/10.4065/83.4.460
4. Pant, R., Firmal, P., Shah, V. K., Alam, A., & Chattopadhyay, S. (2020). Epigenetic regulation of adipogenesis in development of metabolic disorders. *Frontiers in Cell and Developmental Biology, 8*, 619888. https://doi.org/10.3389/fcell.2020.619888
5. Hosseinpour-Niazi, S., Mirmiran, P., Hosseini, S., et al. (2022). Effect of TCF7L2 on the relationship between lifestyle factors and glycemic parameters: A systematic review. *Nutrition Journal, 21*(1), 59. https://doi.org/10.1186/s12937-022-00772-
6. Martemucci, G., Fracchiolla, G., Muraglia, M., Tardugno, R., Dibenedetto, R. S., & D'Alessandro, A. G. (2023). Metabolic syndrome: A narrative review from the oxidative stress to the management of related diseases. *Antioxidants, 12*(12), 2091. https://doi.org/10.3390/antiox12122091
7. Mohamed, S. M., Shalaby, M. A., El-Shiekh, R. A., El-Banna, H. A., Emam, S. R., & Bakr, A. F. (2023). Metabolic syndrome: Risk factors, diagnosis, pathogenesis, and management with natural approaches. *Food Chemistry Advances, 3*, 100335. https://doi.org/10.1016/j.focha.2023.100335
8. Ren, Y., Huang, P., Zhang, L., Tang, Y. F., Luo, S. L., She, Z., Peng, H., Chen, Y. Q., Luo, J. W., Duan, W. X., & Liu, L. J. (2022). Dual regulation mechanism of obesity: DNA methylation and the gut microbiota. *Biomedicines, 10*(8), 1633. https://doi.org/10.3390/biomedicines10081633

9. MDPI. (2024). The role of endocrine disruptors bisphenols and phthalates in metabolic disorders. *International Journal of Molecular Sciences, 25*(1), 675. https://doi.org/10.3390/ijms25010675
10. Nouri, Z., Hajialyani, M., Izadi, Z., Bahramsoltani, R., Farzaei, M. H., & Abdollahi, M. (2020). Nanophytomedicines for the prevention of metabolic syndrome: A pharmacological and biopharmaceutical review. *Frontiers in Bioengineering and Biotechnology, 8*, 425. https://doi.org/10.3389/fbioe.2020.00425
11. Galgani, J. E., & Fernández-Verdejo, R. (2021). Pathophysiological role of metabolic flexibility on metabolic health. *Obesity Reviews, 22*(3), e13131. https://doi.org/10.1111/obr.13131
12. Mohajer, N., Du, C. Y., Checkcinco, C., & Blumberg, B. (2021). Obesogens: How they are identified and molecular mechanisms of action. *Frontiers in Endocrinology, 12*, 780888. https://doi.org/10.3389/fendo.2021.780888
13. PubMed. (2018). TCF7L2 and Wnt signaling pathway in glucose metabolism. *Trends in Endocrinology and Metabolism, 29*(6), 437–445. https://doi.org/10.1016/j.tem.2018.03.003
14. Sharma, A. K. (2023). Current trends in nanotheranostics: A concise review on bioimaging and smart wearable technology. *Nano, 7*(3), 258–269. https://doi.org/10.7150/ntno.82886
15. Thomaz, F. S., John, O. D., Sinha, P., Shafie, S. R., & Worrall, S. (2024). The metabolic syndrome: An overview and proposed mechanisms. *Obesities, 4*(3), 226–255. https://doi.org/10.3390/obesities4030020

Pathophysiology Mechanism in Therapeutics of Metabolic Syndrome

2

Nitish Kumar Singh, Pooja Dubey, Surbhi Singh, and Royana Singh

Abstract

Metabolic syndrome (MetS) is a bunch of interconnected metabolic abnormalities, including central obesity, insulin resistance, dyslipidemia, hypertension, and chronic systemic inflammation. These interrelated conditions significantly increase the risk of type 2 diabetes mellitus, cardiovascular disease, and nonalcoholic fatty liver disease. This chapter examines the intricate pathophysiological mechanisms underlying MetS, including insulin signaling defects, adipose tissue dysfunction, dyslipidemia, oxidative stress, and mitochondrial impairment. It further outlines current and emerging therapeutic strategies, including dietary and physical activity interventions, pharmacologic agents (e.g., metformin, statins, GLP-1 receptor agonists), and novel approaches targeting gut microbiota and metabolic inflammation. A precision medicine approach, integrating genomics and digital health, is highlighted as the future direction in managing MetS. Understanding these mechanisms holistically is essential to designing effective preventive and therapeutic interventions to combat this global health challenge.

N. K. Singh
Multidisciplinary Research Unit, ICMR-DHR, Institute of Medical Sciences, Banaras Hindu University, Varanasi, Uttar Pradesh, India

P. Dubey · S. Singh · R. Singh (✉)
Department of Anatomy, Institute of Medical Science, Banaras Hindu University, Varanasi, Uttar Pradesh, India
e-mail: surbhi@bhu.ac.in; royanasingh@bhu.ac.in

V. K. Chaturvedi et al. (eds.), *Nanotheranostics in Metabolic Syndrome*,
https://doi.org/10.1007/978-981-95-3746-4_2

Keywords

Metabolic syndrome · Insulin resistance · Dyslipidemia · Oxidative stress · Mitochondrial dysfunction · Adipose tissue dysfunction · Lifestyle interventions · GLP-1 receptor agonists · Gut microbiota · Precision medicine

1 Introduction

Metabolic syndrome (MetS) is a complex and growing public health concern defined by a cluster of interrelated metabolic disturbances that significantly increase the likelihood of developing cardiovascular disease, type 2 diabetes, stroke, nonalcoholic fatty liver disease, chronic kidney disease, and premature death [56]. Originally conceptualized by Gerald Reaven in 1988 under the term "Syndrome X," MetS includes five major components: central (abdominal or visceral) obesity, insulin resistance, systemic hypertension, atherogenic dyslipidemia (primarily elevated triglycerides and reduced high-density lipoprotein cholesterol), and impaired glucose tolerance [70]. In recent decades, the incidence of MetS has surged globally, paralleling profound shifts in diet, physical activity, socioeconomic structure, and urbanization. The rapid transition from traditional to Westernized lifestyles has led to caloric excess, high consumption of refined carbohydrates and saturated fats, and diminished physical activity, collectively contributing to the obesity pandemic and a sharp increase in MetS prevalence [150].

Metabolic syndrome (MetS) is not just a diagnostic entity but a manifestation of intricate systemic disturbances affecting multiple organ systems. Central to its pathophysiology is insulin resistance, wherein key tissues like muscle, liver, and adipose tissue fail to respond effectively to insulin. This impairs glucose uptake, enhances hepatic glucose production, and disrupts lipid metabolism, resulting in hyperglycemia, hyperinsulinemia, and dyslipidemia [167]. Dysfunctional adipose tissue further exacerbates this state by secreting proinflammatory cytokines (e.g., TNF-α, IL-6) and reducing anti-inflammatory adipokines such as adiponectin, creating a chronic inflammatory milieu that drives endothelial dysfunction and atherogenesis [27].

Visceral adiposity, more lipolytically active than subcutaneous fat, releases excess free fatty acids (FFAs), promoting lipotoxicity in the liver, muscle, and pancreas. This contributes to a dyslipidemic profile characterized by high triglycerides, low HDL, and small, dense LDL particles [10]. Hypertension in MetS arises from insulin resistance-induced renal sodium retention, sympathetic overdrive, and RAAS activation, compounded by reduced nitric oxide and elevated endothelin-1 [44].

Additionally, mitochondrial dysfunction and oxidative stress amplify insulin resistance and inflammation. Organ crosstalk—particularly involving the gut–liver and gut–brain axes—underscores the role of dysbiosis and neurohormonal pathways in MetS, highlighting its systemic and interconnected nature [48].

1.1 Epidemiology and Global Burden

Metabolic syndrome (MetS) has emerged as one of the most persistent global public health challenges of the twenty-first century. Defined by a collection of interrelated metabolic abnormalities, including central obesity, dyslipidemia, hypertension, and impaired glucose metabolism, MetS significantly elevates the risk of developing cardiovascular disease (CVD), type 2 diabetes mellitus (T2DM), and premature mortality. In recent decades, the incidence of metabolic syndrome (MetS) has risen sharply worldwide, driven by changes in lifestyle patterns, increased urbanization, aging demographics, and evolving dietary habits [77].

1.1.1 Global Prevalence

Estimates suggest that the global prevalence of MetS ranges between 20% and 30% among adults. However, this figure is not static and varies widely depending on several factors, including the diagnostic criteria applied, the population studied, and regional gaps in lifestyle and socioeconomic status [3]. For instance, data derived from the National Health and Nutrition Examination Survey (NHANES) in the United States indicate that as many as 34.7% of adults meet the NCEP ATP III criteria for MetS. The prevalence is even higher among specific subgroups, including older adults, individuals with obesity, and certain ethnic minorities such as Hispanic Americans and African Americans [126].

MetS affects approximately 25–30% of adults in Europe, although variability exists between Western and Eastern European nations due to differences in diet, healthcare access, and socioeconomic development [25]. In the Middle East and North Africa (MENA) region, some studies report prevalence rates exceeding 40%, reflecting high rates of obesity and physical inactivity. Similarly, in South Asia, particularly in countries like India, Pakistan, and Bangladesh, the prevalence of MetS is increasing rapidly, often manifesting at younger ages and lower body mass indices (BMIs), a phenomenon attributed to the unique metabolic phenotype of South Asians characterized by increased visceral adiposity and insulin resistance [3].

In Southeast Asia, Latin America, and Sub-Saharan Africa, the epidemiological burden of MetS is on a sharp upward trajectory [117]. This tendency is largely driven by demographic transitions, rural-to-urban migration, and the adoption of Westernized diets rich in refined carbohydrates, saturated fats, and processed foods. Urban populations in these regions often experience a dual burden of malnutrition and undernutrition in early life, followed by overnutrition in adulthood, strongly associated with insulin resistance and MetS development [34]. As these regions undergo economic development, their healthcare systems are being increasingly strained by the rising tide of noncommunicable diseases (NCDs), with MetS playing a pivotal role in this transformation.

1.1.2 Pediatric and Adolescent MetS

A particularly alarming trend is the emergence of MetS in pediatric and adolescent populations. In the past, MetS was largely viewed as a condition of middle-aged or older adults. However, the global rise in childhood and adolescent obesity has shifted this paradigm [168]. Studies now show that MetS affects 3–10% of adolescents in developed countries, with higher rates observed among those who are overweight or obese [111]. In the United States, for example, approximately one-third of obese adolescents meet the criteria for MetS. The implications are profound, as children with MetS are at increased risk of developing early-onset T2DM, hypertension, dyslipidemia, and CVD conditions traditionally associated with adulthood [68].

Furthermore, exposure to adverse intrauterine environments, such as maternal obesity, gestational diabetes, or undernutrition, may epigenetically predispose offspring to MetS later in life. These findings underscore the need for preventive strategies that begin early in life, even during pregnancy, to curb the intergenerational transmission of metabolic risk [26].

1.1.3 Economic Impact

Metabolic syndrome (MetS) imposes a significant and complex economic burden. Direct healthcare expenditures stem from the diagnosis, treatment, and ongoing management of its core conditions—such as obesity, type 2 diabetes, hypertension, and dyslipidemia as well as associated complications like heart attacks, strokes, chronic kidney disease, and nonalcoholic fatty liver disease (NAFLD). Indirect costs arise from reduced workplace productivity, including absenteeism, diminished performance while on the job (presenteeism), early onset of disability, and premature retirement [32].

1.1.4 Risk Factors

The development of metabolic syndrome (MetS) is driven by a complex interplay of multiple factors, including genetic predisposition, environmental influences, lifestyle behaviors, and hormonal imbalances.

- *Central Obesity, especially visceral fat accumulation*: Excess abdominal fat is considered the primary risk factor for MetS. Visceral adipose tissue is metabolically active and contributes to systemic inflammation and insulin resistance by releasing free fatty acids and inflammatory mediators [10].
- *Physical inactivity*: Sedentary behavior is strongly associated with impaired glucose uptake, reduced insulin sensitivity, increased fat accumulation, and dyslipidemia [31].
- *Dietary habits*: Diets high in saturated fats, trans fats, refined sugars, and sugar-sweetened beverages are linked to weight gain, dyslipidemia, insulin resistance, and endothelial dysfunction. Conversely, diets rich in whole grains, fiber, unsaturated fats, fruits, and vegetables offer protective effects [112].
- *Age and gender*: The prevalence of MetS increases with age. In early adulthood, MetS tends to be more common in men, but postmenopausal women rapidly catch up due to hormonal changes that promote fat redistribution, insulin resistance, and lipid abnormalities [124].
- *Genetic predisposition*: Numerous gene variants have been associated with components of MetS. For instance, polymorphisms in the *FTO*, *PPARG*,

TCF7L2, and *ADIPOQ* genes influence susceptibility to obesity, insulin resistance, and dyslipidemia [138].
- *Epigenetic mechanisms*: Epigenetic modifications such as DNA methylation, histone acetylation, and regulation by microRNAs (miRNAs) are increasingly recognized as key regulators of metabolic gene expression. These modifications can be influenced by prenatal nutrition, environmental exposures, and lifestyle factors, and may partially explain the heritability and intergenerational transmission of MetS [6].

1.2 Diagnostic Criteria

Metabolic syndrome (MetS) represents a convergence of several cardiometabolic risk factors that collectively increase the likelihood of developing type 2 diabetes mellitus (T2DM), cardiovascular disease (CVD), and related complications [150]. In recognition of its growing health burden, several international organizations have developed diagnostic criteria for MetS, aiming to standardize its identification and facilitate early intervention. The three most prominent and widely used definitions are those proposed by: The National Cholesterol Education Program Adult Treatment Panel III (NCEP ATP III) in 2001, The World Health Organization (WHO) in 1999, and The International Diabetes Federation (IDF) in 2005 [56].

1.2.1 NCEP ATP III (2001)

According to the NCEP ATP III guidelines, metabolic syndrome is diagnosed when an individual exhibits three or more out of five specific metabolic risk factors. This framework emphasizes a flexible approach, without mandating central obesity or insulin resistance as prerequisites, as summarized in Table 2.1 [126].

1.2.2 WHO (1999)

The WHO was the first major body to define MetS and emphasized *insulin resistance* as a mandatory criterion. Diagnosis requires the presence of insulin resistance (e.g., T2DM, impaired glucose tolerance) plus two or more additional factors, as summarized in Table 2.2 [102].

Table 2.1 Diagnostic criteria for metabolic syndrome based on NCEP ATP III guidelines

Component	Men	Women
Waist circumference	>102 cm (40 inches)	>88 cm (35 inches)
Triglycerides	≥150 mg/dL (1.7 mmol/L) or on treatment	≥150 mg/dL (1.7 mmol/L) or on treatment
HDL cholesterol	<40 mg/dL (1.03 mmol/L)	<50 mg/dL (1.29 mmol/L)
Blood pressure	≥130/85 mmHg or on antihypertensive meds	≥130/85 mmHg or on antihypertensive meds
Fasting glucose	≥100 mg/dL (5.6 mmol/L) or on glucose meds	≥100 mg/dL (5.6 mmol/L) or on glucose meds

Table 2.2 Diagnostic criteria for metabolic syndrome based on WHO guidelines

Component	Criteria	Sex-specific thresholds
Core requirement	Evidence of insulin resistance or glucose abnormality	Includes type 2 diabetes, impaired fasting glucose, impaired glucose tolerance, or elevated HOMA-IR
Blood pressure	Elevated systolic or diastolic pressure	≥140/90 mmHg (both sexes)
Triglyceride levels	Raised fasting triglyceride concentration	≥150 mg/dL (1.7 mmol/L)
HDL cholesterol	Reduced high-density lipoprotein levels	<35 mg/dL (0.9 mmol/L) in men <39 mg/dL (1.0 mmol/L) in women
Abdominal obesity	Central fat accumulation based on waist-to-hip ratio or BMI	WHR >0.90 in men, >0.85 in women or BMI >30 kg/m^2
Microalbuminuria	Increased urinary protein excretion	Urinary albumin ≥20 μg/min or ACR ≥30 mg/g

Table 2.3 Diagnostic criteria for metabolic syndrome based on IDF (2005) guidelines

Component	Criteria	Sex/ethnicity-specific thresholds
Required: central obesity	Waist circumference adjusted for ethnic background	Europid men: ≥94 cm Europid women: ≥80 cm South Asian/Chinese men: ≥90 cm South Asian/Chinese women: ≥80 cm
Triglycerides	Elevated fasting triglycerides or on treatment for dyslipidemia	≥150 mg/dL (1.7 mmol/L)
HDL cholesterol	Reduced HDL levels or under therapeutic intervention	Men: <40 mg/dL (1.03 mmol/L) Women: <50 mg/dL (1.29 mmol/L)
Blood pressure	Increased blood pressure or receiving antihypertensive therapy	Systolic ≥130 mmHg or diastolic ≥85 mmHg
Fasting blood glucose	Raised fasting glucose or a prior diagnosis of diabetes	≥100 mg/dL (5.6 mmol/L) or diagnosed T2DM

1.2.3 IDF (2005)

The IDF criteria emphasize *central obesity as an essential criterion*, recognizing its pivotal role in the pathogenesis of MetS. The IDF also introduced *ethnicity-specific waist circumference thresholds* to reflect global variations in body fat distribution, as summarized in Tables 2.3 and 2.4 [5].

Table 2.4 Highlight differences in the diagnostic emphasis and flexibility among these definitions

Feature	NCEP ATP III (2001)	WHO (1999)	IDF (2005)
Mandatory criteria	None	Insulin resistance	Central obesity
Number of factors required	Any three out of five	Insulin resistance + two other factors	Central obesity + two other factors
Central obesity emphasis	One of five factors	Considered with BMI/WHR	Essential with ethnic-specific cutoffs
Insulin resistance	Not mandatory	Mandatory	Not mandatory
Microalbuminuria	Not included	Included	Not included
Flexibility in diagnosis	High	Moderate	Moderate

1.3 Ethnic-Specific Considerations

Ethnicity plays a pivotal role in the clinical expression, prevalence, and diagnostic assessment of metabolic syndrome (MetS). A uniform diagnostic criterion does not equally apply to all ethnic groups, as significant heterogeneity exists in body fat distribution, susceptibility to insulin resistance, and metabolic response to environmental stimuli. This necessitates ethnic-specific adaptations in diagnostic thresholds and risk-stratification strategies [3].

1.3.1 Body Composition and Central Obesity

One of the key components of MetS, central obesity, varies markedly across populations. Although widely used, body Mass Index (BMI) and waist circumference often fail to account for the metabolic impact of fat distribution in diverse ethnic groups [68]. For instance, Asians tend to accumulate more visceral adipose tissue (VAT) at lower BMIs compared to Caucasians, predisposing them to metabolic complications despite appearing nonobese by conventional standards. This phenomenon has led to the concept of "metabolically obese normal weight" (MONW) individuals, who carry a higher risk of insulin resistance and type 2 diabetes mellitus (T2DM), as summarized in Table 2.5 [153].

1.3.2 Genetic and Epigenetic Predisposition

Ethnic disparities in the prevalence of MetS are also partly attributable to genetic architecture. Specific single-nucleotide polymorphisms (SNPs) have been linked to increased susceptibility in different populations [120]. For instance:

- *FTO gene variants* are strongly associated with obesity in Europeans but have variable associations in Asians and Africans [25].
- *TCF7L2* polymorphisms have been implicated in glucose dysregulation predominantly in Caucasians and Hispanics [35].
- *APOA5 and CETP variants* show strong links to dyslipidemia among East Asians [104].

Table 2.5 Ethnic-specific waist circumference cut-offs for central obesity (IDF Recommendations)

Population group	Men (waist circumference)	Women (waist circumference)
European descent	94 cm or more	80 cm or more
South Asian origin	90 cm or more	80 cm or more
Chinese	90 cm or more	80 cm or more
Japanese	85 cm or more	90 cm or more
Sub-Saharan African	94 cm or more	80 cm or more
South & Central American	Follow South Asian values or use regional standards	Same as for men

Moreover, epigenetic modifications—such as DNA methylation patterns influenced by early-life nutrition and environmental exposures—also exhibit ethnic specificity, affecting gene expression related to metabolism and inflammation.

1.4 Pathophysiological Hallmarks

Metabolic syndrome (MetS) is not merely a collection of clinical markers but a manifestation of interconnected molecular and cellular dysfunctions. Central to its pathogenesis are insulin resistance, chronic low-grade inflammation, oxidative stress, and hormonal imbalance—each contributing to the syndrome's multisystemic impact [157].

1.4.1 Insulin Resistance: The Cornerstone of MetS

Insulin resistance (IR) refers to a diminished biological response to insulin in target tissues—mainly liver, muscle, and adipose tissue. It is the pivotal mechanism that links the core features of MetS [167].

1.4.1.1 Hepatic Insulin Resistance

- Leads to unchecked gluconeogenesis, contributing to fasting hyperglycemia [162].
- Stimulates de novo lipogenesis and overproduction of very low-density lipoproteins (VLDL), fostering hypertriglyceridemia and hepatic steatosis [79].

1.4.1.2 Skeletal Muscle IR

- Impairs glucose uptake via reduced GLUT4 translocation [46].
- Reduces glycogen synthesis, contributing to postprandial hyperglycemia [2].

1.4.1.3 Adipose Tissue IR

- Enhances lipolysis, increasing free fatty acids (FFAs) in circulation [119].
- FFAs promote hepatic IR, interfere with insulin signaling, and contribute to lipotoxicity [24].

This interplay generates a vicious cycle wherein IR leads to compensatory hyperinsulinemia, which itself exacerbates metabolic disturbances [167].

1.4.2 Inflammation and Adipokine Imbalance

MetS is characterized by chronic low-grade systemic inflammation originating predominantly from dysfunctional adipose tissue. This inflammation is both a driver and consequence of insulin resistance [45].

1.4.2.1 Adipose Tissue Inflammation

- In obesity, especially visceral fat expansion, adipocytes become hypertrophic and hypoxic [27].
- Macrophages infiltrate the adipose tissue, shifting from an anti-inflammatory M2 to a proinflammatory M1 phenotype [27].
- These immune cells release cytokines such as:
  - *TNF-α*: Inhibits insulin receptor signaling [160].
  - *IL-6*: Promotes hepatic C-reactive protein (CRP) synthesis and insulin resistance [160].
  - *MCP-1*: Recruits monocytes, perpetuating inflammation [160].

1.4.2.2 Dysregulated Adipokines

- *Adiponectin*: An insulin-sensitizing adipokine is reduced in MetS [125].
- *Leptin*: Often elevated in obesity, but with leptin resistance blunting satiety signals and promoting sympathetic activation [125].
- *Resistin and visfatin*: Contribute to endothelial dysfunction and atherogenesis [121].

1.4.3 Oxidative Stress and Mitochondrial Dysfunction

Reactive oxygen species (ROS) play a dual role as signaling molecules and mediators of tissue damage. In MetS, ROS production overwhelms antioxidant defenses, promoting cellular injury [40].

- *Sources of ROS in MetS*:
  - Mitochondrial electron transport chain leakage [166].
  - NADPH oxidase activation by cytokines and FFAs [83].
  - Glycation reactions in hyperglycemia [23].

ROS activate redox-sensitive transcription factors (e.g., NF-κB), augmenting the expression of proinflammatory cytokines. Oxidative stress also:

- Damages endothelial cells, contributing to hypertension and atherosclerosis [74].
- Impairs insulin receptor substrates (IRS-1/2), worsening insulin resistance [167].
- Promotes lipid peroxidation and mitochondrial DNA damage, impairing ATP production, shown in the Fig. 2.1 [99].

1.4.4 Neurohormonal Dysregulation and Gut Microbiota

Neuroendocrine and microbial factors play a complementary role in sustaining the MetS phenotype [86].

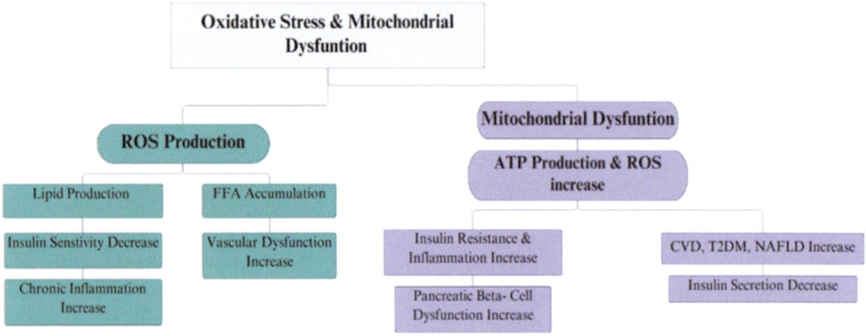

Fig. 2.1 A hierarchical diagram illustrates the role of oxidative stress and mitochondrial dysfunction in triggering ROS production, insulin resistance, inflammation, and the progression of metabolic diseases such as CVD, T2DM, and NAFLD

1.4.4.1 Neurohormonal Activation

- *RAAS Activation*: Angiotensin II promotes vasoconstriction, inflammation, and oxidative stress [13].
- *Sympathetic Nervous System Overactivity*: Elevates blood pressure and promotes lipolysis [148].
- *Cortisol Excess*: Chronic stress or subclinical hypercortisolism (e.g., Cushingoid features) leads to central obesity and insulin resistance [91].

1.4.4.2 Gut Microbiota

- Dysbiosis alteration in microbial composition disrupts intestinal barrier function, allowing lipopolysaccharide (LPS) translocation into circulation (metabolic endotoxemia) [54].
- Short-chain fatty acid (SCFA) imbalance affects energy harvest and lipid metabolism [164].
- Microbial metabolites modulate bile acid signaling via the farnesoid X receptor (FXR), influencing glucose and lipid homeostasis [133].

2 Metabolism and Metabolic Disorders

Metabolism is a fundamental biological process through which organisms convert food and nutrients into energy, structural components, and signaling molecules necessary for cellular function and survival. This dynamic and highly regulated network of biochemical reactions ensures the maintenance of cellular and systemic homeostasis. The human metabolic network encompasses various interdependent pathways involving carbohydrates, lipids, proteins, nucleic acids, vitamins, and minerals. These reactions are compartmentalized within cells and orchestrated by hormones, enzymes, and coenzymes [1].

Metabolism is classically divided into two overarching categories:

- *Anabolism*—the biosynthesis of complex molecules such as proteins, nucleic acids, and lipids from simpler precursors. It is an energy-consuming process that supports growth, repair, and cellular differentiation [128].
- *Catabolism*—the degradation of macromolecules into smaller units, resulting in the release of energy in the form of adenosine triphosphate (ATP). Catabolic pathways include glycolysis, β-oxidation, and amino acid breakdown [128].

2.1 Metabolic Disorders Can Be Categorized into Two Main Groups

1. *Inherited Metabolic Disorders*—Often manifest early in life due to mutations in genes encoding metabolic enzymes or transporters [135].
2. *Acquired Metabolic Disorders*—These result from environmental, dietary, or lifestyle influences and tend to manifest in adolescence or adulthood [62].

One of the most clinically relevant acquired metabolic disorders today is *metabolic syndrome (MetS)*. MetS is not a single disease entity but a cluster of interrelated metabolic disturbances that heighten the risk for type 2 diabetes mellitus (T2DM), cardiovascular disease (CVD), nonalcoholic fatty liver disease (NAFLD), and certain cancers. The primary components of MetS include: Central obesity (visceral adiposity), Insulin resistance (IR), Dyslipidemia (elevated triglycerides, low HDL cholesterol), Hypertension, Impaired glucose metabolism [60].

2.2 Overview of Metabolic Pathways

Metabolic pathways consist of carefully controlled enzymatic steps that convert macronutrients into energy or key molecules needed for biosynthesis. These processes take place within different parts of the cell, including the cytosol, mitochondria, and endoplasmic reticulum, and are regulated by factors such as nutrient levels, hormonal cues, enzyme allosteric effects, and posttranslational enzyme modifications [1].

In the context of MetS, disturbances in three core macronutrient metabolic pathways carbohydrates, lipids, and proteins are central to the syndrome's pathophysiology [31].

2.2.1 Carbohydrate Metabolism

Carbohydrates are the most immediate source of energy for human cells, particularly for the brain and erythrocytes. Glucose, the principal monosaccharide, is absorbed through the intestinal epithelium and transported via the bloodstream to various tissues [87].

Key pathways in carbohydrate metabolism include:

- *Glycolysis*: Occurs in the cytosol; glucose is converted to pyruvate, producing ATP and NADH. Under anaerobic conditions, pyruvate is converted to lactate [122].
- *Glycogenesis*: Excess glucose is stored as glycogen in the liver and skeletal muscles, a process promoted by insulin [122].
- *Glycogenolysis*: The breakdown of glycogen back to glucose, primarily during fasting, is driven by glucagon and epinephrine [122].
- *Gluconeogenesis*: Synthesis of glucose from noncarbohydrate precursors (lactate, alanine, glycerol) in the liver and kidneys, especially during prolonged fasting [122].
- *Pentose Phosphate Pathway (PPP)*: Produces NADPH and ribose-5-phosphate for anabolic reactions and redox balance [122].

2.2.2 Lipid Metabolism

Lipids are hydrophobic molecules crucial for energy storage, cellular architecture, and signaling. The primary pathways include:

- *Lipolysis*: Breakdown of triglycerides in adipose tissue into glycerol and free fatty acids (FFAs), facilitated by hormone-sensitive lipase (HSL) [14].
- *Fatty Acid β-Oxidation*: FFAs are oxidized in mitochondria to produce acetyl-CoA, NADH, and $FADH_2$. Acetyl-CoA enters the tricarboxylic acid (TCA) cycle for further energy production [82].
- *Lipogenesis*: The synthesis of fatty acids from acetyl-CoA, typically in the liver and adipose tissue. It is stimulated by excess carbohydrates and insulin [65].
- *Cholesterol Synthesis*: A multistep process that begins with acetyl-CoA and is regulated by HMG-CoA reductase, the target of statins [14].

2.2.3 Protein Metabolism

Proteins are composed of amino acids, which serve both structural and regulatory roles. After digestion, amino acids enter circulation for use in protein synthesis, energy production, or conversion into other bioactive molecules [42].

Major processes include:

- *Transamination and Deamination*: Amino acids are converted into keto acids, which enter the TCA cycle [42].
- *Urea Cycle*: Detoxifies ammonia produced from amino acid catabolism, converting it into urea for renal excretion [42].
- *Gluconeogenesis*: Glucogenic amino acids such as alanine and glutamine serve as substrates for glucose synthesis during fasting [42].
- *mTOR Pathway Activation*: Senses nutrient availability, particularly branched-chain amino acids (BCAAs), and regulates cellular growth and protein synthesis [42].

2.3 Genetic and Acquired Metabolic Disorders

Metabolic disorders encompass a broad spectrum of conditions arising from disruptions in the body's metabolic processes. These disturbances can result from inherited genetic mutations (inborn errors of metabolism) or from acquired factors such as lifestyle, aging, and organ dysfunction. Together, these disorders highlight the vulnerability of metabolic homeostasis to both intrinsic and extrinsic influences, and understanding their etiologies is central to addressing the broader challenge of metabolic syndrome (MetS) [165].

2.3.1 Inborn Errors of Metabolism (IEMs)

Inborn errors of metabolism (IEMs) are a group of rare, genetically determined disorders resulting from mutations that affect the function of enzymes, cofactors, or transporters involved in key metabolic pathways. These errors often lead to the accumulation of toxic substrates or the deficiency of critical metabolic products. Although individually rare, collectively, IEMs represent a significant burden due to their early onset and often severe clinical manifestations [149].

2.3.1.1 Genetic Basis and Inheritance Patterns

Most IEMs follow an autosomal recessive inheritance pattern, meaning that both parents must carry one copy of the mutated gene for the child to manifest the disorder. X-linked and mitochondrial inheritance patterns are less common but also occur. Modern genetic techniques, including whole-exome sequencing, have facilitated the identification of these mutations, enabling early diagnosis and targeted treatment, as summarized in Table 2.6 [149].

2.3.2 Acquired Metabolic Disorders

In contrast to IEMs, acquired metabolic disorders develop due to interactions between environmental factors and host physiology. These conditions often manifest in adulthood and are linked to lifestyle factors such as diet, physical activity, stress, and exposure to environmental toxins. Unlike genetic disorders, acquired metabolic diseases are typically preventable and reversible, particularly in their early stages [62].

Table 2.6 Key enzymes in metabolic disorders

Disorder	Deficient enzyme	Accumulated metabolite
Phenylketonuria (PKU)	Phenylalanine hydroxylase	Phenylalanine
Maple syrup urine disease	Branched-chain α-keto acid dehydrogenase	Leucine, isoleucine, valine
Galactosemia	Galactose-1-phosphate uridyltransferase	Galactose-1-phosphate
Gaucher disease	Glucocerebrosidase	Glucocerebroside

2.3.2.1 Etiological Factors

Acquired metabolic disorders arise from:

- *Obesity*: Particularly central (visceral) obesity, which promotes insulin resistance and systemic inflammation [155].
- *Sedentary lifestyle*: Reduces muscle glucose uptake and promotes fat accumulation [155].
- *Persistent stress and sleep disturbances*: These conditions disrupt the hypothalamic–pituitary–adrenal (HPA) axis, leading to elevated cortisol levels and subsequent changes in glucose and lipid metabolism [120].
- *Exposure to environmental toxins*: Includes substances that interfere with hormonal systems, such as bisphenol A (BPA), phthalates, and certain heavy metals [91].
- *Environmental exposures*: Including endocrine-disrupting chemicals (e.g., BPA, phthalates) and heavy metals (e.g., lead, cadmium) [153].

2.3.2.2 Common Acquired Disorders Associated with Metabolic Syndrome

1. *Type 2 Diabetes Mellitus (T2DM)*
   (a) A progressive metabolic disorder marked by resistance to insulin and impaired pancreatic β-cell function. It often develops gradually and is frequently diagnosed at advanced stages. T2DM can lead to complications affecting the kidneys (nephropathy), nerves (neuropathy), eyes (retinopathy), and cardiovascular system [150].
2. *Nonalcoholic Fatty Liver Disease (NAFLD)*
   (a) This condition encompasses a range of liver abnormalities starting from simple fat accumulation (steatosis), progressing to inflammation and liver cell injury (nonalcoholic steatohepatitis, NASH), which may further develop into fibrosis and cirrhosis. NAFLD is strongly associated with visceral fat accumulation and insulin resistance, and it has become the leading cause of chronic liver disease worldwide [10, 118].
3. *Dyslipidemia*
   (a) Atherogenic dyslipidemia is a hallmark of MetS: elevated triglycerides, small dense LDL particles, and low HDL cholesterol [67].
   (b) Promotes atherosclerosis, myocardial infarction, and stroke [67].
4. *Hypertension*
   (a) Multifactorial origin: endothelial dysfunction, RAAS activation, increased sympathetic tone [148].
   (b) Often coexists with obesity and insulin resistance [148].
5. *Polycystic Ovary Syndrome (PCOS)*
   (a) Endocrine disorder in women of reproductive age, characterized by hyperandrogenism, ovulatory dysfunction, and polycystic ovaries [142].
   (b) Strongly associated with insulin resistance and obesity [142].
6. *Gout and Hyperuricemia*
   (a) MetS is a major risk factor for elevated serum uric acid levels [18].
   (b) Hyperinsulinemia reduces renal uric acid clearance [18].

2.4 Dyslipidemia in Metabolic Syndrome

Dyslipidemia is a key feature of metabolic syndrome, marked by increased triglyceride and low-density lipoprotein (LDL) levels, along with decreased concentrations of high-density lipoprotein (HDL). Commonly seen in individuals with insulin resistance, type 2 diabetes, and obesity, this lipid imbalance significantly contributes to the risk of developing cardiovascular diseases [67].

2.4.1 Pathophysiological Mechanisms of Dyslipidemia

- *Free Fatty Acid (FFA) Overflow*: In insulin-resistant states, uncontrolled lipolysis in adipose tissue releases an excess of FFAs. These are taken up by the liver, which synthesizes and secretes very-low-density lipoprotein (VLDL) particles, contributing to hypertriglyceridemia [17].
- *Impaired Lipoprotein Lipase (LPL) Activity*: LPL is responsible for hydrolyzing triglycerides in circulating lipoproteins. Insulin resistance suppresses LPL activity, impairing clearance of VLDL and chylomicrons [59].
- *Decreased HDL Synthesis*: Insulin normally enhances apolipoprotein A-I expression and HDL maturation. In MetS, insulin resistance and chronic inflammation reduce HDL production and promote HDL catabolism [137].
- *Increased Cholesteryl Ester Transfer Protein (CETP) Activity*: CETP exchanges triglycerides from VLDL for cholesteryl esters from HDL, resulting in triglyceride-rich, less stable HDL that is rapidly cleared [14].

2.4.2 Clinical Implications

Dyslipidemia in MetS is typically atherogenic, with small dense LDL particles and low HDL levels contributing to endothelial dysfunction and plaque formation. It increases the risk for myocardial infarction, stroke, and peripheral artery disease [67].

2.4.3 Genetic Influences

Although acquired, dyslipidemia often has a genetic basis influencing severity. Polymorphisms in genes like: *APOB* (apolipoprotein B), *LDLR* (LDL receptor), *LPL* (lipoprotein lipase), and *PCSK9* (proprotein convertase subtilisin/kexin type 9) can modulate lipid levels and treatment responses [21].

2.4.4 Management

Therapeutic strategies include:

- *Lifestyle Modification*: Diets rich in omega-3 fatty acids, fiber, and low in trans fats, combined with regular physical activity [130].
- *Pharmacotherapy*: Statins (HMG-CoA reductase inhibitors), fibrates, niacin, ezetimibe, and PCSK9 inhibitors [97].
- *Emerging Approaches*: Targeting ANGPTL3 and CETP inhibitors, shown in the Fig. 2.2 [97].

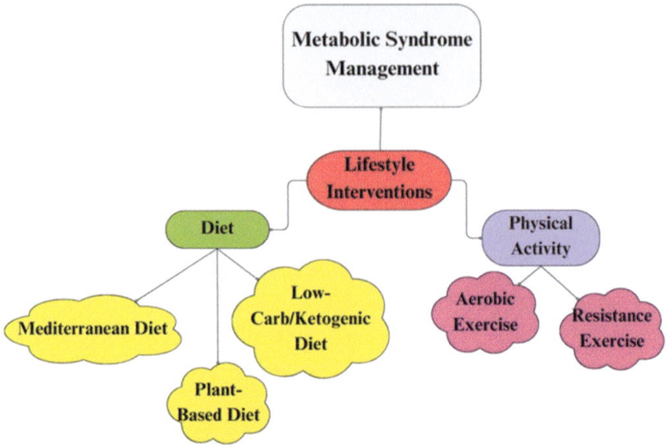

Fig. 2.2 Flowchart illustrating lifestyle interventions for managing metabolic syndrome, including dietary approaches (Mediterranean, plant-based, low-carb/ketogenic) and physical activities (aerobic and resistance exercise)

2.4.5 Genetic Contributions

While lifestyle and environmental factors play dominant roles in acquired dyslipidemia, genetic predisposition cannot be ignored. Several genetic polymorphisms have been identified that modulate lipid metabolism and predispose individuals to dyslipidemia, particularly when compounded by poor dietary habits or physical inactivity [143].

- *APOB (Apolipoprotein B)*: Mutations can impair LDL receptor binding, reducing LDL clearance [131].
- *LDLR (Low-Density Lipoprotein Receptor)*: Variants may impair LDL uptake by hepatocytes, raising serum LDL-C [64].
- *PCSK9 (Proprotein Convertase Subtilisin/Kexin Type 9)*: Gain-of-function mutations enhance LDL receptor degradation, increasing LDL levels [156].
- *LPL (Lipoprotein Lipase)*: Mutations reduce hydrolytic activity on TRLs, contributing to hypertriglyceridemia [72].

2.4.6 Clinical Manifestations and Diagnosis

Most individuals with dyslipidemia are asymptomatic and are diagnosed during routine screening. However, in severe cases or familial dyslipidemias, clinical manifestations may include xanthomas, xanthelasmas, and corneal arcus. Laboratory diagnosis involves fasting or nonfasting lipid profile assessments. Key parameters include such as Total cholesterol (TC), Triglycerides (TG), LDL-C (calculated or direct), HDL-C, nonHDL cholesterol (TC-HDL-C) [143].

2.4.7 Management Strategies

1. *Lifestyle Modifications*: Lifestyle interventions remain the first line of management. Key recommendations include:
   - *Diet*: Reduction of saturated and trans fats, increased intake of omega-3 fatty acids, fiber, and plant sterols [47].
   - *Exercise*: Regular aerobic activity improves insulin sensitivity and promotes favorable lipid changes [151].
   - *Weight Loss*: Even a 5–10% reduction in body weight significantly improves lipid parameters [151].
2. *Pharmacological Therapy*:
   - *Statins* (HMG-CoA reductase inhibitors): Most effective in reducing LDL-C and total cholesterol; have pleiotropic anti-inflammatory effects [47].
   - *Fibrates*: Used for hypertriglyceridemia; activate PPAR-α to increase LPL activity [47].
   - *Niacin*: Increases HDL-C, reduces TG, and LDL-C, but limited by side effects [151].
   - *Omega-3 fatty acids*: Particularly effective in reducing TG levels [130].
   - *PCSK9 inhibitors*: Newer agents that significantly lower LDL-C, particularly useful in familial hypercholesterolemia or statin-intolerant patients [156].
   - *Ezetimibe*: Inhibits cholesterol absorption; often used as adjunct to statins [130].
3. *Novel Therapies and Future Directions*: Advances in understanding lipid metabolism have led to emerging therapies targeting specific molecular pathways, including antisense oligonucleotides against ApoC-III, ANGPTL3 inhibitors, and gene editing approaches [81].

2.4.8 Public Health Implications

Given its widespread prevalence and role in CVD, dyslipidemia is a major public health concern. National programs aimed at cardiovascular risk screening and lipid monitoring are essential for early identification and intervention. Moreover, policies promoting healthier diets, physical activity, and smoking cessation are vital in tackling this metabolic derangement at the population level [145].

2.5 Hyperinsulinemia and Insulin Resistance

Hyperinsulinemia and insulin resistance (IR) are central pathophysiological components of metabolic syndrome (MetS), underlying a range of clinical disorders including type 2 diabetes mellitus (T2DM), polycystic ovarian syndrome (PCOS), nonalcoholic fatty liver disease (NAFLD), hypertension, and cardiovascular diseases. While often considered two sides of the same coin, hyperinsulinemia refers to elevated circulating insulin levels, whereas insulin resistance denotes a diminished response of peripheral tissues—primarily muscle, liver, and adipose tissue—to insulin's biological actions. In acquired metabolic disorders, these phenomena reflect both a cause and consequence of metabolic dysregulation [167].

2.5.1 Physiology of Insulin Action

Insulin is a peptide hormone produced by the β-cells of the pancreas that is essential for regulating both glucose and lipid metabolism. Following the intake of carbohydrates, a rise in blood glucose levels stimulates insulin release. The hormone binds to insulin receptors on target cells, triggering downstream signaling pathways. These include the PI3K–Akt pathway, which mediates key metabolic processes such as glucose transport and glycogen formation, and the MAPK pathway, which influences cellular proliferation and differentiation [76].

2.5.2 Pathogenesis of Insulin Resistance and Hyperinsulinemia

In acquired metabolic disorders, a complex interplay of genetic predisposition, chronic overnutrition, physical inactivity, and ectopic fat deposition culminates in systemic insulin resistance [167]. The initial response of pancreatic β-cells is to increase insulin secretion to compensate for the reduced insulin sensitivity. This leads to compensatory hyperinsulinemia, which maintains normoglycemia in the early stages. However, chronic overstimulation of β-cells ultimately results in β-cell dysfunction, leading to impaired glucose tolerance and overt diabetes [76].

2.5.3 Mechanisms Involved in IR and Hyperinsulinemia Include

2.5.3.1 Adipose Tissue Dysfunction

- Visceral adiposity is metabolically active and secretes proinflammatory adipokines (e.g., TNF-α, IL-6, resistin) and less adiponectin, fostering insulin resistance [121].
- Enlarged adipocytes exhibit enhanced lipolysis, releasing excess FFAs that interfere with insulin signaling in muscle and liver [159].

2.5.3.2 Lipotoxicity and Ectopic Fat Accumulation

- Excess free fatty acids (FFAs) in liver and muscle tissues disrupt insulin signaling by activating protein kinase C (PKC). This activation leads to the phosphorylation of serine residues on the insulin receptor substrate (IRS), which hampers its ability to propagate insulin signals effectively [17].
- Intrahepatic and intramyocellular lipid deposition strongly correlate with systemic insulin resistance [59].

2.5.3.3 Chronic Inflammation

- Low-grade systemic inflammation is a hallmark of obesity-associated IR. Proinflammatory cytokines inhibit insulin receptor signaling pathways and recruit immune cells (macrophages) to adipose tissue [160].
- NF-κB and JNK pathways are activated in inflammatory states, reducing insulin sensitivity [160].

2.5.3.4 Mitochondrial Dysfunction and Oxidative Stress

- Impaired mitochondrial oxidative capacity leads to accumulation of lipid intermediates and reactive oxygen species (ROS), both of which inhibit insulin signaling [99].
- Mitochondrial biogenesis is also reduced in insulin-resistant states, further exacerbating energy dysregulation [48].

2.5.3.5 Endoplasmic Reticulum (ER) Stress

- Nutrient overload induces ER stress in hepatocytes and adipocytes, which activates the unfolded protein response (UPR) and promotes insulin resistance via c-Jun N-terminal kinase (JNK) activation [146].

2.5.3.6 Hyperinsulinemia-Induced Insulin Resistance

- Persistent hyperinsulinemia downregulates insulin receptors and desensitizes downstream signaling pathways [163].
- It also induces hepatic lipogenesis, increasing VLDL secretion and contributing to dyslipidemia and NAFLD [10].

2.5.3.7 Clinical and Biochemical Features

Insulin resistance is a largely asymptomatic condition until it progresses to impaired glucose tolerance or T2DM. However, several clinical markers can suggest its presence:

- Central obesity (waist circumference >102 cm in men or >88 cm in women) [69]
- Acanthosis nigricans: a dermatological sign associated with hyperinsulinemia [69]
- Polycystic ovary syndrome (PCOS) in reproductive-aged women [69]

2.5.3.8 Hyperinsulinemia and Its Systemic Effects

While initially compensatory, chronic hyperinsulinemia has widespread pathophysiological effects:

- *Endothelial Dysfunction*: Insulin promotes nitric oxide production; resistance shifts balance toward vasoconstriction, promoting hypertension [71].
- *Sympathetic Nervous System Activation*: Hyperinsulinemia increases catecholamine release, elevating blood pressure and heart rate [148].
- *Procoagulant State*: Insulin resistance elevates fibrinogen and PAI-1, increasing thrombosis risk [49].
- *Cell Proliferation*: Chronic activation of MAPK pathway contributes to smooth muscle proliferation and atherosclerosis [92].
- *Cancer Risk*: Hyperinsulinemia enhances IGF-1 signaling, linked to increased risk of breast, colorectal, and endometrial cancers [57].

2.5.3.9 Diagnosis and Assessment Tools

Several surrogate indices and dynamic tests assess IR:

- *HOMA-IR*: Most commonly used in epidemiological studies [57].
- *Matsuda Index*: Derived from oral glucose tolerance tests (OGTT) [57].
- *QUICKI*: (Quantitative Insulin Sensitivity Check Index).
- *Euglycemic-hyperinsulinemic clamp*: Gold standard in research settings [57].
- *Triglyceride-glucose (TyG) index*: An emerging marker correlated with IR and cardiovascular risk [100].

2.6 Nonalcoholic Fatty Liver Disease (NAFLD)

Nonalcoholic fatty liver disease (NAFLD) encompasses a range of liver disorders marked by the buildup of fat in liver cells in people who consume minimal or no alcohol. It has become the leading cause of chronic liver disease worldwide, closely linked to the increasing prevalence of obesity, type 2 diabetes, and metabolic syndrome [90]. Nonalcoholic fatty liver disease (NAFLD) includes a spectrum of liver abnormalities, starting with simple fat accumulation and potentially advancing to nonalcoholic steatohepatitis (NASH), followed by fibrosis, cirrhosis, and, in some cases, hepatocellular carcinoma (HCC). As the liver-related component of metabolic syndrome, NAFLD not only results from metabolic disturbances but also exacerbates systemic metabolic imbalance [11].

2.6.1 Epidemiology and Public Health Burden

Globally, NAFLD affects an estimated 25–30% of the adult population, with higher prevalence rates in specific regions such as the Middle East, South America, and parts of Asia. Among individuals with obesity or T2DM, the prevalence of NAFLD can exceed 70% [55]. Notably, NAFLD is increasingly diagnosed in pediatric populations, especially in children with obesity or insulin resistance, raising concerns about early liver-related and metabolic morbidity [154].

In India and other parts of South Asia, the burden of NAFLD is rising alarmingly, often in nonobese individuals due to higher visceral fat deposition and increased genetic susceptibility, illustrating the importance of ethnic-specific considerations. NAFLD has now overtaken viral hepatitis as the leading cause of chronic liver disease and liver transplantation in many Western countries. The socioeconomic cost of NAFLD is substantial due to increased healthcare utilization, loss of productivity, and long-term liver-related complications [152].

2.6.2 Pathogenesis: From Steatosis to Steatohepatitis

The development of NAFLD is multifactorial and complex, involving genetic, metabolic, and environmental contributors. The classical "two-hit" hypothesis has evolved into a more refined "multiple parallel hits" model, which better captures the interplay of various pathogenic factors [11].

1. *Insulin Resistance and Lipid Accumulation*: Insulin resistance is a central driver of NAFLD. In adipose tissue, insulin resistance leads to unregulated lipolysis and the release of free fatty acids (FFAs) into the circulation. These FFAs are taken up by the liver and re-esterified into triglycerides, resulting in hepatic steatosis. Concurrently, insulin fails to suppress hepatic gluconeogenesis and stimulates de novo lipogenesis (DNL), further exacerbating triglyceride accumulation [129].
2. *Mitochondrial Dysfunction and Oxidative Stress*: The liver's attempt to metabolize excess FFAs through β-oxidation results in the generation of reactive oxygen species (ROS). When the antioxidant defenses are overwhelmed, oxidative stress damages cellular structures, proteins, and DNA. This mitochondrial dysfunction not only impairs ATP synthesis but also contributes to hepatocyte injury and inflammation [48].
3. *Inflammation and Cell Death*: Hepatic steatosis sensitizes the liver to inflammatory stimuli. Kupffer cells (resident macrophages) are activated and produce proinflammatory cytokines such as TNF-α, IL-6, and IL-1β. Additionally, injured hepatocytes release damage-associated molecular patterns (DAMPs) that trigger further immune activation. This creates a cycle of chronic inflammation, apoptosis, and necroptosis, distinguishing NASH from benign steatosis [29].
4. *Fibrogenesis and Progression to Cirrhosis*: Persistent inflammation activates hepatic stellate cells, which deposit extracellular matrix components, leading to fibrosis. Over time, fibrotic remodeling disrupts the hepatic architecture, progressing to cirrhosis. NASH-related cirrhosis carries a significant risk of liver failure and hepatocellular carcinoma, even in the absence of significant alcohol consumption [51].
5. *Gut–Liver Axis and Microbiota Dysbiosis*: Emerging evidence implicates the gut microbiome in NAFLD pathogenesis. Increased intestinal permeability ("leaky gut") allows endotoxins such as lipopolysaccharides (LPS) to translocate into portal circulation, activating hepatic inflammation. Dysbiosis may also affect bile acid metabolism and choline availability, impacting lipid homeostasis [37].

2.6.3 Clinical Features and Diagnosis

Most patients with NAFLD are asymptomatic and are often diagnosed incidentally through elevated liver enzymes (ALT, AST) or imaging studies. When symptoms are present, they are typically nonspecific, such as fatigue, right upper quadrant discomfort, or malaise [105].

Diagnostic Modalities

- *Biochemical markers*: Mild elevations in ALT and AST, with ALT often being predominant. In advanced disease, AST may surpass ALT [144].
- *Imaging*:
  - *Ultrasound*: Common screening tool; shows increased echogenicity but lacks sensitivity in mild steatosis [105].
  - *FibroScan (Transient elastography)*: Measures liver stiffness and fat content [105].
  - *MRI-PDFF*: Most accurate noninvasive method to quantify hepatic fat [105].

- *Liver biopsy* remains the gold standard for diagnosing NASH and assessing fibrosis, but is invasive and not routinely performed unless clinically indicated [105].

2.6.4 Association with Metabolic Syndrome and Cardiovascular Risk

NAFLD is tightly linked with all components of MetS—central obesity, insulin resistance, dyslipidemia, and hypertension. It is often considered the hepatic manifestation of MetS. Importantly, NAFLD is not only a marker of systemic metabolic risk but also an active contributor to cardiometabolic disease [20].

Cardiovascular disease (CVD) is the leading cause of mortality in patients with NAFLD, even more than liver-related complications. Hepatic fat accumulation and systemic inflammation contribute to atherogenesis, endothelial dysfunction, and plaque instability [141].

2.6.5 Genetic and Epigenetic Influences

Genetic susceptibility plays a key role in the development and severity of NAFLD. Several genetic variants have been identified:

- *PNPLA3 (patatin-like phospholipase domain-containing 3)*: The I148M variant is strongly associated with hepatic fat accumulation and fibrosis progression [103].
- *TM6SF2 and MBOAT7*: Influence lipid processing and VLDL secretion [134].
- *GCKR and HSD17B13*: Modulate hepatic glucose and lipid metabolism [103].

Epigenetic modifications such as DNA methylation, histone acetylation, and noncoding RNAs further regulate gene expression and may mediate environmental influences like diet and physical activity [134].

2.7 Type 2 Diabetes Mellitus (T2DM)

Type 2 diabetes mellitus (T2DM) is a chronic, progressive metabolic disorder characterized by hyperglycemia resulting from a combination of insulin resistance and impaired insulin secretion. It constitutes more than 90% of all diabetes cases worldwide and is closely linked to other components of the metabolic syndrome, such as obesity, dyslipidemia, and hypertension [110]. Type 2 diabetes mellitus (T2DM) is the end result of chronic metabolic disturbances, starting with insulin resistance and eventually leading to impaired function of the pancreatic β-cells. The disease poses a significant global health challenge, not only because of its widespread prevalence but also due to the chronic microvascular and macrovascular complications that make it a leading cause of illness and death [15].

2.7.1 Epidemiology and Global Trends

The worldwide impact of type 2 diabetes mellitus (T2DM) is immense. According to the International Diabetes Federation (IDF), around 537 million adults were affected by diabetes in 2021, and this figure is expected to increase to 783 million by 2045

[61]. The fastest growth in diabetes cases is taking place in low- and middle-income nations, driven by factors such as urbanization, reduced physical activity, and changes in dietary habits. Regions like South Asia, the Middle East, and Sub-Saharan Africa are witnessing particularly steep rises. Importantly, T2DM is now being diagnosed at younger ages, including in adolescents and young adults, reflecting the global rise in childhood obesity and metabolic risk factors [113].

2.7.2 Pathophysiological Mechanisms

Two major abnormalities underpin the development of T2DM:

1. *Insulin Resistance (IR)*: In the early stages, insulin resistance develops in peripheral tissues, including skeletal muscle, the liver, and adipose tissue, reducing their responsiveness to insulin's effects. In skeletal muscle, the transport of glucose through GLUT4 is diminished, while in the liver, insulin is unable to effectively inhibit gluconeogenesis, leading to increased production of glucose by the liver. Adipose tissue, particularly visceral fat, becomes resistant to insulin-mediated inhibition of lipolysis, increasing the flux of free fatty acids (FFAs) into the liver and muscle, further exacerbating IR [167].
2. *Pancreatic β-Cell Dysfunction*: As insulin resistance progresses, the pancreatic β-cells initially respond by increasing insulin production. Over time, due to glucotoxicity (persistent hyperglycemia), lipotoxicity (accumulated FFAs), and inflammation, β-cells undergo functional exhaustion and apoptosis, leading to insufficient insulin secretion relative to demand [76].

2.7.3 Molecular Mediators and Cellular Pathways

Several molecular pathways are implicated in T2DM:

- *AMPK (AMP-activated protein kinase)*: A vital energy regulator that is reduced in type 2 diabetes, leading to decreased glucose absorption and fatty acid breakdown.
- *mTOR (mammalian target of rapamycin)*: Excessive activation due to nutrient overload results in the development of insulin resistance [108].
- *Inflammatory Cytokines*: TNF-α, IL-6, and MCP-1 produced by macrophages infiltrating adipose tissue disrupt insulin signaling by causing serine phosphorylation of IRS-1 (insulin receptor substrate) [66].
- *Oxidative Stress*: Excess ROS (reactive oxygen species) damages β-cells and disrupts insulin signaling cascades [41].
- *AGEs (Advanced Glycation End Products)*: Formed during chronic hyperglycemia, these compounds trigger inflammatory responses and endothelial dysfunction [43].

2.7.4 Risk Factors

T2DM is a multifactorial disease, and a wide array of risk factors influences its development:

- *Genetic Susceptibility*: Numerous genes including TCF7L2, SLC30A8, KCNJ11, and FTO are associated with increased T2DM risk [73].
- *Obesity*: Particularly central or visceral obesity is the strongest modifiable risk factor [127].
- *Physical Inactivity*: Sedentary behavior reduces insulin sensitivity and muscle glucose uptake [107].
- *Unhealthy Diet*: High intake of refined carbohydrates, saturated fats, and sugar-sweetened beverages promotes IR and β-cell stress [107].
- *Age*: Risk increases with advancing age, partly due to declining β-cell function and increased visceral adiposity [136].
- *Ethnicity*: Higher risk is observed in South Asians, African Americans, Hispanics, and Native Americans, often at lower BMI thresholds [73].
- *Gestational Diabetes*: Women with a history of GDM are at elevated risk for later T2DM [136].
- *Polycystic Ovary Syndrome (PCOS)*: A known state of IR that predisposes women to diabetes [142].

2.7.5 Clinical Presentation

T2DM often remains asymptomatic for years, with hyperglycemia developing gradually. Typical symptoms, when they occur, include frequent urination (polyuria), excessive thirst (polydipsia), increased hunger (polyphagia), fatigue, blurry vision, repeated infections such as candidiasis, delayed wound healing, and unexplained weight loss (in later stages) [22].

2.7.6 Diagnosis

The diagnostic criteria for T2DM, as per the American Diabetes Association (ADA) and World Health Organization (WHO), include:

- *Fasting Plasma Glucose (FPG)*: ≥126 mg/dL (7.0 mmol/L) [61]
- *2-Hour OGTT (Oral Glucose Tolerance Test)*: ≥200 mg/dL (11.1 mmol/L) [61]
- *HbA1c*: ≥6.5% [61]
- *Random Plasma Glucose*: ≥200 mg/dL with symptoms [61]

2.7.7 Complications

T2DM contributes significantly to global morbidity and mortality, mainly due to its complications, which are categorized into:

- *Microvascular*: Retinopathy, nephropathy, neuropathy [96]
- *Macrovascular*: Coronary artery disease, stroke, peripheral vascular disease [96]

These complications arise from chronic hyperglycemia, oxidative stress, endothelial dysfunction, and prothrombotic states [96].

2.7.8 Management Strategies

T2DM management is multidimensional, aiming to achieve glycemic targets, address comorbid conditions (hypertension, dyslipidemia), and prevent complications [7].

1. *Lifestyle Modification*: Cornerstone of treatment includes diet (e.g., Mediterranean, DASH), exercise (150 min/week moderate aerobic), and weight reduction [85].
2. *Pharmacotherapy*: The initial treatment typically involves metformin. Additional medications include sulfonylureas, SGLT2 inhibitors, DPP-4 inhibitors, GLP-1 receptor agonists, and insulin. Therapy should be tailored to the individual, considering age, existing health conditions, cardiovascular risk, and kidney function [75].
3. *Monitoring*:
   - Self-monitoring of blood glucose (SMBG) [7]
   - Regular HbA1c testing [7]
   - Screening for complications (retinal exams, microalbuminuria, foot care) [7]

2.7.9 Emerging Therapies and Innovations

- *Dual GIP/GLP-1 receptor agonists* (e.g., tirzepatide) show promise in superior glycemic and weight control [75].
- *Gene editing and regenerative therapies* aim to restore β-cell mass or function [75].
- *Gut microbiota modulation* via prebiotics, probiotics, or fecal microbiota transplantation is being explored [75].

3 Pathophysiological Mechanisms of Metabolic Syndrome

Metabolic syndrome (MetS) represents a constellation of interrelated pathophysiological disturbances that include insulin resistance, central obesity, dyslipidemia, hypertension, and systemic inflammation. These features are not merely coexistent but intricately linked through molecular and cellular pathways [12]. Understanding the mechanistic underpinnings is essential for developing effective therapeutic strategies that target not just symptoms but the root causes of the syndrome. This section delves into the five central mechanisms: insulin resistance, adipose tissue dysfunction, dyslipidemia, hypertension, and oxidative stress [123].

3.1 Insulin Resistance

Insulin resistance (IR) is broadly acknowledged as the primary pathological process driving the onset and advancement of metabolic syndrome (MetS). It acts as the common underlying factor connecting the key metabolic disturbances of MetS, including elevated blood sugar, abnormal lipid levels, high blood pressure, central obesity, and persistent low-grade inflammation. IR refers to a diminished biological response of insulin-sensitive tissues, particularly skeletal muscle, adipose tissue, and the liver to normal circulating levels of insulin. This resistance disrupts glucose and lipid homeostasis, impairs protein metabolism, and contributes to endothelial dysfunction, thereby driving the metabolic and cardiovascular components of MetS [167].

Under normal conditions, insulin facilitates glucose absorption in skeletal muscle and fat tissue, reduces glucose production in the liver, and controls fat metabolism by suppressing the breakdown of fats (lipolysis) and promoting fat synthesis (lipogenesis). However, in the insulin-resistant state, these processes are profoundly altered [129]. Insulin signaling is initiated when insulin binds to its receptor (IR), setting off a series of phosphorylation events inside the cell. This process involves insulin receptor substrates (IRS), phosphoinositide 3-kinase (PI3K), and protein kinase B (Akt), ultimately resulting in the movement of glucose transporter type 4 (GLUT4) to the cell surface to enable glucose uptake. In IR, a key molecular defect is the serine phosphorylation (rather than tyrosine) of IRS-1, which impairs downstream signaling and GLUT4 translocation, particularly in skeletal muscle—the primary site of postprandial glucose disposal [84].

In adipose tissue, insulin normally inhibits lipolysis and regulates the secretion of adipokines. This antilipolytic effect is blunted in insulin-resistant states, leading to an increased release of free fatty acids (FFAs) into circulation. Increased levels of free fatty acids (FFAs) lead to abnormal fat buildup in the liver, skeletal muscles, and pancreas, causing lipotoxic effects that further intensify insulin resistance [159]. Insulin resistance in the liver reduces the hormone's effectiveness in inhibiting gluconeogenesis, leading to an increase in glucose production by the liver. Simultaneously, the elevated influx of free fatty acids (FFAs) promotes the buildup of diacylglycerol (DAG), which activates specific protein kinase C isoforms like PKCε. This disrupts insulin signaling, contributing to liver insulin resistance and abnormal lipid levels by enhancing the secretion of very-low-density lipoprotein (VLDL) [69].

Prolonged insulin resistance also places excessive stress on pancreatic β-cells, initially prompting compensatory hyperinsulinemia to maintain normoglycemia. Over time, β-cell exhaustion ensues, leading to impaired insulin secretion, glucose intolerance, and eventual progression to type 2 diabetes mellitus (T2DM). Chronic exposure to elevated FFAs and inflammatory mediators contributes to β-cell apoptosis and functional decline [71].

Inflammation and oxidative stress **intensify** insulin resistance. Proinflammatory cytokines, including TNF-α and IL-6, released by enlarged fat cells and invading immune cells, activate stress-related enzymes such as c-Jun N-terminal kinase (JNK) and inhibitor of nuclear factor kappa-B kinase (IKK). These kinases disrupt insulin signaling by increasing serine phosphorylation of IRS-1. At the same time, mitochondrial dysfunction causes an overproduction of reactive oxygen species (ROS), which harms cellular structures and further impairs insulin function [160].

Clinically, IR presents with elevated fasting plasma glucose and insulin levels, central adiposity, elevated triglycerides, low HDL-C, and hypertension—collectively forming the diagnostic framework for MetS. Thus, insulin resistance is not merely a biochemical defect but a central driver of the systemic metabolic derangements seen in MetS. Gaining insight into its intricate molecular and physiological mechanisms is essential for creating targeted treatments that enhance insulin sensitivity and halt the advancement of metabolic diseases [167].

3.1.1 Molecular Mechanisms of Insulin Resistance

At the molecular scale, insulin resistance arises from disturbances in the insulin signaling pathway. Normally, insulin functions by attaching to the extracellular alpha subunit of the insulin receptor (IR), a transmembrane receptor with tyrosine kinase activity. This interaction triggers autophosphorylation of tyrosine residues on the receptor's beta subunit, activating subsequent signaling pathways [116].

One of the primary targets of this activation is the insulin receptor substrate (IRS) proteins, primarily IRS-1 and IRS-2, which act as scaffolding proteins to recruit and activate downstream signaling molecules. Upon tyrosine phosphorylation, IRS binds and activates PI3K, which catalyzes the production of phosphatidylinositol-3,4,5-triphosphate (PIP3). PIP3 facilitates the activation of Akt (protein kinase B), which is crucial for glucose uptake by facilitating the movement of GLUT4 to the cell surface in muscle and fat tissues [132].

In insulin-resistant states, this cascade is disrupted at multiple levels. One primary mechanism is the serine phosphorylation of IRS-1, often mediated by inflammatory kinases such as JNK, IKKβ, and mTOR/S6K1. This modification prevents effective IRS interaction with PI3K, blunting Akt activation. Consequently, GLUT4 fails to translocate efficiently to the membrane, reducing glucose uptake [28].

Another branch of insulin signaling, the mitogen-activated protein kinase (MAPK) pathway, remains relatively intact in insulin resistance. This imbalance can be deleterious, as MAPK activation promotes the expression of inflammatory genes and cellular proliferation, contributing to a proinflammatory and proatherogenic milieu [28].

Mitochondrial dysfunction aggravates insulin resistance by reducing ATP production and boosting the production of reactive oxygen species (ROS). These ROS activate redox-sensitive kinases, which further inhibit insulin signaling. Mitochondrial stress also limits the cell's ability to oxidize FFAs, leading to intracellular lipid accumulation—a condition linked to the activation of PKC isoforms, especially PKCθ in muscle and PKCε in liver. These kinases inhibit insulin signaling via IRS serine phosphorylation and disruption of membrane lipid rafts required for insulin receptor localization [99].

3.2 Adipose Tissue Dysfunction

Adipose tissue functions as an active endocrine organ essential for metabolic regulation, extending beyond its classic role as a simple energy store. It influences systemic energy balance, insulin sensitivity, immune responses, and lipid and glucose metabolism by secretion of adipokines—bioactive molecules that exert endocrine, paracrine, and autocrine effects. In metabolic syndrome (MetS), adipose tissue undergoes a pathological transformation termed *adipose tissue dysfunction*, characterized by adipocyte hypertrophy, hypoxia, immune cell infiltration, fibrosis, and altered adipokine secretion. This dysfunctional state is instrumental in the development of insulin resistance, systemic inflammation, dyslipidemia, and hypertension [101].

Visceral adipose tissue (VAT) has a particularly harmful impact on metabolic syndrome (MetS). Unlike subcutaneous fat, VAT is metabolically active and prone to inflammation. Chronic energy excess leads to adipocyte hypertrophy and reduced oxygen diffusion, triggering hypoxia and the activation of hypoxia-inducible factor-1α (HIF-1α) [30]. This transcription factor drives fibrotic remodeling and inflammatory gene expression. Enlarged adipocytes become insulin-resistant and release excessive free fatty acids (FFAs) due to impaired insulin suppression of lipolysis. These FFAs accumulate in the liver, muscle, and pancreas, exacerbating lipotoxicity and metabolic dysfunction [101].

Additionally, enlarged adipocytes release higher amounts of proinflammatory cytokines like tumor necrosis factor-alpha (TNF-α), interleukin-6 (IL-6), and monocyte chemoattractant protein-1 (MCP-1). At the same time, levels of adiponectin a beneficial adipokine that improves insulin sensitivity and has anti-inflammatory properties, are significantly decreased. This imbalance contributes to widespread insulin resistance and persistent low-grade inflammation [153].

Immune remodeling is another hallmark of adipose tissue dysfunction. In healthy states, adipose tissue is rich in anti-inflammatory immune cells, including M2 macrophages and regulatory T cells. However, obesity induces a shift toward proinflammatory M1 macrophages, CD8+ T cells, and neutrophils. M1 macrophages gather around dying adipocytes in structures known as crown-like formations, where they secrete cytokines that disrupt insulin signaling and help maintain inflammation [30].

Dysfunctional adipose tissue also contributes to dyslipidemia, a key feature of MetS. Increased levels of free fatty acids (FFAs) stimulate the liver to produce more very-low-density lipoproteins (VLDL). At the same time, inflammatory signals impair lipid metabolism, leading to the formation of small, dense LDL particles and a decrease in HDL cholesterol. Impaired lipolysis, caused by abnormal regulation of enzymes such as hormone-sensitive lipase (HSL) and adipose triglyceride lipase (ATGL), worsens lipid metabolism disturbances [36].

Adipokine dysregulation exacerbates these effects. While adiponectin is diminished, leptin levels rise but become ineffective due to leptin resistance. This not only impairs appetite regulation but also promotes endothelial dysfunction and sympathetic nervous system activation, contributing to hypertension [161].

Additionally, adipose extracellular matrix (ECM) remodeling—characterized by excess collagen VI and fibronectin—leads to fibrosis, impeding healthy adipocyte expansion and driving ectopic lipid deposition. Impaired adipogenesis due to dysfunctional mesenchymal stem cells also limits the formation of new adipocytes, reinforcing inflammation and fibrosis [30].

Thus, adipose tissue dysfunction is a central pathogenic mechanism linking obesity to MetS, influencing multiple metabolic and cardiovascular domains. Targeting adipose inflammation, adipokine imbalances, and ECM remodeling offers promising avenues for therapeutic intervention in MetS [30].

3.2.1 Inflammation and Immune Cell Infiltration in Adipose Tissue Dysfunction

Inflammation contributes to and results from adipose tissue dysfunction, playing a central role in the development of metabolic syndrome. As adipose tissue enlarges during obesity, there is a significant alteration in the types and characteristics of immune cells within the tissue. This shift in immune cell populations leads to a sustained state of chronic, low-grade inflammation, fundamentally changing the tissue's environment and contributing to metabolic disturbances. This process of immune remodeling is marked by a heightened influx and activation of immune cells that promote inflammation. At the same time, there is a noticeable decrease in the populations of immune cells that normally work to suppress inflammation. This imbalance between proinflammatory and anti-inflammatory cells contributes to a persistent inflammatory environment within the tissue [78].

In individuals with a lean body composition, adipose tissue is mainly inhabited by M2-type macrophages. These cells produce anti-inflammatory cytokines, including interleukin-10 (IL-10), which help to regulate inflammation, support tissue remodeling, and enhance insulin responsiveness. This environment contributes to maintaining healthy metabolic function and prevents the development of chronic inflammation. In contrast, obese adipose tissue accumulates M1-type macrophages that produce TNF-α, IL-6, and inducible nitric oxide synthase (iNOS). These macrophages surround dead or dying adipocytes in structures known as crown-like structures (CLSs) and drive local and systemic inflammation [19].

The recruitment of M1 macrophages is mediated by increased secretion of chemokines such as MCP-1 from hypertrophied adipocytes and local endothelial cells. MCP-1 binds to its receptor CCR2 on monocytes, guiding them into the adipose tissue where they differentiate into inflammatory macrophages. Numerous studies in animal models have revealed that either genetically deleting or pharmacologically inhibiting the chemokine MCP-1 or its receptor CCR2 leads to significant improvements in insulin sensitivity. Additionally, these interventions reduce the inflammatory response within adipose tissue, highlighting the critical role of the MCP-1/CCR2 signaling pathway in regulating metabolic inflammation and insulin resistance [38].

In addition to the role of macrophages, other types of immune cells also contribute significantly to inflammation within adipose tissue. Among these, CD8+ T cells are known to infiltrate the adipose tissue at early stages of obesity. Once present, these cells promote the recruitment of macrophages and enhance their activation, thereby amplifying the inflammatory response and contributing to metabolic dysfunction in the tissue. CD4+ T helper 1 (Th1) cells and natural killer (NK) cells also contribute to the proinflammatory milieu by secreting IFN-γ. In dysfunctional adipose tissue, there is a notable decline in the populations of regulatory T cells

(Tregs) and Th2 cells. These immune cells usually play a crucial role in dampening inflammatory responses and supporting the maintenance of insulin sensitivity. Their reduced presence contributes to an imbalance in the immune environment, leading to increased inflammation and impaired metabolic regulation within the tissue [16].

B cells also play a role in the immune activity occurring within adipose tissue. They produce proinflammatory cytokines and pathogenic IgG antibodies that may exacerbate local inflammation and insulin resistance. The interplay between various immune cell types creates a vicious cycle of cytokine release, immune cell recruitment, and metabolic disruption [9].

At the mechanistic level, proinflammatory cytokines disrupt insulin signaling by triggering serine kinases like JNK and IKKβ. These kinases phosphorylate IRS-1 on serine sites, hindering its capacity to initiate the downstream pathways essential for insulin action. This molecular interference promotes insulin resistance not only locally in adipocytes but also systemically as cytokines enter the circulation [63].

Furthermore, the activation of inflammasomes, especially the NLRP3 inflammasome, has been recognized as a key factor driving inflammation in adipose tissue. When NLRP3 is activated, it triggers the cleavage and activation of IL-1β and IL-18, powerful inflammatory molecules that worsen insulin signaling and contribute to endothelial dysfunction [63].

3.3 Hypertension

Hypertension is a hallmark of metabolic syndrome (MetS) and a key driver of its cardiovascular complications. Around half of individuals with MetS exhibit elevated blood pressure (BP), and its pathogenesis is multifaceted, distinct from primary (essential) hypertension [40]. It arises from the convergence of insulin resistance, endothelial dysfunction, sympathetic nervous system (SNS) activation, renal sodium retention, and chronic inflammation—all of which are interlinked through metabolic derangements characteristic of MetS [93].

A key factor is insulin resistance. Under normal conditions, insulin facilitates vasodilation by activating the PI3K–Akt–eNOS signaling pathway, which enhances nitric oxide (NO) production in endothelial cells. In insulin-resistant states, this pathway is impaired, reducing NO synthesis and blunting vasodilation. In contrast, the MAPK pathway stays active, encouraging vasoconstriction and the growth of vascular smooth muscle cells. This disruption in signaling causes a rise in peripheral vascular resistance, which plays a role in maintaining high blood pressure [102].

The renin-angiotensin-aldosterone system (RAAS) is excessively activated in metabolic syndrome. Levels of angiotensin II (Ang II), a strong vasoconstrictor, are increased and play a role in promoting oxidative stress, impairing endothelial function, and driving inflammation. Aldosterone, beyond regulating sodium retention, exerts proinflammatory and profibrotic effects on the vasculature and kidneys. Notably, visceral adipose tissue itself becomes an endocrine source of RAAS components, producing Ang II and aldosterone and reinforcing systemic RAAS activation [13].

Sympathetic nervous system activity is significantly heightened in MetS, largely due to hyperinsulinemia and elevated leptin levels. Insulin acts on central neural circuits to increase SNS outflow, which raises heart rate, constricts blood vessels, and enhances renal sodium reabsorption. Leptin, an adipokine secreted by fat tissue, further stimulates the SNS and promotes oxidative stress. Leptin resistance, commonly seen in obesity and MetS, amplifies these pressor effects while diminishing satiety signaling, contributing to continued metabolic dysregulation [148].

Endothelial dysfunction is another cornerstone of MetS-associated hypertension. The endothelium loses its ability to produce adequate NO and prostacyclin due to oxidative stress and inflammation. Increased levels of reactive oxygen species (ROS) scavenge NO, reducing its vasodilatory effect and enhancing vascular tone. This dysfunction facilitates vasoconstriction, inflammation, and thrombosis, creating a prohypertensive vascular environment [40].

Renal contributions are equally important. Hyperinsulinemia promotes sodium reabsorption in the nephron via activation of sodium transporters like NHE3 and ENaC, leading to volume expansion and elevated BP. Moreover, individuals with MetS often exhibit salt-sensitive hypertension, marked by an impaired ability to excrete sodium in response to dietary intake, further promoting BP elevation [163].

Adipose tissue dysfunction amplifies hypertension through the release of proinflammatory cytokines (e.g., TNF-α, IL-6) and adipokines (e.g., leptin, resistin). Reduced adiponectin, a vasoprotective molecule, results in diminished anti-inflammatory and vasodilatory signaling. This inflammatory state accelerates endothelial injury, vascular remodeling, and RAAS activation [30].

Chronic hypertension and inflammation lead to vascular remodeling and arterial stiffness, driven by smooth muscle hypertrophy and increased extracellular matrix deposition. Advanced glycation end-products (AGEs), which build up during prolonged high blood sugar levels, form cross-links with collagen in blood vessels. This process decreases vascular flexibility, leading to elevated systolic blood pressure and increased pulse pressure [160].

Oxidative stress is a unifying mechanism, fueling RAAS overactivity, endothelial dysfunction, and SNS stimulation. Enzymes such as NADPH oxidase, xanthine oxidase, and mitochondrial sources of ROS are upregulated in MetS, exacerbating redox imbalances and vascular damage [83].

Overall, hypertension in MetS results from complex and interdependent mechanisms involving metabolic, neural, renal, and vascular pathways. Effective management calls for an integrated approach that targets multiple underlying factors, including insulin resistance, overactivation of the RAAS, heightened sympathetic nervous system activity, oxidative stress, and chronic inflammation, in order to lower cardiovascular risk and enhance overall health outcomes [40].

3.3.1 Sympathetic Nervous System and RAAS Activation in MetS-Induced Hypertension

The sympathetic nervous system (SNS) and the renin-angiotensin-aldosterone system (RAAS) are interconnected neurohormonal pathways that play essential roles in maintaining cardiovascular balance and stability. In the context of metabolic syndrome (MetS), chronic activation of these systems contributes to sustained hypertension and end-organ damage [40].

Hyperinsulinemia, a hallmark of insulin resistance, directly stimulates the central SNS through hypothalamic insulin receptors. This leads to increased norepinephrine release, vasoconstriction, increased cardiac output, and enhanced renal sodium reabsorption. The resultant elevation in blood pressure is compounded by the blunted baroreflex sensitivity observed in MetS, which fails to mitigate sympathetic overactivity [147].

Adipose tissue, particularly visceral fat, secretes leptin—a hormone known to activate the SNS. In MetS, leptin levels are elevated, yet its anorexigenic effects are blunted (leptin resistance), whereas its sympatho-excitatory actions persist. This selective resistance exacerbates sympathetic drive, contributing to increased vascular resistance and BP elevation [10].

The RAAS is also upregulated in MetS. Angiotensin II (Ang II), the key effector peptide of the RAAS, exerts vasoconstrictive effects and stimulates aldosterone secretion from the adrenal cortex. Both Ang II and aldosterone promote sodium and water retention, vascular hypertrophy, and fibrosis. Moreover, Ang II enhances SNS activity through central and peripheral mechanisms, creating a vicious cycle of neurohormonal activation [13].

Interestingly, adipose tissue itself is a site of local RAAS component synthesis. Adipocytes produce angiotensinogen, renin, and ACE, enabling autonomous production of Ang II. This local RAAS activation contributes to adipose inflammation, fibrosis, and vascular tone regulation [36].

The convergence of SNS and RAAS activation in MetS results in a hyperadrenergic and hypervolemic state, characterized by elevated vascular resistance, sodium retention, and increased BP. This synergy also amplifies oxidative stress and endothelial dysfunction—hallmarks of hypertension in MetS. Oxidative stress enhances Ang II signaling via redox-sensitive kinases and promotes SNS-mediated norepinephrine release [13].

Therapeutically, blocking these pathways has proven beneficial in managing MetS-associated hypertension. RAAS inhibitors (ACE inhibitors, ARBs, and mineralocorticoid receptor antagonists) not only lower BP but also improve insulin sensitivity and reduce albuminuria. Beta-blockers and centrally acting sympatholytics attenuate SNS overactivity, although their metabolic effects vary. Newer agents, such as neprilysin inhibitors, show promise in reducing neurohormonal burden while improving cardiovascular outcomes [36].

3.4 Oxidative Stress

Oxidative stress is a fundamental pathophysiological mechanism contributing to the onset and progression of metabolic syndrome (MetS), a cluster of interrelated conditions including insulin resistance, dyslipidemia, central obesity, hypertension, and endothelial dysfunction. It results from a disruption between the production of reactive oxygen species (ROS) and the ability of the body's antioxidant defenses to counteract them. Although normal ROS levels are important for cell signaling, their overproduction in metabolic syndrome causes extensive molecular damage, promotes inflammation, and contributes to vascular issues [158].

Mitochondrial dysfunction is a major contributor to oxidative stress in metabolic syndrome. Insulin resistance disrupts the normal β-oxidation of free fatty acids (FFAs), leading to electron leakage and the generation of superoxide radicals. Reactive oxygen species (ROS) produced by mitochondria can harm lipids, proteins, and DNA, while also disrupting insulin signaling pathways. This creates a self-reinforcing cycle that sustains and worsens metabolic dysfunction. Moreover, high blood sugar levels contribute to increased ROS production and vascular injury by activating pathways such as the polyol pathway and the formation of advanced glycation end-products (AGEs). AGEs bind to RAGE receptors, initiating inflammatory signaling via NF-κB, thereby exacerbating oxidative stress [99].

NADPH oxidase is another major ROS-producing enzyme, especially in adipocytes, endothelial, and immune cells. Its activation, driven by cytokines, Angiotensin II, and FFAs, reduces nitric oxide (NO) bioavailability by generating peroxynitrite (ONOO−), a cytotoxic species that contributes to endothelial dysfunction. Simultaneously, antioxidant defenses—enzymatic (e.g., SOD, catalase) and nonenzymatic (e.g., glutathione, vitamins C and E)—are diminished in MetS, allowing ROS to accumulate and propagate lipid peroxidation and cellular injury [39].

Visceral fat plays a significant role in oxidative stress. Enlarged fat cells, along with infiltrating M1 macrophages produce reactive oxygen species and inflammatory substances, which exacerbate insulin resistance and widespread inflammation. Within the liver, oxidative stress causes lipid peroxidation and activates enzymes such as JNK, which contribute to the development and advancement of nonalcoholic fatty liver disease (NAFLD) into steatohepatitis (NASH). Pancreatic β-cells, inherently low in antioxidants, are also highly susceptible to ROS, resulting in apoptosis and impaired insulin secretion [10].

Oxidative stress is a major factor contributing to endothelial dysfunction, which serves as an early indicator of cardiovascular disease in metabolic syndrome. ROS reduce NO production, increase vascular tone, and promote vasoconstriction. They also upregulate adhesion molecules such as VCAM-1 and ICAM-1, enhancing leukocyte infiltration and vascular inflammation [33].

Therapeutically, oxidative stress modulation offers promising avenues for MetS management. Lifestyle interventions like caloric restriction, physical activity, and weight loss enhance mitochondrial efficiency and reduce ROS. Pharmacological agents such as metformin, statins, and ARBs have antioxidant effects alongside their metabolic actions. Natural compounds including polyphenols, flavonoids, and omega-3 fatty acids possess ROS-scavenging properties and are under investigation as adjunct therapies. Targeting oxidative stress may thus ameliorate multiple metabolic disturbances in MetS, offering a holistic strategy for prevention and treatment [39].

3.4.1 Redox Signaling and Inflammatory Crosstalk in Metabolic Syndrome

Redox signaling and inflammation are intricately connected in the development and progression of metabolic syndrome (MetS). In normal conditions, small amounts of reactive oxygen species (ROS) help regulate cellular activities such as gene expression and immune system function. In metabolic syndrome, an overproduction of reactive oxygen species (ROS) disturbs this equilibrium, triggering redox-sensitive transcription factors like NF-κB and AP-1. These factors promote the production of proinflammatory cytokines, including TNF-α and IL-6 [94]. This initiates a self-perpetuating cycle of inflammation and oxidative damage. At the same time, the activity of Nrf2—a transcription factor responsible for enhancing antioxidant defenses—is diminished in metabolic syndrome because oxidative changes to its regulator, Keap1, lower the cell's ability to combat oxidative stress [58].

Reactive oxygen species (ROS) stimulate the activation of the NLRP3 inflammasome in both adipose tissue and macrophages, causing the secretion of inflammatory cytokines IL-1β and IL-18, which further aggravate insulin resistance and inflammation in the tissue. Furthermore, an imbalance in redox status leads to endoplasmic reticulum (ER) stress, activating the unfolded protein response (UPR) through the CHOP, PERK, and IRE1 signaling pathways. Prolonged ER stress worsens β-cell impairment and promotes the development of fatty liver disease [95].

Additionally, reactive oxygen species (ROS) affect key metabolic regulators such as AMPK and sirtuins. In metabolic syndrome, reduced activity of AMPK and lower levels of SIRT1 result in compromised mitochondrial function and heightened oxidative stress. These disruptions further impair energy metabolism and promote inflammation [94].

Targeting redox-inflammatory pathways offers therapeutic potential in MetS. Antioxidants that modulate transcription factors, inflammasomes, and metabolic sensors could help restore metabolic balance and reduce systemic inflammation. Thus, addressing redox–inflammation interplay presents a promising strategy for managing MetS and its complications [95].

4 Therapeutic Strategies for Metabolic Syndrome

Metabolic syndrome (MetS) poses a formidable challenge due to its multifactorial etiology and systemic manifestations. Treatment aims to address the underlying causes of the syndrome, such as insulin resistance, obesity, abnormal lipid levels, high blood pressure, and persistent inflammation, rather than just managing the symptoms [77]. Effective management necessitates an integrative and multifaceted approach, encompassing lifestyle modification, pharmacotherapy, surgical interventions, and emerging molecular strategies such as gene editing and targeted biologics. This section comprehensively discusses the current and evolving therapeutic modalities aimed at mitigating the pathophysiological burden of MetS [157].

4.1 Lifestyle Modification and Behavioral Interventions

Lifestyle modification forms the cornerstone of MetS management and prevention. The primary focus is on achieving and maintaining a healthy weight, enhancing insulin sensitivity, reducing cardiovascular risk factors, and improving overall metabolic health. Lifestyle changes have shown efficacy not only in reversing certain MetS components but also in reducing the incidence of type 2 diabetes mellitus (T2DM) and cardiovascular disease (CVD) [77].

4.1.1 Dietary Interventions

Dietary regulation significantly impacts each component of MetS. Recommended interventions include:

- *Mediterranean diet*: Rich in monounsaturated fats, whole grains, fruits, vegetables, and fish, it has demonstrated improvement in insulin sensitivity, lipid profiles, and endothelial function [157].
- *Low-carbohydrate diets*: Successful in lowering insulin levels, supporting weight reduction, and enhancing the triglyceride-to-HDL cholesterol ratio [60].
- *DASH diet (Dietary Approaches to Stop Hypertension)*: Focuses on reducing sodium consumption and increasing potassium-rich foods, which have been shown to decrease blood pressure and lower the risk of cardiovascular disease [57].

Key nutrients implicated in metabolic regulation include:

- *Omega-3 fatty acids*: Anti-inflammatory effects and lipid-lowering properties [57].
- *Dietary fiber*: Enhances glycemic control and satiety [57].
- *Polyphenols*: Found in berries, tea, and cocoa, these have antioxidant and anti-inflammatory effects beneficial in MetS [57].

4.1.2 Physical Activity and Exercise Training

Consistent exercise enhances insulin sensitivity, decreases visceral fat, and improves lipid metabolism. The most effective regimens include:

- *Aerobic exercise*: Engaging in activities like brisk walking, jogging, swimming, or cycling for a minimum of 150 min weekly boosts cardiovascular health and helps regulate glucose metabolism [31].
- *Resistance training*: Enhances muscle mass and insulin receptor density [31].
- *High-intensity interval training (HIIT)*: Offers substantial benefits in shorter time frames; particularly useful for improving VO2 max and reducing adipose tissue inflammation [31].

Exercise also modulates mitochondrial function and oxidative stress, crucial in reversing metabolic derangements. Moreover, physical activity positively influences psychological health, reducing stress, which is a known exacerbator of MetS.

4.2 Pharmacological Interventions

While lifestyle changes form the foundation of metabolic syndrome (MetS) therapy, pharmacological agents are often required to manage individual components when lifestyle alone proves insufficient. Given the heterogeneity of MetS, pharmacological treatment typically involves multiple drug classes aimed at specific abnormalities—insulin resistance, dyslipidemia, hypertension, and proinflammatory states. This section reviews major drug classes and their mechanisms in ameliorating MetS components [151].

4.2.1 Insulin Sensitizers

Biguanides (e.g., Metformin) Metformin is the most widely prescribed insulin-sensitizing agent. It activates AMP-activated protein kinase (AMPK), a key cellular energy sensor, leading to decreased hepatic gluconeogenesis, enhanced peripheral glucose uptake, and improved lipid profiles. Beyond glycemic control, metformin promotes modest weight loss and reduces CVD risk, making it ideal for MetS patients with impaired glucose tolerance or T2DM [50].

Thiazolidinediones (e.g., Pioglitazone, Rosiglitazone) These medications function by activating peroxisome proliferator-activated receptor gamma (PPAR-γ), a receptor that regulates fat cell development and improves the body's response to insulin [80]. They also exert anti-inflammatory effects and redistribute fat from visceral to subcutaneous depots. However, concerns over weight gain, fluid retention, and potential cardiovascular risks limit their widespread use. Emerging agents like *dual PPAR-α/γ agonists* (e.g., Saroglitazar) aim to combine lipid-lowering and insulin-sensitizing effects with improved safety [139].

4.2.2 Lipid-Lowering Agents

Statins (HMG-CoA Reductase Inhibitors) Statins reduce LDL cholesterol levels and offer additional benefits, such as stabilizing arterial plaques and decreasing overall inflammation in the body. Despite concerns about insulin resistance, their cardiovascular benefits often outweigh risks in MetS patients [89].

Fibrates By activating PPAR-α, fibrates lower triglycerides and raise HDL-C, addressing key features of atherogenic dyslipidemia common in MetS. They are particularly effective in patients with elevated triglycerides and low HDL levels [140].

Omega-3 Fatty Acids (EPA/DHA) High-dose omega-3 supplementation reduces hepatic VLDL synthesis and lowers triglycerides. Prescription formulations like

Icosapent ethyl have demonstrated significant reductions in cardiovascular events [164].

Niacin (Vitamin B3) Niacin increases HDL levels and reduces triglycerides, though its use has declined due to flushing side effects and lack of cardiovascular benefit in recent trials [53].

4.3 Antihypertensive Therapy and Anti-Inflammatory Agents

4.3.1 Blood Pressure Control

Hypertension is a key MetS component and a major contributor to cardiovascular risk. Drug selection should consider comorbidities:

- *ACE Inhibitors (e.g., Lisinopril) and ARBs (e.g., Losartan)*: These reduce blood pressure while also improving insulin sensitivity and reducing inflammation. They are first-line in hypertensive MetS patients, particularly those with diabetes or microalbuminuria [115].
- *Calcium Channel Blockers and Diuretics*: Effective in lowering blood pressure, though some thiazide diuretics may worsen insulin resistance and lipid profiles [115].
- *Beta-blockers*: Less favored due to potential adverse effects on glucose metabolism, though vasodilating agents like carvedilol have a more favorable metabolic profile [115].

4.3.2 Anti-Inflammatory and Antioxidant Therapies

Chronic low-grade inflammation is a unifying mechanism in MetS. While traditional anti-inflammatory drugs are not routinely used, several agents have shown metabolic benefits:

- *Aspirin*: Low-dose aspirin may reduce cardiovascular events in high-risk MetS patients [4].
- *Colchicine and IL-1β Inhibitors*: Being explored for cardiometabolic protection in clinical trials (e.g., CANTOS) [4].
- *Natural compounds* like curcumin and resveratrol exhibit antioxidant and anti-inflammatory properties in experimental models [4].

4.4 Novel Pharmacological Agents

4.4.1 GLP-1 Receptor Agonists (e.g., Liraglutide, Semaglutide)

Initially designed to treat type 2 diabetes, GLP-1 receptor agonists stimulate insulin release, inhibit glucagon production, delay gastric emptying, and support weight reduction. They have also been shown to improve various aspects of metabolic syndrome and provide cardiovascular benefits [88].

4.4.2 Dual and Triple Incretin Agonists (e.g., Tirzepatide)

Tirzepatide, a dual GLP-1/GIP agonist, shows superior glycemic and weight reduction compared to GLP-1 monotherapy. Early trials suggest triple agonists may offer even greater therapeutic potential [109].

4.4.3 SGLT2 Inhibitors (e.g., Empagliflozin, Dapagliflozin)

These medications reduce blood sugar levels by increasing the elimination of glucose through the urine. They also reduce weight and blood pressure and have been shown to confer cardio-renal protection, making them attractive in MetS management [109].

4.4.4 FXR and PPAR Modulators

Farnesoid X receptor (FXR) agonists regulate bile acid and glucose metabolism, while selective PPAR modulators target lipid and glucose pathways with fewer side effects than earlier generations [106].

4.5 Gut Microbiota Modulation

The gut microbiota is essential for maintaining metabolic balance. An imbalance in its composition, known as dysbiosis, is linked to increased inflammation, impaired insulin sensitivity, and disrupted lipid metabolism. Therapeutic strategies include:

- *Probiotics and Prebiotics*: Modulate microbial composition to favorably affect host metabolism [133].
- *Fecal Microbiota Transplantation (FMT)*: Investigated for restoring healthy microbial balance [133].
- *Postbiotics and Microbial Metabolites*: Short-chain fatty acids (SCFAs), generated through the fermentation of dietary fiber, help enhance insulin sensitivity and lower inflammation levels [133].

4.5.1 Nutrigenomics and Precision Medicine

Advances in omics technologies allow for individualized treatment plans:

- *Genetic Profiling*: Identifies risk alleles for insulin resistance, obesity, or dyslipidemia, guiding therapeutic decisions [98].
- *Epigenetic Modifiers*: Targeting histone deacetylases and microRNAs that regulate metabolic gene expression may offer new interventions [69].
- *Metabolomics and Proteomics*: Help predict individual responses to dietary and pharmacological interventions [98].

4.5.2 Cell and Gene-Based Therapies

Although in early stages, regenerative approaches offer exciting possibilities:

- *Adipose-Derived Stem Cells (ADSCs)*: Studied for their potential to reduce inflammation and improve insulin sensitivity [52].

- *CRISPR/Cas9 Gene Editing*: Potential to correct genetic mutations associated with metabolic dysregulation [114].
- *Exosome Therapy*: Leveraging extracellular vesicles for targeted delivery of therapeutic molecules [114].

5 Integration of Therapeutic Strategies with Clinical Outcomes

Metabolic syndrome (MetS) is a multifaceted disorder requiring a comprehensive treatment framework. While individual therapeutic strategies target specific pathophysiological mechanisms, their integration is paramount to achieving meaningful clinical outcomes. This section explores how combining lifestyle, pharmacological, surgical, and emerging therapies influences the progression of MetS and related cardiovascular and metabolic complications [68]. Furthermore, it reviews evidence from clinical trials, real-world studies, and guidelines that underscore the importance of personalized and multidisciplinary care in improving patient prognosis.

5.1 Impact of Lifestyle Modification on Clinical Outcomes

Lifestyle interventions are the cornerstone of MetS management, with substantial evidence demonstrating their benefits on both intermediate biomarkers and hard clinical endpoints. Randomized controlled trials (RCTs), Programs like the Diabetes Prevention Program (DPP) and the Finnish Diabetes Prevention Study (DPS) demonstrated that comprehensive lifestyle interventions led to a 58% reduction in the development of type 2 diabetes over a 3-year period among high-risk individuals exhibiting features of metabolic syndrome [130].

Dietary patterns emphasizing whole foods, reduced saturated fats, and increased fiber intake improve lipid profiles, glycemic control, and blood pressure. Physical activity, through mechanisms like enhanced insulin sensitivity and anti-inflammatory effects, decreases visceral adiposity and improves endothelial function [47]. Long-term adherence to lifestyle modification correlates with reduced cardiovascular events, decreased all-cause mortality, and improved quality of life. However, sustaining lifestyle changes remains challenging due to psychosocial and socioeconomic factors. Integration of behavioral counseling, digital health platforms, and community-based programs is critical to improve adherence and amplify clinical benefits [151].

5.2 Pharmacological Therapies and Outcomes in MetS

Medications target the underlying biochemical and physiological disturbances of metabolic syndrome. Blood pressure-lowering drugs help decrease hypertension, which in turn reduces the likelihood of stroke and heart attacks. In addition to lowering lipid levels, statins possess multiple beneficial effects, such as reducing inflammation and stabilizing arterial plaques, which greatly decrease cardiovascular illness and death in individuals with metabolic syndrome [102].

Medications that lower blood glucose, like metformin, enhance insulin sensitivity and slightly decrease the risk of cardiovascular disease. Recent medications such as GLP-1 receptor agonists and SGLT2 inhibitors have shown strong cardiometabolic advantages beyond just controlling blood sugar, including promoting weight loss, lowering blood pressure, and protecting kidney function. Their integration into MetS management protocols represents a paradigm shift toward comprehensive risk reduction [109].

Pharmacotherapy is most effective when combined with lifestyle changes and requires individualized consideration of comorbidities, drug interactions, and patient preferences. Evidence from landmark trials (e.g., LEADER, EMPA-REG OUTCOME) confirms the mortality benefit of these agents in populations with metabolic risk [102].

5.3 Surgical Interventions: Outcomes and Patient Selection

Bariatric surgery provides significant and lasting benefits in reducing weight, improving insulin resistance, and lowering cardiovascular risk factors, which lead to higher rates of diabetes remission, fewer cardiovascular incidents, and reduced mortality. Surgical techniques like Roux-en-Y gastric bypass and sleeve gastrectomy trigger metabolic effects that extend beyond mere weight reduction, including beneficial changes in gut hormone levels and bile acid signaling [8].

Longitudinal studies report that surgical patients experience improvements in hypertension, dyslipidemia, and systemic inflammation, with benefits persisting for over a decade. However, careful patient selection, preoperative evaluation, and postoperative follow-up are essential to optimize outcomes and minimize complications [145].

5.4 Emerging Therapies and Their Translational Impact

Emerging approaches such as precision medicine, microbiome modulation, and gene therapy hold promise for individualized treatment. Early-phase clinical trials suggest that gut microbiota-targeted interventions can favorably modify metabolic profiles and inflammatory status. Metabolomic and genomic profiling enables stratification of patients to tailor therapies, enhancing efficacy and reducing adverse effects. However, translational challenges remain, including validation of biomarkers, cost-effectiveness, and ethical considerations. As evidence accrues, integration of these novel therapies into clinical practice may transform MetS management from a one-size-fits-all to a precision-based approach [67].

5.5 Multidisciplinary Care Models and Real-World Implementation

The complexity of MetS necessitates a multidisciplinary approach involving endocrinologists, cardiologists, dietitians, psychologists, and primary care providers. Coordinated care models improve guideline adherence, patient education, and risk factor control, translating into better clinical outcomes. Digital health tools and telemedicine platforms enhance monitoring, facilitate patient engagement, and support timely therapeutic adjustments. Real-world studies demonstrate that integrated care pathways reduce hospitalizations, emergency visits, and healthcare costs [5].

5.6 Conclusion

Integrating diverse therapeutic strategies to manage metabolic syndrome (MetS) is vital to interrupt the complex and interconnected pathophysiological mechanisms underlying the condition. A multifaceted approach that combines lifestyle modifications, pharmacological agents, surgical interventions, and innovative therapies such as gene editing and microbiome modulation offers the most promise in curbing disease progression and associated comorbidities. Personalizing these interventions within a patient-centered framework enhances adherence, optimizes therapeutic outcomes, and minimizes adverse effects. Furthermore, early intervention and continuous monitoring are essential to adapt treatment plans according to evolving patient needs. As our understanding of MetS deepens through advances in systems biology and precision medicine, integrated therapeutic models will become increasingly effective. Ultimately, a cohesive strategy that targets both metabolic derangements and individual risk profiles holds the key to sustainable management and improved quality of life for patients with MetS.

References

1. Abdulkader, F. (2021). Metabolic pathways and cell signaling. In *Essential aspects of immunometabolism in health and disease* (pp. 13–30). Springer. https://doi.org/10.1007/978-3-030-86684-6_2
2. Adeva-Andany, M. M., González-Lucán, M., Donapetry-García, C., Fernández-Fernández, C., & Ameneiros-Rodríguez, E. (2016). Glycogen metabolism in humans. *BBA Clinical, 5*, 85–100. https://doi.org/10.1016/j.bbacli.2016.02.001
3. Adjei, N. K., Samkange-Zeeb, F., Boakye, D., Saleem, M., Christianson, L., Kebede, M. M., Heise, T. L., Brand, T., Esan, O. B., Taylor-Robinson, D. C., Agyemang, C., & Zeeb, H. (2024). Ethnic differences in metabolic syndrome in high-income countries: A systematic review and meta-analysis. *Reviews in Endocrine and Metabolic Disorders, 25*(4), 727–750. https://doi.org/10.1007/S11154-024-09879-9
4. Agnello, F., & Capodanno, D. (2022). Anti-inflammatory strategies for atherosclerotic artery disease. *Expert Opinion on Drug Safety, 21*(5), 661–672. https://doi.org/10.1080/14740338.2022.2036717

5. Alberti, K. G. M. M., Zimmet, P., & Shaw, J. (2006). Metabolic syndrome – A new world-wide definition. A consensus statement from the International Diabetes Federation. *Diabetic Medicine, 23*(5), 469–480. https://doi.org/10.1111/J.1464-5491.2006.01858.X
6. Alegría-Torres, J. A., Baccarelli, A., & Bollati, V. (2011). Epigenetics and lifestyle. *Epigenomics, 3*(3), 267–277. https://doi.org/10.2217/EPI.11.22
7. Al-Worafi, Y. M. (2024). Diabetes mellitus: Causes and risk factors in developing countries. In *Handbook of medical and health sciences in developing countries* (pp. 1–26). Springer. https://doi.org/10.1007/978-3-030-74786-2_322-1
8. Ammar, W., Basset, H. A., Al Faramawy, A., Hegazy, T., & Sharaf, Y. (2020). Bariatric surgery and cardiovascular outcome. *The Egyptian Heart Journal, 72*(1), 67. https://doi.org/10.1186/S43044-020-00096-8
9. Balcerczyk, A., Eljaafari, A., & Pirola, L. (2024). Adipose stem cells drive T cell infiltration in obesity. *Trends in Endocrinology and Metabolism, 35*(11), 931–933. https://doi.org/10.1016/j.tem.2024.06.013
10. Bansal, S., Vachher, M., Arora, T., Kumar, B., & Burman, A. (2023). Visceral fat: A key mediator of NAFLD development and progression. *Human Nutrition and Metabolism, 33*, 200210. https://doi.org/10.1016/j.hnm.2023.200210
11. Bashir, A., Duseja, A., De, A., Mehta, M., & Tiwari, P. (2022). Non-alcoholic fatty liver disease development: A multifactorial pathogenic phenomena. *Liver Research, 6*(2), 72–83. https://doi.org/10.1016/j.livres.2022.05.002
12. Basil, B., Myke-Mbata, B. K., Eze, O. E., & Akubue, A. U. (2024). From adiposity to steatosis: Metabolic dysfunction-associated steatotic liver disease, a hepatic expression of metabolic syndrome – Current insights and future directions. *Clinical Diabetes and Endocrinology, 10*(1), 39. https://doi.org/10.1186/S40842-024-00187-4
13. Benigni, A., Cassis, P., & Remuzzi, G. (2010). Angiotensin II revisited: New roles in inflammation, immunology and aging. *EMBO Molecular Medicine, 2*(7), 247–257. https://doi.org/10.1002/EMMM.201000080
14. Bhale, A. S., Meilhac, O., d'Hellencourt, C. L., Vijayalakshmi, M. A., & Venkataraman, K. (2024). Cholesterol transport and beyond: Illuminating the versatile functions of HDL apolipoproteins through structural insights and functional implications. *Biofactors, 50*(5), 922–956. https://doi.org/10.1002/BIOF.2057
15. Bhatti, J. S., Sehrawat, A., Mishra, J., Sidhu, I. S., Navik, U., Khullar, N., Kumar, S., Bhatti, G. K., & Reddy, P. H. (2022). Oxidative stress in the pathophysiology of type 2 diabetes and related complications: Current therapeutics strategies and future perspectives. *Free Radical Biology and Medicine, 184*, 114–134. https://doi.org/10.1016/j.freeradbiomed.2022.03.019
16. Blüher, M. (2013). Adipose tissue dysfunction contributes to obesity related metabolic diseases. *Best Practice & Research: Clinical Endocrinology & Metabolism, 27*(2), 163–177. https://doi.org/10.1016/j.beem.2013.02.005
17. Boden, G. (2001). Free fatty acids – The link between obesity and insulin resistance. *Endocrine Practice, 7*(1), 44–51. https://doi.org/10.4158/ep.7.1.44
18. Bowden, R. G., Richardson, K. A., & Richardson, L. T. (2022). Uric acid and metabolic syndrome: Findings from national health and nutrition examination survey. *Frontiers in Medicine, 9*, 1039230. https://doi.org/10.3389/FMED.2022.1039230/FULL
19. Bradley, D., Deng, T., Shantaram, D., & Hsueh, W. A. (2024). Orchestration of the adipose tissue immune landscape by adipocytes. *Annual Review of Physiology, 86*, 199–223. https://doi.org/10.1146/ANNUREV-PHYSIOL-042222-024353
20. Byrne, C. D., & Targher, G. (2015). NAFLD: A multisystem disease. *Journal of Hepatology, 62*(S1), S47–S64. https://doi.org/10.1016/j.jhep.2014.12.012
21. Cai, G., Liu, J., Cai, M., & Shao, L. (2024). Exploring the causal effect between lipid-modifying drugs and idiopathic pulmonary fibrosis: A drug-target Mendelian randomization study. *Lipids in Health and Disease, 23*(1), 237. https://doi.org/10.1186/S12944-024-02218-6
22. Campbell, I. (2000). Epidemiology and clinical presentation of type 2 diabetes. *Value in Health, 3*(Suppl. 1), 3–6. https://doi.org/10.1046/j.1524-4733.2000.36014.x

23. Canton, M., Sánchez-Rodríguez, R., Spera, I., Venegas, F. C., Favia, M., Viola, A., & Castegna, A. (2021). Reactive oxygen species in macrophages: Sources and targets. *Frontiers in Immunology, 12*, 734229. https://doi.org/10.3389/FIMMU.2021.734229/FULL
24. Capurso, C., & Capurso, A. (2012). From excess adiposity to insulin resistance: The role of free fatty acids. *Vascular Pharmacology, 57*(2–4), 91–97. https://doi.org/10.1016/j.vph.2012.05.003
25. Carbone, F., Després, J. P., Ioannidis, J. P. A., Neeland, I. J., Garruti, G., Busetto, L., Liberale, L., Ministrini, S., Vilahur, G., Schindler, T. H., Macedo, M. P., Di Ciaula, A., Krawczyk, M., Geier, A., Baffy, G., Faienza, M. F., Farella, I., Santoro, N., Frühbeck, G., et al. (2025). Bridging the gap in obesity research: A consensus statement from the European Society for Clinical Investigation. *European Journal of Clinical Investigation, 55*(8), e70059. https://doi.org/10.1111/ECI.70059
26. Castro-Rodríguez, D. C., Rodríguez-González, G. L., Menjivar, M., & Zambrano, E. (2020). Maternal interventions to prevent adverse fetal programming outcomes due to maternal malnutrition: Evidence in animal models. *Placenta, 102*, 49–54. https://doi.org/10.1016/j.placenta.2020.04.002
27. Chait, A., & den Hartigh, L. J. (2020). Adipose tissue distribution, inflammation and its metabolic consequences, including diabetes and cardiovascular disease. *Frontiers in Cardiovascular Medicine, 7*, 22. https://doi.org/10.3389/FCVM.2020.00022/FULL
28. Chandrasekaran, P., & Weiskirchen, R. (2024). Cellular and molecular mechanisms of insulin resistance. *Current Tissue Microenvironment Reports, 5*(3), 79–90. https://doi.org/10.1007/S43152-024-00056-3
29. Chen, Z., Yu, R., Xiong, Y., Du, F., & Zhu, S. (2017). A vicious circle between insulin resistance and inflammation in nonalcoholic fatty liver disease. *Lipids in Health and Disease, 16*(1), 203. https://doi.org/10.1186/S12944-017-0572-9
30. Choe, S. S., Huh, J. Y., Hwang, I. J., Kim, J. I., & Kim, J. B. (2016). Adipose tissue remodeling: Its role in energy metabolism and metabolic disorders. *Frontiers in Endocrinology, 7*, 30. https://doi.org/10.3389/FENDO.2016.00030/FULL
31. Chomiuk, T., Niezgoda, N., Mamcarz, A., & Śliż, D. (2024). Physical activity in metabolic syndrome. *Frontiers in Physiology, 15*, 1365761. https://doi.org/10.3389/FPHYS.2024.1365761/FULL
32. Chong, K. S., Chang, Y. H., Yang, C. T., Chou, C. K., Ou, H., & Kuo, S. (2024). Longitudinal economic burden of incident complications among metabolic syndrome populations. *Cardio vascular Diabetology, 23*(1), 246. https://doi.org/10.1186/S12933-024-02335-7
33. Cipriano, A., Viviano, M., Feoli, A., Milite, C., Sarno, G., Castellano, S., & Sbardella, G. (2023). NADPH oxidases: From molecular mechanisms to current inhibitors. *Journal of Medicinal Chemistry, 66*(17), 11632–11655. https://doi.org/10.1021/ACS.JMEDCHEM.3C00770
34. Colozza, D., Wang, Y. C., & Avendano, M. (2023). Does urbanisation lead to unhealthy diets? Longitudinal evidence from Indonesia. *Health and Place, 83*, 103091. https://doi.org/10.1016/j.healthplace.2023.103091
35. Cornelis, M. C., Qi, L., Kraft, P., & Hu, F. B. (2009). TCF7L2, dietary carbohydrate, and risk of type 2 diabetes in US women. *American Journal of Clinical Nutrition, 89*(4), 1256–1262. https://doi.org/10.3945/ajcn.2008.27058
36. Crewe, C., An, Y. A., & Scherer, P. E. (2017). The ominous triad of adipose tissue dysfunction: Inflammation, fibrosis, and impaired angiogenesis. *Journal of Clinical Investigation, 127*(1), 74–82. https://doi.org/10.1172/JCI88883
37. da Silva Rosa, S. C., Nayak, N., Caymo, A. M., & Gordon, J. W. (2020). Mechanisms of muscle insulin resistance and the cross-talk with liver and adipose tissue. *Physiological Reports, 8*(19), e14607. https://doi.org/10.14814/PHY2.14607
38. Dahdah, N., Tercero-Alcázar, C., Malagón, M. M., Garcia-Roves, P. M., & Guzmán-Ruiz, R. (2024). Interrelation of adipose tissue macrophages and fibrosis in obesity. *Biochemical Pharmacology, 225*, 116324. https://doi.org/10.1016/j.bcp.2024.116324

39. Das, S., Gautam, N., Dey, S. K., Maiti, T., & Roy, S. (2009). Oxidative stress in the brain of nicotine-induced toxicity: Protective role of Andrographis paniculata Nees and vitamin E. *Applied Physiology, Nutrition and Metabolism, 34*(2), 124–135. https://doi.org/10.1139/H08-147
40. Das, D., Shruthi, N. R., Banerjee, A., Jothimani, G., Duttaroy, A. K., & Pathak, S. (2023). Endothelial dysfunction, platelet hyperactivity, hypertension, and the metabolic syndrome: Molecular insights and combating strategies. *Frontiers in Nutrition, 10*, 1221438. https://doi.org/10.3389/FNUT.2023.1221438/FULL
41. Diane, A., Al-Shukri, N. A., Bin Abdul Mu-u-min, R., & Al-Siddiqi, H. H. (2022). β-cell mitochondria in diabetes mellitus: A missing puzzle piece in the generation of hPSC-derived pancreatic β-cells? *Journal of Translational Medicine, 20*(1), 163. https://doi.org/10.1186/S12967-022-03327-5
42. Emery, P. (2009). Basic metabolism: Protein. *Surgery, 27*(5), 185–189. https://doi.org/10.1016/j.mpsur.2009.04.005
43. Eringa, E. C., Serne, E. H., Meijer, R. I., Schalkwijk, C. G., Houben, A. J. H. M., Stehouwer, C. D. A., Smulders, Y. M., & Van Hinsbergh, V. W. M. (2013). Endothelial dysfunction in (pre)diabetes: Characteristics, causative mechanisms and pathogenic role in type 2 diabetes. *Reviews in Endocrine and Metabolic Disorders, 14*(1), 39–48. https://doi.org/10.1007/S11154-013-9239-7
44. Ertuglu, L. A., Elijovich, F., Laffer, C. L., & Kirabo, A. (2021). Salt-sensitivity of blood pressure and insulin resistance. *Frontiers in Physiology, 12*, 793924. https://doi.org/10.3389/FPHYS.2021.793924/FULL
45. Esser, N., Legrand-Poels, S., Piette, J., Scheen, A. J., & Paquot, N. (2014). Inflammation as a link between obesity, metabolic syndrome and type 2 diabetes. *Diabetes Research and Clinical Practice, 105*(2), 141–150. https://doi.org/10.1016/j.diabres.2014.04.006
46. Esteves, J. V., Enguita, F. J., & Machado, U. F. (2017). MicroRNAs-mediated regulation of skeletal muscle GLUT4 expression and translocation in insulin resistance. *Journal of Diabetes Research, 2017*, 7267910. https://doi.org/10.1155/2017/7267910
47. Fappa, E., Yannakoulia, M., Pitsavos, C., Skoumas, I., Valourdou, S., & Stefanadis, C. (2008). Lifestyle intervention in the management of metabolic syndrome: Could we improve adherence issues? *Nutrition, 24*(3), 286–291. https://doi.org/10.1016/j.nut.2007.11.008
48. Fernando, R., Wardelmann, K., Deubel, S., Kehm, R., Jung, T., Mariotti, M., Vasilaki, A., Gladyshev, V. N., Kleinridders, A., Grune, T., & Castro, J. P. (2020). Low steady-state oxidative stress inhibits adipogenesis by altering mitochondrial dynamics and decreasing cellular respiration. *Redox Biology, 32*, 101507. https://doi.org/10.1016/j.redox.2020.101507
49. Festa, A., Williams, K., Tracy, R. P., Wagenknecht, L. E., & Haffner, S. M. (2006). Progression of plasminogen activator inhibitor-1 and fibrinogen levels in relation to incident type 2 diabetes. *Circulation, 113*(14), 1753–1759. https://doi.org/10.1161/CIRCULATIONAHA.106.616177
50. Foretz, M., Guigas, B., Bertrand, L., Pollak, M., & Viollet, B. (2014). Metformin: From mechanisms of action to therapies. *Cell Metabolism, 20*(6), 953–966. https://doi.org/10.1016/j.cmet.2014.09.018
51. Fujii, H., & Kawada, N. (2012). Inflammation and fibrogenesis in steatohepatitis. *Journal of Gastroenterology, 47*(3), 215–225. https://doi.org/10.1007/S00535-012-0527-X
52. Gadelkarim, M., Abushouk, A. I., Ghanem, E., Hamaad, A. M., Saad, A. M., & Abdel-Daim, M. M. (2018). Adipose-derived stem cells: Effectiveness and advances in delivery in diabetic wound healing. *Biomedicine and Pharmacotherapy, 107*, 625–633. https://doi.org/10.1016/j.biopha.2018.08.013
53. Ganji, S. H., Kamanna, V. S., & Kashyap, M. L. (2003). Niacin and cholesterol: Role in cardiovascular disease (Review). *Journal of Nutritional Biochemistry, 14*(6), 298–305. https://doi.org/10.1016/S0955-2863(02)00284-X

54. Ghosh, S. S., Wang, J., Yannie, P. J., & Ghosh, S. (2020). Intestinal barrier function and metabolic/liver diseases. *Liver Research, 4*(2), 81–87. https://doi.org/10.1016/j.livres.2020.03.002
55. Golabi, P., Owrangi, S., & Younossi, Z. M. (2024). Global perspective on nonalcoholic fatty liver disease and nonalcoholic steatohepatitis – Prevalence, clinical impact, economic implications and management strategies. *Alimentary Pharmacology and Therapeutics, 59*(S1), S1–S9. https://doi.org/10.1111/APT.17833
56. Grundy, S. M., Brewer, H. B., Cleeman, J. I., Smith, S. C., & Lenfant, C. (2004). Definition of metabolic syndrome: Report of the National Heart, Lung, and Blood Institute/American Heart Association conference on scientific issues related to definition. *Circulation, 109*(3), 433–438. https://doi.org/10.1161/01.CIR.0000111245.75752.C6
57. Grundy, S. M., Hansen, B., Smith, S. C., Cleeman, J. I., & Kahn, R. A. (2004). Clinical management of metabolic syndrome. *Arteriosclerosis, Thrombosis, and Vascular Biology, 24*(2), e19–e24. https://doi.org/10.1161/01.ATV.0000112379.88385.67
58. Harijith, A., Ebenezer, D. L., & Natarajan, V. (2014). Reactive oxygen species at the crossroads of inflammasome and inflammation. *Frontiers in Physiology, 5*, 352. https://doi.org/10.3389/FPHYS.2014.00352/FULL
59. Heeren, J., & Scheja, L. (2021). Metabolic-associated fatty liver disease and lipoprotein metabolism. *Molecular Metabolism, 50*, 101238. https://doi.org/10.1016/j.molmet.2021.101238
60. Hoffman, D. J., Powell, T. L., Barrett, E. S., & Hardy, D. B. (2021). Developmental origins of metabolic diseases. *Physiological Reviews, 101*(3), 739–795. https://doi.org/10.1152/PHYSREV.00002.2020
61. Hossain, M. J., Al-Mamun, M., & Islam, M. R. (2024). Diabetes mellitus, the fastest growing global public health concern: Early detection should be focused. *Health Science Reports, 7*(3), e2004. https://doi.org/10.1002/HSR2.2004
62. Hsu, C. C.-T., & Krings, T. (2019). Acquired metabolic diseases. In *Clinical neuroradiology* (pp. 1459–1498). Springer. https://doi.org/10.1007/978-3-319-68536-6_69
63. Huh, J. Y., Park, Y. J., Ham, M., & Kim, J. B. (2014). Crosstalk between adipocytes and immune cells in adipose tissue inflammation and metabolic dysregulation in obesity. *Molecules and Cells, 37*(5), 365–371. https://doi.org/10.14348/molcells.2014.0074
64. Hulthe, J., Bokemark, L., Wikstrand, J., & Fagerberg, B. (2000). The metabolic syndrome, LDL particle size, and atherosclerosis. The atherosclerosis and insulin resistance (AIR) study. *Arteriosclerosis, Thrombosis, and Vascular Biology, 20*(9), 2140–2147. https://doi.org/10.1161/01.ATV.20.9.2140
65. Hulver, M. W., Berggren, J. R., Cortright, R. N., Dudek, R. W., Thompson, R. P., Pories, W. J., MacDonald, K. G., Cline, G. W., Shulman, G. I., Dohm, G. L., & Houmard, J. A. (2003). Skeletal muscle lipid metabolism with obesity. *American Journal of Physiology-Endocrinology and Metabolism, 284*(4), 741–747. https://doi.org/10.1152/AJPENDO.00514.2002
66. Igoillo-Esteve, M., Marselli, L., Cunha, D. A., Ladrière, L., Ortis, F., Grieco, F. A., Dotta, F., Weir, G. C., Marchetti, P., Eizirik, D. L., & Cnop, M. (2010). Palmitate induces a pro-inflammatory response in human pancreatic islets that mimics CCL2 expression by beta cells in type 2 diabetes. *Diabetologia, 53*(7), 1395–1405. https://doi.org/10.1007/S00125-010-1707-Y
67. Ikewaki, K., & Tohyama, J.. (2004). Dyslipidemia in metabolic syndrome. Nippon Rinsho. Japanese Journal of Clinical Medicine, 62(6), 1099–1103. doi: https://doi.org/10.1007/978-3-319-12125-3_58-1.
68. Islam, M. S., Wei, P., Suzauddula, M., Nime, I., Feroz, F., Acharjee, M., & Pan, F. (2024). The interplay of factors in metabolic syndrome: Understanding its roots and complexity. *Molecular Medicine (Cambridge, Mass.), 30*(1), 279. https://doi.org/10.1186/S10020-024-01019-Y

69. Jahdkaran, M., & Sistanizad, M. (2025). From lipids to glucose: Investigating the role of dyslipidemia in the risk of insulin resistance. *Journal of Steroid Biochemistry and Molecular Biology, 250*, 106744. https://doi.org/10.1016/j.jsbmb.2025.106744
70. Jahić, R., & Fajkić, A. (2025). Blood biomarkers associated with metabolic syndrome: Implications for public health screening. In *Handbook of public health nutrition* (pp. 1–25). Springer. https://doi.org/10.1007/978-3-031-32047-7_94-1
71. Jansson, P. A. (2007). Endothelial dysfunction in insulin resistance and type 2 diabetes. *Journal of Internal Medicine, 262*(2), 173–183. https://doi.org/10.1111/J.1365-2796.2007.01830.X
72. Jiang, S., Ren, Z., Yang, Y., Liu, Q., Zhou, S., & Xiao, Y. (2023). The GPIHBP1-LPL complex and its role in plasma triglyceride metabolism: Insights into chylomicronemia. *Biomedicine and Pharmacotherapy, 169*, 115874. https://doi.org/10.1016/j.biopha.2023.115874
73. Joseph, A., Thirupathamma, M., Mathews, E., & Alagu, M. (2022). Genetics of type 2 diabetes mellitus in Indian and Global Population: A review. *Egyptian Journal of Medical Human Genetics, 23*(1), 135. https://doi.org/10.1186/S43042-022-00346-1
74. Kaneto, H., Katakami, N., Matsuhisa, M., & Matsuoka, T. A. (2010). Role of reactive oxygen species in the progression of type 2 diabetes and atherosclerosis. *Mediators of Inflammation, 2010*, 11. https://doi.org/10.1155/2010/453892
75. Kang, A. Y., Park, S. K., Park, S. Y., Lee, H. J., Han, Y., Lee, S. R., Suh, S. H., Kim, D. K., & Park, M. K. (2011). Therapeutic target achievement in type 2 diabetic patients after hyperglycemia, hypertension, dyslipidemia management. *Diabetes and Metabolism Journal, 35*(3), 264–272. https://doi.org/10.4093/DMJ.2011.35.3.264
76. Karwen, T., Kolczynska-Matysiak, K., Gross, C., Löffler, M. C., Friedrich, M., Loza-Valdes, A., Schmitz, W., Wit, M., Dziaczkowski, F., Belykh, A., Trujillo-Viera, J., El-Merahbi, R., Deppermann, C., Nawaz, S., Hastoy, B., Demczuk, A., Erk, M., Wieckowski, M. R., Rorsman, P., et al. (2023). Platelet-derived lipids promote insulin secretion of pancreatic β cells. *EMBO Molecular Medicine, 15*(9), e16858. https://doi.org/10.15252/EMMM.202216858
77. Kassi, E., Pervanidou, P., Kaltsas, G., & Chrousos, G. (2011). Metabolic syndrome: Definitions and controversies. *BMC Medicine, 9*, 48. https://doi.org/10.1186/1741-7015-9-48
78. Kawai, T., Autieri, M. V., & Scalia, R. (2021). Adipose tissue inflammation and metabolic dysfunction in obesity. *American Journal of Physiology – Cell Physiology, 320*(3), C375–C391. https://doi.org/10.1152/AJPCELL.00379.2020
79. Kawano, Y., & Cohen, D. E. (2013). Mechanisms of hepatic triglyceride accumulation in non-alcoholic fatty liver disease. *Journal of Gastroenterology, 48*(4), 434–441. https://doi.org/10.1007/S00535-013-0758-5
80. Kheniser, K., Aminian, A., & Kashyap, S. R. (2022). Effects of metabolic medicine and metabolic surgery on patient-reported outcomes among patients with type 2 diabetes. *Metabolic Syndrome and Related Disorders, 20*(9), 497–508. https://doi.org/10.1089/MET.2022.0039
81. Kim, J. Y., & Kim, N. H. (2023). New therapeutic approaches to the treatment of dyslipidemia 1: ApoC-III and ANGPTL3. *Journal of Lipid and Atherosclerosis, 12*(1), 23–36. https://doi.org/10.12997/JLA.2023.12.1.23
82. Kissebah, A. H., & Krakower, G. R. (1997). Effects of obesity on lipid metabolism. In *Clinical research in diabetes and obesity* (pp. 73–91). Humana Press. https://doi.org/10.1007/978-1-4757-3906-0_5
83. Kushwaha, P. P., Gupta, S., Singh, A. K., Prajapati, K. S., Shuaib, M., & Kumar, S. (2020). MicroRNA targeting nicotinamide adenine dinucleotide phosphate oxidases in cancer. *Antioxidants and Redox Signaling, 32*(5), 267–284. https://doi.org/10.1089/ARS.2019.7918
84. Laukkanen, O., Pihlajamäki, J., Lindström, J., Eriksson, J., Valle, T. T., Hämäläinen, H., Ilanne-Parikka, P., Keinänen-Kiukaanniemi, S., Tuomilehto, J., Uusitupa, M., & Laakso, M. (2004). Common polymorphisms in the genes regulating the early insulin signalling pathway: Effects on weight and the conversion from impaired glucose tolerance to type 2 diabetes. The Finnish Diabetes Prevention Study. *Diabetologia, 47*(5), 871–877. https://doi.org/10.1007/S00125-004-1395-6

85. Leenders, M., Verdijk, L. B., van der Hoeven, L., Adam, J. J., van Kranenburg, J., Nilwik, R., & Van Loon, L. J. C. (2013). Patients with type 2 diabetes show a greater decline in muscle mass, muscle strength, and functional capacity with aging. *Journal of the American Medical Directors Association, 14*(8), 585–592. https://doi.org/10.1016/j.jamda.2013.02.006
86. Lemche, E., Chaban, O. S., & Lemche, A. V. (2016). Neuroendorine and epigentic mechanisms subserving autonomic imbalance and HPA dysfunction in the metabolic syndrome. *Frontiers in Neuroscience, 10*, 142. https://doi.org/10.3389/FNINS.2016.00142/FULL
87. Leong, S. Y., Duque, S. M., Muhammad Abduh, S. B., & Oey, I. (2019). Carbohydrates. In *Innovative thermal and non-thermal processing, bioaccessibility and bioavailability of nutrients and bioactive compounds* (pp. 171–206). Woodhead Publishing. https://doi.org/10.1016/B978-0-12-814174-8.00006-8
88. Liu, Q. K. (2024). Mechanisms of action and therapeutic applications of GLP-1 and dual GIP/GLP-1 receptor agonists. *Frontiers in Endocrinology, 15*, 1431292. https://doi.org/10.3389/FENDO.2024.1431292/FULL
89. Liu, J. C., Lei, S. Y., Zhang, D. H., He, Q. Y., Sun, Y. Y., Zhu, H. J., Qu, Y., Zhou, S. Y., Yang, Y., Li, C., & Guo, Z. N. (2024). The pleiotropic effects of statins: A comprehensive exploration of neurovascular unit modulation and blood–brain barrier protection. *Molecular Medicine, 30*(1), 256. https://doi.org/10.1186/S10020-024-01025-0
90. Loomba, R., Friedman, S. L., & Shulman, G. I. (2021). Mechanisms and disease consequences of nonalcoholic fatty liver disease. *Cell, 184*(10), 2537–2564. https://doi.org/10.1016/j.cell.2021.04.015
91. Lucassen, E. A., & Cizza, G. (2012). The hypothalamic-pituitary-adrenal axis, obesity, and chronic stress exposure: Sleep and the HPA axis in obesity. *Current Obesity Reports, 1*(4), 208–215. https://doi.org/10.1007/S13679-012-0028-5
92. Luo, L., Fu, C., Bell, C. F., Wang, Y., & Leeper, N. J. (2023). Role of vascular smooth muscle cell clonality in atherosclerosis. *Frontiers in Cardiovascular Medicine, 10*, 1273596. https://doi.org/10.3389/FCVM.2023.1273596/FULL
93. Ma, X., McKie, P. M., Iyer, S. R., Scott, C., Bailey, K., Johnson, B. K., Benike, S. L., Chen, H., Miller, W. L., Cabassi, A., Burnett, J. C., & Cannone, V. (2024). MANP in hypertension with metabolic syndrome: Proof-of-concept study of natriuretic peptide-based therapy for cardiometabolic disease. *JACC Basic to Translational Science, 9*(1), 18–29. https://doi.org/10.1016/j.jacbts.2023.08.011
94. Magnani, N. D., Marchini, T., Calabró, V., Alvarez, S., & Evelson, P. (2020). Role of mitochondria in the redox signaling network and its outcomes in high impact inflammatory syndromes. *Frontiers in Endocrinology, 11*, 568305. https://doi.org/10.3389/FENDO.2020.568305/FULL
95. Manoharan, R. R., Prasad, A., Pospíšil, P., & Kzhyshkowska, J. (2024). ROS signaling in innate immunity via oxidative protein modifications. *Frontiers in Immunology, 15*, 1359600. https://doi.org/10.3389/FIMMU.2024.1359600/FULL
96. Mansour, A., Mousa, M., Abdelmannan, D., Tay, G., Hassoun, A., & Alsafar, H. (2023). Microvascular and macrovascular complications of type 2 diabetes mellitus: Exome wide association analyses. *Frontiers in Endocrinology, 14*, 1143067. https://doi.org/10.3389/FENDO.2023.1143067/FULL
97. Mazur, M., Przytuła, A., Szymańska, M., & Popiołek-Kalisz, J. (2024). Dietary strategies for cardiovascular disease risk factors prevention. *Current Problems in Cardiology, 49*(9), 102746. https://doi.org/10.1016/j.cpcardiol.2024.102746
98. McCarthy, M. I. (2010). Genomics, type 2 diabetes, and obesity. *New England Journal of Medicine, 363*(24), 2339–2350. https://doi.org/10.1056/NEJMRA0906948
99. Mercer, J. R., Cheng, K. K., Figg, N., Gorenne, I., Mahmoudi, M., Griffin, J., Vidal-Puig, A., Logan, A., Murphy, M. P., & Bennett, M. (2010). DNA damage links mitochondrial dysfunction to atherosclerosis and the metabolic syndrome. *Circulation Research, 107*(8), 1021–1031. https://doi.org/10.1161/CIRCRESAHA.110.218966

100. Miller, M., Stone, N. J., Ballantyne, C., Bittner, V., Criqui, M. H., Ginsberg, H. N., Goldberg, A. C., Howard, W. J., Jacobson, M. S., Kris-Etherton, P. M., Lennie, T. A., Levi, M., Mazzone, T., & Pennathur, S. (2011). Triglycerides and cardiovascular disease: A scientific statement from the American Heart Association. *Circulation, 123*(20), 2292–2333. https://doi.org/10.1161/CIR.0B013E3182160726
101. Mitov, M. I. (2024). Endocrine functions of the adipose tissue. In *Adipose tissue* (pp. 125–145). Springer. https://doi.org/10.1007/978-3-031-72570-8_4
102. Mohamed, S. M., Shalaby, M. A., El-Shiekh, R. A., El-Banna, H. A., Emam, S. R., & Bakr, A. F. (2023). Metabolic syndrome: Risk factors, diagnosis, pathogenesis, and management with natural approaches. *Food Chemistry Advances, 3*, 100335. https://doi.org/10.1016/j.focha.2023.100335
103. Mohamed, A. A., Al Dweik, R., Abdelghafour, R. A., Ramadan, A., Abbas, A. M., Samir, H. H., Muharram, N. M., Ahmed Elshiha, R. I., El-Salawy, N., Ghaith, D., Darwish, M. K., Abd El Salam, S. M., Sultan, E. A., Soliman, A. S., Ezz Al Arab, M., Elamir, A. Y., Mohamed, A. A., Hassanin, A. S. A., Abouaggour, A. A. M., et al. (2024). Anthropometry, laboratory, and PNPLA3 polymorphisms in a novel model for early identification and evaluation of nonalcoholic fatty liver disease. *Informatics in Medicine Unlocked, 48*, 101513. https://doi.org/10.1016/j.imu.2024.101513
104. Mozafari, S., Ashoori, M., Emami Meybodi, S. M., Solhi, R., Mirjalili, S. R., Firoozabadi, A. D., & Soltani, S. (2024). Association between APOA5 polymorphisms and susceptibility to metabolic syndrome: A systematic review and meta-analysis. *BMC Genomics, 25*(1), 590. https://doi.org/10.1186/S12864-024-10493-X
105. Muthiah, M. D., Cheng Han, N., & Sanyal, A. J. (2022). A clinical overview of non-alcoholic fatty liver disease: A guide to diagnosis, the clinical features, and complications—What the non-specialist needs to know. *Diabetes, Obesity and Metabolism, 24*(S2), 3–14. https://doi.org/10.1111/DOM.14521
106. Navab, M., Anantharamaiah, G. M., Reddy, S. T., Hama, S., Hough, G., Grijalva, V. R., Yu, N., Ansell, B. J., Datta, G., Garber, D. W., & Fogelman, A. J. (2005). The Farnesoid X receptor: A molecular link between bile acid and lipid and glucose metabolism. *Arteriosclerosis, Thrombosis, and Vascular Biology, 25*(10), 2020–2031. https://doi.org/10.1161/01.ATV.0000178994.21828.A7
107. Nazarko, L. (2022). Healthwise, Part 5. Prevention and treatment of type 2 diabetes. *British Journal of Healthcare Assistants, 16*(1), 18–25. https://doi.org/10.12968/BJHA.2022.16.1.18
108. Nguyen-Tu, M. S., Da Silva Xavier, G., Leclerc, I., & Rutter, G. A. (2018). Transcription factor-7–like 2 (TCF7L2) gene acts downstream of the Lkb1/Stk11 kinase to control mTOR signaling, cell growth, and insulin secretion. *Journal of Biological Chemistry, 293*(36), 14178–14189. https://doi.org/10.1074/jbc.RA118.003613
109. Ni, L., Yuan, C., Chen, G., Chen, G., Zhang, C., Zhang, C., Zhang, C., Zhang, C., & Wu, X. (2020). SGLT2i: Beyond the glucose-lowering effect. *Cardiovascular Diabetology, 19*(1), 98. https://doi.org/10.1186/S12933-020-01071-Y
110. Ning, C., Jiao, Y., Wang, J., Li, W., Zhou, J., Lee, Y. C., Ma, D. L., Leung, C. H., Zhu, R., & David Wang, H. M. (2022). Recent advances in the managements of type 2 diabetes mellitus and natural hypoglycemic substances. *Food Science and Human Wellness, 11*(5), 1121–1133. https://doi.org/10.1016/j.fshw.2022.04.004
111. Noubiap, J. J., Nansseu, J. R., Lontchi-Yimagou, E., Nkeck, J. R., Nyaga, U. F., Ngouo, A. T., Tounouga, D. N., Tianyi, F. L., Foka, A. J., Ndoadoumgue, A. L., & Bigna, J. J. (2022). Global, regional, and country estimates of metabolic syndrome burden in children and adolescents in 2020: A systematic review and modelling analysis. *The Lancet Child & Adolescent Health, 6*(3), 158–170. https://doi.org/10.1016/S2352-4642(21)00374-6
112. Nurkolis, F., Harbuwono, D. S., Taslim, N. A., Soegondo, S., Suastika, K., Sparringa, R. A., Mustika, A., Syam, A. F., Santini, A., Holly, J. M. P., Bhunia, A. K., & Tjandrawinata, R. R. (2025). New insight on dietary strategies to increase insulin sensitivity and reduce diabetes prevalence: An expert perspective and recommendation. *Discover Food, 5*(1), 136. https://doi.org/10.1007/S44187-025-00422-6

113. Ogurtsova, K., Guariguata, L., Barengo, N. C., Ruiz, P. L. D., Sacre, J. W., Karuranga, S., Sun, H., Boyko, E. J., & Magliano, D. J. (2022). IDF diabetes atlas: Global estimates of undiagnosed diabetes in adults for 2021. *Diabetes Research and Clinical Practice, 183*, 109118. https://doi.org/10.1016/j.diabres.2021.109118
114. Parker, V. E. R., Savage, D. B., O'Rahilly, S., & Semple, R. K. (2011). Mechanistic insights into insulin resistance in the genetic era. *Diabetic Medicine, 28*(12), 1476–1486. https://doi.org/10.1111/J.1464-5491.2011.03463.X
115. Patel, B. M., & Mehta, A. A. (2013). Choice of anti-hypertensive agents in diabetic subjects. *Diabetes and Vascular Disease Research, 10*(5), 385–396. https://doi.org/10.1177/1479164113485250
116. Pei, J., Wang, B., & Wang, D. (2022). Current studies on molecular mechanisms of insulin resistance. *Journal of Diabetes Research, 2022*, 1863429. https://doi.org/10.1155/2022/1863429
117. Popkin, B. M., & Reardon, T. (2018). Obesity and the food system transformation in Latin America. *Obesity Reviews, 19*(8), 1028–1064. https://doi.org/10.1111/OBR.12694
118. Pouwels, S., Sakran, N., Graham, Y., Leal, A., Pintar, T., Yang, W., Kassir, R., Singhal, R., Mahawar, K., & Ramnarain, D. (2022). Non-alcoholic fatty liver disease (NAFLD): A review of pathophysiology, clinical management and effects of weight loss. *BMC Endocrine Disorders, 22*(1), 63. https://doi.org/10.1186/S12902-022-00980-1
119. Prentki, M., & Madiraju, M. S. R. (2012). Glycerolipid/free fatty acid cycle and islet β-cell function in health, obesity and diabetes. *Molecular and Cellular Endocrinology, 353*(1–2), 88–100. https://doi.org/10.1016/j.mce.2011.11.004
120. Prone-Olazabal, D., Davies, I., & González-Galarza, F. F. (2023). Metabolic syndrome: An overview on its genetic associations and gene-diet interactions. *Metabolic Syndrome and Related Disorders, 21*(10), 545–560. https://doi.org/10.1089/MET.2023.0125
121. Rabe, K., Lehrke, M., Parhofer, K. G., & Broedl, U. C. (2008). Adipokines and insulin resistance. *Molecular Medicine, 14*(11–12), 741–751. https://doi.org/10.2119/2008-00058.RABE
122. Raddatz, D., & Ramadori, G. (2011). Hepatic carbohydrate metabolism. In *Molecular pathology of liver diseases* (pp. 109–123). Springer. https://doi.org/10.1007/978-1-4419-7107-4_8
123. Rahmawati, N. D., Andriani, H., Wirawan, F., Farsia, L., Waits, A., & Karim Taufiqurahman, K. A. (2024). Body mass index as a dominant risk factor for metabolic syndrome among Indonesian adults: A 6-year prospective cohort study of non communicable diseases. *BMC Nutrition, 10*(1), 43. https://doi.org/10.1186/S40795-024-00856-8
124. Ramezankhani, A., Azizi, F., & Hadaegh, F. (2022). Gender differences in changes in metabolic syndrome status and its components and risk of cardiovascular disease: A longitudinal cohort study. *Cardiovascular Diabetology, 21*(1), 227. https://doi.org/10.1186/S12933-022-01665-8
125. Roy, P. K., Islam, J., & Lalhlenmawia, H. (2023). Prospects of potential adipokines as therapeutic agents in obesity-linked atherogenic dyslipidemia and insulin resistance. *The Egyptian Heart Journal, 75*(1), 24. https://doi.org/10.1186/S43044-023-00352-7
126. Russell, J. B. W., Koroma, T. R., Sesay, S., Samura, S. K., Lakoh, S., Bockarie, A., Abiri, O. A., Conteh, V., Conteh, S., Smith, M., Mahdi, O. Z., & Lisk, D. R. (2024). Prevalence and correlates of metabolic syndrome among adults in freetown, Sierra Leone: A comparative analysis of NCEP ATP III, IDF and harmonized ATP III criteria. *International Journal of Cardiology: Cardiovascular Risk and Prevention, 20*, 200236. https://doi.org/10.1016/j.ijcrp.2024.200236
127. Ruze, R., Liu, T., Zou, X., Song, J., Chen, Y., Xu, R., Yin, X., & Xu, Q. (2023). Obesity and type 2 diabetes mellitus: Connections in epidemiology, pathogenesis, and treatments. *Frontiers in Endocrinology, 14*, 1161521. https://doi.org/10.3389/FENDO.2023.1161521/FULL
128. Saghiv, M. S., & Sagiv, M. S. (2020). Metabolism. In *Basic exercise physiology* (pp. 33–147). Springer. https://doi.org/10.1007/978-3-030-48806-2_2

129. Samuel, V. T., Liu, Z. X., Qu, X., Elder, B. D., Bilz, S., Befroy, D., Romanelli, A. J., & Shulman, G. I. (2004). Mechanism of hepatic insulin resistance in non-alcoholic fatty liver disease. *Journal of Biological Chemistry, 279*(31), 32345–32353. https://doi.org/10.1074/jbc.M313478200
130. Santos, L. (2022). The impact of nutrition and lifestyle modification on health. *European Journal of Internal Medicine, 97*, 18–25. https://doi.org/10.1016/j.ejim.2021.09.020
131. Schaefer, J. R., Scharnagl, H., Baumstark, M. W., Schweer, H., Zech, L. A., Seyberth, H., Winkler, K., Steinmetz, A., & März, W. (1997). Homozygous familial defective apolipoprotein B-100: Enhanced removal of apolipoprotein E-containing VLDLs and decreased production of LDLs. *Arteriosclerosis, Thrombosis, and Vascular Biology, 17*(2), 348–353. https://doi.org/10.1161/01.ATV.17.2.348
132. Schinner, S., Scherbaum, W. A., Bornstein, S. R., & Barthel, A. (2005). Molecular mechanisms of insulin resistance. *Diabetic Medicine, 22*(6), 674–682. https://doi.org/10.1111/J.1464-5491.2005.01566.X
133. Schoeler, M., & Caesar, R. (2019). Dietary lipids, gut microbiota and lipid metabolism. *Reviews in Endocrine and Metabolic Disorders, 20*(4), 461–472. https://doi.org/10.1007/S11154-019-09512-0
134. Seko, Y., Yamaguchi, K., & Itoh, Y. (2018). The genetic backgrounds in nonalcoholic fatty liver disease. *Clinical Journal of Gastroenterology, 11*(2), 97–102. https://doi.org/10.1007/S12328-018-0841-9
135. Shahid, M., Rasool, A., & Anjum, F. (2021). Inherited metabolic disorders: A current status. In *Endocrine disrupting chemicals-induced metabolic disorders and treatment strategies* (pp. 69–82). Springer. https://doi.org/10.1007/978-3-030-45923-9_4
136. Shivananda Nayak, B., Sobrian, A., Latiff, K., Pope, D., Rampersad, A., Lourenço, K., & Samuel, N. (2014). The association of age, gender, ethnicity, family history, obesity and hypertension with type 2 diabetes mellitus in Trinidad. *Diabetes and Metabolic Syndrome: Clinical Research and Reviews, 8*(2), 91–95. https://doi.org/10.1016/j.dsx.2014.04.018
137. Siebel, A. L., Heywood, S. E., & Kingwell, B. A. (2015). HDL and glucose metabolism: Current evidence and therapeutic potential. *Frontiers in Pharmacology, 6*, 258. https://doi.org/10.3389/FPHAR.2015.00258/FULL
138. Siitonen, N., Pulkkinen, L., Lindström, J., Kolehmainen, M., Eriksson, J. G., Venojärvi, M., Ilanne-Parikka, P., Keinänen-Kiukaanniemi, S., Tuomilehto, J., & Uusitupa, M. (2011). Association of ADIPOQ gene variants with body weight, type 2 diabetes and serum adiponectin concentrations: The Finnish Diabetes Prevention Study. *BMC Medical Genetics, 12*, 5. https://doi.org/10.1186/1471-2350-12-5
139. Soccio, R. E., Chen, E. R., & Lazar, M. A. (2014). Thiazolidinediones and the promise of insulin sensitization in type 2 diabetes. *Cell Metabolism, 20*(4), 573–591. https://doi.org/10.1016/j.cmet.2014.08.005
140. Staels, B., Dallongeville, J., Auwerx, J., Schoonjans, K., Leitersdorf, E., & Fruchart, J. C. (1998). Mechanism of action of fibrates on lipid and lipoprotein metabolism. *Circulation, 98*(19), 2088–2093. https://doi.org/10.1161/01.CIR.98.19.2088
141. Stahl, E. P., Dhindsa, D. S., Lee, S. K., Sandesara, P. B., Chalasani, N. P., & Sperling, L. S. (2019). Nonalcoholic fatty liver disease and the heart: JACC state-of-the-art review. *Journal of the American College of Cardiology, 73*(8), 948–963. https://doi.org/10.1016/J.JACC.2018.11.050
142. Su, P., Chen, C., & Sun, Y. (2025). Physiopathology of polycystic ovary syndrome in endocrinology, metabolism and inflammation. *Journal of Ovarian Research, 18*(1), 34. https://doi.org/10.1186/S13048-025-01621-6
143. Taskinen, M. R. (2003). Diabetic dyslipidaemia: From basic research to clinical practice. *Diabetologia, 46*(6), 733–749. https://doi.org/10.1007/S00125-003-1111-Y
144. Thakur, S., Kumar, V., Das, R., Sharma, V., & Mehta, D. K. (2024). Biomarkers of hepatic toxicity: An overview. *Current Therapeutic Research – Clinical and Experimental, 100*, 100737. https://doi.org/10.1016/j.curtheres.2024.100737

145. Thobani, A., & Wenger, N. K. (2022). Dyslipidemia. In *Handbook of outpatient cardiology* (pp. 177–191). Springer. https://doi.org/10.1007/978-3-030-88953-1_11
146. Tsiotra, P. C., & Tsigos, C. (2006). Stress, the endoplasmic reticulum, and insulin resistance. *Annals of the New York Academy of Sciences, 1083*, 63–76. https://doi.org/10.1196/ANNALS.1367.007
147. Vaidya, R. A., Desai, S., Moitra, P., Salis, S., Agashe, S., Battalwar, R., Mehta, A., Madan, J., Kalita, S., Udipi, S. A., & Vaidya, A. B. (2023). Hyperinsulinemia: An early biomarker of metabolic dysfunction. *Frontiers in Clinical Diabetes and Healthcare, 4*, 1159664. https://doi.org/10.3389/FCDHC.2023.1159664/FULL
148. Valensi, P. (2021). Autonomic nervous system activity changes in patients with hypertension and overweight: Role and therapeutic implications. *Cardiovascular Diabetology, 20*(1), 170. https://doi.org/10.1186/S12933-021-01356-W
149. van de Burgt, N., van Doesum, W., Grevink, M., van Niele, S., de Koning, T., Leibold, N., Martinez-Martinez, P., van Amelsvoort, T., & Cath, D. (2023). Psychiatric manifestations of inborn errors of metabolism: A systematic review. *Neuroscience and Biobehavioral Reviews, 144*, 104970. https://doi.org/10.1016/j.neubiorev.2022.104970
150. Vasishta, S., Umakanth, S., Adiga, P., & Joshi, M. B. (2022). Extrinsic and intrinsic factors influencing metabolic memory in type 2 diabetes. *Vascular Pharmacology, 142*, 106933. https://doi.org/10.1016/j.vph.2021.106933
151. Wang, Q., Chair, S. Y., Wong, E. M. L., & Qiu, X. (2022). Actively incorporating lifestyle modifications into daily life: The key to adherence in a lifestyle intervention programme for metabolic syndrome. *Frontiers in Public Health, 10*, 929043. https://doi.org/10.3389/FPUBH.2022.929043/FULL
152. Wang, J., Du, J., Wang, M., Jin, M., Tang, Z., & Mao, Y. (2025). Global, regional, and national burden of NAFLD in youths and young adults aged 15–39 years, 1990–2021, its attributable risk factors, and projections to 2035: A systematic analysis of the Global Burden of Disease Study 2021. *Frontiers in Nutrition, 12*, 1509232. https://doi.org/10.3389/FNUT.2025.1509232/FULL
153. Williams, R., & Periasamy, M. (2021). Genetic and environmental factors contributing to visceral adiposity in Asian populations. *Endocrinology and Metabolism, 35*(4), 681–695. https://doi.org/10.3803/ENM.2020.772
154. Wong, V. W. S., Ekstedt, M., Wong, G. L. H., & Hagström, H. (2023). Changing epidemiology, global trends and implications for outcomes of NAFLD. *Journal of Hepatology, 79*(3), 842–852. https://doi.org/10.1016/j.jhep.2023.04.036
155. Wu, H., & Ballantyne, C. M. (2020). Metabolic inflammation and insulin resistance in obesity. *Circulation Research, 126*(11), 1549–1564. https://doi.org/10.1161/CIRCRESAHA.119.315896
156. Xia, X. D., Peng, Z. S., Gu, H. M., Wang, M., Wang, G. Q., & Zhang, D. W. (2021). Regulation of PCSK9 expression and function: Mechanisms and therapeutic implications. *Frontiers in Cardiovascular Medicine, 8*, 764038. https://doi.org/10.3389/FCVM.2021.764038/FULL
157. Xia, J., Ding, L., & Liu, G. (2025). Metabolic syndrome and dermatological diseases: Association and treatment. *Nutrition and Metabolism, 22*(1), 36. https://doi.org/10.1186/S12986-025-00924-1
158. Yamagata, K. (2018). Metabolic syndrome: Preventive effects of dietary flavonoids. *Studies in Natural Products Chemistry, 60*, 1–28. https://doi.org/10.1016/B978-0-444-64181-6.00001-2
159. Yu, Y. H., & Ginsberg, H. N. (2005). Adipocyte signaling and lipid homeostasis: Sequelae of insulin-resistant adipose tissue. *Circulation Research, 96*(10), 1042–1052. https://doi.org/10.1161/01.RES.0000165803.47776.38
160. Zatterale, F., Longo, M., Naderi, J., Raciti, G. A., Desiderio, A., Miele, C., & Beguinot, F. (2020). Chronic adipose tissue inflammation linking obesity to insulin resistance and type 2 diabetes. *Frontiers in Physiology, 10*, 1607. https://doi.org/10.3389/FPHYS.2019.01607/FULL

161. Zhang, T., Chen, J., Tang, X., Luo, Q., Xu, D., & Yu, B. (2019). Interaction between adipocytes and high-density lipoprotein: New insights into the mechanism of obesity-induced dyslipidemia and atherosclerosis. *Lipids in Health and Disease, 18*(1), 223. https://doi.org/10.1186/S12944-019-1170-9
162. Zhang, X., Yang, S., Chen, J., & Su, Z. (2019). Unraveling the regulation of hepatic gluconeogenesis. *Frontiers in Endocrinology, 9*, 802. https://doi.org/10.3389/FENDO.2018.00802/FULL
163. Zhang, A. M. Y., Xia, Y. H., Lin, J. S. H., Chu, K. H., Wang, W. C. K., Ruiter, T. J. J., Yang, J. C. C., Chen, N., Chhuor, J., Patil, S., Cen, H. H., Rideout, E. J., Richard, V. R., Schaeffer, D. F., Zahedi, R. P., Borchers, C. H., Johnson, J. D., & Kopp, J. L. (2023). Hyperinsulinemia acts via acinar insulin receptors to initiate pancreatic cancer by increasing digestive enzyme production and inflammation. *Cell Metabolism, 35*(12), 2119–2135.e5. https://doi.org/10.1016/j.cmet.2023.10.003
164. Zhang, D., Jian, Y. P., Zhang, Y. N., Li, Y., Gu, L. T., Sun, H. H., Liu, M., Di, Zhou, H. L., Wang, Y. S., & Xu, Z. X. (2023). Short-chain fatty acids in diseases. *Cell Communication and Signaling, 21*(1), 212. https://doi.org/10.1186/S12964-023-01219-9
165. Zhang, K., Ma, Y., Luo, Y., Song, Y., Xiong, G., Ma, Y., Sun, X., & Kan, C. (2023). Metabolic diseases and healthy aging: Identifying environmental and behavioral risk factors and promoting public health. *Frontiers in Public Health, 11*, 1253506. https://doi.org/10.3389/FPUBH.2023.1253506/FULL
166. Zhao, R. Z., Jiang, S., Zhang, L., & Yu, Z. B. (2019). Mitochondrial electron transport chain, ROS generation and uncoupling (Review). *International Journal of Molecular Medicine, 44*(1), 3–15. https://doi.org/10.3892/IJMM.2019.4188
167. Zhao, X., An, X., Yang, C., Sun, W., Ji, H., & Lian, F. (2023). The crucial role and mechanism of insulin resistance in metabolic disease. *Frontiers in Endocrinology, 14*, 1149239. https://doi.org/10.3389/FENDO.2023.1149239/FULL
168. Zong, X., Kelishadi, R., Kim, H. S., Schwandt, P., Matsha, T. E., Mill, J. G., Caserta, C. A., Medeiros, C. C. M., Kollias, A., Whincup, P. H., Pacifico, L., López-Bermejo, A., Zhao, M., Zheng, M., & Xi, B. (2024). A proposed simplified definition of metabolic syndrome in children and adolescents: A global perspective. *Bmc Medicine, 22*(1), 190. https://doi.org/10.1186/S12916-024-03406-Y

Phytochemical-Based Nanomedicine in the Treatment and Management of Metabolic Syndrome

3

Vijay Kumar, Padma Sharma, Abhinav Sharma, Nisha Kharka, Kiran Toppo, Komal Kumar Ingle, Sachitra Kumar Ratha, and Vipin Rai

Abstract

Metabolic syndrome (MetS) comprises a group of interconnected conditions such as obesity, insulin resistance, dyslipidemia and hypertension that substantially raise the chances of developing type 2 diabetes and cardiovascular diseases. The increasing incidence of MetS globally needs immediate attention to efficient and safer therapeutic approaches. Plant extracts serve as a promising alternative for anti-inflammatory, antioxidant properties, insulin-sensitization and other therapeutics. The clinical use of these extracts remains challenging due to their instability and poor bioavailability. Nanotechnology can be used to overcome these problems by using an advanced delivery system. Nanoformulations from plant extracts serve as effective tools for MetS management. Plant extracts have been utilized to develop various nanomedicine delivery systems, including nanoemulsions, solid lipid nanocarriers, liposomes, micelles and green-synthesized metal nanoparticles. These nanoformulations make plant-based extract more soluble and stable and offer targeted delivery as well as controlled release over time. Several studies confirm nanoformulations from curcumin, berberine, resveratrol, quercetin and other plant extracts lower oxidative damage,

V. Kumar · A. Sharma · K. Toppo · K. K. Ingle
Algology Laboratory, CSIR-National Botanical Research Institute, Lucknow, Uttar Pradesh, India

P. Sharma
Department of Environment Studies, Panjab University, Chandigarh, India

N. Kharka · S. K. Ratha (✉)
Algology Laboratory, CSIR-National Botanical Research Institute, Lucknow, Uttar Pradesh, India

Academy of Scientific and Innovative Research, Ghaziabad, Uttar Pradesh, India
e-mail: sachitra.ratha@nbri.res.in

V. Rai
Rutgers University, Newark, NJ, USA

V. K. Chaturvedi et al. (eds.), *Nanotheranostics in Metabolic Syndrome*,
https://doi.org/10.1007/978-981-95-3746-4_3

inflammatory responses, insulin resistance and control lipid profiles. The curcumin, resveratrol, quercetin and berberine-based nanoparticles have stronger anti-inflammatory, antioxidative, enhanced lipid and glucose regulatory effects. Plant-based nanomedicine showed their ability to control MetS-related pathways through AMPK, NF-κB and PPAR-γ systems. Further studies are required in this area to validate the safety, long-term effects and clinical effectiveness. Phytonanomedicine has the power to revolutionize the treatment of MetS while providing better therapeutic effects than conventional medicine. Plant-based nanosystems operate as a link between traditional plants and contemporary pharmaceutical delivery systems and therefore create opportunities for management and treatments of metabolic disorders.

Keywords

Metabolic syndrome · Nanomedicine · Plant extract · Drug delivery · Curcumin · Cardiovascular diseases

1 Introduction

Medical science defines metabolic syndrome (MetS) as a complicated health condition which results from various interconnected medical issues such as obesity, insulin resistance, hypertension, dyslipidemia and increased blood glucose levels (Fig. 3.1). These health conditions create an environment where people develop serious medical issues including type 2 diabetes and cardiovascular diseases (CVDs) [48]. MetS exists as a global healthcare challenge as it affects 20–30% of the global population [91]. MetS diagnosis depends on criteria established by the International Diabetes Federation (IDF) and the National Cholesterol Education Program's Adult Treatment Panel III (NCEP ATP III), which defines parameters for waist circumference, fasting blood glucose, triglycerides, HDL cholesterol and blood pressure levels [106]. The development of MetS results from genetic, environmental and behavioral factors [64].

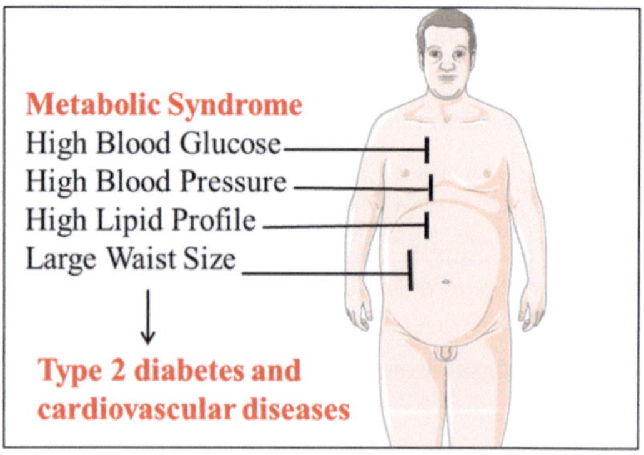

Fig. 3.1 Metabolic syndrome

Insulin resistance, chronic inflammation and oxidative stress drive MetS development through genetic and environmental factors [73]. The impairment of insulin signaling caused by increased free fatty acids creates hyperinsulinemia which disrupts glucose metabolism. The production of adipocytokines including tumor necrosis factor-α (TNF-α), interleukin-6 (IL-6) and leptin becomes abnormal during chronic inflammation which occurs when individuals have visceral obesity [26]. The production of adiponectin decreases through oxidative stress, thus compromising metabolic homeostasis because adiponectin works as a protective adipokine to prevent inflammation and atherosclerosis [87].

The current MetS management strategies involve modifying lifestyles by changing diets, boosting physical activity levels and using medication to manage blood glucose, lipid levels and blood pressure [49]. These treatments along with pharmacological approaches focus on managing each aspect of the metabolic syndrome. However, multiple factors combine to make MetS difficult to manage which requires complete solutions to address its fundamental triggers and connected network pathways [74].

Conventional medications cause unwanted side effects including gastrointestinal discomfort; that lead patients to become noncompliant specifically when receiving extended treatments [58]. The medical community requires quicker development of novel therapies which offer better safety and performance in treatment procedures.

Natural products extracted from medicinal plants have become increasingly popular for treating MetS in recent years. Phytochemical compounds such as curcumin, berberine, resveratrol, quercetin and gymnemic acid show exceptional behavior as anti-inflammatory, antioxidant and insulin-sensitizing capabilities. Medical use of these compounds remains restricted due to their poor absorption rate, low solubility and short lifespan in human tissue [43, 61].

Nanotechnology is a novel approach which addresses existing drug delivery challenges. The use of nanoemulsions, solid lipid nanoparticles (SLNs), nanostructured lipid carriers (NLCs) and polymeric nanoparticles presents distinct advantages through enhanced bioavailability, targeted delivery and decreased systemic adverse effects. The protective encapsulation of phytochemicals within nanoformulations combined with targeted drug release at the target site represents an important property [56, 127].

This chapter gives a complete evaluation of nanoformulated plant extracts as effective MetS treatments along with their management of associated symptoms. It also examines the mechanism of action, therapeutic benefits and technology-driven enhancements to understand innovative approaches that improve the treatment of metabolic syndrome.

2 Types of Phytochemical-Based Nanoformulations

Nanoformulations function as diagnostic tools along with drug delivery systems. Drug formulations using nanoformulations reduce the side effects, act on the target site, control release and enhance the solubility of the plant extracts [116]. Different kinds of nanoformulation like liposomes, polymeric nanoparticles, solid lipid nanoparticles, dendrimers, micelles, nanoemulsions, metallic nanoparticles, carbon-based nanomaterials, quantum dots, exosomes, nanocrystals, magnetic nanoparticles, niosomes, nanogels and protein-based nanoparticles from plant-based extract are used for the treatment of diseases [83]. Curcumin, berberine, resveratrol and quercetin are promising phytochemicals with anti-inflammatory, antioxidant and insulin-sensitizing properties, making them effective in treating metabolic disorders [57].

The therapeutic use of phytochemicals faces significant barriers because they dissolve poorly and have limited uptake by the body and short elimination times [50] Several nanoformulations have been developed as solutions to those challenges. Nanoformulations improve phytochemical delivery by enhancing stability together with solubility and targeting properties to boost their therapeutic value [60]. Table 3.1 describes various nanoformulations containing plant-based compounds.

The pharmaceutical industry has transformed bioactive compound delivery through different types of phytochemical-based nanoformulations [140]. Nanoformulations increase the solubility levels as well as stability and bioavailability of phytochemicals while allowing precise targeted drug delivery processes that maintain prolonged effects. Such nanoformulations address traditional delivery problems which allow them to become a potent tool for treating diabetes, obesity, dyslipidemia and hypertension disorders [131, 132].

3 Mechanisms of Action of Plant Compounds and Their Nanomedicine

Plant-based nanomedicine represents a promising method of treatment for metabolic disorders that include diabetes, obesity, dyslipidemia and hypertension. Researchers have extensively studied the therapeutic elements present in curcumin, berberine, resveratrol and quercetin [62]. Plant-based nanomedicine applies these compounds for metabolic syndrome management because they demonstrate anti-inflammatory action, insulin-modulating and antioxidant properties. However, their poor solubility, low bioavailability and rapid metabolism limit their clinical application [92]. Nanoformulations solve these problems by improving substance solubility and strengthening stability alongside specific delivery methods which enhances both clinical results and treatment benefits [82]. The generalized mechanism of plant-based nanoformulation is shown in Fig. 3.2. The following section outlines the mechanisms through which these phytochemicals function after being formulated as nanoparticles.

Table 3.1 Types of plant-based nanoformulations and their applications

Nanocarrier	Description	Advantages	Applications	Reference
Polymeric nanoparticles	Comprises of biodegradable polymers like chitosan, PLGA and PCL. The polymer matrix encapsulates phytochemicals for controlled release.	High stability and biocompatibility. Protects phytochemicals from degradation. Targeted delivery.	Curcumin-loaded PLGA nanoparticles for diabetic wound healing. Chitosan nanoparticles with Berberine for insulin sensitivity. Quercetin nanoparticles for anti-inflammatory and antioxidant effects.	[4, 59, 118]
Liposomes	Spherical phospholipid bilayers that encapsulate hydrophilic (in core) or hydrophobic (in bilayer) phytochemicals	Improves absorption of hydrophobic compounds; Biocompatible and biodegradable; Functional modifications	Curcumin-loaded liposomes for anti-inflammatory and antioxidant effects. Resveratrol liposomes for insulin sensitivity. Berberine liposomes for lipid-lowering and glucose regulation.	[8, 79, 131, 132]
Solid lipid nanoparticles (SLNs)	Colloidal carriers use solid lipids (e.g., fatty acids, waxes, triglycerides) to encapsulate phytochemicals	High stability and biocompatibility; Protects phytochemicals; Suitable for oral and topical delivery	Curcumin SLNs for wound healing and anti-inflammatory effects. Berberine SLNs for lipid and glucose control. Myricitrin SLNs for antioxidant and antidiabetic effects.	[2, 108, 134]
Nanoemulsions	Oil and water colloidal dispersions stabilized by surfactants, featuring droplet sizes smaller than 1 nanometer	Enhances solubility and absorption of hydrophobic compounds; Improves permeation Versatile delivery	Curcumin nanoemulsions for anti-inflammatory and antioxidant effects. Garlic oil nanoemulsions for lipid lowering. Kenaf seed oil nanoemulsions for liver health and lipid profile improvement.	[3, 33, 96]
Nanostructured lipid carriers (NLCs)	Advanced carriers combine solid and liquid lipids for disordered lipid structures, enabling controlled release	Higher drug loading capacity than SLNs. Improved stability and bioavailability.	Silymarin NLCs for antioxidant and antihyperglycemic effects. Curcumin NLCs for wound healing and anti-inflammatory effects.	[46, 98, 122]

(continued)

Table 3.1 (continued)

Nanocarrier	Description	Advantages	Applications	Reference
		Suitable for hydrophilic and hydrophobic compounds	Quercetin NLCs for lipid and glucose regulation.	
Gold and silver nanoparticles	Phytochemicals are attached to or embedded within metallic nanoparticles.	High drug-loading capacity; Biocompatible and biodegradable; Intrinsic antioxidant and antimicrobial properties	Curcumin-conjugated gold nanoparticles for anti-inflammatory and antioxidant effects. Green-synthesized silver nanoparticles for antidiabetic and antimicrobial effects. Resveratrol-gold nanoparticles for insulin sensitivity.	[22, 40, 112]
Dendrimers	Tree-like nanostructures with well-defined size and shape. Phytochemicals are encapsulated within or conjugated to the surface.	High drug-loading capacity; Precise control over size and surface properties; Targeted delivery	Curcumin-loaded dendrimers for anti-inflammatory and antioxidant effects. Quercetin dendrimers for lipid profile and glucose regulation.	[14, 51]
Micelles	Self-assembled nanostructures formed by amphiphilic molecules in aqueous solutions. Hydrophobic phytochemicals reside in the core, while the hydrophilic shell provides stability.	Enhances solubility and bioavailability of hydrophobic compounds; Compact structure for tissue acceptance; Targeted delivery	Curcumin-loaded micelles for anti-inflammatory and antioxidant effects. Oleanolic acid micelles for insulin sensitivity and glucose regulation.	[13, 131, 132]
Nanofibers	Nanometer-scale fibers are used for topical and transdermal drug delivery. Phytochemicals are incorporated into the nanofiber matrix.	High surface area for drug incorporation; Effective for wound healing and tissue regeneration; Sustained release	Curcumin-loaded nanofibers for treating diabetic wounds and resveratrol nanofibers for promoting skin health and providing antioxidant benefits.	[86, 128]

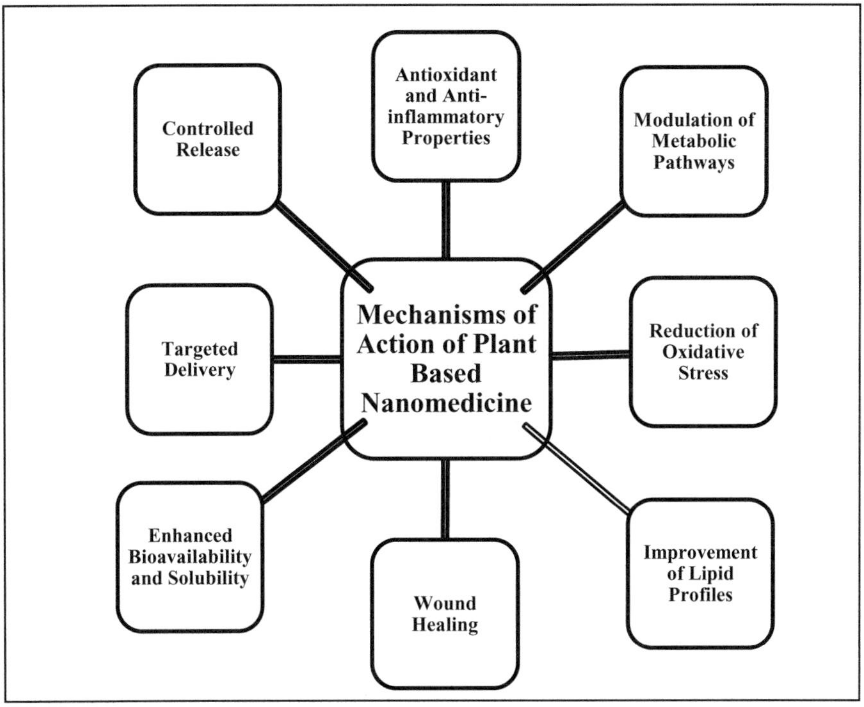

Fig. 3.2 Generalized mechanisms of action of plant-based nanomedicine

3.1 Curcumin

Turmeric-derived polyphenolic compound curcumin serves as a known medicinal ingredient because it demonstrates anti-inflammatory, antioxidant and anti-hyperglycemic properties. Nanoformulations containing curcumin-loaded nanoparticles, liposomes and nanoemulsions enhance curcumin availabilities through improved absorption and extend its release durations. The mechanism of action of curcumin and curcumin-associated nanoformulations is shown in Fig. 3.3.

3.2 Berberine

The plant-derived alkaloid berberine functions as an anti-hyperglycemic and anti-inflammatory substance and it is present in *Berberis aristata* and related plants [31, 32]. The bioavailability and effectiveness of berberine are improved through its nanoformulation with both solid lipid nanoparticles (SLNs) and liposomes. Researchers have used SLN nanoformulations containing berberine for improved treatment outcomes in diabetic and dyslipidemic diseases. The mechanism of action of berberine and berberine-associated nanoformulations is shown in Fig. 3.4.

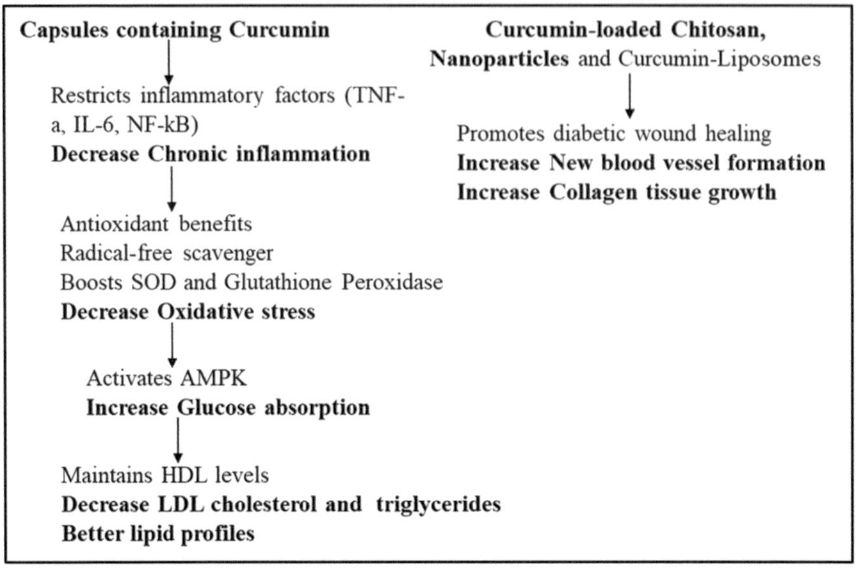

Fig. 3.3 Mechanism of action curcumin and curcumin-associated nanoformulations

Fig. 3.4 Mechanism of action of berberine and berberine-associated nanoformulations

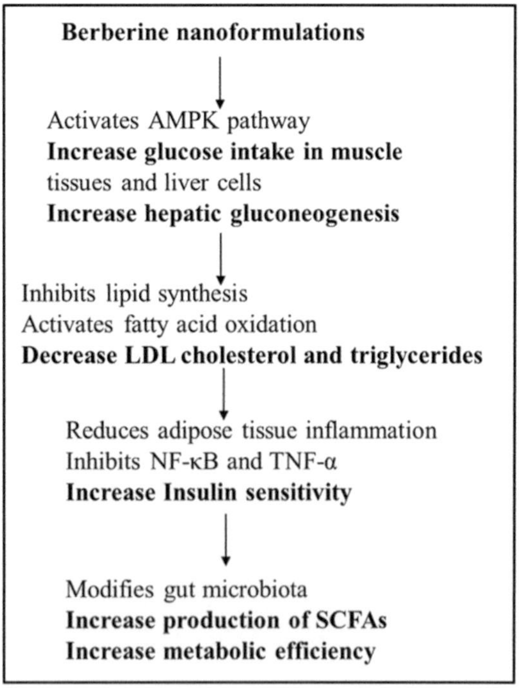

3.3 Resveratrol

The natural plant compound resveratrol occurs in grapes and berries where it serves as an anti-inflammatory, antioxidant and insulin-sensitization regulator. The use of nanoformulations including resveratrol-loaded nanoparticles alongside nanoliposomes improves both stability and make-up of this compound [124]. The mechanism of action of resveratrol and resveratrol-associated nanoformulations is shown in Fig. 3.5.

3.4 Quercetin

Fruits and vegetables contain the flavonoid quercetin which shows antioxidant, anti-inflammatory and anti-hyperglycemic properties. The bioavailability and solubility of quercetin get better when it is loaded into nanoparticles along with nanoemulsion formulation [11]. The mode of action of quercetin and quercetin-associated nanoformulations is shown in Fig. 3.5.

The mechanism of action of plant-based medicine acts on metabolic syndromes through different signaling pathways by anti-inflammatory action, antioxidant,

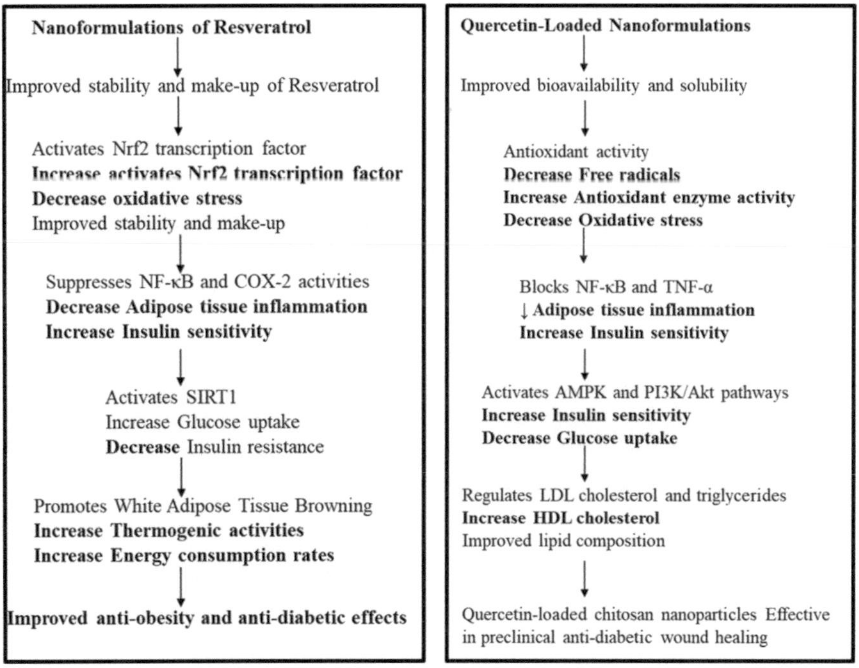

Fig. 3.5 Mechanism of action of resveratrol and quercetin and their associated nanoformulations for metabolic syndrome management

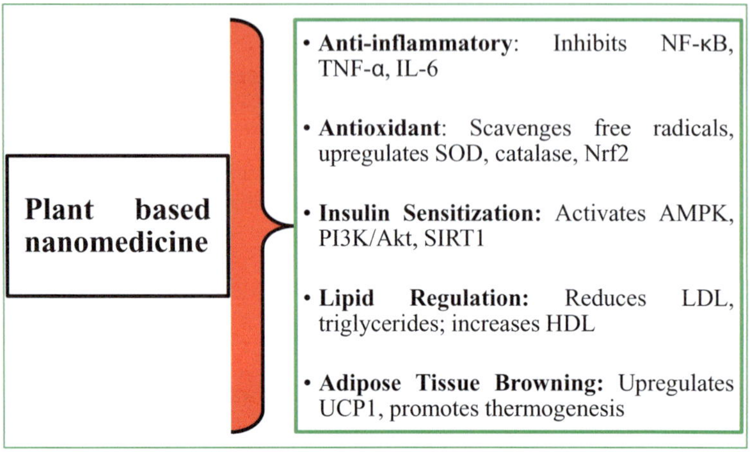

Fig. 3.6 Mechanism of action of phytonanomedicines

insulin sensitization, lipid regulation, adipose tissue browning, etc. The mode of action of phytonanomedicines is shown in Fig. 3.6.

4 Plant-Based Nanomedicine in Metabolic Diseases

4.1 Diabetes Mellitus

The field of plant-based nanomedicine develops as a modern therapeutic approach for diabetes mellitus management through the combination of plant bioactive compounds with nanotechnology precision. The combination of natural antidiabetic agents with this strategy delivers improved bioavailability, enhanced stability and target-specific delivery to patients [113]. This approach provides health practitioners with an effective treatment option beyond traditional medical methods.

Nanoliposomes serve as an effective carrier for betalain delivery to diabetic rats induced by streptozotocin. The delivery method used sustained release properties and has antioxidant and anti-hyperglycemic benefits that resulted in better hyperglycemia control, improved oxidative stress regulation and lipid profile together with decreased tissue damage [12]. The delivery of selenium nanoparticles (MPE-SeNPs) of mulberry and *Pueraria lobata* extracts (MPE) into diabetic rats showed controlled drug release, stability improvements, hypoglycemic response, lower blood sugar levels and an ameliorated overall hypoglycemic effects [37].

Laboratory results showed that curcumin contained in liposomes given to streptozotocin-diabetic rats has hepatoprotective benefits along with antioxidant protection. It decreased oxidative stress together with better glycemic management [28]. Quercetin delivered through core-shell nanoparticles using chitosan and alginate to diabetic rats showed strong results against blood glucose elevation and high lipid levels in the blood. It lowered blood glucose levels while simultaneously improving lipid profile [90]. Nanocomposite medications made from *Bambusa bambos* accelerated the wound-healing process and promoted tissue regeneration in diabetic rats [119–121]. The detailed list of plant-based nanoformulations for addressing the diabetic mellitus is shown in Table 3.2.

The natural and sustainable plant-based nanomedicine technology provides targeted therapies for diabetes management. The plant-based nanomedicine decreases the risks of synthetic drug side effects while improving the therapeutic benefits of plant compounds [39]. The ongoing research shows great promise to develop individualized treatments for diabetes mellitus that simultaneously tackle hyperglycemia and its related conditions.

4.2 Obesity

The merger of plant bioactive compounds with nanotechnology provides a new treatment strategy for obesity through plant-based nanomedicine applications. This dual technology enhances therapeutic agent availability, targeted delivery, and stability, making it a suitable approach for managing obesity and its associated medical conditions [62]. Nanoencapsulation of curcumin enhances its bioavailability and enables targeted delivery to adipose tissue, where it modulates lipid metabolism and reduces inflammation, thereby supporting effective weight management [71]. The administration of catechin and gingerol-bearing nanomedicine proves effective for obesity treatment through improvements in metabolic markers and fat tissue reduction [72]. The treatment of gold nanoparticles obtained from *Smilax glabra* rhizome in diabetic rats led to a reduction in blood glucose levels, better lipid profile performance and complete tissue recovery in both liver and cardiac areas [19]. Silver nanoparticles extracted from *Argyreia nervosa* inhibit carbohydrate metabolism enzymes, antibacterial activity and lower carbohydrate metabolism activity [109]. Gold nanoparticles extracted from *Dendropanax morbifera* lower triglycerides and PPAR-γ while decreasing Jak2, STAT3, CEBPα and ap2 protein expression which resulted in improved lipid metabolism and decreased adipocytes [135]. The detailed list of plant-based nanoformulations for the management of obesity is shown in Table 3.2.

Plant-based nanomedicine presents a natural sustainable and effective method for targeting obesity treatment. This approach reduces the unwanted effects of synthetic pharmaceutical substances and it improves the therapeutic efficiency of plant-derived compounds. Research developments show significant prospects to produce targeted obesity therapies that simultaneously treat both obesity as well as its associated metabolic effects [115].

Table 3.2 List of some plant-based nanosystems for the treatment of metabolic syndrome

Disease condition	Phytonanomedicine	Type of nanoformulation	Model organism	Mechanism of action	Therapeutic benefits	References
Diabetes mellitus type II	Thymoquinone	Polymeric nanocapsules (gum rosin)	Diabetic rats	Sustained release, anti-hyperglycemic and lipid-lowering effects	Lowered blood sugar level, enhanced lipid profile and comparable efficacy to metformin.	[104]
Diabetes mellitus type II	Oleanolic acid (OA)	Polygalacturonic acid (PGA)-OA micelles	Diabetes mellitus type II model rats	Improved gastrointestinal permeability, enhanced drug absorption and sustained release	Regulated blood sugar levels, enhanced insulin sensitivity and sustained benefits.	[137]
Diabetes	Curcumin nanoparticles	Polymeric nanoparticles (e.g., PLGA)	Diabetic rats/mice	Anti-inflammatory, antioxidant and insulin-sensitizing properties	Lowers blood glucose levels, enhances pancreatic β-cell function and prevents complications such as neuropathy.	[125]
Diabetes	Berberine-loaded nanoparticles	Liposomes	Diabetic rats	Enhances glucose uptake, reduces insulin resistance and inhibits gluconeogenesis	Improves glycemic control and reduces diabetic complications.	[88]
Diabetes	Quercetin nanoformulations	Solid lipid nanoparticles (SLNs)	Diabetic rats	Antioxidant and anti-inflammatory effects	Protects against oxidative stress and improves insulin sensitivity.	[45]
Diabetes mellitus	*Pouteria sapota*	Green-synthesized silver nanoparticles	Streptozotocin-induced rats	Antioxidant, antiglycosylation and α-amylase inhibition	Improved superoxide dismutase and catalase activity, reduced blood glucose	[100]

Diabetes mellitus	Betalain	Nanoliposomes	Streptozotocin-induced rats	Sustained release, antioxidant and anti-hyperglycemic effects	Improved hyperglycemia, oxidative stress and hyperlipidemia; reduced tissue damage	[12]
Diabetes mellitus	Mulberry leaf and *Pueraria Lobata* extracts (MPE)	Selenium-layered nanoparticles (MPE-SeNPs)	Diabetic rats	Controlled release, enhanced stability and hypoglycemic effects	Improved intestinal permeability, reduced blood glucose and enhanced hypoglycemic effect	[37]
Diabetes mellitus type II	Myricitrin	Solid lipid nanoparticles	Streptozotocin–nicotinamide-induced diabetic mice	Antioxidant and antiapoptotic effects	Improved hyperglycemia and diabetes complications	[2]
Diabetes mellitus	Naringenin	Alginate/chitosan core–shell nanoparticles	Streptozotocin-induced mice	Sustained release, improved solubility and bioavailability	Reduced blood glucose, enhanced hypoglycemic effect and nontoxicity	[84]
Diabetes mellitus	*Salvia sclarea*	SLNs	High-fed diet (HFD)-triggered diabetic mice	Improved glycemic control and lipid profile	Increased brown adipose tissue weight, improved HDL levels and glycemic control	[30]
Diabetes mellitus	*Solanum nigrum*	Silver nanoparticles	Diabetic rats	Antioxidant and anti-hyperglycemic effects	Improved dyslipidemia and blood glucose levels	[110]
Diabetes mellitus	Curcumin	Nanoparticles	Diabetic albino rats	Antioxidant and glucose-lowering effects	Reduced blood glucose, improved insulin levels and antioxidant effects	[1]

(continued)

Table 3.2 (continued)

Disease condition	Phytonanomedicine	Type of nanoformulation	Model organism	Mechanism of action	Therapeutic benefits	References
Diabetes mellitus	*Catathelasma ventricosum*	Selenium nanoparticles	Diabetic rats induced with streptozotocin	Antioxidant and antidiabetic effects	Improved body weight, antioxidant activity and lipid levels	[80, 81]
Diabetes mellitus	Silymarin	Nanostructured lipid carriers	Diabetic rats induced with streptozotocin	Antihyperglycemic and antihyperalgesic effects	Reduced blood glucose, triglycerides and oxidative stress	[98]
Diabetes mellitus	Quercetin	Core–shell nanoparticles (chitosan/alginate)	Diabetic rats	Reduce high blood sugar and lipid level	Reduced blood glucose and improved lipid profile	[90]
Diabetes mellitus	*Talinum portulacifolium*	SLNs	Streptozotocin- and HFD-induced diabetic rats	Anti-hyperglycemic and lipid-lowering effects	Reduced blood glucose, improved serum insulin and triglyceride levels	[25]
Diabetic wound healing	Quercetin (QCT)	QCT-loaded silver nanoparticles (QCT-AgNPs) hydrogel	Diabetic wound model	Antimicrobial, anti-inflammatory and antioxidant effects	Enhanced re-epithelialization, reduced wound gap and improved wound healing	[20]
Diabetic wound healing	Curcumin (CUR) and resveratrol (RES)	Hyaluronic acid-functionalized chitosan nanoparticles	In vitro and diabetic wound model	Sustained release, anti-inflammatory and proangiogenic effects	Improved wound closure, enhanced tissue regeneration and reduced inflammation	[63]
STZ-induced diabetes	Curcumin	Pluronic nanomicelles	Diabetic rats induced with STZ	Increased Pancreatic and Duodenal Homeobox 1 (Pdx-1), NK Homeobox 6.1 (NKx6.1)	Improved pancreatic function and insulin production	[44]

Diabetes complications	Curcumin + CoQ10	PLGA-based NPs	STZ-induced diabetic rats	Reduced C-reactive protein (CRP), Interleukin-6 (IL-6) and total cholesterol levels lowered plasma triglycerides and increased HDL.	Reduced inflammation and improved lipid metabolism	[38]
Diabetic wound healing	Curcumin (CUR)	Curcumin-loaded chitosan nanoparticles (CUR-CS NPs) in collagen scaffold	Diabetic wound model	Sustained release, improved solubility and biocompatibility	Robust granulation tissue formation and full epithelialization	[69]
Diabetic wound healing	*Chamaecostus cuspidatus* (insulin plant)	Green-synthesized gold nanoparticles	Streptozotocin-induced diabetic rats	Hypoglycemic and free-radical scavenging effects	Improved wound recovery, controlled blood glucose and reduced oxidative stress	[99]
Diabetic wound healing	Curcumin (CUR)	Thermo-sensitive hydrogel with gelatin microspheres	Streptozotocin-induced diabetic mice	Matrix metalloproteinase-responsive release and enhanced bioavailability	Promoted wound healing, improved tissue regeneration and reduced inflammation	[80, 81]
Diabetic wound healing	*Dendrocalamus hamiltoni*	Nanocomposite	Streptozotocin-induced diabetic rats	Epithelial regeneration and collagen deposition	Improved wound healing and tissue regeneration	[119–121]
Diabetic wound healing	Curcumin (CUR)	Chitosan nanoparticles	Streptozotocin-induced diabetic rats	Macrophage-induced inflammation and angiogenesis	Reduced inflammation and improved angiogenesis	[78]

(continued)

Table 3.2 (continued)

Disease condition	Phytonanomedicine	Type of nanoformulation	Model organism	Mechanism of action	Therapeutic benefits	References
Diabetic wound healing	*Syzygium cumini*	Nanocomposites	Diabetic rats induced with STZ	Fast Epithelial regeneration neovascularization and anti-inflammatory action	Enhanced recovery of wounds and alleviated inflammation	[119–121]
Diabetic wound healing	*Acalypha indica*	Gold nanoparticles	BALB/c mice	Epithelial regeneration and collagen deposition	Enhanced wound healing and tissue regeneration	[27]
Diabetes	*Leonotis leonurus*	NLCs (nanostructured lipid carriers)	INS1 pancreatic β cells, Chang liver cells	Stimulated insulin release and glucose uptake	Improved glycemic control	[94]
Diabetes	*Ficus religiosa*	SLNs (solid lipid nanoparticles)	STZ and fructose-induced diabetic rats	Hypoglycemic and insulin-sensitizing effects	Reduced blood glucose, improved insulin sensitivity	[101]
Diabetes	*Plicosepalus acaciae* & *P. curviflorus*	SLNs	STZ + high-fat diet (HFD)-induced diabetic rats	Reduced blood glucose, insulin resistance and glycated hemoglobin; increased antioxidant enzymes	Improved glycemic control and reduced oxidative stress.	[7]
Diabetes	*Momordica charantia*	Nanoemulsion	Alloxan (ALX)-induced diabetic rats	Reduced blood sugar, enhanced antioxidant enzymes	Improved glycemic control, reduced oxidative stress	[97]
Diabetes	*Syzygium jambolanum*	PLGA nanoparticles	L6 rat skeletal muscle cells, arsenic-induced diabetic rats	Increased glucose uptake, GLUT4 expression; reduced ROS, NF-κB and iNOS	Improved glucose uptake, reduced inflammation	[107]

Diabetes	Stevioside	Pluronic-F-68 copolymer-based PLA nanoparticles	In vitro release study	Enhanced intestinal absorption, bioavailability, biocompatibility and controlled release	Enhanced bioavailability and controlled drug release	[141]
Diabetes	*Argyreia nervosa*	Silver nanoparticles (AgNPs)	In vitro (α-amylase and α-glucosidase inhibition)	Enhanced antidiabetic activity through enzyme inhibition and free radical scavenging	Improved glycemic control, Reduced carbohydrate metabolism, strong antibacterial activity	[109]
Diabetes	*Eysenhardtia polystachya*	Silver nanoparticles (AgNPs)	Zebrafish, INS1 pancreatic β cells	Improved insulin secretion, reduced hyperglycemia and dyslipidemia; antioxidant effects	Improved glycemic control, reduced dyslipidemia	[54]
Diabetes	Quercetin	Quercetin nanorods	Alloxan-induced diabetic rats	Decreased fasting blood glucose (FBG), glucose-6-phosphatase (G6Pase), fructose-1,6-bisphosphatase (FBPase), superoxide dismutase (SOD), catalase (CAT), glutathione (GSH), aspartate aminotransferase (AST), alkaline phosphatase (ALP) and alanine aminotransferase (ALT) levels	Improved blood glucose levels and reduced oxidative stress	[5]

(continued)

Table 3.2 (continued)

Disease condition	Phytonanomedicine	Type of nanoformulation	Model organism	Mechanism of action	Therapeutic benefits	References
Diabetic neuropathic pain	Emodin	PEGMA-DMAEMA-MAMMA nano-macromolecule	T2DM rats (HF-diet + low-dose STZ)	Reduced levels of TNF-α, decreased P2X3 receptor expression and lowered ERK1/2 phosphorylation in the dorsal root ganglion (DRG).	Reduced neuropathic pain and inflammation	[77]
Diabetic peripheral neuropathy	Curcumin	PEGMA-DMAEMA-MAO	Diabetic rats induced with STZ	Reduced levels of IL-1β and phosphorylated Akt (p-Akt), along with the presence of P2Y12 receptors on satellite glial cells (SGCs) in the dorsal root ganglion (DRG)	Reduced neuropathic pain and inflammation	[66]
Diabetic nephropathy	Quercetin	Quercetin (QUE/P) NP	Diabetic rats	Downregulation of ICAM-1 expression	Reduced nephropathy and inflammation	[130]
Diabetes	*Cinnamomum litseifolium*	Nanoemulsion	Inhibition of α-amylase and α-glucosidase	Inhibited carbohydrate-digesting enzymes	Potential hypoglycemic agent	[123]
Diabetes	*Costus speciosus*	PLGA nanoparticles	Diabetic rats induced with STZ	Improved insulin secretion, GLUT4 expression; reduced blood glucose and lipid profile	Improved glycemic control and reduced dyslipidemia	[6]

Diabetes mellitus	Resveratrol	Nanoliposomes	STZ-induced diabetic cells	Upregulated reactive oxygen species (ROS)-inactivating enzymes, including glutathione peroxidase (GSH-Px) and superoxide dismutase (SOD)	Reduced oxidative stress and improved cellular health	[136]
Diabetes	*Stevia rebaudiana*	TiO_2 nanomaterial, chitosan nanoparticles	ALX-induced diabetic rats	Reduced blood glucose, HbA1c and lipid profile; improved insulin secretion and antioxidant activity	Improved glycemic control, reduced dyslipidemia	[75]
Diabetes	*Gymnema sylvestre*	Gold nanoparticles (AuNPs)	ALX-induced diabetic rats	Reduced inflammatory mediators, blood glucose and HbA1c; improved lipid profile	Improved glycemic control, reduced inflammation, reduced dyslipidemia	[70]
Diabetes	*Marsilea quadrifolia*	Gold nanoparticles (AuNPs)	3 T3 L1 adipocytes	Enhanced glucose utilization and GLUT4 translocation	Improved glucose utilization	[35]
Diabetes	*Sambucus nigra*	Gold nanoparticles (AuNPs)	Diabetic rats induced with streptozotocin (STZ)	Reduced inflammation and oxidative stress	Reduced inflammation, reduced oxidative stress	[95]
Diabetes	*Vaccinium arctostaphylos*	Zinc oxide nanoparticles (ZnO NPs)	Diabetic rats induced with ALX	Lowered blood glucose levels and enhanced lipid profile.	Improved glycemic control, reduced dyslipidemia	[23]
Diabetes	*Musa paradisiaca*	Silver nanoparticles (AgNPs)	Diabetic rats induced with streptozotocin (STZ)	Increased insulin levels, reduced blood glucose and HbA1c; improved glucose uptake	Improved glycemic control	[15]

(continued)

Table 3.2 (continued)

Disease condition	Phytonanomedicine	Type of nanoformulation	Model organism	Mechanism of action	Therapeutic benefits	References
Diabetes	*Hibiscus subdariffa*	Zinc oxide nanoparticles (ZnO NPs)	Diabetic rats induced with streptozotocin (STZ)	Reduced blood glucose, proinflammatory cytokines; increased anti-inflammatory cytokines	Improved glycemic control, reduced inflammation	[21]
Diabetes	Quercetin	PLGA NPs	Diabetic rats induced with streptozotocin (STZ)	↑CAT and SOD levels, ↓Drug doses	Reduced oxidative stress and improved therapeutic efficacy	[34]
Diabetes	*Cassia fistula*	Gold nanoparticles (AuNPs)	Diabetic rats induced with streptozotocin (STZ)	Improved lipid profile, reduced blood glucose, HbA1c and renal markers	Improved glycemic control, reduced dyslipidemia, improved renal function	[36]
Diabetes	*Silybum marianum*	Zinc oxide nanoparticles (ZnO NPs)	ALX-induced diabetic rats	Reduced blood glucose, improved lipid profile and insulin levels; antibacterial activity	Improved glycemic control, reduced dyslipidemia, antibacterial activity	[18]
Diabetes	*Zingiber officinale*	Silver nanoparticles (AgNPs)	STZ-induced diabetic rats	Hypoglycemic effects; potential anti-inflammatory and antioxidant properties	Improved glycemic control, reduced inflammation, reduced oxidative stress	[55]
Diabetes	*Syzygium cumini*	Silver nanoparticles (AgNPs)	H9C2 cardiomyocytes (rat heart cells)	Reduced oxidative stress and lipid peroxidation	Improved cardioprotective effects	[19]

Diabetes	*Punica granatum*	Gold nanoparticles (AuNPs)	STZ-induced diabetic rats	Reduced glomerular sclerosis, renal fibrosis and inflammation; increased antioxidant activity via Nrf2	Improved renal function, reduced inflammation, reduced oxidative stress	[85]
Diabetes	*Morus alba*	Silver nanoparticles (AgNPs)	Aluminum-intoxicated diabetic rats	Alleviated retinal cell layer deterioration	Potential for treating diabetic retinopathy	[133]
Obesity	*Salacia chinensis*	Gold nanoparticles (SC-AuNPs)	Obese rats	Reduces body weight, resistin, leptin and inflammatory markers; increases HDL and adiponectin	Decreased body weight, improved lipid profile, reduced liver damage	[53]
Obesity	*Smilax glabra* Rhizome	Gold nanoparticles	Streptozotocin-induced rats	Improves insulin sensitivity, lipid profile and liver markers; restores liver and cardiac tissues	Reduced blood glucose, improved lipid profile, restored liver and cardiac tissues	[16]
Obesity	Resveratrol	Ligand-fabricated resveratrol-loaded nanoparticles (L-Rnano)	Obese C57BL/6 J mice	Promotes the transformation of adipose stromal cells into beige fat cells	Reduced fat mass, inflammation and improved glucose homeostasis	[139]
Obesity	Curcumin	Poly-vinyl alcohol gelatin nanofibers	Model rats	Transdermal delivery reduces subcutaneous adipose tissue volume	Reduced adipose tissue volume by 4–7%	[17]
Obesity	*Glycine max*	Nanophytosomes (thermogel method)	High-fed diet-induced mice	Enhances permeation and release of soy extract; reduces adipose cell size	Reduced adipose cell size, improved release pattern and slight reduction in triglycerides	[47]

(continued)

Table 3.2 (continued)

Disease condition	Phytonanomedicine	Type of nanoformulation	Model organism	Mechanism of action	Therapeutic benefits	References
Obesity	*Oleoresin capsicum*	Monolayer and bilayer nanoemulsion with alginate	Rats subjected to a high-fat diet (HFD) and 3 T3-L1 preadipocyte cell lines	Lowers lipid levels, triglycerides (TGs) and messenger RNA (mRNA) levels of peroxisome proliferator-activated receptor gamma (PPAR-γ) and fatty acid-binding protein	Improved lipid profile, reduced adipocyte differentiation	[76]
Obesity	*Citrus sinensis*	Nanovesicles	High-fat high-sugar diet (HFHSD) mice	Inhibits chylomicron synthesis, reduces TGs and plasma lipids	Reduced lipid levels, improved villi size	[24]
Dyslipidemia	*Ziziphus jujube* (Jujube)	Gold nanoparticles	Diabetic rats induced with STZs	Antioxidant, anti-inflammatory and hepatoprotective effects	Reduced triglycerides, cholesterol and liver oxidative markers; improved insulin levels	[65]
Dyslipidemia	Black Currant	Selenium nanoparticles	Galactose-treated rats	Hypolipidemic and antioxidant activities	Reduced lipid levels, improved antioxidant capacity	[10]
Dyslipidemia	*Nigella sativa*	Silver nanoparticles	Male adult rats	Reduces cholesterol absorption, LDL-c and oxidative stress; increases HDL-c	Improved lipid profile, reduced oxidative stress	[9]
Dyslipidemia	*Allium sativum*	Nanoemulsion	HFD-induced dyslipidemic rats	Reduced lipid profile, liver enzymes and hepatic steatosis	Reduced dyslipidemia, improved liver function	[103]

Dyslipidemia	*Hibiscus cannabinus*	Nanoemulsion	HCD-induced dyslipidemic rats	Reduced TC, LDL-C and hepatic fat accumulation; improved antioxidant activity and liver regeneration	Reduced dyslipidemia, reduced hepatic steatosis, reduced oxidative stress	[33]
High LDL cholesterol	Berberine	PLGA-PEG-PLGA block copolymers	HepG2 cells	Modulation of PCSK-9 mRNA	Reduced LDL cholesterol levels	[93]
Hypertension	Curcumin	Poly (lactic-co-glycolic acid) (PLGA) nanoparticles	High-fed diet (HFD) Wistar rats	Reduces oxidative stress, inflammation and fibrosis; improves vascular function	Normalized blood pressure, decreased liver fat, reduced heart inflammation and improved heart fibrosis	[41]
High glucose stress-induced apoptosis	Berberine	O-hexadecyl-dextran nanoparticles	Primary hepatocytes from Sprague-Dawley rats	Reduced reactive oxygen species (ROS), decreased caspase activation and diminished oxidative stress	Reduced hepatocyte apoptosis and oxidative stress	[67]
Hypertension	Resveratrol	Nanocapsule	HF-diet-induced diabetic mice	Regulation blood pressure	Improved blood pressure control	[111]
Hypertension	Curcumin	Nanoparticles	Male Sprague Dawley rats	Reduces oxidative stress and inflammation; improves vascular remodeling	Decreased thickness of the right ventricular wall, enhanced right ventricle weight-to-body ratio and lower oxidative stress	[105]

(continued)

Table 3.2 (continued)

Disease condition	Phytonanomedicine	Type of nanoformulation	Model organism	Mechanism of action	Therapeutic benefits	References
Hypertension	Copaiba oil	Nanocapsules	Wistar rats	Antioxidant and anti-inflammatory effects; enhances Nrf2 and superoxide dismutase activity	Reduced right ventricular hypertrophy, improved oxidative stress markers, no effect on pulmonary vascular resistance	[29]
Hypertension & Hypercholesterolemia	Curcumin	Nanoemulsion	In vitro study	Inhibition of HMGR along with ACE	Reduced hypertension and cholesterol levels	[102]

4.3 Dyslipidemia

Dyslipidemia is a condition in which there is an abnormal lipid level in the blood and is a significant threat to cardiovascular diseases. Statins which function as traditional therapeutic agents face limitations because they produce different responses among patients and result in multiple side effects [42]. Nanomedicine using plant compounds has proven successful in drug delivery since it optimizes bioactive compound utilization through nanotechnology-based formulations. Compounds derived from plants, such as polyphenols, flavonoids and phytosterols have lipid-lowering, anti-inflammatory and antioxidant effects. The placement of these compounds within nanoparticles enhances their stability and improves solubility and controlled release properties while providing better dyslipidemia management [117].

Curcumin can decrease LDL cholesterol levels, triglycerides and reduce inflammation. The therapeutic application of curcumin is restricted by its poor rate of absorption in the human body. The absorption of curcumin gets improved through liposomes or polymeric nanoparticle formulations while these nanoformulations specifically deliver the compound toward liver cells which carry out lipid metabolism. Scientific investigations prove that nanocurcumin enhances lipid profile benefits while minimizing oxidative stress in models with dyslipidemia [56]. The flavonoid compound quercetin which exists in onions, apples and berries demonstrates strong capacities to decrease blood fats and serve as an antioxidant. The substance lacks clinical value because it demonstrates limited solubility properties combined with fast elimination from the body. Nanostructured lipid carriers (NLCs) combined with silica nanoparticles protect quercetin by improving its bioavailability as they extend the natural circulation period of the compound. Studies on dyslipidemia animal models demonstrate that nanoparticles containing quercetin lower total cholesterol levels, LDL count and triglycerides but elevate HDL levels [129]. The treatment of kenaf seed oil nanoemulsion reduced cholesterol levels, improved liver oxidation status and bioavailability measurements in the rats fed a high-cholesterol. Various tests on rats showed that kenaf seed oil nanoemulsion treatment created three main effects including reduced cholesterol levels, decreased liver fat and enhanced liver cell growth [33]. The garlic oil nanoemulsion treatment for dyslipidemic Wistar showed lower amounts of cholesterol, lesser hepatic lipid deposits and enhanced functionalities in lipid metabolism [103]. The detailed list of plant-based nanoformulations for the management of dyslipidemia is shown in Table 3.2.

Nanotechnology when combined with plant-based compounds creates a new therapeutic methodology for dyslipidemia treatment. The strategy uses natural bioactive molecules in combination with nanotechnology to provide better treatment results with fewer adverse effects on lipid disorders. Additional clinical tests must validate these results before converting them into effective medical treatments [62].

4.4 Hypertension

Plant-based nanomedicine technology is a promising approach for hypertension as it is a sustainable, biocompatible and targeted treatment method. Hypertension is a serious cardiovascular disease that requires long-term treatment and the standard medications have effectiveness and side effects issues [126]. The antihypertensive antioxidant and anti-inflammatory properties of flavonoids, polyphenols and alkaloids from plants form the base of plant-based nanomedicine therapy. These bioactive molecules function with nanoparticles to improve their stability and bioavailability targeting potential [60].

Plant-based therapeutic delivery systems include nanocarriers made of exosomes, polymeric nanoparticles and liposomes which direct plant-based therapeutics to vascular endothelium, kidneys and other target sites to control blood pressure regulation. Studies revealed that nanoparticles containing curcumin reduce oxidative stress and improve endothelial function while resveratrol nanoparticles display anti-inflammatory and vasorelaxant properties. Plant-derived exosomes demonstrate the ability to control cellular signaling mechanisms that affect hypertension [68]. Different studies were carried out related to the plant-based nanoformulations for the treatment of hypertension-related issues. Studies show that the administration of 15 mg/kg of stevioside nanoemulsions to hypertensive rats results in the reduction of blood pressure and improved vascular system functioning [74]. The application of copaiba oil nanocapsules to pulmonary arterial hypertension in Wistar rats showed strong antioxidant and anti-inflammatory properties by boosting the activity of Nrf2 and superoxide dismutase [29] PLGA nanoparticles encapsulating curcumin minimize the oxidative stress, inflammation and fibrosis in Wistar rats fed with a high-fat diet. It regulates blood pressure, reduces hepatic fat accumulation, ameliorates ventricular inflammation and fibrosis [41] The nanoencapsulated *Copaifera* spp. oil reduces oxidative stress and right ventricular hypertrophy by enhancing antioxidant action by Nrf2 in pulmonary arterial hypertension in rats monocrotaline exposure [29].

Plant-based nanomedicine reduces toxicity along with minimizing environmental impact since it substitutes synthetic drugs. This field requires additional effort to resolve scalability problems as well as obtain regulatory approval and demonstrate the long-term safety of this approach. The combination of natural compounds with nanotechnology has made plant-based nanomedicine an innovative and comprehensive approach to hypertension treatment [126]. The detailed list of plant-based nanoformulations for the management of hypertension is shown in Table 3.2.

5 Synergistic Effects of Nanoformulations

The bioavailability of phytochemicals increases through nanoformulations while the formulations direct delivery to particular tissues which minimizes effects that do not target the specific site [31, 32]. For example, nanoemulsions along with liposomes increase hydrophobic compounds such as curcumin and resveratrol by enhancing their stability and improving their solubility [138]. The therapeutic outcomes of berberine and quercetin receive enhancement from sustained release through solid lipid nanoparticles (SLNs) and nanostructured lipid carriers (NLCs) [89]. The medication transport system involving polymeric nanoparticles of chitosan and PLGA functions both as a delivery vehicle and a targeted release controller for affected tissues. Controlled release along with targeted tissue delivery during inflammatory conditions can be achieved through polymeric nanoparticles such as chitosan and PLGA [114].

The therapeutic power of curcumin, berberine, resveratrol and quercetin as plant-derived nanomedical agents' focuses on treating various metabolic syndrome mechanisms [52].

The effectiveness of metabolic syndrome management with nanoformulated drugs becomes more robust because they boost solubility, bioavailability and targeting properties for managing inflammatory, oxidative stress-related conditions insulin resistance and lipid dysregulation treatment [127]. This novel therapeutic approach provides strong potential for metabolic disorder treatment by providing safer and more effective results than established standard treatments.

6 Conclusions and Future Directions

Phytochemical-based nanomedicine shows promising potential for treating metabolic syndrome (MetS). Traditional treatment methods are inadequate due to their weak therapeutic effects, harmful side effects and reduced patient compliance. The application of phytochemicals for therapeutics is also limited due to their low bioavailability and poor stability properties. Nanoformulations from plant extract increase their bioavailability, solubility, stability targeted delivery and biocompatibility with minimal toxic effects. Plant-based nanomedicine helps in the treatment of metabolic syndrome by different pathways. The implementation of plant-based nanomedicine faces several obstacles, like cost-efficiency, scalability, long-term safety, regulatory demands and lack of standardized delivery methods.

Phytochemical-based nanomedicines for MetS therapy need more advancement to achieve their full therapeutic benefits. Future research should be focused on optimizing nanocarrier design systems which enhance the bioavailability and stability while achieving targeted delivery of phytochemicals. The process of developing new nanoformulation design can be accelerated by using computational modeling and simulation techniques. Diverse clinical trials with excellent design must be implemented to establish the security and effectiveness of phytochemical nanomedicine treatments among broad MetS patient groups. The testing procedure requires strict methods together with standardized results assessment with appropriate comparison groups. The analysis of distinct subgroups should determine which patient segments exhibit positive responses and negative responses through stratification using MetS severity levels together with age and gender characteristics. The genomics, transcriptomics and proteomics technologies enable scientists to gain important knowledge about these mechanisms. Scientists must perform additional studies to comprehend how phytochemical-based nanomedicines enrich therapeutic outcomes. Researchers need to examine phytochemical molecular target interactions and signaling pathway responses as well as cellular process effects related to MetS.

References

1. Abu-Taweel, G. M., Attia, M. F., Hussein, J., Mekawi, E. M., Galal, H. M., Ahmed, E. I., & El-Naggar, M. E. (2020). Curcumin nanoparticles have potential antioxidant effect and restore tetrahydrobiopterin levels in experimental diabetes. *Biomedicine & Pharmacotherapy, 131*, 110688. https://doi.org/10.1016/j.biopha.2020.110688
2. Ahangarpour, A., Oroojan, A. A., Khorsandi, L., Kouchak, M., & Badavi, M. (2018). Solid lipid nanoparticles of myricitrin have antioxidant and antidiabetic effects on streptozotocin-nicotinamide-induced diabetic model and myotube cell of male mouse. *Oxidative Medicine and Cellular Longevity, 2018*(1), 7496936. https://doi.org/10.1155/2018/7496936
3. Ahmad, N., Ahmad, R., Al-Qudaihi, A., Alaseel, S. E., Fita, I. Z., Khalid, M. S., & Pottoo, F. H. (2019). Preparation of a novel curcumin nanoemulsion by ultrasonication and its comparative effects in wound healing and the treatment of inflammation. *RSC Advances, 9*(35), 20192–20206. https://doi.org/10.1039/C9RA03102B
4. Ak, A. G., Turan, I., Ozacmak, H. S., & Karatas, A. (2023). Chitosan nanoparticles as promising tool for berberine delivery: formulation, characterization and in vivo evaluation. *Journal of Drug Delivery Science and Technology, 80*, 104203.
5. Alam, M. M., Abdullah, K. M., Singh, B. R., Naqvi, A. H., & Naseem, I. (2016). Ameliorative effect of quercetin nanorods on diabetic mice: mechanistic and therapeutic strategies. *RSC Advances, 6*(60), 55092–55103. https://doi.org/10.1039/C6RA04821H
6. Alamoudi, E. F., Khalil, W. K., Ghaly, I. S., Hassan, N. H., & Ahmed, E. S. (2014). Nanoparticles from of Costus speciosus extract improves the antidiabetic and antilipidemic effects against STZ-induced diabetes mellitus in Albino rats. *International Journal of Pharmaceutical Sciences Review and Research, 29*(1), 279–288.
7. Aldawsari, H. M., Hanafy, A., Labib, G. S., & Badr, J. M. (2014). Antihyperglycemic activities of extracts of the mistletoes Plicosepalus acaciae and P. curviflorus in comparison to their solid lipid nanoparticle suspension formulations. *Zeitschrift für Naturforschung C, 69*(9–10), 391–398. https://doi.org/10.5560/znc.2014-0047
8. Alhazzani, K., Alrewily, S. Q., Alanzi, A. R., Aljerian, K., Raish, M., Hawwal, M. F., & Alanazi, A. Z. (2024). Therapeutic effects of liposomal resveratrol in the mitigation of diabetic nephropathy via modulating inflammatory response, oxidative stress, and apoptosis. *Applied Biochemistry and Biotechnology*, 1–20. https://doi.org/10.1007/s12010-024-05092-1
9. Ali, Z. S., & Khudair, K. K. (2019). Synthesis, characterization of silver nanoparticles using Nigella sativa seeds and study their effects on the serum lipid profile and DNA damage on the rats' blood treated with hydrogen peroxide. *The Iraqi Journal of Veterinary Medicine, 43*(2), 23–37. https://doi.org/10.30539/iraqijvm.v43i2.526
10. Al-Kurdy, M. J. J. (2020). The effect of black currant selenium nanoparticles on dyslipidemia and oxidant-antioxidant status in D-galactose treated rats. *Kufa Journal for Veterinary Medical Sciences, 11*(1), 23–38. https://doi.org/10.36326/kjvs/2020/v11i13300
11. Alsabeelah, N., & Kumar, V. (2023). Formulation and optimization of quercetin nanoemulsion for enhancing its dissolution rate, bioavailability and cardioprotective activity. *Journal of Cluster Science, 34*(4), 1893–1906. https://doi.org/10.1007/s10876-022-02351-1
12. Amjadi, S., Abbasi, M. M., Shokouhi, B., Ghorbani, M., & Hamishehkar, H. (2019). Enhancement of therapeutic efficacy of betanin for diabetes treatment by liposomal nanocarriers. *Journal of Functional Foods, 59*, 119–128. https://doi.org/10.1016/j.jff.2019.05.015

13. An, J. Y., Yang, H. S., Park, N. R., Koo, T. S., Shin, B., Lee, E. H., & Cho, S. H. (2020). Development of polymeric micelles of oleanolic acid and evaluation of their clinical efficacy. *Nanoscale Research Letters, 15*, 1–14. https://doi.org/10.1007/s10876-016-1047-2
14. An, H., Deng, X., Wang, F., Xu, P., & Wang, N. (2023). Dendrimers as nanocarriers for the delivery of drugs obtained from natural products. *Polymers, 15*(10), 2292. https://doi.org/10.3390/polym15102292
15. Anbazhagan, P., Murugan, K., Jaganathan, A., Sujitha, V., Samidoss, C. M., Jayashanthani, S., & Benelli, G. (2017). Mosquitocidal, antimalarial and antidiabetic potential of Musa paradisiaca-synthesized silver nanoparticles: in vivo and in vitro approaches. *Journal of Cluster Science, 28*, 91–107. https://doi.org/10.1007/s10876-016-1047-2
16. Ansari, S., Bari, A., Ullah, R., Mathanmohun, M., Veeraraghavan, V. P., & Sun, Z. (2019). Gold nanoparticles synthesized with Smilax glabra rhizome modulates the anti-obesity parameters in high-fat diet and streptozotocin induced obese diabetes rat model. *Journal of Photochemistry and Photobiology B: Biology, 201*, 111643. https://doi.org/10.1016/j.jphotobiol.2019.111643
17. Ariamoghaddam, A. R., Ebrahimi-Hosseinzadeh, B., Hatamian-Zarmi, A., & Sahraeian, R. (2018). In vivo anti-obesity efficacy of curcumin loaded nanofibers transdermal patches in high-fat diet induced obese rats. *Materials Science and Engineering: C, 92*, 161–171. https://doi.org/10.1016/j.msec.2018.06.030
18. Arvanag, F. M., Bayrami, A., Habibi-Yangjeh, A., & Pouran, S. R. (2019). A comprehensive study on antidiabetic and antibacterial activities of ZnO nanoparticles biosynthesized using *Silybum marianum* L seed extract. *Materials Science and Engineering: C, 97*, 397–405. https://doi.org/10.1016/j.msec.2018.12.058
19. Atale, N., Saxena, S., Nirmala, J. G., Narendhirakannan, R. T., Mohanty, S., & Rani, V. (2017). Synthesis and characterization of Sygyzium cumini nanoparticles for its protective potential in high glucose-induced cardiac stress: a green approach. *Applied Biochemistry and Biotechnology, 181*, 1140–1154. https://doi.org/10.1007/s12010-016-2274-6
20. Badhwar, R., Mangla, B., Neupane, Y. R., Khanna, K., & Popli, H. (2021). Quercetin loaded silver nanoparticles in hydrogel matrices for diabetic wound healing. *Nanotechnology, 32*(50), 505102. https://doi.org/10.1088/1361-6528/ac2536
21. Bala, N., Saha, S., Chakraborty, M., Maiti, M., Das, S., Basu, R., & Nandy, P. (2015). Green synthesis of zinc oxide nanoparticles using Hibiscus subdariffa leaf extract: effect of temperature on synthesis, anti-bacterial activity and anti-diabetic activity. *RSC Advances, 5*(7), 4993–5003. https://doi.org/10.1039/C4RA12784F
22. Barabadi, H., Hosseini, O., Jounaki, K., Sadeghian-Abadi, S., Ashouri, F., Alrikabi, A. M. A., & Mostafavi, E. (2023). Bioinspired green-synthesized silver nanoparticles: in vitro physicochemical, antibacterial, biofilm inhibitory, genotoxicity, antidiabetic, antioxidant, and anticoagulant performance. *Materials Advances, 4*(14), 3037–3054. https://doi.org/10.1039/D3MA00089C
23. Bayrami, A., Parvinroo, S., Habibi-Yangjeh, A., & Rahim Pouran, S. (2018). Bio-extract-mediated ZnO nanoparticles: microwave-assisted synthesis, characterization and antidiabetic activity evaluation. *Artificial Cells, Nanomedicine, and Biotechnology, 46*(4), 730–739. https://doi.org/10.1080/21691401.2017.1337025
24. Berger, E., Colosetti, P., Jalabert, A., Meugnier, E., Wiklander, O. P., Jouhet, J., & Rome, S. (2020). Use of nanovesicles from orange juice to reverse diet-induced gut modifications in diet-induced obese mice. *Molecular Therapy Methods & Clinical Development, 18*, 880–892. https://doi.org/10.1016/j.omtm.2020.08.009
25. Bindu, R. H., Lakshmi, S. M., Himaja, N., Nirosha, K., & Pooja, M. (2014). Formulation characterization and antidiabetic evaluation of *Talinum portulacifolium* (Forssk.) loaded solid lipid nanoparticles in Streptozotocin and high fat diet induced diabetic rats. *Journal of Global Trends in Pharmaceutical Sciences, 5*(4), 2108–2114.

26. Boden, G. (2011). Obesity, insulin resistance and free fatty acids. *Current Opinion in Endocrinology, Diabetes and Obesity, 18*(2), 139–143. https://doi.org/10.1097/MED.0b013e3283444b09
27. Boomi, P., Ganesan, R., Prabu Poorani, G., Jegatheeswaran, S., Balakumar, C., Gurumallesh Prabu, H., & Saravanan, M. (2020). Phyto-engineered gold nanoparticles (AuNPs) with potential antibacterial, antioxidant, and wound healing activities under in vitro and in vivo conditions. *International Journal of Nanomedicine*, 7553–7568. https://doi.org/10.2147/IJN.S257499
28. Bulboacă, A. E., Porfire, A. S., Tefas, L. R., Boarescu, P. M., Bolboacă, S. D., Stănescu, I. C., & Dogaru, G. (2019). Liposomal curcumin is better than curcumin to alleviate complications in experimental diabetic mellitus. *Molecules, 24*(5), 846. https://doi.org/10.3390/molecules24050846
29. Campos, C., De Castro, A. L., Tavares, A. M. V., Fernandes, R. O., Ortiz, V. D., Barboza, T. E., & Belló-Klein, A. (2017). Effect of free and nanoencapsulated copaiba oil on monocrotaline-induced pulmonary arterial hypertension. *Journal of Cardiovascular Pharmacology, 69*(2), 79–85. https://doi.org/10.1097/FJC.0000000000000442
30. Cerri, G. C., Lima, L. C. F., Lelis, D. d. F., Barcelos, L. d. S., Feltenberger, J. D., & Mussi, S. V. (2019). Sclareol-loaded lipid nanoparticles improved metabolic profile in obese mice. *Life Sciences, 218*, 292–299. https://doi.org/10.1016/j.lfs.2018.12.063
31. Chauhan, A., Gaur, R., & Kanta, C. (2024a). Nano formulation of Berberine (Brb): Transforming a natural alkaloid into advanced therapeutics. *Journal of Applied Pharmaceutical Science, 14*(12), 035–041. https://doi.org/10.7324/JAPS.2024.657635
32. Chauhan, D., Yadav, P. K., Sultana, N., Agarwal, A., Verma, S., Chourasia, M. K., & Gayen, J. R. (2024b). Advancements in nanotechnology for the delivery of phytochemicals. *Journal of Integrative Medicine, 22*(4), 385–398. https://doi.org/10.1016/j.joim.2024.04.005
33. Cheong, A. M., Jessica Koh, J. X., Patrick, N. O., Tan, C. P., & Nyam, K. L. (2018). Hypocholesterolemic effects of kenaf seed oil, macroemulsion, and nanoemulsion in high-cholesterol diet induced rats. *Journal of Food Science, 83*(3), 854–863. https://doi.org/10.1111/1750-3841.14038
34. Chitkara, D., Nikalaje, S. K., Mittal, A., Chand, M., & Kumar, N. (2012). Development of quercetin nanoformulation and in vivo evaluation using streptozotocin induced diabetic rat model. *Drug Delivery and Translational Research, 2*, 112–123. https://doi.org/10.1007/s13346-012-0063-5
35. Chowdhury, A., Kunjiappan, S., Bhattacharjee, C., Somasundaram, B., & Panneerselvam, T. (2017). Biogenic synthesis of *Marsilea quadrifolia* gold nanoparticles: a study of improved glucose utilization efficiency on 3T3-L1 adipocytes. *In Vitro Cellular & Developmental Biology-Animal, 53*, 483–493. https://doi.org/10.1007/s11626-017-0136-3
36. Daisy, P., & Saipriya, K. (2012). Biochemical analysis of *Cassia fistula* aqueous extract and phytochemically synthesized gold nanoparticles as hypoglycemic treatment for diabetes mellitus. *International Journal of Nanomedicine*, 1189–1202. https://doi.org/10.2147/IJN.S26650
37. Deng, W., Wang, H., Wu, B., & Zhang, X. (2019). Selenium-layered nanoparticles serving for oral delivery of phytomedicines with hypoglycemic activity to synergistically potentiate the antidiabetic effect. *Acta Pharmaceutica Sinica B, 9*(1), 74–86. https://doi.org/10.1016/j.apsb.2018.09.009
38. Devadasu, V. R., Wadsworth, R. M., & Kumar, M. R. (2011). Protective effects of nanoparticulate coenzyme Q 10 and curcumin on inflammatory markers and lipid metabolism in streptozotocin-induced diabetic rats: a possible remedy to diabetic complications. *Drug Delivery and Translational Research, 1*, 448–455. https://doi.org/10.1007/s13346-011-0041-3
39. Dewanjee, S., Chakraborty, P., Mukherjee, B., & De Feo, V. (2020). Plant-based antidiabetic nanoformulations: the emerging paradigm for effective therapy. *International Journal of Molecular Sciences, 21*(6), 2217. https://doi.org/10.3390/ijms21062217

40. Dharman, S., Maragathavalli, G., Shanmugam, R., & Shanmugasundaram, K. (2023). Curcumin mediated gold nanoparticles and analysis of its antioxidant, anti-inflammatory, antimicrobial activity against oral pathogens. *Pesquisa Brasileira Em Odontopediatria e Clínica Integrada, 23*, e220068. https://doi.org/10.1590/pboci.2023.073
41. Du Preez, R., Pahl, J., Arora, M., Ravi Kumar, M. N. V., Brown, L., & Panchal, S. K. (2019). Low-dose curcumin nanoparticles normalise blood pressure in male wistar rats with diet-induced metabolic syndrome. *Nutrients, 11*(7), 1542. https://doi.org/10.3390/nu11071542
42. Du, Z., & Qin, Y. (2023). Dyslipidemia and cardiovascular disease: current knowledge, existing challenges, and new opportunities for management strategies. *Journal of Clinical Medicine, 12*(1), 363. https://doi.org/10.3390/jcm12010363
43. El-Far, Y. M., Zakaria, M. M., Gabr, M. M., El Gayar, A. M., El-Sherbiny, I. M., & Eissa, L. A. (2016). A newly developed silymarin nanoformulation as a potential antidiabetic agent in experimental diabetes. *Nanomedicine, 11*(19), 2581–2602. https://doi.org/10.2217/nnm-2016-0204
44. El-Far, Y. M., Zakaria, M. M., Gabr, M. M., El Gayar, A. M., Eissa, L. A., & El-Sherbiny, I. M. (2017). Nanoformulated natural therapeutics for management of streptozotocin-induced diabetes: potential use of curcumin nanoformulation. *Nanomedicine, 12*(14), 1689–1711. https://doi.org/10.2217/nnm-2017-0106
45. Elhassan, G. O., & Mohamed, J. M. M. (2024). Quercetin-loaded solid lipid nanoparticles: Preparation and characterization for biomedical application. *Frontiers in Health Informatics, 15*, 234–245. https://doi.org/10.3389/fhealth.2024.00123
46. Elkhateeb, O., Badawy, M. E., Tohamy, H. G., Abou-Ahmed, H., El-Kammar, M., & Elkhenany, H. (2023). Curcumin-infused nanostructured lipid carriers: a promising strategy for enhancing skin regeneration and combating microbial infection. *Bmc Veterinary Research, 19*(1), 206. https://doi.org/10.1186/s12917-023-03774-2
47. El-Menshawe, S. F., Ali, A. A., Rabeh, M. A., & Khalil, N. M. (2018). Nanosized soy phytosome-based thermogel as topical anti-obesity formulation: an approach for acceptable level of evidence of an effective novel herbal weight loss product. *International Journal of Nanomedicine, 13*, 307–318.
48. Fahed, G., Aoun, L., Bou Zerdan, M., Allam, S., Bou Zerdan, M., Bouferraa, Y., & Assi, H. I. (2022). Metabolic syndrome: updates on pathophysiology and management in 2021. *International Journal of Molecular Sciences, 23*(2), 786. https://doi.org/10.3390/ijms23020786
49. Fappa, E., Yannakoulia, M., Pitsavos, C., Skoumas, I., Valourdou, S., & Stefanadis, C. (2008) Lifestyle intervention in the management of metabolic syndrome: could we improve adherence issues? *Nutrition, 24*(3), 286–291. https://doi.org/10.1016/j.nut.2007.11.008
50. Francini-Pesenti, F., Spinella, P., & Calò, L. A. (2019). Potential role of phytochemicals in metabolic syndrome prevention and therapy. *Diabetes, Metabolic Syndrome and Obesity: Targets and Therapy, 12*, 1987–2002.
51. Gallien, J., Srinageshwar, B., Gallo, K., Holtgrefe, G., Koneru, S., Otero, P. S., & Rossignol, J. (2021). Curcumin loaded dendrimers specifically reduce viability of glioblastoma cell lines. *Molecules, 26*(19), 6050. https://doi.org/10.3390/molecules26196050
52. Gangwar, T., Poonia, N., Subudhi, R. N., & Arora, V. (2025). Therapeutic potential and underlying mechanisms of phytoconstituents: emphasizing on resveratol, curcumin, quercetin, berberine, and hesperidin in ulcerative colitis. *Naunyn-Schmiedeberg's Archives of Pharmacology*, 1–18. https://doi.org/10.1007/s00210-025-03811-x
53. Gao, L., Hu, Y., Hu, D., Li, Y., Yang, S., & Dong, X. (2020). Anti-obesity activity of gold nanoparticles synthesized from Salacia chinensis modulates the biochemical alterations in high-fat diet-induced obese rat model via AMPK signaling pathway. *Arabian Journal of Chemistry, 13*(8), 6589–6597. https://doi.org/10.1016/j.arabjc.2020.06.015
54. Garcia Campoy, A. H., Perez Gutierrez, R. M., Manriquez-Alvirde, G., & Muñiz Ramirez, A. (2018). Protection of silver nanoparticles using Eysenhardtia polystachya in peroxide-induced pancreatic β-cell damage and their antidiabetic properties in zebrafish. *International Journal of Nanomedicine, 13*, 2601–2612.

55. Garg, A., Pandey, P., Sharma, P., & Shukla, A. (2016). Synthesis and characterization of silver nanoparticle of ginger rhizome (*Zingiber officinale*) extract: synthesis, characterization and anti diabetic activity in streptozotocin induced diabetic rats. *European Journal of Biomedical and Pharmaceutical Sciences, 3*, 605–611.
56. Gera, M., Sharma, N., Ghosh, M., Lee, S. J., Min, T., Kwon, T., & Jeong, D. K. (2017). Nanoformulations of curcumin: an emerging paradigm for improved remedial application. *Oncotarget, 8*(39), 66680. https://doi.org/10.18632/oncotarget.19164
57. Grabarczyk, M., Justyńska, W., Czpakowska, J., Smolińska, E., Bielenin, A., Glabinski, A., & Szpakowski, P. (2024). Role of plant phytochemicals: resveratrol, curcumin, luteolin and quercetin in demyelination, neurodegeneration, and epilepsy. *Antioxidants, 13*(11), 1364. https://doi.org/10.3390/antiox13111364
58. Grundy, S. M. (2016). Metabolic syndrome update. *Trends in Cardiovascular Medicine, 26*(4), 364–373. https://doi.org/10.1016/j.tcm.2015.10.004
59. Guan, F., Wang, Q., Bao, Y., & Chao, Y. (2021). Anti-rheumatic effect of quercetin and recent developments in nano formulation. *RSC Advances, 11*(13), 7280–7293. https://doi.org/10.1039/D0RA08817J
60. Han, H. S., Koo, S. Y., & Choi, K. Y. (2022). Emerging nanoformulation strategies for phytocompounds and applications from drug delivery to phototherapy to imaging. *Bioactive Materials, 14*, 182–205. https://doi.org/10.1016/j.bioactmat.2021.11.027
61. Hassani, F. V., Shirani, K., & Hosseinzadeh, H. (2016). Rosemary (*Rosmarinus officinalis*) as a potential therapeutic plant in metabolic syndrome: a review. *Naunyn-Schmiedeberg's Archives of Pharmacology, 389*, 931–949. https://doi.org/10.1007/s00210-016-1256-0
62. Hu, F., Sun, D. S., Wang, K. L., & Shang, D. Y. (2022). Nanomedicine of plant origin for the treatment of metabolic disorders. *Frontiers in Bioengineering and Biotechnology, 9*, 811917. https://doi.org/10.3389/fbioe.2021.811917
63. Hussain, Z., Pandey, M., Choudhury, H., Ying, P. C., Xian, T. M., Kaur, T., et al. (2020). Hyaluronic acid functionalized nanoparticles for simultaneous delivery of curcumin and resveratrol for management of chronic diabetic wounds: Fabrication, characterization, stability and in vitro release kinetics. *Journal of Drug Delivery Science and Technology, 57*, 101747. https://doi.org/10.1016/j.jddst.2020.101747
64. Islam, M. S., Wei, P., Suzauddula, M., Nime, I., Feroz, F., Acharjee, M., & Pan, F. (2024). The interplay of factors in metabolic syndrome: understanding its roots and complexity. *Molecular Medicine, 30*(1), 279. https://doi.org/10.1186/s10020-024-01019-y
65. Javanshir, R., Honarmand, M., Hosseini, M., & Hemmati, M. (2020). Anti-dyslipidemic properties of green gold nanoparticle: Improvement in oxidative antioxidative balance and associated atherogenicity and insulin resistance. *Clinical Phytoscience, 6*(1), 1–10. https://doi.org/10.1186/s40816-020-00224-6
66. Jia, T., Rao, J., Zou, L., Zhao, S., Yi, Z., Wu, B., & Li, G. (2018). Nanoparticle-encapsulated curcumin inhibits diabetic neuropathic pain involving the P2Y12 receptor in the dorsal root ganglia. *Frontiers in Neuroscience, 11*, 755. https://doi.org/10.3389/fnins.2017.00755
67. Kapoor, R., Singh, S., Tripathi, M., Bhatnagar, P., Kakkar, P., & Gupta, K. C. (2014). O-hexadecyl-dextran entrapped berberine nanoparticles abrogate high glucose stress induced apoptosis in primary rat hepatocytes. *PLoS One, 9*(2), e89124. https://doi.org/10.1371/journal.pone.0089124
68. Karimi, A., Jazani, A. M., Darzi, M., Azgomi, R. N. D., & Vajdi, M. (2023). Effects of curcumin on blood pressure: A systematic review and dose-response meta-analysis. *Nutrition, Metabolism and Cardiovascular Diseases, 33*(11), 2089–2101. https://doi.org/10.1016/j.numecd.2023.07.003
69. Karri, V. V. S. R., Kuppusamy, G., Talluri, S. V., Mannemala, S. S., Kollipara, R., & Wadhwani, A. D. (2016). Curcumin loaded chitosan nanoparticles impregnated into collagen-alginate scaffolds for diabetic wound healing. *International Journal of Biological Macromolecules, 93*, 1519–1529. https://doi.org/10.1016/j.ijbiomac.2016.05.038

70. Karthick, V., Kumar, V. G., Dhas, T. S., Singaravelu, G., Sadiq, A. M., & Govindaraju, K. (2014). Effect of biologically synthesized gold nanoparticles on alloxan-induced diabetic rats-an in vivo approach. *Colloids and Surfaces B: Biointerfaces, 122*, 505–511. https://doi.org/10.1016/j.colsurfb.2014.07.022
71. Karthikeyan, A., Senthil, N., & Min, T. (2020). Nanocurcumin: A promising candidate for therapeutic applications. *Frontiers in Pharmacology, 11*, 487. https://doi.org/10.3389/fphar.2020.00487
72. Kazmi, I., Imam, S. S., Al-Abbasi, F. A., Afzal, M., Nadeem, M. S., & Alshehri, S. (2022). Environment friendly green synthesis method based natural bioactive functional "catechin and gingerol" loaded nanomedicine for the management of obesity. *International Journal of Pharmaceutics, 628*, 122340. https://doi.org/10.1016/j.ijpharm.2022.122340
73. Keane, K. N., Cruzat, V. F., Carlessi, R., de Bittencourt, P. I. H., Jr., & Newsholme, P. (2015). Molecular events linking oxidative stress and inflammation to insulin resistance and β-cell dysfunction. *Oxidative Medicine and Cellular Longevity, 2015*(1), 181643. https://doi.org/10.1155/2015/181643
74. Kim, H. L., Chung, J., Kim, K. J., Kim, H. J., Seo, W. W., Jeon, K. H., & Kim, S. H. (2022). Lifestyle modification in the management of metabolic syndrome: statement from Korean Society of CardioMetabolic Syndrome (KSCMS). *Korean Circulation Journal, 52*(2), 93–109. https://doi.org/10.4070/kcj.2021.0328
75. Langle, A., González-Coronel, M. A., Carmona-Gutiérrez, G., Moreno-Rodríguez, J. A., Venegas, B., Muñoz, G., & Díaz, A. (2015). Stevia rebaudiana loaded titanium oxide nanomaterials as an antidiabetic agent in rats. *Revista Brasileira de Farmacognosia, 25*, 145–151. https://doi.org/10.1016/j.bjp.2015.03.004
76. Lee, M.-S., Jung, S., Shin, Y., Lee, S., Kim, C.-T., & Kim, I.-H. (2017). Lipolytic efficacy of alginate double-layer nanoemulsion containing Oleoresin capsicum in differentiated 3T3-L1 adipocytes. *Food & Nutrition Research, 61*, 1339553. https://doi.org/10.1080/16546628.2017.1339553
77. Li, L., Sheng, X., Zhao, S., Zou, L., Han, X., Gong, Y., & Liang, S. (2017). Nanoparticle-encapsulated emodin decreases diabetic neuropathic pain probably via a mechanism involving P2X3 receptor in the dorsal root ganglia. *Purinergic Signalling, 13*, 559–568. https://doi.org/10.1007/s11302-017-9583-2
78. Li, F., Shi, Y., Liang, J., & Zhao, L. (2019). Curcumin-loaded chitosan nanoparticles promote diabetic wound healing via attenuating inflammation in a diabetic rat model. *Journal of Biomaterials Applications, 34*(4), 476–486. https://doi.org/10.1177/0885328219860929
79. Li, K. X., Yuan, H., Zhang, J., Peng, X. B., Zhuang, W. F., Huang, W. T., & Qin, S. L. (2024). Curcumin-loaded long-circulation liposomes ameliorate insulin resistance in type 2 diabetic mice. *International Journal of Nanomedicine*, 12099–12110. https://doi.org/10.2147/IJN.S487519
80. Liu, J., Chen, Z., Wang, J., Li, R., Li, T., Chang, M., & Wang, Y. (2018a). Encapsulation of curcumin nanoparticles with MMP9-responsive and thermos-sensitive hydrogel improves diabetic wound healing. *ACS Applied Materials & Interfaces, 10*(19), 16315–16326. https://doi.org/10.1021/acsami.8b03868
81. Liu, Y., Zeng, S., Liu, Y., Wu, W., Shen, Y., & Zhang, L. (2018b). Synthesis and antidiabetic activity of selenium nanoparticles in the presence of polysaccharides from Catathelasma ventricosum. *International Journal of Biological Macromolecules, 114*, 632–639. https://doi.org/10.1016/j.ijbiomac.2018.03.161
82. Liu, Y., Liang, Y., Yuhong, J., Xin, P., Han, J. L., Du, Y., & Ma, Y. (2024). Advances in nanotechnology for enhancing the solubility and bioavailability of poorly soluble drugs. *Drug Design, Development and Therapy*, 1469–1495. https://doi.org/10.2147/DDDT.S447496
83. Lu, H., Zhang, S., Wang, J., & Chen, Q. (2021). A review on polymer and lipid-based nanocarriers and its application to nano-pharmaceutical and food-based systems. *Frontiers in Nutrition, 8*, 783831. https://doi.org/10.3389/fnut.2021.783831

84. Maity, S., Mukhopadhyay, P., Kundu, P. P., & Chakraborti, A. S. (2017). Alginate coated chitosan core-shell nanoparticles for efficient oral delivery of naringenin in diabetic animals – An in vitro and in vivo approach. *Carbohydrate Polymers, 170*, 124–132. https://doi.org/10.1016/j.carbpol.2017.04.066
85. Manna, K., Mishra, S., Saha, M., Mahapatra, S., Saha, C., Yenge, G., & Das Saha, K. (2019). Amelioration of diabetic nephropathy using pomegranate peel extract-stabilized gold nanoparticles: assessment of NF-κB and Nrf2 signaling system. *International Journal of Nanomedicine*, 1753–1777. https://doi.org/10.2147/IJN.S176013
86. Mascarenhas-Melo, F., Araújo, A. R., Rodrigues, M., Mathur, A., Gonçalves, M. B. S., Tanwar, K., & Paiva-Santos, A. C. (2023). Dermatological bioactivities of resveratrol and nanotechnology strategies to boost its efficacy – An updated review. *Cosmetics, 10*(3), 68. https://doi.org/10.3390/cosmetics10030068
87. Matsuda, M., & Shimomura, I. (2014). Roles of adiponectin and oxidative stress in obesity-associated metabolic and cardiovascular diseases. *Reviews in Endocrine and Metabolic Disorders, 15*, 1–10. https://doi.org/10.1007/s11154-013-9271-7
88. Mishra, R., & Aher, A. (2024). Lipid-based berberine loaded lyophilized nanomicelles with enhanced antioxidant effect: Design and characterization. *Journal of Applied Pharmaceutical Science, 14*(6), 126–134. https://doi.org/10.7324/JAPS.2024.154408
89. Mishra, R. V., & Dhole, S. N. (2021). Enhanced pharmacological efficacy of berberine hydrochloride loaded lipid based pellets for the treatment of metabolic diseases. *Biomedical and Pharmacology Journal, 14*(2), 993–1005. https://doi.org/10.13005/bpj/2201
90. Mukhopadhyay, P., Maity, S., Mandal, S., Chakraborti, A. S., Prajapati, A. K., & Kundu, P. P. (2018). Preparation, characterization and in vivo evaluation of pH sensitive, safe quercetin-succinylated chitosan-alginate core-shell-corona nanoparticle for diabetes treatment. *Carbohydrate Polymers, 182*, 42–51. https://doi.org/10.1016/j.carbpol.2017.10.098
91. Noubiap, J. J., Nansseu, J. R., Lontchi-Yimagou, E., Nkeck, J. R., Nyaga, U. F., Ngouo, A. T., & Bigna, J. J. (2022). Geographic distribution of metabolic syndrome and its components in the general adult population: A meta-analysis of global data from 28 million individuals. *Diabetes Research and Clinical Practice, 188*, 109924. https://doi.org/10.1016/j.diabres.2022.109924
92. Nouri, Z., Hajialyani, M., Izadi, Z., Bahramsoltani, R., Farzaei, M. H., & Abdollahi, M. (2020). Nanophytomedicines for the prevention of metabolic syndrome: a pharmacological and biopharmaceutical review. *Frontiers in Bioengineering and Biotechnology, 8*, 425. https://doi.org/10.3389/fbioe.2020.00425
93. Ochin, C. C., & Garelnabi, M. (2018). Berberine encapsulated PLGA-PEG nanoparticles modulate PCSK-9 in HepG2 cells. *Cardiovascular & Haematological Disorders-Drug Targets (Formerly Current Drug Targets-Cardiovascular & Hematological Disorders), 18*(1), 61–70. https://doi.org/10.2174/1871529X18666180201130340
94. Odei-Addo, F., Shegokar, R., Müller, R. H., Levendal, R. A., & Frost, C. (2017). Nanoformulation of Leonotis leonurus to improve its bioavailability as a potential antidiabetic drug. *3 Biotech, 7*, 1–9. https://doi.org/10.1007/s13205-017-0986-0
95. Opris, R., Tatomir, C., Olteanu, D., Moldovan, R., Moldovan, B., David, L., & Filip, G. A. (2017). The effect of *Sambucus nigra* L. extract and phytosinthesized gold nanoparticles on diabetic rats. *Colloids and Surfaces B: Biointerfaces, 150*, 192–200. https://doi.org/10.1016/j.colsurfb.2016.11.033
96. Orugun, O. A., Uronnachi, E. M., Issac, J., Yahaya, Z. S., & Nnamani, N. D. (2023). Formulation and in vitro trypanocidal evaluation of garlic oil nanoemulsions. *Journal of Research in Pharmacy, 27*, 1431–1437. https://doi.org/10.29228/jrp.430
97. Paul, D., Dey, T. K., Mukherjee, S., Ghosh, M., & Dhar, P. (2014). Comparative prophylactic effects of α-eleostearic acid rich nano and conventional emulsions in induced diabetic rats. *Journal of Food Science and Technology, 51*, 1724–1736. https://doi.org/10.1007/s13197-014-1257-2

98. Piazzini, V., Micheli, L., Luceri, C., D'Ambrosio, M., Cinci, L., & Ghelardini, C. (2019). Nanostructured lipid carriers for oral delivery of silymarin: Improving its absorption and in vivo efficacy in type 2 diabetes and metabolic syndrome model. *International Journal of Pharmaceutics, 572*, 118838. https://doi.org/10.1016/j.ijpharm.2019.118838
99. Ponnanikajamideen, M., Rajeshkumar, S., Vanaja, M., & Annadurai, G. (2019). In vivo type 2 diabetes and wound-healing effects of antioxidant gold nanoparticles synthesized using the insulin plant *Chamaecostus cuspidatus* in albino rats. *Canadian Journal of Diabetes, 43*(2), 82–89. https://doi.org/10.1016/j.jcjd.2018.05.006
100. Prabhu, S., Vinodhini, S., Elanchezhiyan, C., & Rajeswari, D. (2018). Evaluation of antidiabetic activity of biologically synthesized silver nanoparticles using Pouteria sapota in streptozotocin-induced diabetic rats. *Journal of Diabetes, 10*(1), 28–42. https://doi.org/10.1111/1753-0407.12554
101. Priyanka, K., Sahu, P. L., & Singh, S. (2018). Optimization of processing parameters for the development of *Ficus religiosa* L. extract loaded solid lipid nanoparticles using central composite design and evaluation of antidiabetic efficacy. *Journal of Drug Delivery Science and Technology, 43*, 94–102. https://doi.org/10.1016/j.jddst.2017.08.006
102. Rachmawati, H., Soraya, I. S., Kurniati, N. F., & Rahma, A. (2016). In vitro study on antihypertensive and antihypercholesterolemic effects of a curcumin nanoemulsion. *Scientia Pharmaceutica, 84*(1), 131. https://doi.org/10.3797/scipharm.ISP.2015.05
103. Ragavan, G., Muralidaran, Y., Sridharan, B., Nachiappa Ganesh, R., & Viswanathan, P. (2017). Evaluation of garlic oil in nano-emulsified form: Optimization and its efficacy in high-fat diet induced dyslipidemia in Wistar rats. *Food and Chemical Toxicology, 105*, 203–213. https://doi.org/10.1016/j.fct.2017.04.019
104. Rani, R., Dahiya, S., Dhingra, D., Dilbaghi, N., Kim, K.-H., & Kumar, S. (2018). Improvement of antihyperglycemic activity of nano-thymoquinone in rat model of type-2 diabetes. *Chemico-Biological Interactions, 295*, 119–132. https://doi.org/10.1016/j.cbi.2018.02.006
105. Rice, K. M., Manne, N. D., Kolli, M. B., Wehner, P. S., Dornon, L., Arvapalli, R., & Blough, E. R. (2016). Curcumin nanoparticles attenuate cardiac remodeling due to pulmonary arterial hypertension. *Artificial Cells, Nanomedicine, and Biotechnology, 44*(8), 1909–1916. https://doi.org/10.3109/21691401.2015.1111235
106. Saif-Ali, R., Kamaruddin, N. A., Al-Habori, M., Al-Dubai, S. A., & Ngah, W. Z. W. (2020). Relationship of metabolic syndrome defined by IDF or revised NCEP ATP III with glycemic control among Malaysians with Type 2 Diabetes. *Diabetology & Metabolic Syndrome, 12*, 1–7. https://doi.org/10.1186/s13098-020-00575-7
107. Samadder, A., Das, S., Das, J., Paul, A., & Khuda-Bukhsh, A. R. (2012). Ameliorative effects of *Syzygium jambolanum* extract and its poly (lactic-co-glycolic) acid nano-encapsulated form on arsenic-induced hyperglycemic stress: a multi-parametric evaluation. *Journal of Acupuncture and Meridian Studies, 5*(6), 310–318. https://doi.org/10.1016/j.jams.2012.09.001
108. Sandhu, S. K., Kumar, S., Raut, J., Singh, M., Kaur, S., Sharma, G., & Kaur, I. P. (2021). Systematic development and characterization of novel, high drug-loaded, photostable, curcumin solid lipid nanoparticle hydrogel for wound healing. *Antioxidants, 10*(5), 725. https://doi.org/10.3390/antiox10050725
109. Saratale, G. D., Saratale, R. G., Benelli, G., Kumar, G., Pugazhendhi, A., & Kim, D. S. (2017). Anti-diabetic potential of silver nanoparticles synthesized with Argyreia nervosa leaf extract high synergistic antibacterial activity with standard antibiotics against foodborne bacteria. *Journal of Cluster Science, 28*, 1709–1727. https://doi.org/10.1007/s10876-017-1179-z
110. Sengottaiyan, A., Aravinthan, A., Sudhakar, C., Selvam, K., Srinivasan, P., & Govarthanan, M. (2016). Synthesis and characterization of Solanum nigrum-mediated silver nanoparticles and its protective effect on alloxan-induced diabetic rats. *Journal of Nanostructure in Chemistry, 6*, 41–48. https://doi.org/10.1007/s40097-015-0178-6
111. Shahraki, A., Bahadorikhalili, S., Hashemzaei, M., Hajinezhad, M., Afsharimoghaddam, A., Sarani, F., & Tajrobekar, O. (2017). Resveratrol nanocapsule as an efficient tool for blood

pressure regulation: a study on metabolic syndrome induced mice. *Bioscience Biotechnology Research Communications, 10*(4), 623–630. https://doi.org/10.21786/bbrc/10.4/4

112. Shahraki, A., Bahadorikhalili, S., Hashemzaei, M., Shahraki, J., & Tajrobekar, O. (2020). Effects of resveratrol nanocapsules on the quantitative insulin sensitivity check index in insulin resistance: a study on metabolic syndrome induce mice. *SN Applied Sciences, 2*(5), 962. https://doi.org/10.1007/s42452-020-2759-3
113. Shahzad, N., Alzahrani, A. R., Ibrahim, I. A. A., Shahid, I., Alanazi, I. M., Falemban, A. H., & Arulselvan, P. (2024). Therapeutic strategy of biological macromolecules based natural bioactive compounds of diabetes mellitus and future perspectives: A systemic review. *Heliyon, 10*(2), e24207. https://doi.org/10.1016/j.heliyon.2024.e24207
114. Sharma, S., Gauba, P., Tyagi, A., & Dang, S. (2025). Chitosan-modified polymeric nanoparticles for the nose-to-brain drug delivery of paroxetine: an in vitro and in vivo evaluation. *Nanoscale, 17*(3), 1687–1702. https://doi.org/10.1039/D4NR04250F
115. Sibuyi, N. R. S., Moabelo, K. L., Meyer, M., Onani, M. O., Dube, A., & Madiehe, A. M. (2019). Nanotechnology advances towards development of targeted-treatment for obesity. *Journal of Nanobiotechnology, 17*, 1–21.
116. Sim, S., & Wong, N. K. (2021). Nanotechnology and its use in imaging and drug delivery (review). *Biomedical Reports, 14*(5), 42.
117. Sindhu, R. K., Goyal, A., Algın Yapar, E., & Cavalu, S. (2021). Bioactive compounds and nanodelivery perspectives for treatment of cardiovascular diseases. *Applied Sciences, 11*(22), 11031.
118. Singha, A., Harshitha, M., Kalladka, K., Chakraborty, G., Maiti, B., Satyaprasad, A. U., & Sil, S. K. (2024). Exploring the potential of curcumin-loaded PLGA nanoparticles for angiogenesis and antioxidant proficiency in zebrafish embryo (Danio rerio). *Future Journal of Pharmaceutical Sciences, 10*(1), 167.
119. Singla, R., Soni, S., Kulurkar, P. M., Kumari, A., Mahesh, S., Patial, V., & Yadav, S. K. (2017a). In situ functionalized nanobiocomposites dressings of bamboo cellulose nanocrystals and silver nanoparticles for accelerated wound healing. *Carbohydrate Polymers, 155*, 152–162.
120. Singla, R., Soni, S., Patial, V., Kulurkar, P. M., Kumari, A. S. M., et al. (2017b). In vivo diabetic wound healing potential of nanobiocomposites containing bamboo cellulose nanocrystals impregnated with silver nanoparticles. *International Journal of Biological Macromolecules, 105*, 45–55. https://doi.org/10.1016/j.ijbiomac.2017.06.109
121. Singla, R., Soni, S., Patial, V., Kulurkar, P. M., Kumari, A. S. M., et al. (2017c). Cytocompatible anti-microbial dressings of *Syzygium cumini* cellulose nanocrystals decorated with silver nanoparticles accelerate acute and diabetic wound healing. *Scientific Reports, 7*(1), 1–13. https://doi.org/10.1038/s41598-017-08897-9
122. Sonawane, D., & Pokharkar, V. (2024). Quercetin-loaded nanostructured lipid carrier in situ gel for brain targeting through intranasal route: Formulation, in vivo pharmacokinetic and pharmacodynamic studies. *AAPS PharmSciTech, 25*(2), 30.
123. Sriramavaratharajan, V., & Murugan, R. (2019). Evaluation of chemical composition, antioxidant and anti-hyperglycemic activities of the essential oil based nanoemulsions of Cinnamomum litseifolium. *Natural Product Research, 33*(16), 2430–2433.
124. Summerlin, N., Soo, E., Thakur, S., Qu, Z., Jambhrunkar, S., & Popat, A. (2015). Resveratrol nanoformulations: challenges and opportunities. *International Journal of Pharmaceutics, 479*(2), 282–290. https://doi.org/10.1016/j.ijpharm.2015.01.003
125. Sun, S., Hou, X., Li, K., Huang, C., Rong, Y., Bi, J., & Wu, D. (2024). Curcumin and metformin infinite coordination polymer nanoparticles for combined therapy of diabetic mice via intraperitoneal injections. *Journal of Functional Biomaterials, 15*(12), 388. https://doi.org/10.3390/jfb15120388

126. Syahputra, R. A., Dalimunthe, A., Utari, Z. D., Halim, P., Sukarno, M. A., Zainalabidin, S., & Harahap, U. (2024). Nanotechnology and flavonoids: Current research and future perspectives on cardiovascular health. *Journal of Functional Foods, 120*, 106355. https://doi.org/10.1016/j.jff.2024.106355
127. Taghipour, Y. D., Hajialyani, M., Naseri, R., Hesari, M., Mohammadi, P., Stefanucci, A., & Abdollahi, M. (2019). Nanoformulations of natural products for management of metabolic syndrome. *International Journal of Nanomedicine*, 5303–5321. https://doi.org/10.2147/IJN.S213831
128. Tamilarasi, G. P., Krishnan, M., Sabarees, G., Gouthaman, S., Alagarsamy, V., & Solomon, V. R. (2022). Emerging trends in curcumin embedded electrospun nanofibers for impaired diabetic wound healing. *Applied Nano, 3*(4), 202–232. https://doi.org/10.3390/applnano3040015
129. Tomou, E. M., Papakyriakopoulou, P., Saitani, E. M., Valsami, G., Pippa, N., & Skaltsa, H. (2023). Recent advances in nanoformulations for quercetin delivery. *Pharmaceutics, 15*(6), 1656. https://doi.org/10.3390/pharmaceutics15061656
130. Tong, F., Liu, S., Yan, B., Li, X., Ruan, S., & Yang, S. (2017). Quercetin nanoparticle complex attenuated diabetic nephropathy via regulating the expression level of ICAM-1 on endothelium. *International Journal of Nanomedicine*, 7799–7813. https://doi.org/10.2147/IJN.S146978
131. Wang, M., Xia, H., Yang, X., Zhang, Q., Li, Y., Wang, Y., & Xie, Z. (2023a). Berberine hydrochloride-loaded liposomes gel: Preparation, characterization and antioxidant activity. *Indian Journal of Pharmaceutical Education and Research, 57*, 74–82. https://doi.org/10.5530/001954642087
132. Wang, X., Wang, Y., Tang, T., Zhao, G., Dong, W., Li, Q., & Liang, X. (2023b). Curcumin-loaded RH60/F127 mixed micelles: Characterization, biopharmaceutical characters and anti-inflammatory modulation of airway inflammation. *Pharmaceutics, 15*(12), 2710. https://doi.org/10.3390/pharmaceutics15122710
133. Xu, L., Li, W., Shi, Q., Li, H., Yang, Z., Liao, D., & Zhang, J. (2019). Synthesis of mulberry leaf extract mediated gold nanoparticles and their ameliorative effect on Aluminium intoxicated and diabetic retinopathy in rats during perinatal life. *Journal of Photochemistry and Photobiology, B: Biology, 196*, 111502. https://doi.org/10.1016/j.jphotobiol.2019.04.011
134. Xue, M., Yang, M. X., Zhang, W., Li, X. M., Gao, D. H., Ou, Z. M., & Yang, S. Y. (2013). Characterization, pharmacokinetics, and hypoglycemic effect of berberine loaded solid lipid nanoparticles. *International Journal of Nanomedicine, 8*, 4677–4687.
135. Yi, M. H., Simu, S. Y., Ahn, S., Aceituno, V. C., Wang, C., Mathiyalagan, R., et al. (2020). Anti-obesity effect of gold nanoparticles from *Dendropanax morbifera* Léveille by suppression of triglyceride synthesis and downregulation of PPARγ and CEBPα signaling pathways in 3T3-L1 mature adipocytes and HepG2 cells. *Current Nanoscience, 16*(2), 196–203. https://doi.org/10.2174/1573413716666200116124822
136. Yücel, Ç., Karatoprak, G. Ş., & Aktaş, Y. (2018). Nanoliposomal resveratrol as a novel approach to treatment of diabetes mellitus. *Journal of Nanoscience and Nanotechnology, 18*(6), 3856–3864. https://doi.org/10.1166/jnn.2018.15247
137. Zhang, Y., Li, J., Wang, Z., Xu, M. Z., Zeng, Z., Huang, J. P., & Guan, Y. Q. (2020). Natural plant-derived polygalacturonic acid-oleanolic acid assemblies as oral-delivered nanomedicine for insulin resistance treatment. *Chemical Engineering Journal, 390*, 124630. https://doi.org/10.1016/j.cej.2020.124630
138. Zhou, H., Zheng, B., & McClements, D. J. (2021). Encapsulation of lipophilic polyphenols in plant-based nanoemulsions: Impact of carrier oil on lipid digestion and curcumin, resveratrol and quercetin bioaccessibility. *Food & Function, 12*(8), 3420–3432. https://doi.org/10.1039/D1FO00275A

139. Zu, Y., Zhao, L., Hao, L., Mechref, Y., Zabet-Moghaddam, M., Keyel, P. A., & Wang, S. (2021). Browning white adipose tissue using adipose stromal cell-targeted resveratrol-loaded nanoparticles for combating obesity. *Journal Of Controlled Release, 333*, 339–351. https://doi.org/10.1016/j.jconrel.2021.03.022
140. Parvin, N., Aslam, M., Joo, S. W., & Mandal, T. K. (2025). Nano-phytomedicine: harnessing plant-derived phytochemicals in nanocarriers for targeted human health applications. *Molecules*, 30(15), 3177. https://doi.org/10.3390/molecules30153177
141. Barwal, I., Sood, A., Sharma, M., Singh, B., & Yadav, S. C. (2013). Development of stevioside Pluronic-F-68 copolymer based PLA-nanoparticles as an antidiabetic nanomedicine. *Colloids and Surfaces B: Biointerfaces, 101*, 510–516. https://doi.org/10.1016/j.colsurfb.2012.07.005

Role of Nanobiosensor in the Diagnostics of Metabolic Syndrome

4

Pankaj Kalia, Rajesh Kumar, Ashutosh Kumar, and Brijesh Kumar

Abstract

Metabolic syndrome is a multifactorial disorder associated with insulin resistance, obesity, hypertension and dyslipidemia, in which increased risk factors promote cardiovascular diseases and diabetes. Early diagnosis is essential for effective management and efforts toward prevention of complications. The appearance of the nanobiosensors is considered as an innovative approach to develop novel diagnostic tools which utilize features of the nanomaterials to achieve high sensitivity, specificity and rapid detection of multiple metabolic biomarkers at a time. Compared to many other methods, these sensors are simple to introduce, fast and allow real-time monitoring and are also possible to integrate with wearable devices and digital health platforms to provide a personalized and continuous health management. Challenges to nanobiosensor development include biocompatibility, sensor stability and regulatory hurdles which are being addressed thanks to progress in nanotechnology, materials science and data analytics. In the end, nanobiosensors hold tremendous promise to revolutionize metabolic syndrome diagnostics by providing opportunity for early detection, precise monitoring and personalized therapeutic intervention which will improve patient outcomes and decrease disease burden.

P. Kalia · R. Kumar (✉)
Maharaja Agrasen School of Pharmacy, Maharaja Agrasen University, Solan, Himachal Pradesh, India

A. Kumar
Department of Ophthalmology, University of California, Los Angeles, Los Angeles, CA, USA

B. Kumar
Department of Pharmacology, Institute of Medical Sciences, Banaras Hindu University, Varanasi, India

V. K. Chaturvedi et al. (eds.), *Nanotheranostics in Metabolic Syndrome*,
https://doi.org/10.1007/978-981-95-3746-4_4

Keywords

Nanobiosensors · Metabolic syndrome · Biomarkers · Early diagnosis · Nanomaterials

1 Introduction

When someone has metabolic syndrome (MetS), they have central obesity, insulin resistance, dyslipidemia and hypertension together which greatly increase the chances of developing cardiovascular diseases and type 2 diabetes mellitus [1]. Diagnosing metabolic syndrome early and correctly is very important so that patients can be treated and prevented on time. The usual way to diagnose includes measuring fasting glucose, blood pressure, various lipids and a patient's physical measurements. Nonetheless, these traditional methods usually are not sensitive, specific or quick enough to spot delicate changes in body chemistry at the beginning of the syndrome. As a result, nanobiosensors provide a strong analytical solution, being able to detect small levels of biomarkers precisely, quickly and with less harm to the body [2]. Enzymes, antibodies, nucleic acids or aptamers are combined with gold nanoparticles, carbon nanotubes, quantum dots or graphene in nanobiosensors. Because of their electrical, optical and mechanical features, these nanomaterials substantially boost and improve how the signal transduction processes function (Fig. 4.1). For people with metabolic syndrome, nanobiosensors can detect free glucose, insulin, adiponectin, leptin and inflammatory lipoproteins, each of which plays a role in the underlying causes of the illness. Take, for example, a type of biosensor called an electrochemical nanobiosensor which is very accurate at identifying blood glucose and insulin. This, in turn, makes it possible to keep tabs on insulin resistance that may lead to MetS [3].

Nanobiosensors are miniaturized in nature such that they can be integrated into portable, point-of-care, diagnostic devices for decentralized healthcare delivery and continuous patient monitoring. This is especially so in low resource settings where centralized laboratory facilities are limited [4]. Additionally, nanobiosensors can multiplex and detect several biomarkers simultaneously which is necessary for representing the whole chemistry of the metabolic syndrome. Other optical nanobiosensors, including those based on surface plasmon resonance (SPR) or fluorescence resonance energy transfer (FRET), also have the promise to quantify adipokines and cytokines with high resolution and hence provide novel insights regarding the patient's inflammatory and metabolic state [5]. Some other advantages of nanobiosensors are they can be integrated in wearable technology and digital health platforms to allow for remote diagnostics and diagnostic tools for personalized medicine [6]. Wearable nanobiosensors continuously monitor physiological parameters and send real-time data which are used for the proactive management of diseases and changes in lifestyles based on dynamic health indicators. Moreover, advances in the nanofabrication, biofunctionalization and data analytics

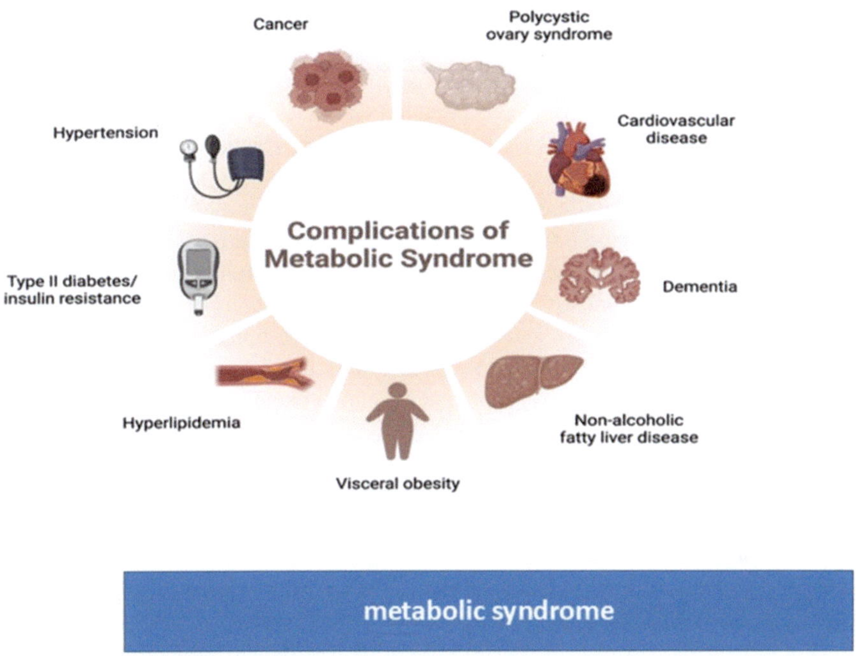

Fig. 4.1 Introduction to complications of metabolic syndrome

are further pushing the frontier of next generation nanobiosensors having greater biocompatibility, stability and analytical throughput [7].

2 Fundamentals of Nanobiosensors

A nanobiosensor uses both a biological part and a nanomaterial part to spot and detect certain biochemical or physiological substances at a molecular level. The main idea behind nanobiosensors is to change a biological reaction such as enzyme-substrate, antigen-antibody or nucleic acid hybridization into a measurable signal. It comes from physical changes, either to electrical, optical, thermal or mechanical properties and is enhanced by the specific qualities of nanomaterials [8]. Thanks to their ability to detect small differences with accuracy, these sensors respond fast, are beneficial for medicine, pollution control and safe foods. The most important elements in a nanobiosensor include a bioreceptor, a transducer and a signal processor [9]. The bioreceptor is a biological substance, including an enzyme, antibody, DNA strand or aptamer, that can specifically interact with what it is looking for [10]. As the gas reaction takes place, it changes the environment inside which the transducer can detect. Transducer systems tend to use nanomaterials such as gold nanoparticles, graphene, quantum dots and carbon nanotubes since they are large in surface area, highly conductive and have unique lighting properties which make them more sensitive to weak signals and help detect minute amounts of substances [11].

Some of the transduction approaches commonly used in nanobiosensor devices are electrochemical, optical, piezoelectric and thermal. They detect current, potential or impedance changes that happen due to a biochemical reaction such sensors detect biochemical interactions by measuring changes in how light is handled, for example through absorbance, fluorescence or SPR [12]. Piezoelectric sensors can sense changes in weight or vibrations on a crystal and thermal sensors spot temperature changes that arise from biochemical reactions. Nanotechnology also makes it possible to add selective ligands or chemical groups to nanobiosensor surfaces which in turn improves how precisely and steadily the sensors' function. Besides, linking nanobiosensors to microfluidics and digital electronics helps create lab-on-a-chip devices and wearable sensors for quick and accurate diagnostics at any time [13].

3 Bioreceptor, Transducer, and Signal Processor

Nanobiosensors are used to analyze and find particular biological molecules in a more precise way than traditional methods. Such receptors operate with the combined actions of the bioreceptor, transducer and signal processor. Thanks to each of these components, nanobiosensors can measure biological interactions which is useful for finding illnesses like metabolic syndrome, infections and cancers [14]. The bioreceptor is the vital part of the nanobiosensor since it binds to and selects the target analyte. Examples of bioreceptors are enzymes, antibodies, nucleic acids, aptamers or whole cells. Because of their specificity, the sensor only detects the right molecule regardless of other factors present in a biological sample [15]. A good example is using glucose oxidase as the bioreceptor for the detection of glucose. In these biosensors, immobilizing the bioreceptors on gold and carbon nanostructures helps to boost the interaction between the sensor and the substance the sensor is trying to detect [16] (Table 4.1).

The transducer changes the biorecognition event into a measurable form. Graphene, quantum dots and nanowires which are common nanomaterials in transducers, have excellent properties relating to electricity, light and what they can withstand [17]. They improve the sensitivity of the sensor because they make one molecular signal much larger. Depending how they detect changes, transducers are varied as electrochemical, optical, piezoelectric and thermal [18]. As an example, if an antigen connects with its antibody on a surface covered by gold nanoparticles, the conductivity or refractive index will be altered which can be measured to quantify the interaction [19]. The signal processor reads the transducer's signal and converts it into a form that is understandable to the user. It requires steps such as amplification, filtering and digitizing the signal for good accuracy and clarity [20]. Often, nanobiosensor systems now have signal processors, microcontrollers and display units combined, using wireless networking to let the monitoring be done instantly and remotely [21].

Table 4.1 Bioreceptor, transducer and signal processor as nanobiosensors

Component	Type/Example	Function	Applications
Bioreceptor	Enzymes (e.g., glucose oxidase)	Recognizes specific analyte (e.g., glucose)	Glucose monitoring in diabetes; metabolic syndrome detection
	Antibodies	Binds specifically to antigens (e.g., insulin, cytokines)	Detection of inflammatory markers, insulin levels
	Aptamers	Synthetic oligonucleotides binding to target molecules	Detection of small molecules (e.g., lipids, cholesterol)
	DNA/RNA probes	Hybridize with complementary nucleic acids	Genetic marker detection; mutation analysis
	Molecularly imprinted polymers (MIPs)	Artificial recognition sites for target molecules	Detection of hormones, metabolites
Transducer	Electrochemical (e.g., amperometric, potentiometric)	Converts biorecognition event into electrical signal	Blood glucose meters, wearable biosensors
	Optical (e.g., fluorescence, surface plasmon resonance)	Converts recognition event into optical signal	Cancer biomarker detection, cholesterol sensing
	Piezoelectric (e.g., quartz crystal microbalance)	Measures mass change upon target binding	Detection of microbial pathogens, proteins
	Thermal (e.g., calorimetric sensors)	Measures heat released/ absorbed during reaction	Enzyme-substrate reaction monitoring
Signal processor	Microcontrollers/ Processors (e.g., Arduino, ARM chips)	Processes and interprets signals into readable format	Real-time monitoring devices; smartphone-integrated biosensors
	Wireless modules (e.g., Bluetooth, NFC, Wi-Fi)	Transmit processed data for remote monitoring	Telehealth applications, continuous patient monitoring
	AI/ML algorithms	Enhance signal quality, reduce noise, enable predictive analytics	Early detection of metabolic syndromes, intelligent diagnostics

4 Nanomaterials

Because nanomaterials have excellent physicochemical properties, they help make nanobiosensors more accurate, sensitive and compact. Due to their size, typically smaller than 100 nanometers, nanomaterials have special features like a large area for reaction, excellent electrical conduction, interaction with light and strong ability to work as catalysts which are vital for biosensing tasks. Most nanobiosensors depend on nanomaterials acting as transducers or offering a place to fix bioreceptors, making it easy to detect biomolecules present at extremely low amounts [22]. Examples of nanomaterials in nanobiosensors are metallic nanoparticles (like gold and silver), carbon-based nanostructures (such as carbon nanotubes and graphene), semiconductor particles (such as quantum dots) and polymeric types [23]. Thanks to their good biocompatibility, manageable chemical properties and unique SPR, gold nanoparticles (AuNPs) find broad use. It is common to use them in optical and colorimetric biosensors for tracking biomolecular interactions by looking at changes in color. Thanks to their top electrical conductivity and strength, carbon nanotubes (CNTs) and graphene are great picks for constructing electrochemical biosensors that sense current or voltage upon an analyte being recognized. The big surface of these nanoparticles is useful for immobilizing enzymes, antibodies or nucleic acids [24].

Thanks to their high stability and adjustable emission, quantum dots as semiconducting nanocrystals allow fluorescence-based biosensors to detect things much more sensitively. They perform best in situations where several biomarkers must be spotted in one test. Using magnetic fields, magnetic nanoparticles can be easily handled, helping to get the sample ready for measurement so that there is a better chance of detecting even small amounts. Using nanomaterials in biosensors has led to the invention of wearable, implantable and lab-on-a-chip diagnostics. Because of these improvements, it is now possible to closely monitor a patient's condition using small wearable devices [25].

5 Nanobiosensors

These small instruments have become important tools for checking for metabolic syndrome which is a health problem that comes with abdominal obesity, insulin resistance, high blood pressure, high blood sugar and poor cholesterol levels. It is important to identify metabolic syndrome at an early stage to stop it from leading to heart problems and diabetes mellitus. Often, traditional ways to diagnose require extra time, intrusive procedures and they fail to detect very weak changes at the molecular level [26]. On the flip side, nanobiosensors allow for quick, very sensitive and minimally painful tests, helping to discover several biomarkers at very low levels and thus enabling greater accuracy in diagnosing diseases. They are created using biorecognition elements alongside nanomaterials such as enzymes, antibodies, aptamers, gold nanoparticles, carbon nanotubes or graphene. Thanks to nanomaterials, the presence of glucose, insulin, leptin, adiponectin, cholesterol and inflammatory cytokines can be detected more clearly [27]. This means that electrochemical nanosensors based on carbon materials can effectively measure blood glucose and insulin and their findings are important for spotting insulin resistance, an important sign of metabolic syndrome. Furthermore, optical nanobiosensors made with quantum dots or plasmonic nanoparticles can distinguish adipokines and proinflammatory cytokines that take part in the condition's pathophysiology [28].

A significant benefit of nanobiosensors is that they can give instant and nearby medical testing. They can be integrated with wearables or transportation options which allows constant tracking of vital body measurements [29]. This proves especially beneficial for those at high risk since it allows them to respond to health changes and make the necessary lifestyle changes. Furthermore, it is possible to design nanobiosensors to spot various biomarkers in a single go (multiplexing) which aids in precisely identifying metabolic syndrome, thereby enhancing both detection and sorting of patients [30].

6 Biomarkers in Metabolic Syndrome

It is important to rely on biomarkers for the accurate detection, outlook and changes of metabolic syndrome (MetS) which features a combination of conditions such as obesity around the waist, resistance to insulin, high blood pressure, abnormal cholesterol levels and changes in glucose processing [31]. Using these biomarkers, doctors can find out early about a disease, understand a patient's risk and check how well therapy is working. As metabolic syndrome is very complex and variable, it is necessary to locate and measure meaningful biomarkers to provide timely and effective intervention for each person [32]. A key biomarker for metabolic syndrome is fasting blood glucose which describes the efficiency of glucose management in the body. High glucose shows that a person's body does not respond well to insulin which is one of the main traits of MetS. Insulin sensitivity can also be estimated by looking at a person's insulin levels and HOMA-IR. Blood tests that measure hemoglobin A1c (HbA1c) are key, since they offer a glimpse of glucose control over several months which is necessary for monitoring the progress of the disease [33]. Triglycerides, HDL cholesterol and LDL cholesterol are all basic biomarkers in metabolic syndrome. MetS is identified when triglycerides are high and HDL-C is low and too much LDL-C is a risk factor for atherosclerosis. They point out problems in how the body processes fats and are related to heart problems linked to the syndrome [34] (Table 4.2).

Adipose tissue produces adipokines like leptin and adiponectin which are used as helpful biomarkers [35]. People with obesity usually have high leptin levels due to resistance, whereas those with metabolic syndrome usually have lower adiponectin levels and fewer of these are linked to insulin sensitivity and inflammation. Inflammatory molecules, known as proinflammatory cytokines which consist of TNF-α, IL-6 and CRP, are increased in MetS and show there is a problem with constant low-level inflammation linked to metabolic problems [36]. Blood pressure is also a key biomarker and if it remains high for a long time, it serves as a criterion for diagnosis and raises the risk of cardiovascular disease. Moreover, studies are looking into resistin, ghrelin and plasminogen activator inhibitor-1 (PAI-1) to see how they could play a part in developing MetS [37].

Table 4.2 Biomarkers in metabolic syndrome with examples and applications

Biomarker category	Biomarker example	Biological role	Applications in metabolic syndrome
Glucose metabolism	Fasting plasma glucose (FPG)	Indicator of insulin resistance and glucose homeostasis	Diagnosis of diabetes and impaired glucose tolerance
	HbA1c	Reflects average blood glucose over 2–3 months	Long-term glucose control assessment
Lipid profile	Triglycerides (TG)	Reflects dyslipidemia	Diagnosis of hypertriglyceridemia and atherogenic risk
	HDL-C (high-density lipoprotein)	Protective role in lipid transport	Low HDL indicates risk for cardiovascular complications
	LDL-C (low-density lipoprotein)	Proatherogenic cholesterol	Elevated in metabolic syndrome; CVD risk factor
Inflammatory markers	C-reactive protein (CRP)	Acute-phase reactant indicating inflammation	Early detection of systemic inflammation and CVD risk
	IL-6, TNF-α	Proinflammatory cytokines	Indicators of chronic low-grade inflammation in metabolic syndrome
Adipokines	Leptin	Regulates appetite and energy balance	Elevated in obesity and insulin resistance
	Adiponectin	Anti-inflammatory and insulin-sensitizing hormone	Low levels indicate higher risk of metabolic syndrome
Oxidative stress markers	Malondialdehyde (MDA)	Lipid peroxidation byproduct	Indicates oxidative damage linked with obesity and insulin resistance
	Superoxide dismutase (SOD)	Antioxidant enzyme	Low SOD associated with oxidative stress in metabolic syndrome
Liver enzymes	Alanine aminotransferase (ALT)	Marker of liver function	Elevated in nonalcoholic fatty liver disease (NAFLD) linked to MetS
	Gamma-glutamyl transferase (GGT)	Involved in oxidative stress	Associated with insulin resistance and liver dysfunction
Hormonal markers	Insulin	Regulates blood glucose	Elevated fasting insulin indicates insulin resistance
	Cortisol	Stress hormone affecting metabolism	Chronic elevation linked to obesity and metabolic imbalance

7 Multiplexed Biomarker Detection

With multiplexed biomarker detection, it is possible to both find and measure several biomarkers at the same time in one assay. This technique plays an essential role in dealing with metabolic syndrome (MetS), since it involves a mix of factors like excess belly fat, insulin resistance, high blood pressure, lipid problems and low-level inflammation [38]. Being diagnosed with metabolic syndrome is best done through multiplexed detection which uses many biomarkers to accurately set a patient's risk and assess the condition [39]. Standard ways of testing tend to screen only one metabolite each time, so they can take a lot of time, are not always cost-effective and might overlook small interactions between parameters. On the other hand, multiplexed platforms, mostly those created with nanobiosensors, can check a group of substances like glucose, insulin, adiponectin, leptin, CRP, IL-6, TNF-α, triglycerides and HDL cholesterol at the same time. It gives a complete picture of the patient's metabolism and supports early and reliable detection of related problems [40].

Using gold nanoparticles, quantum dots, graphene and carbon nanotubes has allowed nanomaterial-based biosensors to carry out more tests simultaneously. Sensitive and specific detections of different analytes in a small volume can be achieved because of their large surface, electric conductivity and optical properties [41]. Metabolic syndrome diagnostics in the clinic are being done with the help of microarray-based biosensors, lab-on-a-chip devices and electrochemical arrays. Also, combining multiplexed nanobiosensors with digital health technologies means patients can be managed remotely, since data is picked up, transmitted wirelessly and interpreted upon receipt using algorithms. It becomes especially important in measuring biomarker levels when people change their habits or take medicines [42].

8 Nanobiosensor Technologies for Metabolic Syndrome

With the help of nanobiosensor technologies, early detection and monitoring of MetS have been much improved since MetS combines key characteristics like central obesity, insulin resistance, abnormalities in fats and cholesterol, high blood pressure and chronic inflammation [43]. In recent years, traditional ways of diagnosing illnesses are not sensitive enough, quick enough or able to connect to the modern management systems needed for immediate health monitoring [44, 45]. On the other hand, nanobiosensors are suited for fast and precise detection of numerous low-level metabolic biomarkers since they have large surfaces, are biocompatible and intensify the signals (Fig. 4.2). Advanced sensors make use of gold nanoparticles, carbon nanotubes, graphene, magnetic nanoparticles and quantum dots to increase how well signals are transmitted and recognized [46, 47]. There are electrochemical, optical, piezoelectric and thermal platforms for nanobiosensor technologies, depending on the way they work. Nanobiosensors detect alterations in electrical signals that arise when analytes (such as glucose or insulin) connect to

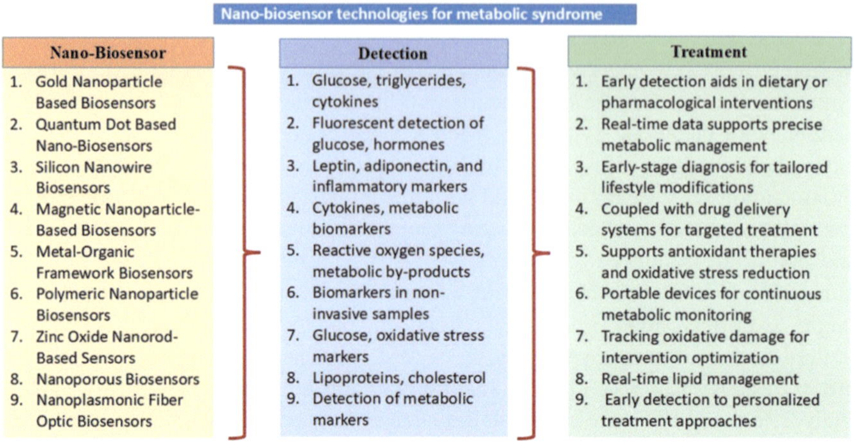

Fig. 4.2 Nanobiosensor technologies for metabolic syndrome

bioreceptors that are placed on nanoscale electrodes. Using quantum dots or SPR, optical nanobiosensors can detect both adipokines and inflammatory cytokines that are important signals of metabolic syndrome [48, 49].

A significant achievement in this area is the building of multiplexed nanosensors that are able to detect several biomarkers simultaneously, for instance, glucose, leptin, adiponectin, CRP, IL-6 and triglycerides [50]. It is important to have this holistic diagnosis since MetS affects a complex biochemical system [51]. Furthermore, the small size of nanobiosensors now supports lab-on-a-chip systems and wearable gadgets that allow constant tracking of metabolic health at the moment it is needed [52]. Besides, linking nanobiosensors with artificial intelligence (AI), microfluidic systems and wireless transmissions is helping to develop smart diagnostics and individualized healthcare [53]. They can keep an eye on metabolic biomarker changes and offer prompt advice which helps people prevent diseases and deal with them better. Portable detectors of glucose and noninvasive monitors that use saliva or sweat for metabolic syndrome are examples of lately developed nanobiosensors [54].

9 Integration with Digital Health Platforms

When nanobiosensors are combined with digital health systems, it results in a better method of managing metabolic syndrome which involves several risk factors such as obesity, insulin resistance, high blood pressure and dyslipidemia [55, 56]. Apps and other digital technologies in health, like mHealth, wearable devices, storage in the cloud and AI, help us watch over patients' health, analyze information and provide personal health care [57]. Having nanobiosensors in addition to microsystems offers an innovative way to notice MetS early, track progress and handle it actively. Scientists can use nanobiosensors to find glucose, insulin, leptin, adiponectin, cholesterol and inflammatory cytokines with very high accuracy and minimal error [58]. Biosensors that are used in wearable or portable devices permit obtaining biological fluids such as sweat, saliva or materials from between the skin. These devices are capable of keeping tabs on your body and immediately sending information to mobile devices or online services which helps both you and your doctor stay up to date [59, 60].

Data visualization and trend analysis are made easier on digital platforms, so users can see their metabolic health trends in the past. With the help of artificial intelligence and machine learning, these systems predict the course of a disease, notice changes from normal markers and suggest ways to change one's life or treatment. As an example, CGMs that share information with mobile apps measure sugar fluctuations and give advice about both meals and drugs. Besides, the use of electronic health records makes it possible for clinicians to use recent information from a patient which aids in accurate diagnosis and solid decision-making [61, 62]. They provide chances for patients to be more involved by reminding them about appointments and giving them information and tools for following their treatment plans and making healthy changes. Joining forces with cloud computing and the Internet of Things makes it easier for nanobiosensors to be used on a large scale. In rural or underserved regions, they offer people continuous care without the requirement for constant visits to health centers [63, 64].

10 Recent Advances in Nanobiosensor

Progress in nanobiosensor technology has now made them more sensitive, more accurate, portable and usable in real time, making them very useful for both modern diagnostics and personal healthcare. Nanobiosensors have seen many innovations in materials, manufacturing and combining with digital and microfluidic tools which led them to work as multifunctional instruments at home and for medical use [65, 66]. Because of these developments, it is now easier to discover conditions such as metabolic syndrome, cancer, infections and neurological problems [67, 68]. One main progress is the development of nanomaterials that have more than one function, for example, graphene, carbon nanotubes, gold nanoparticles, quantum dots and metal-organic frameworks (MOFs) [69]. Because of their excellent electrical, optical and catalytic properties, these materials make it possible to detect biomarkers with high sensitivity and selectivity. As an illustration, graphene electrochemical sensors can now find analytes even at concentrations as low as femtomolar which is much better than traditional biosensors. Using nanostructures has helped improve how signals are amplified. Plasmonic nanoparticles are added to optical biosensors to increase SPR signals and make it possible for labels to be avoided in detection. Also, biosensors that use nanozymes are coming to replace natural enzymes, since they are more stable and less costly, making the biosensors better suited for various situations [46, 70].

The use of flexible biosensors capable of being worn is becoming an important development. Many smartwatches or textiles are now equipped with new devices that can check glucose, lactate, cortisol and hydration level in sweat or interstitial fluid, letting you know about your health right away. Many of these devices are paired with Bluetooth or NFC pieces to guarantee easy data sharing with smartphones or the cloud. Because of artificial intelligence (AI) and machine learning (ML) with nanobiosensor data, new opportunities have come up for detecting diseases in advance [71]. AI tools examine the data from biosensors to spot patterns, know the chances of disease development and choose the best treatment methods. Using lab-on-a-chip models which bring together microfluidics and small biosensors, now allows for more advanced and repetitive testing that needs less sample input [72, 73].

11 Nanobiosensors in Clinical or Preclinical Use

Nanobiosensors, which used to be just for research, are now commonly used in diagnosis, monitoring and therapies in many medical fields. Because they are sensitive, specific and require just little biomaterial, they are strongly desired for use in medicine, where fast and correct analysis of biomarkers is very important. Researchers in the preclinical field use nanobiosensors to get detailed molecular-scale information which helps them explain disease development and develop better drug therapies [74]. Nanobiosensors are used in making devices used to spot biomarkers related to heart conditions, diabetes, cancer, infectious diseases and metabolic syndrome in clinical analysis. Sensors based on nanomaterials like graphene and gold nanoparticles are popularly used for ongoing glucose monitoring (CGM) by diabetic patients. With these tools, people are able to control their blood sugar better, lowering the risk of related problems [75]. With the help of nanobiosensors, pathologists are able to discover circulating tumor DNA, microRNAs or different protein markers with great accuracy and usually sooner than when using standard approaches.

Nanobiosensors are tested before human trials on animals or cells to find out how a disease advances and reacts to certain treatments. These sensors make it possible to detect different biochemical values together, giving a thorough look at how body conditions affect health. Using nanobiosensors, investigators can study how glucose, lipids, adipokines and inflammatory cytokines change in the body of those with metabolic syndrome [76]. Many nanobiosensor devices for commercial use show the difference they make in clinical settings. Apart from blood sugar tests, nanobiosensors able to detect troponin and CRP are increasingly being used in quick emergency situations. The speed with which rapid antigen tests and biosensors were developed for COVID-19 using nanotechnology showed that they can be helpful in controlling infectious diseases. Still, it is difficult to use nanobiosensors routinely in healthcare due to getting approvals, standardization and long-lasting stability [77]. Nonetheless, new advancements in biocompatibility, miniaturization and combining nanotechnology with digital healthcare are solving these problems. Also, by adding artificial intelligence and machine learning to nanobiosensors, we improve how we read data and help both diagnostics and personalized care [78].

12 Challenges and Limitations

Even though nanobiosensors show great potential for tracking and monitoring metabolic syndrome, some issues and limitations stop them from being widely used in clinical care. A main worry is that nanomaterials in sensor fabrication may not be safe for living organisms and could be toxic. Certain metal nanoparticles, quantum dots and carbon nanotubes, if they touch biological tissue or bloodstream, can start oxidative stress, trigger inflammation or become harmful to cells [79]. Because of this, safety is a big issue when it comes to continuous use of nanobiosensors inside or on the body for monitoring metabolic markers. How nanomaterials may stay in the body for long periods is still not well understood which makes it necessary to study their safety in depth before their use in medical treatments [80]. Keeping nanobiosensors steady and ensuring their repeatability is also a major issue. Over time, nanomaterials could break down or combine which results in a decrease in the sensitivity and reliability of sensors. Problems controlled by environmental conditions like temperature, pH and ionic strength might result in errors in the sensor readings. This unpredictability makes it hard to get sensors ready which is needed for dependable results in diagnosing metabolic syndrome [81].

Nanobiosensor function is further complicated due to the biological environment of complex metabolic syndrome. However, biological fluids such as blood or saliva have many interfering substances and proteins that can bind nonspecifically to the sensor surface which causes fouling and reduces specificity. A technical hurdle remains in trying to improve sensor accuracy with real clinical samples, namely the design of antifouling coatings and selective bioreceptors [82].

From a regulatory and manufacturing point of view, the translation of nanobiosensors from lab prototypes toward market ready products raises regulatory scrutiny. Standardized protocols are lacking for the characterization (e.g., quantification and purity verification), quality control and clinical validation of nanomaterials; approvals are delayed and development costs increase. Additionally, mass fabrication of consistent quality for a cost that is affordable continues to be a challenge as well [83].

Additional concerns in regard to data security and user acceptance are introduced with data privacy and integration with digital health platforms. Ensuring patient trust and regulating compliance by safe transmission and storage of sensitive health data is of paramount importance [84].

13 Toxicity Impact

Because of nanobiosensors, it is now possible to detect metabolic syndrome quickly and with great sensitivity. Still, because nanomaterials may be harmful, many experts are worried about using these sensors over a long period of time. Most of the toxic issues linked to nanobiosensors result from the key features of nanomaterials, like size, form, surface charge, composition and how they are modified [85].

Nanobiosensors normally work with metal nanoparticles (for example, gold and silver), carbon-based nanomaterials (such as carbon nanotubes and graphene), quantum dots and metal oxides. Such components, although making sensors more efficient, might trigger cytotoxicity, oxidative stress, inflammation and change the genetic code in cells when they come into touch with them [86]. To give an example, metal nanoparticles can result in the formation of ROS which damages different parts of cells including lipids, proteins and DNA. Some carbon nanotubes may cause problems in lung tissue or result in inflammation when they are inhaled [87].

When it comes to diagnosing metabolic syndrome, nanotechnologies are commonly made as wearable or implanted devices which may lead to long-term contamination of the body. Drawing close contacts with skin or the blood vessels and internal organs for a long time may release harmful chemicals which could cause issues for the metabolic system rather than help in diagnosis. When nanomaterials are released into the environment because a device has been degraded or thrown out, they may be toxic to living organisms [88].

The environment of the body and the immune system's reaction to these materials should be carefully analyzed too. Immune system activation that was not meant to occur can affect the body or a local area, adding difficulty to managing some diseases, mainly in patients with metabolic syndrome who often live with low-grade inflammation. So, both the exterior and innermost structures of nanoparticles are altered and protected to help them interact better with the body [89].

To deal with toxicity, it is necessary to perform both in vitro and in vivo testing prior to using drugs in people. Documents are being developed to determine how to limit negative outcomes by using biodegradable materials and healthy methods of making nanoparticles. Authorities tend to require standard testing methods to determine how safe nanomaterials are in medical devices [90].

14 Future Prospects

Because of progress in nanotechnology, materials science and digital health, the future of nanobiosensors in this field is very promising. Because metabolic syndrome targets different markers in bodily fluids such as glucose, lipids and inflammatory cytokines future nanobiosensors are predicted to be advanced and sensitive enough to measure various markers at once with outstanding speed and accuracy [91].

An especially interesting direction is finding ways to put nanobiosensors in our clothing and bodies. With these devices, people can quickly and comfortably check their metabolic markers from sweat, saliva, in between the cells of the skin or even in breath. New advancements in nanomaterials will increase how comfortable, strong and long-lasting sensors are, enabling better and unnoticed tracking of chronic diseases. With these tools, patients could manage their own health quickly and get personalized suggestions to maintain normal blood sugar, lipid levels and reduce inflammation [92].

With the use of AI and ML, nanobiosensors will play even greater roles. With the help of large data sets, AI algorithms can discover subtle changes in metabolism, warn about early symptoms of a disease, predict its development and set up personalized treatments. Teaming nanobiosensors with smart phones and the internet will allow patients to receive remote care and better health services, especially where medical resources are limited [93].

Moreover, using lab-on-a-chip systems featuring microfluidics and nanobiosensors will allow fast testing of various samples at once with a small amount of fluid. Because of this, it can help diagnose metabolic syndrome faster and allow early treatment which improves how patients fare and saves money on care for related issues like cardiovascular disease and diabetes [94].

Moreover, technological progress in nanomaterial engineering will result in sensors that choose their signals more accurately, detect those signals with higher sensitivity and resist pollution from the fluids they collect. New materials including 2D nanomaterials, hybrid structures and nanocomposites inspired by nature are expected to perform better and be safer [95]. Research that continues to examine the effects of nanomaterials on the environment and so far, as their health impacts will improve the safety and sustainability of future nanobiosensors [96].

15 Conclusion

Nanobiosensors are proving to be very helpful in the detection of metabolic syndrome, a condition marked by many metabolic problems. At the same time, they can find a wide range of biomarkers with a high degree of accuracy and specificity which is required for early and correct identification of problems. Thanks to the special features of nanomaterials such as an enhanced surface, nanobiosensors can speedily, safely and continuously measure the blood levels of glucose, lipids, inflammatory cytokines and a range of relevant metabolic signs.

The use of nanobiosensors with modern tools like wearables, tiny fluid systems and online health systems allows care to be given in real-time and according to patients' needs. Because of these innovations, it will be easier to control and treat the factors of metabolic syndrome early on which should protect patients from diseases that are linked to them, like heart problems and diabetes.

Although sensor reliability, body tolerance and government approval remain issues now, recent progress is removing these obstacles. Very soon, these nanobiosensors might play an important role in daily hospital care and health programs, transforming the way metabolic syndrome is diagnosed and leading to new methods that focus on early detection, constant monitoring and customized treatments.

References

1. Khan, R. S., Khurshid, Z., & Asiri, F. Y. I. (2017). Advancing point-of-care (PoC) testing using human saliva as liquid biopsy. *Diagnostics, 7*(3). https://doi.org/10.3390/diagnostics7030039
2. Nag, S., et al. (2024). Autophagy and mitophagy as potential therapeutic targets in diabetic heart condition: Harnessing the power of nanotheranostics. *Asian Journal of Pharmaceutical Sciences, 19*(3). https://doi.org/10.1016/j.ajps.2024.100927
3. Manoharan Nair Sudha Kumari, S., & Thankappan Suryabai, X. (2024). Sensing the future—frontiers in biosensors: Exploring classifications, principles, and recent advances. *ACS Omega, 9*(50), 48918–48987. https://doi.org/10.1021/acsomega.4c07991
4. Liu, Y., et al. (2023). Nano-biosensor for SARS-CoV-2/COVID-19 detection: methods, mechanism and interface design. *RSC Advances, 13*(26), 17883–17906. https://doi.org/10.1039/d3ra02560h
5. Kumar, A., et al. (2025). Analysis of potential neuropharmacological activity and attenuating effect in chronic constriction induced neuropathic pain using Calotropis procera (Aiton) Dryand flower ethanol extract. *Pharmacological Research – Modern Chinese Medicine, 15*(February), 100624. https://doi.org/10.1016/j.prmcm.2025.100624
6. Singh, R. D., et al. (2018). Quantum dot based nano-biosensors for detection of circulating cell free miRNAs in lung carcinogenesis: From biology to clinical translation. *Frontiers in Genetics, 9*(December), 1–23. https://doi.org/10.3389/fgene.2018.00616
7. Fang, W., et al. (2023). Diagnosis of invasive fungal infections: challenges and recent developments. *Journal of Biomedical Science, 30*(1), 1–35. https://doi.org/10.1186/s12929-023-00926-2
8. Yadav, A. K., et al. (2022). Nanomaterial-based electrochemical nanodiagnostics for human and gut metabolites diagnostics: Recent advances and challenges. *Biosensors, 12*(9). https://doi.org/10.3390/bios12090733
9. Kumar, R., Kumar, A., Kumar, B., & Goyal, A. (2025). Phytochemical investigation and toxicity profiling of Premna latifolia Roxb. leaves extract in rodents. *Comparative Clinical Pathology*. https://doi.org/10.1007/s00580-025-03663-0
10. Khademi, R., Mohammadi, Z., Khademi, R., Saghazadeh, A., & Rezaei, N. (2023). Nanotechnology-based diagnostics and therapeutics in acute lymphoblastic leukemia: a systematic review of preclinical studies. *Nanoscale Advances, 5*(3), 571–595. https://doi.org/10.1039/d2na00483f
11. Nayab, D. E., et al. (2023). Nano biomaterials based strategies for enhanced brain targeting in the treatment of neurodegenerative diseases: an up-to-date perspective. *Journal of Nanobiotechnology, 21*(1), 1–29. https://doi.org/10.1186/s12951-023-02250-1
12. Dubey, D., et al. (2017). Neurological autoantibody prevalence in epilepsy of unknown etiology. *JAMA Neurology, 74*(4), 397–402. https://doi.org/10.1001/jamaneurol.2016.5429
13. Tiwari, S., Sharma, V., Mujawar, M., Mishra, Y. K., Kaushik, A., & Ghosal, A. (2019). Biosensors for epilepsy management: State-of-art and future aspects. *Sensors (Switzerland), 19*(7). https://doi.org/10.3390/s19071525
14. Hussain, K. K., et al. (2020). Biosensors and diagnostics for fungal detection. *Journal of Fungi, 6*(4), 1–26. https://doi.org/10.3390/jof6040349
15. Shandilya, R., Mishra, P. K., Pathak, N., Lohiya, N. K., & Sharma, R. S. (2020). Nanotechnology in reproductive medicine: Opportunities for clinical translation. *Clinical and Experimental Reproductive Medicine, 47*(4), 245–262. https://doi.org/10.5653/cerm.2020.03650
16. Ng, S., Strunk, T., Jiang, P., Muk, T., Sangild, P. T., & Currie, A. (2018). Precision medicine for neonatal sepsis. *Frontiers in Molecular Biosciences, 5*(July). https://doi.org/10.3389/fmolb.2018.00070
17. Piplani, M., Kumar, R., Bhateja, P., Pandey, P., & Yadav, K. (2025, February). Free radicals and AIDS. In *Role of free radicals in pathology* (pp. 581–611). https://doi.org/10.1201/9781003560579-24

18. Barani, M., et al. (2021). Nanomaterials in the management of gram-negative bacterial infections. *Nanomaterials, 11*(10), 1–26. https://doi.org/10.3390/nano11102535
19. Kumar, R., Dutt, S., Tripathi, A. D., Singh, A. K., Chaturvedi, V. K., & Singh, S. K. (2024). Navigating safety and toxicity challenges in nanomedicine: Strategies, assessment, and mitigation BT. In V. Gautam, R. Kumar, K. Das Manandhar, & S. C. Kamble (Eds.), *Nanomedicine: Innovations, applications, and breakthroughs in the quest for health and medicine's future* (pp. 15–37). Springer Nature Switzerland. https://doi.org/10.1007/978-3-031-72467-1_2
20. Upreti, P., Devhare, L. D., Abdulmageed, L. H., Kumar, Y. G., Kumar, R., & Dharmamoorthy, G. (2024). Combatting antibiotic resistance: Leveraging fecal microbial transplantation for gut health BT. In M. Gangwar & G. Nath (Eds.), *Emerging paradigms for antibiotic-resistant infections: Beyond the pill* (pp. 211–232). Springer Nature Singapore. https://doi.org/10.1007/978-981-97-5272-0_10
21. Zhang, F., Xu, W., Deng, Z., & Huang, J. (2024). A bibliometric and visualization analysis of electrochemical biosensors for early diagnosis of eye diseases. *Frontiers in Medicine, 11*-(January), 1–17. https://doi.org/10.3389/fmed.2024.1487981
22. Hamid, A., et al. (2024, August). Artificial intelligence (AI) based colonoscopy. In *Nanobiotechnology and artificial intelligence in gastrointestinal diseases*. https://doi.org/10.1088/978-0-7503-6134-7CH12
23. Mauriz, E. (2020). Clinical applications of visual plasmonic colorimetric sensing. *Sensors (Switzerland), 20*(21), 1–31. https://doi.org/10.3390/s20216214
24. Abbasi, R., Hu, X., Zhang, A., Dummer, I., & Wachsmann-Hogiu, S. (2024). Optical image sensors for smart analytical chemiluminescence biosensors. *Bioengineering, 11*(9), 912. [Online]. Available: https://www.mdpi.com/2306-5354/11/9/912
25. Nanda, B. P., Rani, P., Paul, P., Aman, Subrahmanya S, G., & Bhatia, R. (2024). Recent trends and impact of Localized Surface Plasmon Resonance (LSPR) and Surface-Enhanced Raman Spectroscopy (SERS) in modern analysis. *Journal of Pharmaceutical Analysis, 14*(11), 100959. https://doi.org/10.1016/j.jpha.2024.02.013
26. Sadeghi, S., Rahaie, M., & Ostad-Hasanzadeh, B. (2022). Nanostructures in non-invasive prenatal genetic screening. *Biomedical Engineering Letters, 12*(1), 3–18. https://doi.org/10.1007/s13534-021-00208-6
27. Kumar, R., Dutt, S., Goyal, A., Kumar, A., & Kumar, B. (2024, August). Nanotechnology in gastrointestinal endoscopy. In *Nanobiotechnology and artificial intelligence in gastrointestinal diseases*. https://doi.org/10.1088/978-0-7503-6134-7CH5
28. Li, Y., et al. (2023). Needs, challenges and countermeasures of SARS-CoV-2 surveillance in cold-chain foods and packaging to prevent possible COVID-19 resurgence: A perspective from advanced detections. *Viruses, 15*(1). https://doi.org/10.3390/v15010120
29. Kumar, R., Singh, Y., Piplani, M., Bhateja, P., Garg, R., & Kumar, B. (2024). The chemistry and pharmacological study of Copaifera spp. oils. *The Chemistry inside Spices & Herbs: Research and Development, 4*, 79–116. https://doi.org/10.2174/9789815196832124040006
30. Umapathy, V. R., et al. (2022). Emerging biosensors for oral cancer detection and diagnosis—A review unravelling their role in past and present advancements in the field of early diagnosis. *Biosensors, 12*(7). https://doi.org/10.3390/bios12070498
31. Singh, A., et al. (2024). Synthesis, crystal structure and screening for anticonvulsant and antianxiety activities of three new Ni(II), Cu(II), and Zn(II) dithiocarbamate complexes. *Journal of Molecular Structure, 1298*(P2), 137052. https://doi.org/10.1016/j.molstruc.2023.137052
32. Khondakar, K. R., & Kaushik, A. (2023). Role of wearable sensing technology to manage long COVID. *Biosensors, 13*(1). https://doi.org/10.3390/bios13010062
33. de Araujo, W. R., Lukas, H., Torres, M. D. T., Gao, W., & de la Fuente-Nunez, C. (2024). Low-cost biosensor technologies for rapid detection of COVID-19 and future pandemics. *ACS Nano, 18*(3), 1757–1777. https://doi.org/10.1021/acsnano.3c01629
34. Wright, J. M., & Jones, G. B. (2018). Harnessing the digital exhaust: Incorporating wellness into the pharma model. *Digital Biomarkers, 2*(1), 31–46. https://doi.org/10.1159/000488132

35. Kumar, A., Singh, P., Kumar, R., & Dutt, S. (2024, August). Novel drug delivery systems for inflammatory bowel disease. In *Nanobiotechnology and artificial intelligence in gastrointestinal diseases*. https://doi.org/10.1088/978-0-7503-6134-7CH4
36. Macovei, D. G., Irimes, M. B., Hosu, O., Cristea, C., & Tertis, M. (2023). Point-of-care electrochemical testing of biomarkers involved in inflammatory and inflammatory-associated medical conditions. *Analytical and Bioanalytical Chemistry. Springer Berlin Heidelberg, 415*(6). https://doi.org/10.1007/s00216-022-04320-z
37. Ahmad, S., Lohiya, S., Taksande, A., Meshram, R. J., Varma, A., & Vagha, K. (2024). A comprehensive review of innovative paradigms in microbial detection and antimicrobial resistance: Beyond traditional cultural methods. *Cureus, 16*(6), 1–8. https://doi.org/10.7759/cureus.61476
38. Singh, A., et al. (2024). Dithiocarbamate-based novel anti-histaminic agents: synthesis, characterization, crystal structure and thermal study. *Dalton Transactions*. https://doi.org/10.1039/d4dt01777c
39. Faiyazuddin, M., et al. (2023). Virulence traits and novel drug delivery strategies for mucormycosis post-COVID-19: a comprehensive review. *Frontiers in Immunology, 14-*(September), 1–21. https://doi.org/10.3389/fimmu.2023.1264502
40. Samson, R., & Dharne, M. (2022). COVID-19 associated mucormycosis: evolving technologies for early and rapid diagnosis. *3 Biotech, 12*(1), 1–17. https://doi.org/10.1007/s13205-021-03080-4
41. Singh, A., Prasad, L. B., Shiv, K., Kumar, R., & Garai, S. (2023). Synthesis, characterization, and in vitro antibacterial and cytotoxic study of Co(II), Ni(II), Cu(II), and Zn(II) complexes of N-(4-methoxybenzyl) N-(phenylethyl) dithiocarbamate ligand. *Journal of Molecular Structure, 1288*(May), 135835. https://doi.org/10.1016/j.molstruc.2023.135835
42. Kim, B., Yang, Q., Chan, L. W., Bhatia, S. N., Ruoslahti, E., & Sailor, M. J. (2021). Fusogenic porous silicon nanoparticles as a broad-spectrum immunotherapy against bacterial infections. *Nanoscale Horizons, 6*(4), 330–340. https://doi.org/10.1039/d0nh00624f
43. Singh, M., et al. (2023, July). Pharmacokinetic screening to estimate the drug likeliness characteristics of selected Herbal Anticancer Drugs. *Research Journal of Pharmacy and Technology, 16*(7), 3422–3426. https://doi.org/10.52711/0974-360X.2023.00566
44. Bonini, A., et al. (2022). Emerging biosensing technologies towards early sepsis diagnosis and management. *Biosensors, 12*(10), 1–22. https://doi.org/10.3390/bios12100894
45. Garg, R., et al. (2022). COVID-19: A challenge towards the sustainability of health in platform era. *NeuroQuantology, 20*(11), 2503–2519. https://doi.org/10.14704/NQ.2022.20.11.NQ66250
46. Kumar, A., et al. (2022). Assessment of pain-ameliorative potential of ethanolic extract of Calotropis procera and its active components characterization by HPTLC and GC-MS analysis. *Systematic Reviews in Pharmacy, 13*(09), 581–586. https://doi.org/10.31858/0975-8453.13.6.659-664
47. Dhiman, R., Kumar, B., Kumar, D., & Kumar, A. (2016). Liquid-liquid extraction method for Ziprasidone (ZRS) bioanalysis by using ZRS-D 8 (stable isotope) as internal standard. *Journal of Chemical and Pharmaceutical Research, 8*(3), 722–727.
48. Boselli, L., Pomili, T., Donati, P., & Pompa, P. P. (2021). Nanosensors for visual detection of glucose in biofluids: Are we ready for instrument-free home-testing? *Materials (Basel), 14*(8). https://doi.org/10.3390/ma14081978
49. Kumar, R., et al. (2021). HPTLC and GC-MS analyses of biologically active extracts and fractions from Premna latifolia Roxb leaves reveal anti-inflammatory and analgesic effects. *Systematic Reviews in Pharmacy, 12*(10), 811–818.
50. Kumar, A., et al. (2022). Acute and subacute toxicity study of ethanolic extract of Calotropis procera (Aiton) Dryand flower in Swiss albino mice. *Phytomedicine Plus, 2*(2), 100224. https://doi.org/10.1016/j.phyplu.2022.100224
51. Dhiman, R., Kumar, D., Kumar, B., & Pandey, B. L. (2016). Quantitative determination of atorvastatin, ortho-hydroxy atorvastatin, para-hydroxy atorvastatin in human plasma using

rosuvastatin as internal standard by LC-MS/MS. *International Journal of Pharmaceutical Chemistry Science, 4*(4), 487–500.

52. Vural, B., Uludag, I., Ince, B., Ozyurt, C., Ozturk, F., & Sezginturk, M. K. (2023). Fluid-based wearable sensors: A turning point in personalized healthcare. *Turkish Journal of Chemistry, 47*(5), 944–967. https://doi.org/10.55730/1300-0527.3588
53. Kumar, A., et al. (2022, February). Toxicity (acute and subacute) assessment and in-vivo antiurolithiatic activity of ethanolic extract of Caesalpinia bonducella seed in albino Wistar rat. *Journal of Applied Pharmaceutical Science, 12*(2), 187–197. https://doi.org/10.7324/JAPS.2022.120220
54. Kim, J. H., et al. (2021). Technological advances in electrochemical biosensors for the detection of disease biomarkers. *Biomedical Engineering Letters, 11*(4), 309–334. https://doi.org/10.1007/s13534-021-00204-w
55. Kumar, R., Kumar, B., Kumar, A., Kumar, A., & Singh, M. (2022). GC-MS analysis of phytocomponents in the methanol extract of Premna latifolia Roxb. *Pharmacognosy Research, 14*(1), 19–23. https://doi.org/10.5530/pres.14.1.4
56. Vijaya Durga, R. D. (2015). Determination of S-amlodipine in human plasma using tizanidine as internal standard by Lc/Ms/Ms method. *Journal of Bioanalysis & Biomedicine, 07*(03), 91–96. https://doi.org/10.4172/1948-593x.1000130
57. Gacche, R., Birge, R. B., & Chandra, D. (2024). Biology beyond boundaries: Mitochondrial insights, computational breakthroughs, and clinical transformations: International conference hosted by Savitribai Phule Pune University, Pune, India, January 29–31, 2024. *Cell Communication and Signaling, 22*(1). https://doi.org/10.1186/s12964-024-01764-x
58. Kumar, A., et al. (2021). Antiurolithiasis, antioxidant, anti-inflammatory, analgesic, and diuretic activity of ethanolic extract of seeds of Caesalpinia bonducella. *International Journal of Pharmaceutical Investigation, 11*(3), 306–311. https://doi.org/10.5530/ijpi.2021.3.54
59. Zhang, Q., Zhao, G., Wang, S., Song, Y., & Sun, Y. (2025). Polymer-based electrochemical sensors for the diagnosis of neurodegenerative diseases. *Cellular and Molecular Neurobiology. Springer US, 45*(1). https://doi.org/10.1007/s10571-025-01570-0
60. Dhiman, R., Durga, V., & Rio, A. J. (2015). Bioanalytical method for lornoxicam determination in human plasma by using piroxicam as internal standard by LC-MS/MS. *Journal of Chemical and Pharmaceutical Research, 7*(8), 54–60.
61. Azab, M. Y., Hameed, M. F. O., & Obayya, S. S. A. (2023). Overview of optical biosensors for early cancer detection: Fundamentals, applications and future perspectives. *Biology (Basel), 12*(2). https://doi.org/10.3390/biology12020232
62. Kumar, R., Kumar, B., Kumar, A., Kumar, A., & Singh, M. (2021). HPTLC and GC-MS analyses of biologically active extracts and fractions from Premna Latifolia Roxb leaves. *Journal of Neurological Disorders, 9*, 454.
63. El-Safty, S. A., & Shenashen, M. A. (2020). Nanoscale dynamic chemical, biological sensor material designs for control monitoring and early detection of advanced diseases. *Materials Today Bio, 5*(February), 100044. https://doi.org/10.1016/j.mtbio.2020.100044
64. Dhiman, R., Durga, V., Jayasankar, K., & Raj, S. (2015). Solid phase extraction: Potential carbocisteine bioanalysis using carbocisteine 13c3 stable isotope: LC-MS/MS technique with pharmacokinetics on healthy human subjects. *Der Pharmacia Lettre, 7*(8), 126–134.
65. Gomez-Villalba, L. S., Salcines, C., & Fort, R. (2023). Application of inorganic nanomaterials in cultural heritage conservation, risk of toxicity, and preventive measures. *Nanomaterials, 13*(9). https://doi.org/10.3390/nano13091454
66. Dhiman, R., et al. (2015). Novel LC-MS/MS bioanalytical method validation for gabapentin in human plasma in accordance with current regulatory guidances. *International Journal of Pharmaceutical and Chemical Sciences, 4*(4), 462–468. [Online]. Available: http://www.ijpcsonline.com/files/08-10-15/07-1035.pdf
67. Pandey, H., Srivastava, S., Kumar, A., Kumar, R., & Tripathi, Y. B. (2019). Pharmacokinetics study of puerarin absorption in blood after consumption of Pueraria tuberosa water extract (PTWE) by rats. *Exploration of Animal Medicine Research, 9*(2), 188–196.

68. Pandey, H., Srivastava, S., Kumar, R., & Tripathi, Y. B. (2018). Preclinical acute and repeated dose toxicity of Pueraria tuberosa (PTWE) on Charles Foster rats. *International Journal of Pharmaceutical Sciences and Research, 9*(11), 4572–4581. https://doi.org/10.13040/IJPSR.0975-8232.9(11).4572-81
69. Barhoum, A., et al. (2022). Review on natural, incidental, bioinspired, and engineered nanomaterials: History, definitions, classifications, synthesis, properties, market, toxicities, risks, and regulations. *Nanomaterials, 12*(2). https://doi.org/10.3390/nano12020177
70. Dahlan, N. A., et al. (2022). Role of nanomaterials in the fabrication of bioNEMS/MEMS for biomedical applications and towards pioneering food waste utilisation. *Nanomaterials, 12*(22). https://doi.org/10.3390/nano12224025
71. Harun-Ur-Rashid, M., Jahan, I., Foyez, T., & Bin Imran, A. (2023). Bio-inspired nanomaterials for micro/nanodevices: A new era in biomedical applications. *Micromachines, 14*(9). https://doi.org/10.3390/mi14091786
72. Shao, Y., et al. (2021). Shining light on chiral inorganic nanomaterials for biological issues. *Theranostics, 11*(19), 9262–9295. https://doi.org/10.7150/thno.64511
73. Kumar, R., Kumar, B., & Kumar, A. (2017). Pharmacokinetic profile of Antitubercular drugs and their clinical applications in therapeutic drug monitoring. *Asian Journal of Pharmacy and Pharmacology, 3*(5), 145–154. [Online]. Available: http://ajpp.in/uploaded/p71.pdf
74. Anžlovar, A., & Žagar, E. (2022). Cellulose structures as a support or template for inorganic nanostructures and their assemblies. *Nanomaterials, 12*(11). https://doi.org/10.3390/nano12111837
75. Parvin, N., Kumar, V., Joo, S. W., & Mandal, T. K. (2024). Emerging trends in nanomedicine: Carbon-based nanomaterials for healthcare. *Nanomaterials, 14*(13). https://doi.org/10.3390/nano14131085
76. Dediu, V., Ghitman, J., Gradisteanu Pircalabioru, G., Chan, K. H., Iliescu, F. S., & Iliescu, C. (2023). Trends in photothermal nanostructures for antimicrobial applications. *International Journal of Molecular Sciences, 24*(11), 1–41. https://doi.org/10.3390/ijms24119375
77. Ferrone, E., Araneo, R., Notargiacomo, A., Pea, M., & Rinaldi, A. (2019). ZnO nanostructures and electrospun ZnO–polymeric hybrid nanomaterials in biomedical, health, and sustainability applications. *Nanomaterials, 9*(10), 1–33. https://doi.org/10.3390/nano9101449
78. Zhang, D., Chen, Y., Hao, M., & Xia, Y. (2024). Putting hybrid nanomaterials to work for biomedical applications. *Angewandte Chemie International Edition, 63*(16). https://doi.org/10.1002/anie.202319567
79. Hekmat, A., Saso, L., Lather, V., Pandita, D., Kostova, I., & Saboury, A. A. (2022). Recent advances in nanomaterials of group XIV elements of periodic table in breast cancer treatment. *Pharmaceutics, 14*(12). https://doi.org/10.3390/pharmaceutics14122640
80. Ranjha, M. M. A. N., et al. (2022). Biocompatible nanomaterials in food science, technology, and nutrient drug delivery: Recent developments and applications. *Frontiers in Nutrition, 8*(January), 1–15. https://doi.org/10.3389/fnut.2021.778155
81. Anastasiadis, S. H., Chrissopoulou, K., Stratakis, E., Kavatzikidou, P., Kaklamani, G., & Ranella, A. (2022). How the physicochemical properties of manufactured nanomaterials affect their performance in dispersion and their applications in biomedicine: A review. *Nanomaterials, 12*(3). https://doi.org/10.3390/nano12030552
82. Hu, X., et al. (2021). Tailor-made nanomaterials for diagnosis and therapy of pancreatic ductal adenocarcinoma. *Advancement of Science, 8*(7). https://doi.org/10.1002/advs.202002545
83. Boros, B. V., & Ostafe, V. (2020). Evaluation of ecotoxicology assessment methods of nanomaterials and their effects. *Nanomaterials, 10*(4). https://doi.org/10.3390/nano10040610
84. Zhu, L., et al. (2021). Applications of aptamer-bound nanomaterials in cancer therapy. *Biosensors, 11*(9). https://doi.org/10.3390/bios11090344
85. McCourt, K. M., et al. (2022). Potential environmental and health implications from the scaled-up production and disposal of nanomaterials used in biosensors. *Biosensors, 12*(12). https://doi.org/10.3390/bios12121082

86. Kumar, S., et al. (2023). Optically active nanomaterials and its biosensing applications—A review. *Biosensors, 13*(1), 1–38. https://doi.org/10.3390/bios13010085
87. Dastgheib, Z. S., Abolmaali, S. S., Farahavar, G., Salmanpour, M., & Tamaddon, A. M. (2024). Gold nanostructures in melanoma: Advances in treatment, diagnosis, and theranostic applications. *Heliyon, 10*(15), e35655. https://doi.org/10.1016/j.heliyon.2024.e35655
88. Harish, V., et al. (2022). Review on nanoparticles and nanostructured materials: Bioimaging, biosensing, drug delivery, tissue engineering, antimicrobial, and agro-food applications. *Nanomaterials, 12*(3). https://doi.org/10.3390/nano12030457
89. Butler, J., Handy, R. D., Upton, M., & Besinis, A. (2023). Review of antimicrobial nanocoatings in medicine and dentistry: Mechanisms of action, biocompatibility performance, safety, and benefits compared to antibiotics. *ACS Nano, 17*(8), 7064–7092. https://doi.org/10.1021/acsnano.2c12488
90. Krishnan, S. K., Singh, E., Singh, P., Meyyappan, M., & Nalwa, H. S. (2019). A review on graphene-based nanocomposites for electrochemical and fluorescent biosensors. *RSC Advances, 9*(16), 8778–8781. https://doi.org/10.1039/c8ra09577a
91. McInnes, S. J. P., & Voelcker, N. H. (2009). Silicon-polymer hybrid materials for drug delivery. *Future Medicinal Chemistry, 1*(6), 1051–1074. https://doi.org/10.4155/fmc.09.90
92. Govindan, B., Sabri, M. A., Hai, A., Banat, F., & Haija, M. A. (2023). A review of advanced multifunctional magnetic nanostructures for cancer diagnosis and therapy integrated into an artificial intelligence approach. *Pharmaceutics, 15*(3). https://doi.org/10.3390/pharmaceutics15030868
93. Kyriakides, T. R., et al. (2021). Biocompatibility of nanomaterials and their immunological properties. *Biomedical Materials, 16*(4), 1–49. https://doi.org/10.1088/1748-605X/abe5fa
94. Safarkhani, M., et al. (2024). Advancements in aptamer-driven DNA nanostructures for precision drug delivery. *Advancement of Science, 11*(26), 1–30. https://doi.org/10.1002/advs.202401617
95. Tsukanov, A. A., Turk, B., Vasiljeva, O., & Psakhie, S. G. (2022). Computational indicator approach for assessment of nanotoxicity of two-dimensional nanomaterials. *Nanomaterials, 12*(4). https://doi.org/10.3390/nano12040650
96. Wan, T., et al. (2023). Tissue-engineered nanomaterials play diverse roles in bone injury repair. *Nanomaterials, 13*(9). https://doi.org/10.3390/nano13091449

Nanomaterials in Targeted Drug Delivery for Metabolic Syndrome

5

Gitali Sen, Pradeep Kumar Patel, Soki Daeme Malang, Pooja Kathait, Ankur, Navneet Yadav, Sunday O. Otimenyin, and Alakh N. Sahu

Abstract

The prevalence of metabolic syndrome is rising significantly, driven by factors such as urbanization, movement-deficient lifestyles, and excessive energy consumption. Metabolic syndrome, characterized by multifactorial conditions including obesity, insulin resistance, dyslipidaemia, and hypertension, poses significant challenges in treatment due to its multifactorial nature and systemic effects. Conventional therapeutic approaches for metabolic syndrome frequently face challenges of low bioavailability, off-target effects, and poor patient compliance. Nanoparticles are utilized to enhance treatment effectiveness, reduce side effects, and minimize drug dosage. Their small size, high permeability, and stability enable effective absorption by target organs. Various nanoparticles, differing in usage and particle size, have been studied for this purpose. Advanced nanocarriers, including lipid-based carrier systems like liposomes, polymeric nanoparticles, dendrimers, and lipid-based nanostructures, micelles, and niosomes, reported good efficacy in delivering bioactive molecules to specific tissues, particularly adipose tissue, liver, and vascular endothelium which are the key sites of metabolic syndrome pathology. Functionalization of nanocarriers with ligands like peptides, and antibodies, enhances their ability to selectively

G. Sen · P. K. Patel · S. D. Malang · P. Kathait · Ankur · A. N. Sahu (✉)
Phytomedicine Research Laboratory, Department of Pharmaceutical Engineering & Technology, IIT (BHU), Varanasi, India
e-mail: ansahu.phe@iitbhu.ac.in

N. Yadav
Department of Mechanical Engineering, Faculty of Science and Engineering, Swansea University, Sketty, Swansea, UK

S. O. Otimenyin
Department of Pharmacology, Faculty of Pharmaceutical Sciences, University of Jos, Jos, Plateau, Nigeria

V. K. Chaturvedi et al. (eds.), *Nanotheranostics in Metabolic Syndrome*,
https://doi.org/10.1007/978-981-95-3746-4_5

attach to overexpressed receptors on diseased cells. Moreover, stimuli-responsive nanomaterials enable controlled drug release in response to pH, temperature, or enzymatic activity, further improving precision. In conclusion, the inclusion of nanotechnology advanced in enhancing the efficacy of various medications. By preventing drug degradation and enabling controlled sustained release, nanotechnology significantly improves drug performance. These advancements not only reduce the required dosage but also minimize side effects, making nanotechnology a valuable tool in the treatment of complex conditions like metabolic syndrome.

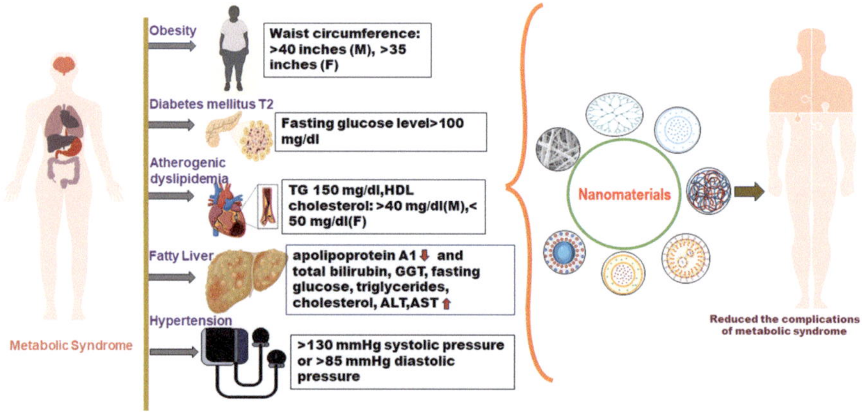

Keywords

Metabolic syndrome · Nanoparticle · Nanocarrier · Nanotechnology · Bioavailability

1 Introduction

1.1 Overview of Metabolic Syndrome

Metabolic syndrome (MetS) is a group of interconnected metabolic abnormal characteristics with high socioeconomic costs that altogether promote the risk of cardiovascular disease (CVD), type 2 diabetes mellitus (T2DM), and other health-related complications. It is marked by multiple risk factors making it a complicated disorder. It has a significant economic impact and is a growing global issue [1]. Visceral obesity, abnormal cholesterol levels, elevated blood sugar, elevated blood pressure, and other characteristics of metabolic syndrome have emerged as one of the world's most remarkable public health matters. While experts have known about these risk factors for over 80 years, the clustering was not given much attention until 1988, when Reaven introduced 'syndrome X', which includes elevated very low density lipoproteins (VLDL) triglycerides, resistance of insulin, hyperglycaemia, hypertension, along with low high density lipoproteins (HDL) cholesterol as key features. Interestingly, he left out obesity, particularly visceral obesity in his description, which is now considered a crucial component of the syndrome [2]. Non-alcoholic fatty liver disease (NAFLD) is characterized by an accumulation of too much fat in the liver, which is detected through several imaging techniques or a liver biopsy [3]. In individuals who drink little to no alcohol, it is the term used to describe a variety of cytologic abnormalities, from benign steatosis to non-alcoholic steatohepatitis (NASH). NAFLD has a severe issue of developing into hepatocellular carcinoma, cirrhosis, end-stage liver disease, liver-related death, and potential reappearance even after a liver transplant [4]. A particular limit for alcohol consumption could not be established, so the term 'fatty liver' was first used in 1836. In contrast, the term 'NAFLD' was first introduced in 1980 and was used for 40 years until 2020, when it was re-examined for a better reflection that it relates to metabolic issues showing the presence of metabolic dysfunction rather than the absence of other conditions, like alcohol intake [5]. NAFLD was ultimately again classified as metabolic-associated fatty liver disease (MAFLD) to more accurately represent its underlying pathophysiology and metabolic origins [3]. The easiest way to explain the metabolic syndrome is to think of abdominal adipose fats as an endocrine organ that releases excess damaging free fatty acid (FFA), angiotensin II, and adipokines into the bloodstream. First, the elevated blood FFA blocks muscles from taking up glucose. Angiotensin II and excess FFA harm the pancreas. The paradox of fasting hyperglycaemia despite elevated plasma insulin levels, known as insulin resistance, is explained by the fact that even if the pancreas secrets more insulin, it is not enough to combat the high blood sugar, which leads to the issue of high blood sugar levels even with more insulin present this is called insulin resistance. Because of the vasoconstrictive properties, angiotensin II raises blood pressure. Additionally, substances like tumour necrosis factor-α, inflammatory cytokines, and other cytokines (interleukins) also reduce the effectiveness of insulin and may increase blood pressure. The high blood sugar and elevated levels of FFA modulate the right conditions for the liver to produce more triglycerides. As triglycerides increase in the blood, lipoproteins end up carrying more triglycerides and less HDL cholesterol [6]. Individuals are diagnosed with (MetS) if they exhibit at least three of the succeeding risk factors which include abdominal obesity (an accumulation of excess fat around the waist), high triglyceride levels, high blood pressure, low levels of HDL cholesterol, or impaired glucose tolerance [7].

1.2 Challenges in Current Drug Delivery Systems

Traditional drug delivery systems (TDDS), including tablets, capsules, syrups, ointments, and injectables, have long served as the predominant modalities for pharmaceutical administration because of their accessibility and cost-effectiveness. On the other hand, these conventional approaches exhibit several inherent limitations, such as poor bioavailability, fluctuations in systemic drug concentrations, and the necessity for frequent dosing to maintain therapeutic efficacy. These restrictions frequently lead to worse effective therapeutic results, more adverse effects, and lower patient adherence to medication. Maintaining drug concentrations within the therapeutic range is challenging with TDDS since these do not have mechanisms for prolonged release. These have limitations including poor absorption and lack of target specificity, even though they offer ease and precise dosage [8]. Solubility and permeability issues are often observed in commercially available anti-diabetic medications, especially sulfonylureas, which leads to lower bioavailability. This results in the need for more frequent dosing, increasing the likelihood of patient non-compliance and the chance of missing a dose [9]. The drawbacks associated with other anti-diabetic medications, such as thiazolidinediones like pioglitazone, include a short half-life (3–5 hours), which leads to low bioavailability and diminished therapeutic effectiveness [9, 10].

1.3 Importance of Nanomaterials in Drug Delivery

The unique properties of nanomaterials such as enhanced targeting, increased bioavailability, controlled release, reduced side effects etc. make them a promising tool for enhancing the efficiency and safety of drug delivery systems. Natural product-based nanoformulations should be nontoxic, biodegradable, and biocompatible. Additionally, they should have a high capacity to penetrate cells, act quickly, and remain stable after oral administration [11]. When developing nanodrugs, there are multiple strategies to increase the therapeutic efficacy. Examples include the encapsulation of drug molecules in nanoparticles (NPs) to shield them from harsh physiological conditions, and using surface conjugation to direct drugs to specific tissues. This method promotes more therapeutic concentrations at the targeted site even with reduced doses while taking advantage of the small size of NPs to navigate significant physiological obstacles, such as the immune system, renal clearance, and enzymatic or mechanical degradation, among others [12].

2 Understanding Metabolic Syndrome

2.1 Definition and Diagnostic Criteria

MetS, also familiar with other terms such as insulin resistance syndrome (IRS) and the 'Deadly Quartet', is a clinical condition defined by abdominal obesity (involving visceral and retroperitoneal fat) along with a combination of cardiovascular risk factors. These include impaired glucose metabolism, high-density lipoprotein cholesterol (HDL-C) reduction, high arterial blood pressure, elevated triglyceride levels, and hyperinsulinaemia driven by primary insulin resistance. MetS is a multifactorial ailment characterized by a constellation of mutually dependent risk factors which tend to manifest concurrently rather than as a singular pathological condition. Its diagnosis serves as a pivotal marker of heightened cardiovascular and metabolic risk, underscoring the necessity for early therapeutic interventions to mitigate the progression of severe health complications [13].

Over the past few decades, multiple diagnostic criteria have been proposed for metabolic syndrome. In 2001, the National Cholesterol Education Program (NCEP) Adult Treatment Panel III (ATP III) established a straightforward diagnostic framework utilizing commonly assessed clinical markers, such as waist circumference, triglyceride levels, HDL-C level, arterial blood pressure, and fasting glucose levels. MetS is diagnosed when abnormalities are detected in at least three of these five indicators. The ATP III criteria have been mostly adopted in both clinical practice and research due to their practicality and comprehensive approach, avoiding attention on a single cause. Given the lack of strong scientific evidence for significant changes, both the American Heart Association (AHA) and the National Heart, Lung, and Blood Institute (NHLBI) support the continued use of the ATP III criteria with minor modifications and additional clarifications [14].

2.2 Pathophysiology of Metabolic Syndrome

The primary markers of MetS in an individual include elevated blood pressure, hyperglycaemia, and increased triglyceride levels. Under normal circumstances, blood glucose levels are typically maintained, ranging between 80 and 90 mg/dL in a fasting condition before breakfast. Postprandial, these levels rise to about 120–140 mg/dL within an hour. However, feedback mechanisms work to restore normal glucose concentrations, although the baseline may be slightly elevated in individuals with diabetes. Persistently high blood glucose levels can affect various tissues, mostly blood vessels, potentially leading to conditions such as heart attacks, strokes, and kidney disease [15] (Fig. 5.1 and Table 5.1).

Although not all individuals with overweight or obesity experience metabolic dysfunction, the majority exhibit insulin resistance [1]. Central obesity is recognized as a key contributor to metabolic syndrome, as visceral adipose tissue secretes various bioactive molecules, known as adipocytokines. These include tumour necrosis factor-alpha (TNF-α), leptin, interleukin-6 (IL-6), and angiotensin II, all of which contribute to the development of insulin resistance. Additionally, plasminogen activator inhibitor-1 (PAI-1), a protein associated with thrombogenic vascular disorders, is elevated in this state. Conversely, adiponectin, a crucial adipocytokine with protective properties against T2DM, inflammation, hypertension, and atherosclerotic vascular diseases, is markedly reduced in individuals with excess visceral fat. This decline may contribute to the pathogenesis of MetS [16].

In insulin resistance, the PI3K-Akt signalling pathway is impaired, whereas the MAP kinase pathway remains unaffected, leading to a dysregulation between the two signalling cascades. Suppression of the PI3K-Akt pathway reduces endothelial nitric oxide (NO) production, contributing to endothelial dysfunction and impairing the translocation of GLUT4. This, in turn, lowers glucose uptake in skeletal muscle and in adipose tissue. Conversely, the MAP kinase pathway remains active, promoting endothelin-1 (ET-1) production, upregulating vascular cell adhesion molecule

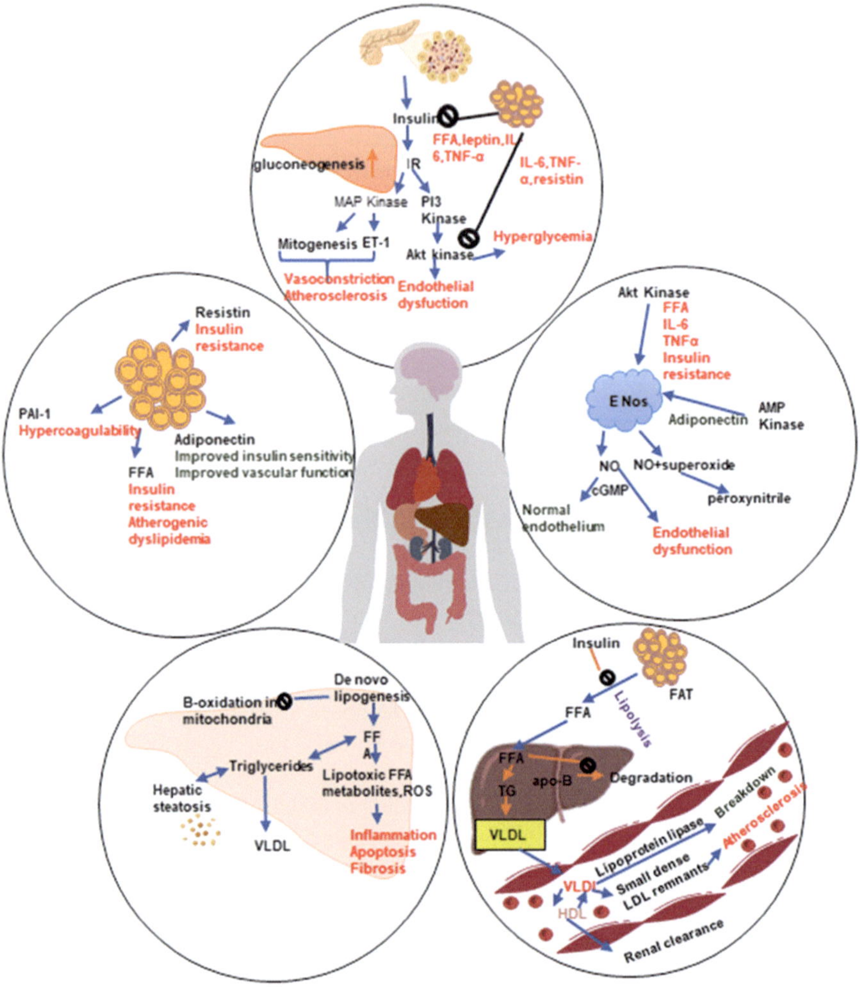

Fig. 5.1 Pathophysiology of metabolic syndrome

expression, and delivering mitogenic signals to vascular smooth muscle cells. Collectively, these alterations induced by insulin resistance contribute to vascular dysfunction and elevate the risk of atherosclerosis [17].

Atherogenic dyslipidaemia is marked by increasing serum triglyceride levels, decreased HDL cholesterol, and rise in small, dense low-density lipoprotein (LDL). This lipid profile is predominantly associated with insulin resistance and increased visceral adiposity [17]. Insulin resistance contributes to atherogenic dyslipidaemia through multiple mechanisms. In normal conditions, insulin suppresses lipolysis in adipocytes, but when insulin signalling is hindered, lipolysis is augmented, leading to elevated FFA levels. In the liver, FFAs provide substrates for triglyceride synthesis and promote the production of apolipoprotein B (apoB), the primary protein

Table 5.1 Criteria/Definition of metabolic syndrome

Disorder	NCEP ATP III (2005 revision)	WHO (1998)	EGIR (1999)	IDF (2005)
Visceral obesity	Waistline measurement: >40 inches (male), >35 inches (female)	Waist/hip index: >0.90 (Male), >0.85 (Female) and Body mass index >30 kg/m^2	Waistline measurement: ≥94 cm (male), ≥80 cm (female)	Visceral obesity is needed
Hyperglycaemia	Fasting blood sugar level ≥100 mg/dL	Insulin resistance is mandatory	Insulin resistance is already required	Fasting blood sugar level ≥100 mg/dL
Dyslipidaemia	Serum triglycerides ≥150 mg/dL	Serum triglyceride ≥150 mg/dL and serum high-density lipoprotein-cholesterol level: <35 mg/dL (male), <39 mg/dL (female)	Serum triglyceride ≥177 mg/dL and high-density lipoprotein-cholesterol <39 mg/dL	Blood serum triglyceride ≥150 mg/dL
Atherogenic dyslipidaemia	Serum high-density lipoprotein cholesterol: <40 mg/mL (male), <50 mg/dL (female)			Blood serum high-density lipoprotein cholesterol level: <40 mg/dL (male), <50 mg/dL (female)
Hypertension	>130 mmHg systolic and >85 mmHg diastolic	≥140/90 mmHg	≥140/90 mmHg	>130 mmHg systolic and >85 mmHg diastolic

molecule of VLDL particles, thereby increasing VLDL secretion. Additionally, insulin typically facilitates the breakdown of apoB via PI3K-dependent pathways, and insulin resistance disrupts this process, further driving the overproduction of VLDL. Lastly, Insulin helps regulate the function of lipoprotein lipase, the key enzyme responsible for clearing VLDL. Consequently, hypertriglyceridemia associated with insulin resistance arises from both the elevation in the production of VLDL and the decreased clearance of VLDL [18].

Hypertension is associated with at least four vascular abnormalities that contribute to its pathophysiology. The first abnormality involves elevated levels of vasoconstrictor hormones, such as catecholamines, angiotensin II, and vasopressin act collectively in alterations of vascular function that modulate vasoconstriction while impairing vasodilation. Several studies on isolated vessels have indicated that both conduit arteries and arterioles from hypertensive animal models exhibit heightened vasoconstrictor responses to various stimuli. Genes involved in regulating G protein-coupled receptor (GPCR) signalling, particularly those encoding regulators of G protein signalling (RGS) proteins, play a critical role in controlling GTP hydrolysis and terminating G protein signalling. RGS proteins 1 and 2 are particularly associated with GPCRs that amplify the vascular response to vasoconstrictors, such as angiotensin II, thromboxane, and norepinephrine [19].

Lipotoxicity and glucotoxicity, resulting from excessive exposure of hepatocytes to high lipid and carbohydrate levels, respectively, play a crucial role in the onset of simple steatosis (SS) and its progression to NASH. In this context, a high-fat, high-carbohydrate diet, commonly associated with obesity, increases the risk of developing SS and NASH. The underlying pathophysiological mechanisms linking lipotoxicity and glucotoxicity to SS and NASH involve mitochondrial dysfunction, endoplasmic reticulum stress, and oxidative stress [20]. Impaired mitochondrial function leads to hindered fatty acid oxidation, therefore enhancing lipid accumulation. Excess FFAs undergo peroxidation and producing reactive oxygen species (ROS), which in turn damage hepatocytes. ROS activate kupffer cells and inflammatory cytokines, contributing to hepatocellular injury [21].

2.3 Current Therapeutic Approaches

Polypharmacy for metabolic disorders involves a diverse array of medications, including lipid-lowering agents, antihypertensive agents, anti-diabetic drugs, heart failure medications, and anti-obesity treatments. A widely employed strategy for managing metabolic diseases involves the use of statins to reduce LDL cholesterol levels. Statins act as inhibitors of 3-hydroxy-3-methyl-glutaryl-coenzyme A (HMG-CoA) reductase, an enzyme critical to the cholesterol biosynthesis pathway [22]. Another method for reducing lipid levels in metabolic disorders is the use of ezetimibe [22, 23] which prevents the absorption of cholesterol in the intestines. Nicotinic acid is also an efficient lipid-lowering medication that lowers all apoB-containing lipoproteins while increasing HDL cholesterol levels, as well as it appears to lower the risk of CVD in patients with MetS and T2DM. Additionally, angiotensin system inhibitors are frequently prescribed to manage hypertension associated with metabolic diseases. Angiotensin-converting enzyme (ACE) inhibitors and angiotensin receptor blockers (ARBs) not only aid in reducing blood pressure but also slower the progression of cardiovascular and renal complications in diabetes. For treating heart failure in metabolic conditions, diuretics, beta-adrenergic blockers, or calcium channel blockers are commonly implemented [23, 24]. Anti-diabetic treatments include PPARγ agonists, which help regulate glucose metabolism, DPP-4 inhibitors, which enhance insulin secretion, GLP-1 receptor agonists, which improve insulin response and slow digestion, and SGLT-2 inhibitors, which lower blood sugar by increasing glucose excretion through urine [7].

3 Nanomaterials

3.1 Overview of Nanomaterials

Nanomaterials are tiny particles that usually measure between 1 and 100 nm in size. Their compact size and Surface area in relation to volume give them special physical, chemical, and biological properties. This makes them useful in a variety of fields, including healthcare, electronics, and environmental science.

3.2 Advantages of Nanomaterials for Targeted Drug Delivery

The unique properties of nanomaterials such as enhanced targeting, increased bioavailability, controlled release, reduced side effects etc. make them a promising tool for improving the effectiveness and safety of drug delivery systems. Natural product-based nanoformulations should be nontoxic, biodegradable, and biocompatible. Additionally, they should have a high capacity to penetrate cells, act quickly, and remain stable after oral administration [11]. When developing nanodrugs, there are multiple strategies to positively modulate the therapeutic efficacy. Examples include the encapsulation of drug molecules in NPs to shield them from harsh physiological environments, utilizing surface conjugation to direct drugs to specific tissues, which allows for high therapeutic concentrations at the target site even with reduced doses, and leveraging the small size of NPs to navigate significant physiological challenges, such as the immune system, renal clearance, and enzymatic or mechanical degradation, among others [12].

3.3 Types of Nanomaterials

Introduction to Types of Nanoparticles

NPs can be generally, categorized according to their structure, composition, and function. They are primarily classified into three categories: hybrid, inorganic, and organic, each offering specific advantages based on its application. Because of their biocompatibility, organic nanoparticles like liposomes and polymeric nanoparticles are mostly utilized in drug delivery and biomedicine. In contrast, inorganic NPs like gold, silica, or quantum dots are mostly utilized and have potential in imaging and diagnostics because of their stability and optical characteristics. Hybrid NPs combine both organic and inorganic components to capitalize on the strengths of each type.

In addition to their composition, NPs can differ in shape, size, and surface modifications, which further influence their properties and applications. Understanding the different types of NPs and their distinctive features is critical for developing innovative solutions across diverse scientific and industrial domains.

3.3.1 Liposomes

Lipid NPs or liposomes are nanoformulations characterized by their spherical shape and a range of heterogeneous sizes, typically measuring from a few hundred to several thousand nanometres in diameter. These structures consist of a lipid bilayer containing cholesterol, enabling them to effectively deliver hydrophilic drugs in an aqueous environment while encapsulating hydrophobic drugs within the lipid layer. The distinctive properties of liposomes, such as biocompatibility due to the use of naturally derived, safe lipids, their nanoscale size, and the ability to tailor both their hydrophilic and hydrophobic properties, enhance tissue specificity in drug delivery [25]. The structure of liposomes features water-soluble groups facing inwards, while the hydrophilic groups are oriented outwards, creating a double-layer spherical membrane. Liposomes offer several advantages, including non-toxicity, biodegradability, minimal immune reactivity, uniform particle size distribution, stability and the capability to encapsulate a broad spectrum of substances. The phospholipid most frequently utilized in liposome formulations is phosphatidylcholine, which made up of a hydrophilic polar head group, phosphocholine, a glycerol backbone, and a hydrophobic fatty acid hydrocarbon tail [15]. Liposomes coated with chitosan have been demonstrated to improve insulin absorption in the gastrointestinal tract (GIT) through their mucoadhesive properties. Additionally, a novel nanoformulation of liposomes, functionalized with folic acid and coated with polyacrylic acid and polyallylamine hydrochloride, has been developed for the oral delivery of insulin [26].

3.3.2 Solid Lipid Nanoparticles

Solid lipid nanoparticles (SLNs) are particles having a lipid nucleus that exhibit size variation from 50 to 1000 nm. They are mostly employed in oral drug delivery applications owing to their excellent biodegradability and biocompatibility [26]. SLNs possess several advantageous properties, which include small size, high surface area, more drug loading capacity in addition to the ability to interact at the interface of different phases, making them appealing for improving the efficacy of pharmaceutical compounds. Recent developments in surface-modified solid lipid nanoparticles (SMSLNs) have incorporated materials like albumin, polyethylene glycol, heparin and polysaccharides to enhance the oral delivery of phytobioactive compounds. Chitosan, known for its biodegradability, low immunogenicity, and ability to support controlled release under different pH conditions, proves especially effective for the oral administration of these compounds. The behaviour of both SLNs and modified SLNs (MSLNs) in oral delivery systems has been thoroughly investigated. Coating SLNs with chitosan, together with additional modifications to its structure, offers numerous advantages, including sustained release of bioactive compounds along with an enhanced positive surface charge, which helps reduce the initial burst release of SMSLNs [27].

3.3.3 Polymeric Nanoparticles

Polymeric NPs can be made in multiple structures such as solid, hollow, or dense porous configurations. For instance, nanospheres and nanorods are examples of porous structures, whereas nanocapsules and nanoshells are examples of hollow structures [25]. NPs are colloidal drug delivery systems with particle diameters ranging from 10 to 1000 nm. They can exist as a matrix system, where the drug is uniformly dispersed within the particles, known as a nanosphere, or as a reservoir system, where the drug is enclosed within a cavity surrounded by a polymeric membrane, referred to as a nanocapsule [9]. A single polymer membrane surrounds a cavity, within which the drug is encapsulated by NPs, forming vesicular systems. These systems degrade into biocompatible components through hydrolysis, facilitating controlled drug release at the target site. During this process, the matrix undergoes analysis, where it is uniformly degraded at the polymer level, with the degradation rate being linked to the surface area. The polymer disintegrates into lactic acid and glycolic acid, there are further metabolized into carbon dioxide and water [15].

3.3.4 Nanofibers

Nanofibers have many applications in important domains because of their small diameter (between 1 and 100 nm), large specific surface area, and significant porosity. Nanofibers can be generally categorized into three types based on their composition: polymeric, inorganic, and organic/inorganic composites. Among these, polymer nanofibers are especially favoured as drug delivery carriers because of their high porosity, large specific surface area-to-volume ratio, excellent biocompatibility and stability, as well as their resemblance to the extracellular matrix (ECM) [28]. At present, polymeric nanofibers can be encompassed several proteins [29], polysaccharides [30], and growth factors, and they have also been utilized to encapsulate hydrophobic drugs like ibuprofen and paclitaxel [31].

3.3.5 Dendrimers

Dendrimers are composed of a single molecule with a central core and a series of macromolecular branches, which are gradually formed by attaching various distinct units. Their distinctive structure makes them ideal for drug delivery applications, as they can display multiple copies of functional groups on their surface [32]. The unimolecular micelle structure of dendrimers arises from their hydrophilic outer and inner regions. They are capable of forming both covalent and non-covalent complexes with drug molecules and hydrophobic substances, contributing significantly to their solubilization properties. Polyamidoamine (PAMAM) dendrimers are anticipated to offer a positive outlook in improving the solubility of drugs within drug delivery systems [33].

A special class of polymeric macromolecules known as dendrimers have many types extending from a core, creating a nearly flawless three-dimensional geometric pattern. These can be formulated by two primary approaches a) divergent methods and b) convergent methods. These two methods of dendrimer synthesis vary among the direction of growth, either inwards towards the core or outwards from it. The divergent synthesis approach, first introduced by Tomalia in the 1980s, involves the creation of three-dimensional PAMAM dendrimers by progressively growing divisions that extend radially from a central core to the periphery. Furthermore, the presence of tertiary amines and amide linkages in PAMAM dendrimers facilitates the binding of various targeted and guest molecules [34].

3.3.6 Micelle

The structure and composition of micellar NPs can be precisely adjusted to achieve various particle sizes, drug-loading capacities, and release profiles. Micelles, which are self-assembling NPs made from polymeric amphiphiles, can be designed for the slow and controlled delivery of hydrophobic medications. They feature an internal hydrophobic core that effectively encapsulates drugs that have low solubility in water, while their external surface possesses enough polarity to enable dissolution in aqueous environments [12]. The hydrophobic core of micelles allows for the encapsulation and release of bioactive medicinal compounds, diagnostic agents, or a combination of both. Conjugating moieties like targeting ligands to the micelle's outer surface can enhance active binding to cells and tissues associated with the disorders [25]. Micelles have gained significant attention due to their ability to encapsulate drug compounds that are poorly soluble in water or aqueous environments. Commonly known as 'polymeric micelles', these structures are efficient in delivering drugs with low water solubility. They have been applied in cancer treatments targeting mitochondria. This drug delivery approach also contributes to enhancing the drug's bioavailability, which can be further improved by adding external surfactants to the micelle formulation [35].

3.3.7 Niosomes

These are artificial vesicles stabilized by cholesterol and created by non-ionic surfactants arranged in concentric bilayers. They are recognized for their drug pool, safety, and capacity to administer drugs continuously. The presence of cholesterol prevents drug leakage and increases the rigidity of vesicles. Additionally, its 'multi-philic' (lipophilic, hydrophilic, and amphiphilic) components improve drug solubility [26]. Niosomes can be classified based on their size into three types: small unilamellar vesicles (SUV) ranging from 10 to 100 nm, large unilamellar vesicles (LUV) ranging from 100 to 3000 nm, and multilamellar vesicles (MLV), which consist of multiple bilayers [36]. Niosomes are generally osmotically active, chemically stable, and have a longer shelf life compared to liposomes. They can encapsulate hydrophobic drug molecules within their bilayer membranes and hydrophilic drugs in aqueous environments. By shielding drug molecules from the biological environment, niosomes enhance therapeutic efficacy, leading to improved bioavailability and controlled drug release. Additionally, they facilitate targeted drug delivery by confining the drug's effects to specific cells while reducing clearance from systemic circulation in sustained-release drug delivery systems [37]. Niosomes can be prepared as drug delivery and gene therapy vectors through several techniques. Also, for the sake of targeting, enhancing circulation and target searching time, these can be coated by various types of substances including polyethylene glycol (PEG)

[38, 39], hyaluronic acid [40], antibodies [41], etc. for distinct applications. Surfactants, when combined with an additive component like cholesterol, can generate various types of non-ionic surfactant vesicles.

3.4 Mechanisms of Targeted Drug Delivery for Metabolic Syndrome

3.4.1 Passive Targeting Through Enhanced Permeability and Retention (EPR) Effect

Passive targeting refers to the accumulation of nanomaterials at the plaque site, either through the increased permeability of leaky vasculature or via phagocytosis by the mononuclear phagocyte system. Similar to cancer, the enhanced permeability and retention (EPR) effect may also play a role in atherosclerosis due to extensive neovascularization and the irregular architecture of endothelial cells. Notably, passive targeting is significantly affected by the physicochemical properties of nanomaterials, such as their size, shape, and surface characteristics [42].

3.4.2 Active Targeting Mechanisms

Active targeting in metabolic syndrome involves directing therapeutic agents specifically to tissues or cells affected by the syndrome, such as adipose tissue, liver, and muscle. This strategy modulates treatment potential and reduces side effects by targeting molecular markers or receptors which are overexpressed in affected tissues. Active targeting by nanomaterials shows a promising approach to address the challenges. Nanomaterials, such as liposomes, dendrimers, and SLNs, are designed to deliver drugs directly to targeted tissues or cells involved in MetS. Therefore, they increase the drug's potency. This targeting is established by functionalizing the NPs with respective ligands that recognize and bind to specific target receptors or molecular markers which are overexpressed in the affected tissues, including insulin receptors, inflammatory cytokines, or adipose tissue markers [43]. In the case of MetS, inflammation shows a vital role in the pathogenesis of insulin resistance and other metabolic disorders. Nanomaterials can be formulated to deliver anti-inflammatory drugs to the sites of inflammation, which are mostly present in adipose tissue, liver, and muscle. For example, NPs functionalized with ligands which target pro-inflammatory cytokines like IL-6 or TNF-α, which can help minimize inflammation and improve insulin sensitivity [44].

3.4.3 Ligand–Receptor Interactions

In MetS, certain tissues and cells, like adipocytes, endothelial cells, hepatocytes, and muscle cells, are found to show altered expression of receptors involved in several metabolic pathways, including glucose uptake, lipid metabolism, and inflammation. By functionalizing nanoparticles with specific ligands which recognize and bind to these receptors, nanomaterials can increase the targeting of therapeutic molecules directly to these tissues. This decreases off-target effects and promote therapeutic efficacy. Insulin receptor (IR) and GLUT4 transporter IR are overexpressed in skeletal muscle and adipose tissue in individuals with insulin resistance. The glucose transporter GLUT4 is responsible for the uptake of glucose into these cells. NPs functionalized with insulin or insulin-mimetic ligands can specifically bind to the IR, facilitating the uptake of the NP-encapsulated drug into insulin-resistant tissues [45]. This enhances glucose uptake, thereby improving insulin sensitivity and metabolic control. TNF-α is a key pro-inflammatory cytokine involved in insulin resistance and inflammation. Its receptor (TNFR) is overexpressed on adipocytes, macrophages, and hepatocytes in individuals with MetS. Nanomaterials functionalized with TNF-α inhibitors or antibodies targeting TNFR can reduce pro-inflammatory signalling and improve insulin sensitivity by blocking the inflammatory cascade [46]. Angiotensin II Receptor (AT2R): The angiotensin II type 1 receptor (AT1R) plays a principal role in regulating blood pressure and vascular function. Overactivation of AT1R in the hepatocytes and adipose tissue contributes to the elevated blood pressure and vascular dysfunction associated with MetS. NPs functionalized with angiotensin receptor blockers (ARBs) or other antihypertensive agents can selectively target AT1R, reducing vascular inflammation and improving blood pressure regulation [47].

3.4.4 Antibody–Drug Conjugates

A commercial-grade available lipid-based formulation containing the cationic lipid DOGS (dioctadecylamidoglycylspermine) has been utilized for the delivery of antibodies (such as anti-connexin 32 mAb, anti-actin antibody and rabbit IgG), along with proteins and peptides. Antibody-loaded DOGS lipid NPs engage with cell surfaces by electrostatically binding to negatively charged heparan sulphate proteoglycans, leading to endocytosis [48, 49]. Upon endosomal escape, which is likely facilitated by the cationic amine groups and amphiphilic nature of DOGS, NP dissociation occurs, allowing the release of intact cargo molecules into the cytoplasm. In a separate study, Bayele et al. examined a class of dendrimer-like peptide lipids containing multiple lysine residues and linear alkyl chains. Using a Cy3-labelled c-myc monoclonal antibody (mAb), they observed that these peptide lipids could enter HeLa cells via a receptor-independent mechanism. Their results indicated that the lipid (C14)3(L)7(NH2)8, composed of seven lysine residues and eight primary amine groups, exhibited superior intracellular antibody delivery compared to a similar lipid containing fifteen lysine residues and sixteen primary amine groups [50].

4 Characterization Techniques for Nanomaterials

4.1 Scanning Electron Microscopy (SEM)

One of the principal approaches for characterizing nanomaterials is the SEM. Scanning an electron probe across a surface and detecting the emitted secondary electrons is likely to generate secondary electron images of both organic as well as inorganic materials having nanoscale resolution. SEMs are employed for various tasks, such as high-speed compositional analysis, cryogenic research, and ordinary morphological examinations [51].

4.2 Transmission Electron Microscopy (TEM)

TEM is one of the widely applied techniques employed for characterizing nanomaterials in electron microscopy. A variety of data analysis approaches can be employed along with TEM in combination for multiple applications; For example, by combining energy dispersive X-ray spectroscopy with chemical investigation from electron energy loss spectroscopy, the electronic structure and chemical configuration of nanomaterials can be quantitatively analysed. The size, distribution, aggregation, and movement of nanoformulations in an aqueous phase are determined by the TEM wet test method [52].

4.3 Atomic Force Microscopy (AFM)

It is composed of a micro-machined cantilever with a sharp tip at the end, because of electrostatic repulsion, van der Waals interaction and traction between atoms on the measured surface as well as at the tip. Unlike existing electron microscopic approaches, AFM can characterize nanomaterial dynamics in biological contexts, such as real-time observation of nanomaterials' interaction with hold-up lipid bilayers [52].

4.4 Dynamic Light Scattering (DLS)

DLS is a conventional light scattering method that uses a monochromatic light source, like a laser, the technique helps for determination of the particle size distribution of nanomaterials in a solution or suspension by observing changes in Rayleigh scattering at a fixed angle [53]. Some of the benefits of the DLS method include shorter measurement times, more accurate measurement results, the competence to measure a range of concentrations analysis, diluted samples along repeatable measurements [51, 54].

4.5 Entrapment Efficiency (EE)

EE measures how effectively a drug is encapsulated in NPs, showing the percentage of the initial drug successfully trapped. It's crucial for evaluating encapsulation efficiency and potential drug availability. EE is calculated using the formula:

$$\%\text{EE} = \frac{(T - C)}{T} \times 100 \tag{5.1}$$

Where T is the total drug and C is the free drug in the supernatant. High EE values indicate efficient drug encapsulation, improving drug stability and controlled release. Optimization of EE is a prominent issue for developing powerful nanomaterial-based drug delivery systems [55].

4.6 Loading Capacity (LC)

Drug loading is a key process to increase the bioavailability and dissolution rate of poorly soluble drugs using nanomaterials. It involves drug adsorption onto the NPs surface or within its pores, influenced by factors like surface area, pore volume, and the drug's solubility. The two main methods are solvent-based, which uses a solution to load the drug, and solvent-free, which is more environmentally friendly. Drug loading efficiency is calculated by the following formula [56]:

$$\text{LC}\% = \left(\frac{\text{Drug loaded}}{\text{Carrier weight}}\right) \times 100 \tag{5.2}$$

4.7 Drug Release

The release of drugs from nanoformulations provides a major role in the determination of the therapeutic efficacy, particularly in controlled and targeted drug delivery systems. The process of transferring active pharmaceutical ingredients from the site of administration to a specific target organ within a biological system is commonly known as 'drug delivery'. These studies enable the optimization of nanoformulations for a sustained, controlled, or stimuli-responsive release of drugs that ensure better treatment efficacy with reduced side effects [56].

4.8 Pharmacokinetics

Nanomaterials have tremendous prospects in biomedical applications, especially for drug delivery, imaging, magnetic hyperthermia, and tissue regeneration. Pharmacokinetics the study of absorption, distribution, metabolism, and excretion (ADME) is relevant to understanding how drugs behave on the molecular level. Nanomaterials can alter ADME profiles as a result of enabling targeted drug delivery. The pharmacokinetics of these nanomaterials' hinges on protein adsorption. Hence, elucidating the relationship between the properties of nanomaterials and biological behaviour is central to the rational design of an effective drug delivery system and a forecast of responses in vivo [57].

4.9 X-Ray Diffraction (XRD)

It is an advanced method for identifying the phase of a crystalline material. X-rays have wavelengths ranging from 0.2 to 10 nm. Since the size of X-rays is similar to the distances between layers in crystals, X-ray crystallography is an ideal technique and non-detrimental technique for nanomaterial characterization as XRD data can be utilized for identifying the macroscopic features of the tested specimens as well as nanometre parameters. It is also possible for calculation of the crystallite size of the material using the Scherrer equation and peak broadening, that is, the whole width at half of the highest intensity. Diffraction is the result of the interaction between the sample and the X-ray beam and is ruled by Bragg's law: $n\lambda = 2d\sin(\theta)$ [51].

4.10 Differential Scanning Calorimetry (DSC)

DSC is a thermo-analytical approach that supplies valuable insights on the behaviour of lipid melting and crystallization. DSC detects a variety of thermal transitions related to physicochemical alteration when NPs are heated or cooled under controlled conditions. Similarly, it will also provide a lot of information concerning glass transition temperatures, eutectic systems, polymorphism, and crystal structure [58].

4.11 Thermogravimetric Analysis (TGA)

TGA is a technique which monitors the mass of a sample as a function of time or temperature. It is commonly employed for the quantitative analysis of ligand adsorption onto NP surfaces. This method has been employed to calculate the ligand packing density and quantify the number of ligands adsorbed to the surface of gold and silver NP surfaces. TGA has been used to describe carbon nanotube (CNT) oxidation [59].

4.12 Drug–Excipient Compatibility

Compatibility between drug and excipient is very significant to ensure the stability and efficacy of nanoformulations. Fourier transform infrared spectroscopy (FTIR) detects chemical interactions through the comparison of vibrational peaks of the drug, excipient, and their physical mixtures within a specified wavelength range. DSC assesses thermal behaviour through the analysis of endothermic or exothermic transitions that may indicate incompatibility. High performance thin layer chromatography (HPTLC) provides a fingerprint analysis for confirmation of stability by consistent retention factor (Rf) values. These studies are very imperative in formulation development to prevent adverse interactions and ensure drug stability [60].

4.13 In Vitro and In Vivo Therapeutic Activity

The special features of nanomaterials, like nanoscale size and large surface area, offer great therapeutic potential. Their in vitro activity has been studied for various uses, including antimicrobial, enzymatic activity, anticancer, and anti-inflammatory therapies. Metallic NPs, particularly gold and silver, exhibit potent in vitro activity through the generation of ROS, disruption of microbial membranes, or inhibition of cancer cell proliferation. Lipid-based nanocarriers, like liposomes and SLNs, enhance drug solubility, protect active agents from degradation, and facilitate sustained release.

In vivo, nanomaterials enhance bioavailability, enable targeted drug delivery, and reduce systemic toxicity. Many NPs leverage the EPR effect for passive targeting in tumours, while others are engineered for active targeting by incorporating ligands that are specific to disease biomarkers associated with MetS. Their biocompatibility assures minimum immune response, while their ability to load multiple drugs or diagnostic agents simultaneously broadens their therapeutic scope. Long-term toxicity, biodegradation, and regulatory concerns remain some of the challenges that persist. In conclusion, nanomaterials are creating a revolution in in vitro and in vivo therapeutic strategies with prominent advancements in the field of precision medicine [61].

5 Overview of Therapeutic Strategies

Nanomaterials have emerged as a novel therapeutic strategy in the management of MetS, aiming to address its components, like obesity, insulin resistance, dyslipidaemia, and hypertension, at the cellular and molecular levels.

5.1 Management of Metabolic Syndrome

5.1.1 Diabetes Management

Nanocarriers generally mention particles with a size range from 1 to 100 nm. Several nanocarriers for insulin delivery have been documented including polymeric nanoparticles, liposomes, SLNs, PAMAM dendrimer niosomes and micelles [26, 62]. PAMAM dendrimers are effective for delivering genes and hormones that help in regulating elevated blood sugar levels. By lowering elevated blood sugar levels, polyamidoamine G4 has mimicked blood glucose-lowering medications [63]. Several studies have demonstrated that the intraperitoneal route (IP) of administration results in a higher pronounced hypoglycaemic effect, while sustained hyperglycaemic responses are standardized through both subcutaneous (SC) and IP routes [64].

The bioavailability of niosomes containing metformin-loaded demonstrated a prolonged hypoglycaemic effect and sustained metformin release. When compared with the traditional dose form, niosome-containing repaglinide (RPG) exhibited superior hypoglycaemic effects. Niosomes are very superior for improved bioavailability, reduced dosage frequency, and long-term hypoglycaemic effects [65, 66]. Chitosan, a naturally occurring polymer, consists of N-acetyl-D-glucosamine and deacetylated glucosamine. Its excellent permeability, biocompatibility, biodegradability, strong adhesiveness in the GIT, and safety make it highly suitable for a wide range of applications [26]. Chitosan-based NPs incorporating oleic acid appear to show protection to insulin against degradation by enzymes. This indicates that these NPs may increase the stability and effectiveness of insulin in therapeutic applications [67]. To enhance mucoadhesiveness, hydrophilicity as well as the permeability in alkaline environments, derivatives of chitosan like carboxylated chitosan, thiolated chitosan, and trimethyl chitosan have been utilized in the production of NPs [68].

Compared to traditional liposomes, bile salt-based liposomes have been found to ameliorate insulin absorption and its bioavailability [26]. Chitosan-coated liposomes have been found to reduce insulin absorption in the GIT due to their mucoadhesive property [69]. A liposome-based nanoformulation, functionalized with folic acid and coated with polyacrylic acid and polyallylamine hydrochloride, has likely been developed for the oral delivery of insulin [70].

Polymeric NPs offer a strategy to enhance insulin uptake in the GIT. The release of insulin and its subsequent pharmacological effects are regulated by synthetic or natural polymeric materials. Insulin-loaded NPs, made from biodegradable polymers like poly(lactide-co-glycolide), polyanhydride, and polyalkyl cyanoacrylate, facilitate transport of insulin across the intestinal mucosa, improving absorption through intestinal epithelial cells [62].

5.1.2 Treatment of Atherosclerosis

In the context of atherosclerosis, non-metallic NPs including carbon-based nanomaterials and mesoporous silica are frequently utilized. With a wide surface area, regular pore channels, a distinctive mesh-like pore structure, and an adjustable pore size, MSNs range in size from 10 to 600 nm. Its ability to act as a nanocarrier to enhance the water solubility of medications and lessen negative effects is made possible by its high cargo loading and tunable central pore structure [42]. Due to the lipid composition, which improves intestinal permeability and lymphatic transport, enhanced, prolonged retention, mucoadhesion, and increased stability in the GIT, SLNs are mostly utilized to treat atherosclerosis [71]. Various carbon-based nanomaterials own exclusive structures and properties and have more photothermal conversion efficiency with distinct fluorescence effect (FL effect), these are acceptable for photothermal therapy (PTT), PA imaging, and FL tracing (e.g. CNTs [72], carbon nanocages [73], graphene oxide) in the treatment of atherosclerosis [74].

Lipid-based nanomaterials primarily consist of liposomes and reconstituted HDL NPs (rHDL NPs) [42]. Primarily, rHDL NPs demonstrate antioxidant properties, exert anti-inflammatory effects, and promote reverse cholesterol transport in foam cells through ABCA1, ABCG1, and SR-B1. This process helps reduce lipid accumulation, making rHDL NPs particularly effective for treating atherosclerosis [75].

5.1.3 Obesity Management

Adipose tissue (AT), often regarded as an inert fat storage organ, has been recognized as a principal player in maintaining the body's homeostasis. It represents less than 30% of total body mass in lean individuals and can comprise up to 80% in obese individuals. The body contains two primary types of AT, each differing in function and location: brown adipose tissue (BAT) and white adipose tissue (WAT) [76]. WAT plays a crucial role in the development of obesity and its related disorders. As a result of chronic excess energy consumption, WAT expands through both an increase in the number of pre-adipocytes (hyperplasia) and an enlargement of adipocytes (hypertrophy) [77]. Obesity disrupts the activity of adipokine expression, along with impacting the activities of other main organs, including the brain, liver, kidney, heart, and muscles. Obesity also impacts WAT functions and homeostasis. Secondary disorders emerge when the physiological functions of these organs are changed. For example, hepatic steatosis can occur due to the persistent assembling of ectopic fat in the liver.

Strategies targeting WAT to reduce its size, remove hypertrophic adipocytes, convert WAT into BAT, or inhibit adipogenesis could greatly improve treatments for obesity and its related multimorbidity [78].

To cure obesity, Chen et al. developed a novel rapid-dissolving microneedle patch that contains rosiglitazone NPs. By successfully piercing the skin and delivering the medication in the dermis, this technique demonstrated notable efficacy in obese mice. In addition to enhancing inflammatory markers and reducing body weight and fat, the medication also stimulated the browning of white adipocytes, a process in which WAT is converted into metabolically active brown or beige AT. Remarkably, it elevated the levels of the anti-inflammatory cytokine Interleukin 10 (IL-10) in inguinal WAT (iWAT), while lowering the mRNA expression of proinflammatory factors likeTNF-α, interleukin 1 beta (IL-1β), and IL-6 [79]. To improve targeted delivery to AT, Liang et al. created an inventive microneedle patch delivery system for 2,4-dinitrophenol (DNP) by employing tetra decanoic acid-DNP ester nanomicelles. When compared to oral administration, this method enhanced biosafety and exhibited notable anti-obesity effects in a mouse model fed a high-fat diet. Enhancement of Uncoupling protein 1 (UCP1) expression and browning, better lipid droplet reduction, and improved preadipocyte endocytosis were all observed in the system. Notably, it reduced systemic toxicity by achieving increased accumulation in iWAT with lesser off-target dispersion. The treatment enhanced metabolism and lipid profile as well as reduced body weight and fat mass [80]. Inorganic NPs are effective in addressing obesity, either due to their physicochemical characteristics or their role as carriers for drug delivery. Prominent examples include iron oxide, gold, mesoporous silica, and calcium phosphate nanoparticles [26]. It has been reported that cerium oxide NPs inhibit adipogenesis by modulating negatively the mRNA transcription of key genes and the assembling the triglycerides. This, in turn, leads to a reduction in body weight, as well as lowered levels of insulin, glucose, and triglycerides in the blood [81]. Lipid NPs are commonly utilized for delivering drugs that have low solubility in aqueous environments. A nanoformulation of trans-resveratrol has been reported to exhibit greater solubility and stability. Additionally, it was noted to enhance the browning effects of resveratrol on WAT, suggesting an improved treatment for obesity [82].

5.1.4 Treatment of Hypertension

Many hypotensive agents, such as atenolol (ATN) and furosemide (FUR), exhibit physicochemical properties that render them suboptimal for oral administration. Their poor aqueous solubility significantly limits their bioavailability, while their potential role as P-glycoprotein substrates and susceptibility to extensive first-pass metabolism further compromise systemic drug levels. Additionally, these drugs often possess a short elimination half-life, necessitating frequent dosing to maintain therapeutic plasma concentrations. These pharmacokinetic limitations have constrained their clinical utility in antihypertensive (AH) therapy [83]. To optimize their therapeutic efficacy, the development of advanced drug delivery systems capable of sustaining and modulating drug release kinetics is imperative. Nanotechnology-based delivery platforms, in particular, offer a promising strategy to enhance drug bioavailability, prolong circulation time, and mitigate dose-dependent adverse effects. Notably, nanoscale drug carriers can circumvent first-pass metabolism and P-glycoprotein-mediated efflux by facilitating direct systemic absorption. Furthermore, the incorporation of these agents into nanocarriers enables targeted drug delivery, thereby improving therapeutic outcomes. Given these advantages, -based drug delivery systems are emerging as a viable alternative to traditional AH treatment strategies [84]. Felodipine (FD), a dihydropyridine calcium-channel antagonist, is employed either alone or in combination with angiotensin-converting enzyme (ACE) inhibitors for the treatment of chronic stable angina, hypertension and Prinzmetal's variant angina. As a Biopharmaceutics Classification System (BCS) Class II drug, FD has poor water solubility and undergoes significant hepatic first-pass metabolism, resulting in a low oral bioavailability of about 15% [85]. The enhanced pharmacodynamic response observed with NP-based formulations, compared to conventional drug suspensions, is primarily attributed to improved bioavailability, facilitated by direct uptake via Peyer's patches in the intestinal lymphatic system. Poly(lactic-co-glycolic acid) (PLGA)-based FD-loaded nanoparticles offer a promising alternative to existing AH and anti-anginal therapies by enhancing bioavailability, prolonging drug release, and maintaining therapeutic plasma concentrations for up to 3 days [86]. Nitrendipine, a dihydropyridine calcium channel antagonist, is commonly taken as an AH agent. However, its clinical effectiveness is constrained by its poor water solubility and less oral bioavailability (10–20%), primarily due to extensive hepatic first-pass metabolism, which produces four pharmacologically inactive metabolites. To overcome this limitation, alternative administration routes such as transdermal, buccal, rectal, and parenteral delivery have been investigated. Nevertheless, the oral route remains the most preferred due to patient compliance and ease of administration, although factors such as gastrointestinal pH, residence time, and solubility significantly influence drug absorption. To enhance the bioavailability of nitrendipine, SLNs have been investigated as a promising delivery system. SLNs can be formulated using various lipid matrices, including tripalmitin (a triglyceride), cetyl palmitate (a wax), and glyceryl monostearate (a monoglyceride), to improve drug solubility and sustain release. Several techniques, including high-pressure homogenization and solvent emulsification/evaporation, have been disclosed in the literature for the formulation of SLNs, offering a potential strategy to overcome the pharmacokinetic limitations of nitrendipine and optimize its therapeutic efficacy [87].

5.1.5 Fatty Liver Disease

Nanoemulsions have a major role in the domain of nanotechnology for treating NAFLD. The high interfacial area and nanoscale size of the emulsions make them ideal for drug encapsulation and release. Nanoemulsions can be utilized to deliver therapeutic medicines to the liver with precision in the context of NAFLD. By encapsulating the medication in a nanoemulsion, the drug's bioavailability can be enhanced and its effective dosage can be lowered, along with minimizing adverse effects. A naturally available polyphenolic chemical with strong anti-inflammatory and antioxidant effects, curcumin has a low bioavailability because of its quick metabolism and systemic excretion. However, by enhancing its bioavailability, curcumin-loaded nanoemulsions have demonstrated an improved therapeutic impact in NAFLD [88]. The most popular nanocarriers in the pharmaceutical industry for creating controlled, sustained, and burst drug delivery systems are polymer nanoparticles (P-NPs), whose release characteristics are closely linked to the polymeric shell that was built. In addition to providing improved bioavailability, higher stability of any volatile materials with non-immunogenicity, and a low to nontoxicity character, they can be used to mask the taste and shield drugs and substances from stomach and environmental degradation [3]. In the design of polymeric nanoparticles (P-NPs) for MAFLD therapy, the use of polymers is primarily limited to chitosan and PLGA. Notably, chitosan alone has demonstrated promising results as an excipient, exhibiting synergistic anti-MAFLD effects. There are additional polymers, such as catechin polymers synthesized via acyl-catalysed methods, that have also shown potential for anti-MAFLD therapy, though they have not yet been incorporated into P-NP formulations for this purpose [89], e.g., pyrrole-based polymers [90], Hericium Erinaceus exo-polymers [91], and beta-cyclodextrin [92, 93].

5.2 Examples of Nanomaterial Applications

5.2.1 Drug-Loaded Nanoparticles for Insulin Delivery

Glucose oxidase (GOx) is the most widely utilized enzyme in glucose sensors due to its remarkable specificity, stability, and high catalytic efficiency. GOx facilitates the oxidation of glucose in the presence of oxygen, resulting in gluconic acid and hydrogen peroxide. This enzymatic reaction leads to a localized decrease in pH, which is advantageous for biosensing applications. This pH change can be assessed quantitatively, making it ideal for glucose detection. Notably, GOx remains stable within the physiological pH range, with optimal stability at a pH of 5. However, it starts to degrade when the pH drops below 2 or exceeds 8, further highlighting its suitability for glucose-sensing applications [94]. A GOx system [95] in which a glucose-responsive insulin delivery system utilizing pH-responsive hydrogel nanoparticles was developed, incorporating GOx, catalase, and insulin. The NPs were composed of poly (diethyl aminoethyl methacrylate-g-ethylene glycol) [P(DEAEM-g-EG)]. As GOx catalysed the transformation of glucose into gluconic acid, the resulting decrease in pH caused the cationic hydrogel to swell, enabling the controlled release of insulin. Meanwhile, catalase within the NPs converted the hydrogen peroxide by-product of the GOx reaction into oxygen, hindering oxygen depletion and enhancing the system's overall efficiency. This glucose-responsive system demonstrated effective insulin release and holds significant potential for advancing closed-loop insulin delivery methods. A dual-responsive polymeric micelle system for insulin delivery, designed to respond to both glucose and hydrogen peroxide, was engineered using the block copolymer poly (ethylene glycol)-block-poly (amino phenylboronic ester) (PEG-b-PAPBE). This system enables controlled insulin release by leveraging the dynamic interactions of the phenylboronic ester with glucose and its oxidative responsiveness to hydrogen peroxide. The copolymer was produced through Michael addition polymerization, and the micelles were formed by the controlled dropwise addition of tetrahydrofuran as a surfactant. While the micelles exhibited intrinsic glucose responsiveness, GOx was encapsulated within them to enhance their glucose sensitivity. At elevated glucose levels, the PAPBE units in the micelles reacted with glucose, resulting in the cleavage of the phenylboronic ester (PBE) groups and the subsequent formation of phenylboronic acid-glucose (PBA-glucose) complexes, facilitating insulin release [96].

5.2.2 Nanocarriers for Antidiabetic Agents

Drug nanoencapsulation is increasingly being recognized as a promising strategy to minimize the adverse effects of drug administration when enabling controlled release and targeted delivery [97]. Polymeric materials are widely regarded as the optimal material for encapsulating various target molecules, including drugs, due to their biodegradability and biocompatibility. They facilitate controlled cargo release through mechanisms such as pH variation, temperature shifts, surface modifications, and molecular conjugation [98, 99]. Metformin is commonly used to regulate blood glucose levels; however, its high-dose administration is associated with adverse side effects. Encapsulating metformin within polymeric nanocapsules enables sustained release and enhances therapeutic efficacy, allowing for effective treatment at lower doses. Metformin-loaded alginate nanoparticles (MLANs) were fabricated via emulsion cross-linking technology, utilizing alginate's hydrogel-forming capability in aqueous environments to improve metformin's bioavailability and therapeutic efficacy. Calcium-induced gelation facilitated the formation of biocompatible, biodegradable, and cost-effective nanoformulations with high drug entrapment efficiency. In streptozotocin-induced diabetic and hyperglycaemic rat models, MLANs effectively lowered blood glucose levels (BGLs) and mitigated postprandial hyperglycaemia. Oral administration of MLANs at a reduced dose exhibited significant hypoglycaemic effects comparable to a threefold dose of free metformin, confirming their potential for improved diabetes management with lower drug concentrations [97]. An optimal drug carrier ensures the accurate administration of therapeutic molecules to the target site at the appropriate dosage. MSNs function as highly adaptable nanocarriers, characterized by their unique physicochemical and morphological attributes. Their exceptionally high surface area enables the efficient loading and transport of substantial drug quantities. Surface functionalization allows MSNs to achieve stimuli-responsive drug release. Studies have demonstrated their capability to deliver a wide range of therapeutic compounds, including insulin, incretin mimetics (e.g. exenatide and liraglutide), tolbutamide, and vitamin K [100]. Exenatide (EXT), the first GLP-1 receptor agonist, is certified as an adjunct treatment for type 2 diabetes; however, its short half-life necessitates frequent subcutaneous administration. To enhance its sustained delivery, EXT was encapsulated within orderly arranged hexagonal mesoporous silica nanostructures (EXT-SBA-15). In vitro release studies demonstrated an initial rapid drug release, followed by a controlled and sustained delivery over 14 days. In diabetic mice, subcutaneous administration of EXT-SBA-15 effectively reduced blood glucose levels, indicating its potential for prolonged glycaemic control [101]. Dextran-capped, PBA-functionalized magnetic mesoporous silica (MMS) was employed for glucose-sensitive delivery of the poorly aqueous-soluble sulfonylurea drug, tolbutamide. The competitive interaction between glucose 1,2-diol groups and PBA in the presence of glucose resulted in the relocation of the dextran polymer from the NP surface, thereby triggering drug release. In vitro studies demonstrated that approximately 80% of tolbutamide was released upon subject to a glucose concentration of 50 mg/mL, highlighting the system's potential for glucose-triggered drug delivery in diabetes management [102]. SLNs are colloidal dispersions consisting of solid lipids at room temperature, stabilized by non-ionic emulsifiers. These nanocarriers have been utilized for the delivery of poorly water-soluble antidiabetic phytocompounds. Xue et al. formulated berberine-loaded SLNs, achieving an EE of 58% and a LC of 4.2%. Oral administration of this formulation at a dose of 100 mg/kg in mice considerably enhanced berberine bioavailability compared to an equivalent dose of free berberine [103]. In a further investigation, myricetin-loaded SLNs (EE: 56.2%, LC: 5.62%) were orally administered at a dose of 10 mg/kg per day, resulting in a significant decrease in blood glucose levels. This hypoglycaemic effect was equivalent to that achieved with 200 mg/kg of metformin [104].

5.2.3 Lipid-Based Nanocarriers for Statins

Rosuvastatin (RSV), a statin-class drug, is employed for prevention of the cardiovascular disorders by lowering LDL cholesterol levels. Its pharmacological action is mediated through the competitive inhibition of the enzyme 3-hydroxy-3-methylglutaryl-CoA (HMG-CoA) reductase. However, following oral administration, rosuvastatin undergoes extensive pre-systemic metabolism, resulting to a relatively low oral bioavailability of approximately 20%. Peak plasma concentrations of RSV are established within 3–5 hours post-administration.

Additionally, the drug exhibits a high degree of plasma protein binding, with approximately 88% bound primarily to serum albumin [105]. Various pharmaceutical formulation approaches have been explored to increase the bioavailability of drugs with poor water solubility. These approaches include the development of colloidal drug delivery systems, like SLNs, polymeric NPs, lipid-based NPs, microemulsions, and self-microemulsifying drug delivery systems (SMEDDS), all of which have been extensively studied and documented in the literature [106]. Flexible liposomes, referred to as 'Transfersomes', represent a subclass of lipid-based NPs composed of phospholipids and a single-chain surfactant [107]. The incorporation of a surfactant, also known as an edge activator, enhances the flexibility of these nanoparticles by minimizing the rigidity of the phospholipid bilayer, thereby transforming them into ultra-deformable vesicles. Owing to their effectiveness in delivering a broad range of pharmacologically active compounds, these highly deformable nanoparticles have attracted considerable research interest. Chitosomes, as a nanoformulation, show promise as a nanocarrier for RSV cytotoxicity, while liposomal nanoparticles serve as an effective carrier to improve RSV bioavailability in the treatment of hyperlipidaemia [105]. The creation and characterization of chitosan nanoparticles were investigated to improve the effectiveness of pravastatin on erythrocyte redox status. The integration of chitosan-based nanogels with erythrocytes presents a promising multifunctional carrier system, providing substantial progress in drug delivery technology. Chitosan nanogel-loaded erythrocytes demonstrated sustained release of pravastatin, and erythrocytes exhibited increased pravastatin uptake when delivered as pravastatin-chitosan nanoparticles compared to pravastatin in solution form. Additionally, Chitosan NPs and pravastatin-loaded chitosan NPs demonstrated a more significant impact on erythrocytes' redox status than free pravastatin. This suggests that pravastatin-loaded chitosan NPs hold potential as a well-performing drug delivery mechanism for the targeted transport of pravastatin to erythrocytes [108]. A study developed atorvastatin calcium-loaded chitos NPs using several drug-to-polymer ratios of 1:1, 1:2, 1:3, and 1:4. Among these, the NPs with a 1:4 ratio exhibited the foremost encapsulation proficiency. These drug-loaded NPs illustrated sustained drug release and improved bioavailability [109]. However, despite the prolonged systemic exposure to atorvastatin calcium from the NPs, its intestinal absorption remained limited due to poor water solubility. To address this, water-soluble chitosan was proposed to enhance intestinal absorption [110].

6 Challenges and Limitations

6.1 Challenges with Lipid-Based Delivery Systems

Lipid-based delivery systems, like liposomes, are extensively researched, and utilized in clinical treatments because of their high biocompatibility along with effective delivery capabilities. In addition to intravenous administration, liposomes can also be taken orally, enabling absorption into the bloodstream despite the harsh conditions of the GIT, including exposure to gastric acid and bile salts. However, relating to other nanocarriers, the complex processes involved in liposome preparation pose challenges for large-scale production. Moreover, batch-to-batch variability remains a significant issue. To integrate between preclinical studies and clinical applications, ongoing advancements in manufacturing techniques and scalable production methods are essential [111].

When foreign particles like liposomes enter inside the body, they overcome across various protective mechanisms intended to recognize, neutralize, and also eliminate them from the body. These protective mechanisms generally include immune responses, opsonization, and the reticuloendothelial system (RES) [112]. These challenges need to be addressed for the effective use of liposomes. However, certain physiological phenomena, like the improved permeability and retention effect, can be utilized to improve drug delivery efficient delivery of drugs [113].

6.2 Challenges with Polymeric Nanoparticles

Over recent years, biopolymers have become a promising drug delivery option owing to their greater EE, non-toxic nature, and controlled release capabilities. The use of both natural and synthetic biopolymers in delivery systems has opened up a wide range of novel administration routes. Research has particularly focused on oral, nasal, and transdermal drug delivery methods to enhance patient adherence. Nevertheless, there remains significant potential to explore additional biodegradable polymers for drug delivery applications. Developing functionalized designs and innovative biomaterials, still presents significant challenges, as it requires an exhaustive understanding of interactions between the biological environment and polymer while ensuring both the clinical effectiveness and safety [111].

The safety and possible toxicity associated with the polymeric NPs often represent a significant challenge. The factors such as residual materials present on the NPs, aggregation of toxic monomers, compatibility issues, and harmful degradation by-products are the reasons for toxicity associated with them [114]. The most common toxicological concerns linked to the utilization of polymeric NPs are immune responses and excessive production of inflammatory mediators. For instance, PLGA-polyethylene oxide NPs have been shown to elevate levels of a pro-inflammatory mediator—granulocyte-macrophage colony-stimulating factor (GM–CSF) [115]. Furthermore, the use of hydrophilic coating polymers such as polyethylene glycol has demonstrated certain limitations. Research has shown that PEG chains can trigger hypersensitivity reactions and about 25% of the population shows the presence of anti-PEG antibodies. Additionally, PEG is a non-biodegradable polymer, which impairs excretion via the kidney and results in its accumulation in hepatic tissues as well as lysosomes of healthy cells [116, 117].

Also, frequent administration of PEGylated formulations via the intravenous route, triggers the production of anti-PEG IgM from the spleen leading to a phenomenon called as 'accelerated blood clearance, that reduces the benefit of long circulation of drugs [118]. In addition, large-scale production and continuous manufacturing of polymeric NPs remain significant obstacles [119].

6.3 Challenges with Solid Lipid Nanoparticles

To achieve commercial availability and regulatory approval for therapeutic applications, lipid NP must overcome several challenges, including large-scale manufacturing processes, optimization of tailored strategies, sterilization methods, and stability concerns. Furthermore, before proceeding with large-scale production and commercialization of solid lipid NPs, comprehensive clinical evaluation and environmental safety are important. The preparation of standardized protocols is vital to assess the promising risks associated with the exposure of nanomaterials, along with the establishment of a robust regulatory framework to ensure safety and efficacy. In addition, although oral administration is the most convenient and commonly preferred method for conventional drugs, the poor absorption of nanocarriers like SLNs through this route remains a major obstacle to the advancement and acceptance of this technology [120].

6.4 Challenges with Dendrimers

Despite their extensive applications in pharmaceuticals, the biological application of dendrimers is limited due to their inherent toxicity owing to their nanoscale dimensions [121]. The toxicity arises from the interaction between the in vivo biological membrane and the dendrimer surface having negative and cationic charges respectively. Such interactions disrupt the membrane through various mechanisms like the formation of nanohole, thinning, erosion of the membrane, formation of free radicals such as ROS, and cytokine release, which altogether support cellular damage and eventually cell death [121]. In biological systems, dendrimers often show toxicity in the form of cytotoxicity, haemolytic toxicity, and haematological toxicity. An example is amino-terminated PAMAM dendrimers, with reported haemolytic effects on rat blood cell solutions and cytotoxicity on human intestinal adenocarcinoma (Caco-2) cells [122]. To address this issue, two key approaches have been employed: first, biocompatible dendrimers designing and synthesis, and second, the modification of surface charges through surface engineering. Chemical surface modification has a vital role in mitigating the toxicity associated with dendrimers [123].

6.5 Challenges with Nanofibers

Nanofibers have demonstrated significant potential for drug delivery applications; however, several challenges must be addressed before electrospun nanofibers can be effectively utilized as drug delivery systems. Some key obstacles comprise of active ingredient stability, drug LC, initial burst release of drugs, residual solvent content, and the feasibility of industrial large-scale production. Although some studies report improved drug loading and extended-release with specific polymers, the uneven distribution of drugs within nanofibers often leads to an initial burst release, requiring further optimization to refine this technology for mainstream drug delivery.

The scalability of nanofiber production using the electrospinning technique remains a significant challenge. While laboratory-scale electrospinning machines typically handle small volumes in the millilitre range, transitioning to industrial-scale production involving volumes in litres presents numerous technical difficulties. These challenges highlight the need for further advancements in both process optimization and machinery design [124]. Therefore, it is crucial to thoroughly understand and address these issues to enable the utilization of electro-spun nanofibers in mainstream delivery systems for drugs, ensuring an optimal release profile and LC of drugs [125].

6.6 Challenges with Niosome

Niosome aqueous solutions owing to fusion, aggregation, drug leakage, and hydrolysis of entrapped drug substances have a limited shelf life. Sonication and extrusion are tedious processes that need dedicated processing equipment to generate MLV. Niosomes also exhibit problems with physical stability. Niosomes may accumulate, fuse, leak drugs, or hydrolyse stored medications while the dispersion is being stored. Furthermore, sterilizing niosomes is a laborious process. Dry heat and steam sterilization are not suitable for lipid-based formulations with a relatively low gel-to-liquid transition temperature, as the high temperatures involved in these processes can cause significant drug leakage from the bilayer vesicles [126, 127]. Niosomes should not be subjected to membrane filtration or heat sterilization. Therefore, to create commercially available niosomal preparations, these challenges require more research [126, 128].

6.7 Challenges with Polymeric Micelles

Despite having the advantages of smaller size, high water solubility, structural stability, and low toxicity; polymeric micelles possess certain limitations. The AB-type block copolymer is among the preferred structures for forming polymeric micelles having a well-defined size, shape, and diameter. The AB-type block copolymer has a simple architecture, consisting of two polymer blocks connected sequentially. However, its synthesis is generally more challenging compared to random polymers, where various units are arranged randomly along the polymer chain. In the polymeric micelle studies, high levels of understanding of polymer chemistry are required. Due to the synthesis-linked problem, the option of monomer units and their possible combinations are significantly restricted. In some instances, where laboratories can synthesize the required block copolymer, it becomes difficult to synthesize the block copolymer on a large industrial scale with high reproducibility. Also, advanced expertise in the field of polymer chemistry is important for synthesizing controlled chemical structures with precise chain lengths. The scale-up issue is more critical than synthesizing polymer since physical parameters such as rate of solvent exchange and diffusion are greatly influenced by the scale-up of the drug incorporation processes including the exchange of solvent through a dialysis membrane. Thus, further research is essential to achieve significant advancements in the incorporation of technology [128].

7 Future Directions and Perspectives

The future of nanotechnology and advanced delivery systems for managing MetS highlights several important areas for advancements that can transform the healthcare system [11]. Fabricating materials at the nanoscale enables researchers to develop NPs with distinct properties, such as a large surface area and the capability to be functionalized with various ligands for targeted drug delivery to specific cells or tissues [129], allowing a controlled release, improving both treatment precision and effectiveness while minimizing side effect, reducing drug dosage and toxicity, improving target site concentration and ensuring safer, more efficient treatments [130]. Biodegradable NPs have demonstrated potential for delivering proteins, peptides, and natural compounds at minimal doses [11]. Stimuli-responsive nanocarriers and theranostic platforms were found to enable targeted therapy with real-time monitoring [130]. Also, the incorporation of personalized medicine, powered by pharmacogenomics, AI, and nanotechnology, has the future of transforming metabolic disorder management by enabling precise diagnosis, personalized treatments, and reduced side effects [131]. AI-driven profiling [132] and pharmacokinetic modelling [131] enhance biomarker discovery, while nanotechnology aids in predicting, preventing, and managing these conditions effectively [133]. These advancements address challenges like biocompatibility, toxicity, and regulatory barriers, paving the way for more effective, personalized, and accessible solutions for managing metabolic syndrome [134].

However, MetS studies often overlook the central nervous system (CNS), despite its role in obesity, hypertension, and inflammation. Leptin's effects on renal activity and oxidative stress, along with inflammatory mediators like TNF-α and IL-6, highlight the complex interplay driving MetS [11]. Also, there is slow progress in managing MetS due to challenges such as the lack of robust preclinical models, and the need for detailed studies on lipid metabolism. So there is a greater need for developing biodegradable, biocompatible, and orally stable nanoformulations with rapid action and cellular uptake while addressing the various challenges through these innovative therapies [11]. Nanotechnology also has a futuristic that will transform gene and RNA-based therapies by enabling the safe delivery of tools like CRISPR and siRNA to address molecular dysfunctions.

8 Summary and Conclusion

8.1 Summary of Key Insights

- Nanomaterials present transformative potential in addressing various dimensions of MetS, offering new avenues for treatment, prevention, and monitoring.
- Nanomaterials for targeted delivery such as liposomes, polymeric NPs, dendrimers, carbon-based nanomaterials, and SLNs play a vital role in drug delivery to the targeted site for treating MetS.

8.2 Outlook on the Future of Nanotechnology in Metabolic Syndrome Treatment

- Nanotechnology offers significant potential in the treatment of metabolic syndrome, a collection of interrelated conditions which elevate the risk of CVD, stroke, diabetes mellitus, and fatty liver disease.
- As researchers explore innovative applications, the outlook for nanotechnology in this field is multifaceted in drug delivery systems and the targeted approach can enhance the efficacy of treatments by minimizing side effects, particularly in managing insulin resistance and lowering cholesterol levels.
- Diagnostic tools can detect biomarkers at low concentrations, which can be crucial for monitoring glucose levels and other metabolic indicators.
- NPs as therapeutic agents can be engineered to enhance the bioavailability and stability of drugs used in MetS management.
- Despite its potential, the integration of nanotechnology in medical practice faces issues like regulatory hurdles, safety as well as toxicity concerns, and therefore, there is a need for further research to establish effective protocols.
- Public perception and acceptance of nanomedicine also play a crucial role in its adoption. Continued research and innovation will be essential for translating these technologies into practical applications that improve patient outcomes.

Authors' Contributions Gitali Sen, Pradeep Kumar Patel, Soki Daeme Malang, Pooja Kathait, Ankur, Navneet Yadav and Sunday O. Otimenyin made the conceptualization, designed the content and wrote the original draft. Alakh N. Sahu provided the scientific input, reviewed the final version and incorporated appropriate content in the final version of book chapter.

References

1. Kassi, E., et al. (2011). Metabolic syndrome: Definitions and controversies. *BMC Medicine, 9*, 1–13.
2. Alberti, K. G. M., Zimmet, P., & Shaw, J. (2005). The metabolic syndrome—a new worldwide definition. *The Lancet, 366*(9491), 1059–1062.
3. Abou Assi, R., Abdulbaqi, I. M., & Siok Yee, C. (2021). The evaluation of drug delivery nanocarrier development and pharmacological briefing for metabolic-associated fatty liver disease (MAFLD): An update. *Pharmaceuticals, 14*(3), 215.

4. Basaranoglu, M., & Neuschwander-Tetri, B. A. (2006). Nonalcoholic fatty liver disease: Clinical features and pathogenesis. *Gastroenterology & Hepatology, 2*(4), 282.
5. Åberg, F., et al. (2018). Interaction between alcohol consumption and metabolic syndrome in predicting severe liver disease in the general population. *Hepatology, 67*(6), 2141–2149.
6. Samson, S. L., & Garber, A. J. (2014). Metabolic syndrome. *Endocrinology and Metabolism Clinics, 43*(1), 1–23.
7. Lillich, F. F., Imig, J. D., & Proschak, E. (2021). Multi-target approaches in metabolic syndrome. *Frontiers in Pharmacology, 11*, 554961.
8. Abbasi, M., Boka, D. A., & DeLoit, H. (2024). Nanomaterial-enhanced microneedles: Emerging therapies for diabetes and obesity. *Pharmaceutics, 16*(10), 1344.
9. Uppal, S., et al. (2018). Nanoparticulate-based drug delivery systems for small molecule anti-diabetic drugs: An emerging paradigm for effective therapy. *Acta Biomaterialia, 81*, 20–42.
10. Rafiee, M. H., et al. (2023). Oral pioglitazone HCl-loaded solid lipid microparticles: Formulation design and bioactivity studies. *Journal of Applied Pharmaceutical Science, 13*(2), 161–174.
11. Taghipour, Y. D., et al. (2019). Nanoformulations of natural products for management of metabolic syndrome. *International Journal of Nanomedicine, 14*, 5303–5321.
12. Ventola, C. L. (2017). Progress in nanomedicine: Approved and investigational nanodrugs. *Pharmacy and Therapeutics, 42*(12), 742.
13. Magliano, D. J., Shaw, J. E., & Zimmet, P. Z. (2006). How to best define the metabolic syndrome. *Annals of Medicine, 38*(1), 34–41.
14. Liberopoulos, E., Mikhailidis, D., & Elisaf, M. (2005). Diagnosis and management of the metabolic syndrome in obesity. *Obesity Reviews, 6*(4), 283–296.
15. Bahadori, E., et al. (2019). Recent advances in nanotechnology for the treatment of metabolic syndrome. *Diabetes & Metabolic Syndrome: Clinical Research & Reviews, 13*(2), 1561–1568.
16. Matsuzawa, Y. (2010). Adiponectin: A key player in obesity related disorders. *Current Pharmaceutical Design, 16*(17), 1896–1901.
17. Huang, P. L. (2009). A comprehensive definition for metabolic syndrome. *Disease Models & Mechanisms, 2*(5–6), 231–237.
18. Semenkovich, C. F. (2006). Insulin resistance and atherosclerosis. *The Journal of Clinical Investigation, 116*(7), 1813–1822.
19. Heximer, S. P., et al. (2003). Hypertension and prolonged vasoconstrictor signaling in RGS2-deficient mice. *The Journal of Clinical Investigation, 111*(4), 445–452.
20. Mota, M., et al. (2016). Molecular mechanisms of lipotoxicity and glucotoxicity in nonalcoholic fatty liver disease. *Metabolism, 65*(8), 1049–1061.
21. Polyzos, S. A., Kountouras, J., & Mantzoros, C. S. (2019). Obesity and nonalcoholic fatty liver disease: From pathophysiology to therapeutics. *Metabolism, 92*, 82–97.
22. Grundy, S. M. (2006). Drug therapy of the metabolic syndrome: Minimizing the emerging crisis in polypharmacy. *Nature Reviews Drug Discovery, 5*(4), 295–309.
23. Bianchi, C., et al. (2007). Treating the metabolic syndrome. *Expert Review of Cardiovascular Therapy, 5*(3), 491–506.
24. Dandona, P., et al. (2005). Metabolic syndrome: A comprehensive perspective based on interactions between obesity, diabetes, and inflammation. *Circulation, 111*(11), 1448–1454.
25. Pang, A. S.-R., et al. (2024). Nanoparticles as drug delivery systems for the targeted treatment of atherosclerosis. *Molecules, 29*(12), 2873.
26. Adetunji, C. O., et al. (2022). Potentialities of nanomaterials for the management and treatment of metabolic syndrome: A new insight. *Materials Today Advances, 13*, 100198.
27. Ganesan, P., et al. (2018). Recent developments in solid lipid nanoparticle and surface-modified solid lipid nanoparticle delivery systems for oral delivery of phyto-bioactive compounds in various chronic diseases. *International Journal of Nanomedicine, 13*, 1569–1583.
28. Duan, X., Chen, H.-l., & Guo, C. (2022). Polymeric nanofibers for drug delivery applications: A recent review. *Journal of Materials Science: Materials in Medicine, 33*(12), 78.

29. Zhao, Y., et al. (2008). Controlled release and interaction of protein using self-assembling peptide RATEA16 nanofiber hydrogels. *Journal of Polymer Science Part A: Polymer Chemistry, 46*(14), 4927–4933.
30. Kim, Y.-J., Ebara, M., & Aoyagi, T. (2012). Temperature-responsive electrospun nanofibers for 'on–off' switchable release of dextran. *Science and Technology of Advanced Materials, 13*(6), 064203.
31. Fang, R., et al. (2015). Nanoscale drug delivery systems: A current review on the promising paclitaxel formulations for future cancer therapy. *Nano, 10*(05), 1530004.
32. Hawker, C. J., & Frechet, J. M. (1990). Preparation of polymers with controlled molecular architecture. A new convergent approach to dendritic macromolecules. *Journal of the American Chemical Society, 112*(21), 7638–7647.
33. Patel, H., & Patel, P. (2013). Dendrimer applications–A re view. *International Journal of Pharma and Bio Sciences, 4*(2), 454–463.
34. Bharali, D. J., et al. (2009). Nanoparticles and cancer therapy: A concise review with emphasis on dendrimers. *International Journal of Nanomedicine, 4*, 1–7.
35. Buchke, S., et al. (2022). Mitochondria-targeted, nanoparticle-based drug-delivery systems: Therapeutics for mitochondrial disorders. *Life, 12*(5), 657.
36. Kaur, I. P., et al. (2004). Vesicular systems in ocular drug delivery: An overview. *International Journal of Pharmaceutics, 269*(1), 1–14.
37. Moghassemi, S., & Hadjizadeh, A. (2014). Nano-niosomes as nanoscale drug delivery systems: An illustrated review. *Journal of Controlled Release, 185*, 22–36.
38. Liu, T., & Guo, R. (2005). Preparation of a highly stable niosome and its hydrotrope-solubilization action to drugs. *Langmuir, 21*(24), 11034–11039.
39. Liu, T., et al. (2007). Structure behaviors of hemoglobin in PEG 6000/Tween 80/Span 80/H2O niosome system. *Colloids and Surfaces A: Physicochemical and Engineering Aspects, 293*(1–3), 255–261.
40. Pouyani, T., & Prestwich, G. D. (1994). Functionalized derivatives of hyaluronic acid oligosaccharides: Drug carriers and novel biomaterials. *Bioconjugate Chemistry, 5*(4), 339–347.
41. Marianecci, C., et al. (2010). Non-ionic surfactant vesicles in pulmonary glucocorticoid delivery: Characterization and interaction with human lung fibroblasts. *Journal of Controlled Release, 147*(1), 127–135.
42. Cheng, J., et al. (2023). Nanomedicine for diagnosis and treatment of atherosclerosis. *Advanced Science, 10*(36), 2304294.
43. Cornier, M.-A., et al. (2008). The metabolic syndrome. *Endocrine Reviews, 29*(7), 777–822.
44. Welty, F. K., Alfaddagh, A., & Elajami, T. K. (2016). Targeting inflammation in metabolic syndrome. *Translational Research, 167*(1), 257–280.
45. El-Daly, S. M., et al. (2021). Stimulatory effect of docosahexaenoic acid alone or loaded in zinc oxide or silver nanoparticles on the expression of glucose transport pathway. *Prostaglandins & Other Lipid Mediators, 155*, 106566.
46. Lo, J., et al. (2007). Effects of TNF-α neutralization on adipocytokines and skeletal muscle adiposity in the metabolic syndrome. *American Journal of Physiology-Endocrinology and Metabolism, 293*(1), E102–E109.
47. Juillerat-Jeanneret, L. (2020). The other angiotensin II receptor: AT2R as a therapeutic target. *Journal of Medicinal Chemistry, 63*(5), 1978–1995.
48. Weiss, A., et al. (2011). Intracellular peptide delivery using amphiphilic lipid-based formulations. *Biotechnology and Bioengineering, 108*(10), 2477–2487.
49. Weill, C. O., Biri, S., & Erbacher, P. (2008). Cationic lipid–mediated intracellular delivery of antibodies into live cells. *Biotechniques, 44*(sup7), vii–xi.
50. Bayele, H. K., et al. (2006). Protein transduction by lipidic peptide dendrimers. *Journal of Pharmaceutical Sciences, 95*(6), 1227–1237.
51. Salame, P. H., Pawade, V. B., & Bhanvase, B. A. (2018). Characterization tools and techniques for nanomaterials. In *Nanomaterials for green energy* (pp. 83–111). Elsevier.

52. Lin, P.-C., et al. (2014). Techniques for physicochemical characterization of nanomaterials. *Biotechnology Advances, 32*(4), 711–726.
53. Sapsford, K. E., et al. (2011). Analyzing nanomaterial bioconjugates: A review of current and emerging purification and characterization techniques. *Analytical Chemistry, 83*(12), 4453–4488.
54. Brar, S. K., & Verma, M. (2011). Measurement of nanoparticles by light-scattering techniques. *TrAC Trends in Analytical Chemistry, 30*(1), 4–17.
55. Song, X., et al. (2008). Dual agents loaded PLGA nanoparticles: Systematic study of particle size and drug entrapment efficiency. *European Journal of Pharmaceutics and Biopharmaceutics, 69*(2), 445–453.
56. Khalbas, A. H., et al. (2024). Drug loading methods and kinetic release models using of mesoporous silica nanoparticles as a drug delivery system: A review. *South African Journal of Chemical Engineering, 50*, 261.
57. Somvanshi, S. B., et al. (2021). Nanomaterials and pharmacokinetics. In *Nano-pharmacokinetics and theranostics* (pp. 1–14). Elsevier.
58. Nsairat, H., et al. (2021). Lipid nanostructures for targeting brain cancer. *Heliyon, 7*(9), e07994.
59. Mudalige, T., et al. (2019). Characterization of nanomaterials: Tools and challenges. In *Nanomaterials for food applications* (pp. 313–353).
60. Pires, F. Q., et al. (2017). Use of mixture design in drug-excipient compatibility determinations: Thymol nanoparticles case study. *Journal of Pharmaceutical and Biomedical Analysis, 137*, 196–203.
61. Jain, A. K., & Thareja, S. (2019). In vitro and in vivo characterization of pharmaceutical nanocarriers used for drug delivery. *Artificial Cells, Nanomedicine, and Biotechnology, 47*(1), 524–539.
62. Alai, M. S., Lin, W. J., & Pingale, S. S. (2015). Application of polymeric nanoparticles and micelles in insulin oral delivery. *Journal of Food and Drug Analysis, 23*(3), 351–358.
63. Thanou, M., Verhoef, J., & Junginger, H. (2001). Oral drug absorption enhancement by chitosan and its derivatives. *Advanced Drug Delivery Reviews, 52*(2), 117–126.
64. Labieniec-Watala, M., et al. (2014). Can metabolic impairments in experimental diabetes be cured with poly (amido) amine (PAMAM) G4 dendrimers?–In the search for minimizing of the adverse effects of PAMAM administration. *International Journal of Pharmaceutics, 464*(1–2), 152–167.
65. Souto, E. B., et al. (2019). Nanoparticle delivery systems in the treatment of diabetes complications. *Molecules, 24*(23), 4209.
66. Hasan, A. A., Madkor, H., & Wageh, S. (2013). Formulation and evaluation of metformin hydrochloride-loaded niosomes as controlled release drug delivery system. *Drug Delivery, 20*(3–4), 120–126.
67. Li, X., et al. (2013). Intestinal mucosa permeability following oral insulin delivery using core shell corona nanolipoparticles. *Biomaterials, 34*(37), 9678–9687.
68. Wong, C. Y., Al-Salami, H., & Dass, C. R. (2017). Potential of insulin nanoparticle formulations for oral delivery and diabetes treatment. *Journal of Controlled Release, 264*, 247–275.
69. Kamei, N., et al. (2008). Usefulness of cell-penetrating peptides to improve intestinal insulin absorption. *Journal of Controlled Release, 132*(1), 21–25.
70. Zhu, X., et al. (2016). Polymeric nanoparticles amenable to simultaneous installation of exterior targeting and interior therapeutic proteins. *Angewandte Chemie International Edition, 55*(10), 3309–3312.
71. Rizvi, S. Z. H., et al. (2019). Simvastatin-loaded solid lipid nanoparticles for enhanced anti-hyperlipidemic activity in hyperlipidemia animal model. *International Journal of Pharmaceutics, 560*, 136–143.
72. Gifani, M., et al. (2021). Ultraselective carbon nanotubes for photoacoustic imaging of inflamed atherosclerotic plaques. *Advanced Functional Materials, 31*(37), 2101005.

73. Liu, J., et al. (2021). SR-A-targeted nanoplatform for sequential photothermal/photodynamic ablation of activated macrophages to alleviate atherosclerosis. *ACS Applied Materials & Interfaces, 13*(25), 29349–29362.
74. Oh, B., & Lee, C. H. (2015). Development of Man-rGO for targeted eradication of macrophage ablation. *Molecular Pharmaceutics, 12*(9), 3226–3236.
75. Jebari-Benslaiman, S., et al. (2022). Boosting cholesterol efflux from foam cells by sequential administration of rHDL to deliver microRNA and to remove cholesterol in a triple-cell 2D atherosclerosis model. *Small, 18*(13), 2105915.
76. Vargas-Castillo, A., et al. (2017). Understanding the biology of thermogenic fat: Is browning a new approach to the treatment of obesity? *Archives of Medical Research, 48*(5), 401–413.
77. Nassir, F., et al. (2015). Pathogenesis and prevention of hepatic steatosis. *Gastroenterology & Hepatology, 11*(3), 167.
78. Sibuyi, N. R. S., et al. (2019). Nanotechnology advances towards development of targeted-treatment for obesity. *Journal of Nanobiotechnology, 17*, 1–21.
79. Chen, S., et al. (2024). A quick paster type of soluble nanoparticle microneedle patch for the treatment of obesity. *Biomaterials, 311*, 122687.
80. Liang, S., et al. (2023). Non-cardiotoxic tetradecanoic acid-2, 4-dinitrophenol ester nanomicelles in microneedles exert potent anti-obesity effect by regulating adipocyte Browning and lipogenesis. *Small, 19*(39), 2301751.
81. Rocca, A., et al. (2015). Pilot in vivo investigation of cerium oxide nanoparticles as a novel anti-obesity pharmaceutical formulation. *Nanomedicine: Nanotechnology, Biology and Medicine, 11*(7), 1725–1734.
82. Zu, Y., et al. (2018). Resveratrol liposomes and lipid nanocarriers: Comparison of characteristics and inducing browning of white adipocytes. *Colloids and Surfaces B: Biointerfaces, 164*, 414–423.
83. Bialik, M., et al. (2023). Dual-controlled delivery of furosemide and atenolol using a biodegradable nanosystem for antihypertensive therapy. *Journal of Drug Delivery Science and Technology, 89*, 105006.
84. Alam, T., et al. (2017). Nanocarriers as treatment modalities for hypertension. *Drug Delivery, 24*(1), 358–369.
85. Todd, P. A., & Faulds, D. (1992). Felodipine: A review of the pharmacology and therapeutic use of the extended release formulation in cardiovascular disorders. *Drugs, 44*, 251–277.
86. Shah, U., Joshi, G., & Sawant, K. (2014). Improvement in antihypertensive and antianginal effects of felodipine by enhanced absorption from PLGA nanoparticles optimized by factorial design. *Materials Science and Engineering: C, 35*, 153–163.
87. Mehnert, W., & Mäder, K. (2012). Solid lipid nanoparticles: Production, characterization and applications. *Advanced Drug Delivery Reviews, 64*, 83–101.
88. Moghtadaie, A., et al. (2023). Emerging role of nanotechnology in treatment of non-alcoholic fatty liver disease (NAFLD). *Excli Journal, 22*, 946.
89. Jeon, S.-Y., & Imm, J.-Y. (2014). Lipase inhibition and cholesterol-lowering activities of laccase-catalyzed catechin polymers. *Food Science and Biotechnology, 23*, 1703–1707.
90. S Burnham, B., et al. (2011). Lipid-lowering effects of polymers derived from halophenyl pyrroles. *Letters in Drug Design & Discovery, 8*(6), 544–549.
91. Yang, B.-K., Park, J.-B., & Song, C.-H. (2002). Hypolipidemic effect of exo-polymer produced in submerged mycelial culture of five different mushrooms. *Journal of Microbiology and Biotechnology, 12*(6), 957–961.
92. Alonso, L., Calvo, M. V., & Fontecha, J. (2019). The influence of β-cyclodextrin on the reduction of cholesterol content in egg and duck liver pâté. *Foods, 8*(7), 241.
93. Alonso, L., et al. (2018). Effect of beta cyclodextrin on the reduction of cholesterol in Ewe's milk manchego cheese. *Molecules, 23*(7), 1789.
94. Wilson, R., & Turner, A. (1992). Glucose oxidase: An ideal enzyme. *Biosensors and Bioelectronics, 7*(3), 165–185.

95. Elsherif, M., et al. (2018). Glucose sensing with phenylboronic acid functionalized hydrogel-based optical diffusers. *ACS Nano, 12*(3), 2283–2291.
96. Mohanty, A. R., Ravikumar, A., & Peppas, N. A. (2022). Recent advances in glucose-responsive insulin delivery systems: Novel hydrogels and future applications. *Regenerative Biomaterials, 9*, rbac056.
97. Kumar, S., et al. (2017). Metformin-loaded alginate nanoparticles as an effective antidiabetic agent for controlled drug release. *Journal of Pharmacy and Pharmacology, 69*(2), 143–150.
98. Kumar, S., et al. (2013). Novel approaches for enhancement of drug bioavailability. *Reviews in Advanced Sciences and Engineering, 2*(2), 133–154.
99. Kumar, S., et al. (2016). Synthesis and optimization of ceftriaxone-loaded solid lipid nanocarriers. *Chemistry and Physics of Lipids, 200*, 126–132.
100. Sarkar, S., et al. (2023). Mesoporous silica nanoparticles: Drug delivery vehicles for antidiabetic molecules. *ChemBioChem, 24*(7), e202200672.
101. Chen, C., et al. (2017). Sustained-release study on Exenatide loaded into mesoporous silica nanoparticles: In vitro characterization and in vivo evaluation. *Daru Journal of Pharmaceutical Sciences, 25*, 1–8.
102. Sinha, A., Chakraborty, A., & Jana, N. R. (2014). Dextran-gated, multifunctional mesoporous nanoparticle for glucose-responsive and targeted drug delivery. *ACS Applied Materials & Interfaces, 6*(24), 22183–22191.
103. Xue, M., et al. (2013). Characterization, pharmacokinetics, and hypoglycemic effect of berberine loaded solid lipid nanoparticles. *International Journal of Nanomedicine, 8*, 4677–4687.
104. Ahangarpour, A., et al. (2018). Solid lipid nanoparticles of myricitrin have antioxidant and antidiabetic effects on streptozotocin-nicotinamide-induced diabetic model and myotube cell of male mouse. *Oxidative Medicine and Cellular Longevity, 2018*(1), 7496936.
105. Ahmed, T. A. (2020). Development of rosuvastatin flexible lipid-based nanoparticles: Promising nanocarriers for improving intestinal cells cytotoxicity. *BMC Pharmacology and Toxicology, 21*, 1–12.
106. Krishnaiah, Y. S. (2010). Pharmaceutical technologies for enhancing oral bioavailability of poorly soluble drugs. *Journal of Bioequivalence & Bioavailability, 2*(2), 28–36.
107. Cevc, G., & Blume, G. (2001). New, highly efficient formulation of diclofenac for the topical, transdermal administration in ultradeformable drug carriers, transfersomes. *Biochimica et Biophysica Acta (BBA)-Biomembranes, 1514*(2), 191–205.
108. Harisa, G. I., et al. (2018). Influence of pravastatin chitosan nanoparticles on erythrocytes cholesterol and redox homeostasis: An in vitro study. *Arabian Journal of Chemistry, 11*(8), 1236–1246.
109. Etemad, L., et al. (2023). An overview on nanoplatforms for statins delivery: Perspective study for safe and effective therapy methods. *Environmental Research, 234*, 116572.
110. Bathool, A., et al. (2012). Development and characterization of atorvastatin calcium loaded chitosan nanoparticles for sustain drug delivery. *Advanced Materials Letters, 3*(6), 466–470.
111. Li, T., et al. (2021). Recent developments in delivery of microRNAs utilizing nanosystems for metabolic syndrome therapy. *International Journal of Molecular Sciences, 22*(15), 7855.
112. Forssen, E., & Willis, M. (1998). Ligand-targeted liposomes. *Advanced Drug Delivery Reviews, 29*(3), 249–271.
113. Sawant, R. R., & Torchilin, V. P. (2012). Challenges in development of targeted liposomal therapeutics. *The AAPS Journal, 14*, 303–315.
114. Singh, N., et al. (2017). Drug delivery: Advancements and challenges. In *Nanostructures for drug delivery* (pp. 865–886). Elsevier.
115. Guadagnini, R., et al. (2015). Toxicity screenings of nanomaterials: Challenges due to interference with assay processes and components of classic in vitro tests. *Nanotoxicology, 9*(sup1), 13–24.
116. Thomas, O. S., & Weber, W. (2019). Overcoming physiological barriers to nanoparticle delivery—Are we there yet? *Frontiers in Bioengineering and Biotechnology, 7*, 415.

117. Garay, R. P., et al. (2012). *Antibodies against polyethylene glycol in healthy subjects and in patients treated with PEG-conjugated agents* (pp. 1319–1323). Taylor & Francis.
118. Suk, J. S., et al. (2016). PEGylation as a strategy for improving nanoparticle-based drug and gene delivery. *Advanced Drug Delivery Reviews, 99*, 28–51.
119. Colby, A. H., et al. (2021). Pilot-scale production of expansile nanoparticles: Practical methods for clinical scale-up. *Journal of Controlled Release, 337*, 144–154.
120. Scioli Montoto, S., Muraca, G., & Ruiz, M. E. (2020). Solid lipid nanoparticles for drug delivery: Pharmacological and biopharmaceutical aspects. *Frontiers in Molecular Biosciences, 7*, 587997.
121. Li, X., et al. (2022). Safety challenges and application strategies for the use of dendrimers in medicine. *Pharmaceutics, 14*(6), 1292.
122. Jevprasesphant, R., et al. (2003). The influence of surface modification on the cytotoxicity of PAMAM dendrimers. *International Journal of Pharmaceutics, 252*(1–2), 263–266.
123. Jain, K., et al. (2010). Dendrimer toxicity: Let's meet the challenge. *International Journal of Pharmaceutics, 394*(1–2), 122–142.
124. Persano, L., et al. (2013). Industrial upscaling of electrospinning and applications of polymer nanofibers: A review. *Macromolecular Materials and Engineering, 298*(5), 504–520.
125. Thakkar, S., & Misra, M. (2017). Electrospun polymeric nanofibers: New horizons in drug delivery. *European Journal of Pharmaceutical Sciences, 107*, 148–167.
126. Ag Seleci, D., et al. (2016). Niosomes as nanoparticular drug carriers: Fundamentals and recent applications. *Journal of Nanomaterials, 2016*(1), 7372306.
127. Abdelbary, G., & El-Gendy, N. (2008). Niosome-encapsulated gentamicin for ophthalmic controlled delivery. *AAPS PharmSciTech, 9*, 740–747.
128. Masjedi, M., & Montahaei, T. (2021). An illustrated review on nonionic surfactant vesicles (niosomes) as an approach in modern drug delivery: Fabrication, characterization, pharmaceutical, and cosmetic applications. *Journal of Drug Delivery Science and Technology, 61*, 102234.
129. Emeihe, E. V., et al. (2024). Revolutionizing drug delivery systems: Nanotechnology-based approaches for targeted therapy. *International Journal of Life Science Research Archive, 7*(1), 040–058.
130. Rocha, S., et al. (2021). Nano-based drug delivery systems used as vehicles to enhance polyphenols therapeutic effect for diabetes mellitus treatment. *Pharmacological Research, 169*, 105604.
131. Alqahtani, A. A., et al. (2023). 3D printed pharmaceutical systems for personalized treatment in metabolic syndrome. *Pharmaceutics, 15*(4), 1152.
132. Ramírez-Mejía, M. M., & Méndez-Sánchez, N. (2023). Advances in biomedical engineering for understanding and managing metabolic syndrome: A comprehensive review. *BME Horizons, 1*(2), 68–68.
133. Singh, S., et al. (2023). Unveiling the future of metabolic medicine: Omics technologies driving personalized solutions for precision treatment of metabolic disorders. *Biochemical and Biophysical Research Communications, 682*, 1.
134. Radhakrishnan, A., Kuppusamy, G., & Subramaniam, T. (2022). Personalized nanotools for the treatment of metabolic disorders. *Recent Innovations in Chemical Engineering (Formerly Recent Patents on Chemical Engineering), 15*(1), 3–13.

Recent Advances in Nanotechnology-Based Approaches for Diagnosis and Treatment of Metabolic Syndrome

6

Sutapa Biswas Majee, Rachayeeta Bera, Hassan Raza, and Arshad Warsi

Abstract

Metabolic syndrome is an aggregate of altered metabolic and biochemical phenomena, precipitating in diabetes, obesity, dyslipidemia, and cardiovascular diseases. Their diagnosis and treatment should be initiated early to prevent their worsening. The current chapter reviews PubMed, Scopus, and Elsevier databases on the latest scientific approaches in nanosensors and nanotherapeutics in metabolic syndrome. The research articles are past 2020, and the recent works are focused. Hyperglycemia and obesity are increasing at an alarming rate primarily due to sedentary lifestyles, high-fat diets, and calorie-rich foods, which are preventable. Dyslipidemia is slowly engulfing with its consequences and is one of the primary causes of atherosclerosis, plaque formation, and other cardiovascular diseases. Obesity can lead to diabetes, dyslipidemia, and cardiovascular symptoms. The design of robust, precise, reusable, specific, sensitive, and accurate nanosensors will minimize the burden of metabolic syndrome, afflicting the majority of the population globally. However, the role of nanosensors is limited to detection of the condition. Nanotherapeutics offers versatile opportunities for designing novel therapeutic platforms for diabetes, obesity, and dyslipidemia. The therapeutic delivery systems promise superior efficiency, minimally invasive or noninvasive techniques, and cost-effectiveness. They promote controlled and sustained drug release with drugs of varying physicochemical properties. But nanotechnology might provide benchmark therapeutic novelties in metabolic

S. B. Majee (✉) · R. Bera · H. Raza
NSHM College of Pharmaceutical Technology, NSHM Knowledge Campus, Kolkata-Group of Institutions, Kolkata, West Bengal, India

A. Warsi
Gomal Center of Biochemistry and Biotechnology (GCBB), Gomal University, Dera Ismail Khan, Pakistan

V. K. Chaturvedi et al. (eds.), *Nanotheranostics in Metabolic Syndrome*,
https://doi.org/10.1007/978-981-95-3746-4_6

syndrome management, only after human clinical trials are completed. Scale-up and regulatory approval are mandatory requirements for their commercialization. No study on metabolic syndrome has been done on humans. Thus, the ultimate challenge of nanosensors and nanotherapeutics for metabolic syndrome is far ahead, and the future scope lies in maximizing their use and benefits for clinical applications in humans.

Keywords

Diabetes · Dyslipidemia · Nanosensor · Nanotherapeutics · Obesity

1 Introduction

Metabolic syndrome is a conglomerate of altered cellular metabolic and biochemical pathways, leading to suboptimal energy production in the body, with chronic or acute clinical manifestations of a specific disease symptom. These disorders are widespread across various sections of the population, especially those with a sedentary lifestyle and imbalanced nutritional status. They include diabetes and related complications, obesity, dyslipidemia, and cardiovascular diseases. Pathophysiology of metabolic syndromes is a vast topic and can occur due to malfunctioning or upregulation of specific genetic components. However, interested persons may go through peer-reviewed articles published in reputed international publications [22, 72].

Early diagnosis and treatment of metabolic syndromes is crucial to overcoming future risks and increasing the quality of life span. The demerits of conventional diagnostic procedures are the multistep sample treatment process, lack of appropriate instrumentation facilities, prolonged time for analysis (>30 min), specificity, accuracy, and sensitivity. Recent literature surveys, mainly from 2020 onward, reveal endeavors in the field of nanotechnology to enable the detection and management of metabolic disorders with facile instrument fabrication, minute sample volume, easy operation, high sensitivity, selectivity, reproducibility, and precision, culminating in personalized healthcare solutions [40, 87].

The global health landscape is facing an almost insurmountable task of managing metabolic diseases. Diabetes mellitus is estimated to be the leading cause of mortality in developed nations in the next 20 years. Hyperglycemia seems to be an alarming precedence in metabolic disorders with far-reaching consequences, rising to almost 11% by 2045 [23, 100]. The development of easily manufacturable, robust, precise, reusable, and accurate diagnostics will optimize the surveillance of high-risk patients with utmost necessity. The financial losses and inherent inefficiencies of the current antidiabetic medicines cannot be overlooked.

Obesity is a long-term metabolic disorder endangering adipocyte differentiation, characterized by excess caloric intake and fat accumulation, leading to fatty tissue expansion and increasing the propensity for various hypertrophic and hyperplastic conditions. The primary causes are sedentary lifestyles, high-fat and calorie-rich food, increased salt intake, and biochemical, hormonal, genetic, epigenetic, microbial, sociocultural, and environmental influences. Among others, it enhances the risk of several chronic metabolic disorders such as type II diabetes mellitus, insulin resistance, and pathophysiological conditions leading to hyperlipidemia, hypertension, cardiovascular complications, coronary artery diseases, inflammation or thrombosis, nonalcoholic fatty liver disease, gall stones, etc. It can precipitate obesity-induced inflammation of vital organs including the heart, brain, skeletal muscle, brain, adipose, pancreas, and liver [5, 88, 98]. Elevation of lipids and glucose can increase vascular resistance, the propensity for vascular hypertrophy and heart enlargement. Increased levels of low-density lipoprotein cholesterol with simultaneous reduction of high-density lipoprotein (HDL-C) can trigger atherosclerosis. Bariatric surgery and antiobesity drugs are restricted by their nonspecificity, unpredictable weight reduction effects, and low efficacy ($<10\%$ weight loss in 12 months). Other side effects of commonly used antiobesity drugs include cardiac valvular anomalies, stroke, pulmonary hypertension, psychotic disorders, addiction susceptibility, and even death. No drug has been able to reduce a 10% loss in weight over a sustained 12-month period [89]. The incidence of obesity is increasing at an alarming rate posing a risk to the current global scenario. Recent global data show the epidemic nature of obesity and its consequences on healthcare status with repeated warnings on altered lifestyle and dietary intake. Thus, efforts are essential for revolutionizing the novel field of nanotechnology in managing obesity.

Dyslipidemia occurs because of cholesterol imbalance in the body. Cholesterol constitutes a crucial element in mammalian cell membrane physiology and is essential for cellular integrity and flexibility. However, hypercholesterolemia or dyslipidemia may result in hepatic oxidative stress, atherosclerosis, arterial hypertension, coronary artery diseases, peripheral artery disease, cerebrovascular disease, hepatic inflammation, and advanced hepatic failures. Atherosclerotic cardiovascular disease therefore constitutes primary risk for hypercholesteremia and is one of the primary causes of vascular diseases worldwide [54–56, 65, 73].

Cardiovascular diseases (CVDs) constitute serious health risks that include heart and blood vessels. Its etiology is related to hyperlipidemia, hypertension, diabetes, obesity, and family history. CVD outpaced cancer and other life-threatening diseases and ranked first in terms of morbidity and mortality. Atherosclerosis complicates the risk of precise localization of unstable or rupture-prone plaques before stroke or myocardial infarction. Therefore, there is an urgent need to investigate biomarkers for predicting risks of CVD and ensuring safe and effective therapeutic modalities [27].

Diabetes, obesity, dyslipidemia, and cardiovascular diseases are precursors to unhealthy lifestyle diseases and increase the susceptibility to various other metabolic syndromes and ailments. Understanding the cellular mechanism of metabolic disease progression and suitable manipulation technologies leads to the development of nanoscale sensors or nanosensors for their diagnosis. Rational design of multifunctional nanosensors, wearable devices, electronic devices, laboratory-on-a-chip devices etc. with nanoscale sensors for metabolic syndrome will enhance their morphological suitability, biocompatibility, environmental stability, pharmacokinetic profile, in vivo stability, flexibility, and shelf-life and produce surrogate endpoints for diagnostic purposes.

The nanosensor market was estimated to be more than USD 750 million in 2023 and is expected to reach around 7.5% CAGR in 2032. This need is primarily due to the increasing demand for real-time monitoring and detection facilities for lifestyle diseases. The nanosensors are virtually moving toward miniaturization and integration into compact and portable devices [32].

Harnessing the unique properties of cutting-edge functional nanocarriers, nanotechnology offers unprecedented and versatile opportunities to design novel therapeutic platforms and revolutionize delivery strategies in diabetes, obesity, hypercholesterolemia, and CVD. These targeted nanotherapeutic strategies can be developed into more efficient, minimally invasive, and cost-effective approaches. They promote controlled distribution and deposition of molecules of varying physicochemical properties in specific biological tissues and can be very beneficial in therapeutic interventions of lifestyle diseases [2, 84]. Drug-polymer ratio is the most vital aspect of such nanocarriers affecting various properties of nanotherapeutic systems such as particle size distribution, surface properties, on-demand release profile, and lastly, stability of the nanocarriers, thereby maximizing therapeutic efficacy and minimizing the incidences of off-target adverse effects [103].

The chapter aims to shed light on the current literature database of PubMed, Scopus, and Elsevier, on the latest research in nanotechnology in the diagnosis and drug delivery of metabolic syndrome, with emerging nanosensors and nanotherapeutics. The inter-relationship between diabetic complications, obesity, dyslipidemia, and cardiovascular disorders is a rapidly evolving archetype. Evidence was taken from studies done before February 2025 till 2020 with few studies before 2020. The search terms included "diabetes," "hypercholesterolemia," "dyslipidemia," "cardiovascular disease," "lipid-lowering therapies," "obesity," "nanosensors," "nanotechnology," "nanotheranostics," "nanotherapeutics," and others. The searches were restricted to English-language articles. The inclusion criteria involved primarily peer-reviewed research articles, meta-analyses, and clinical trials.

The chapter focuses on the applications of nanotechnology as sensing platforms and platforms for therapeutic management of metabolic syndrome, and challenges in the design and delivery of novel nanotechnological approaches for the management of lifestyle diseases, afflicting the global population.

2 Nanosensors and Nanotherapeutics in Diabetes Care

The global health landscape is marked by paramount drift with a significant number of people suffering from immense risk of diabetes. Diabetes is bound to endanger the world population with a risk of more than $2 trillion being diagnosed with the disease [50]. Dysregulation of blood glucose homeostasis, chronic kidney complications, cardiovascular diseases, and blindness are some markers. Glucose monitoring devices should be manufacturable at a fast speed in sufficient numbers with a high sensitivity, and accuracy to prevent premature mortality. Moreover, they should be stable and lasting enough for easy estimation. A better understanding of the etiology and pathophysiology of diabetes is essential for analyzing the cellular and molecular cascades of the dreaded disease. Thus, stimuli-sensitive nanosensors with minimum off-target cytotoxicity will suffice as highly efficient and beneficial devices for glucose monitoring [23, 68].

In diabetes research, nanotechnology has envisioned nanosensors as the glucose-sensing devices of the time ahead. Clinical diagnosis of diabetes is initiated by the development of novel functional and target-specific smart nanosensors and progress in instrumental methods of analysis. Recent progress in nanosensors is creating immense potential for nanomaterial application in diabetes diagnosis over conventional methods [100]. They can test diabetes by determining the blood glucose content of an individual through skin and sweat, thereby permitting continuous, real-time monitoring in minimally invasive or noninvasive manners. Nanodimensions, extremely low surface area, light-weight, biocompatibility, superior mechanical strength, improved reaction area for enzymatic catalysis or nonenzymatic reaction, superior electrochemical attributes, tunable electric conductivity, sensitivity, specificity, speed, suitability at a quantum level, precision, and reproducibility have radically transformed diabetes diagnosis. Nanosensors are bound to counteract market competition in the future [69].

The application of nanosensors in the electrochemical monitoring of blood glucose has led to significant strides in diabetes detection. Although enzymatic amperometric nanosensors, employing primarily electrochemical methods of sensing, offer superior speed, selectivity, and sensitivity, they have their limitations such as the intrinsic nature of immobilized glucose oxidase on the solid electrode, the precise quantity of enzyme present, restricted operation facility, low stability with temperature, pH, humidity variation, and reproducibility. Thus, nonenzymatic glucose nanosensors find wide applications and are marked by easy modification process, superior shelf-life, sensitivity, specificity, and low detection capacity. Nonenzymatic devices act as high-performance electrocatalysts and promote direct electro-oxidation of the minute quantity of glucose in human body fluids accessing the electrode surface [71, 82]. Metallic electrochemical nanosensors of platinum, gold, silver, etc. serve a crucial role and can help in glucose detection at even nano or picoconcentration, but at a higher price and with risk of toxicity. Thus, inexpensive metal oxides or sulfides such as copper and nickel have found wide applications in delivering nonenzymatic nanosensors for glucose at optimum cost [57]. Analysis and readout of the biological signal consist of three elements, mainly, detector for recognition of biological elements such as enzymes, receptors, antibodies, lectins, signal transducer, or nanosensor that transduces the recognized signal in vivo into a quantifiable, output one, and finally, a reading device that converts the measurable signal into an appropriate readable form by the physician or the patient [23]. There have been efforts to produce nanoprobes with superior β-cell specificity and, thus, high imaging potential. This may offer excellent opportunities for noninvasive quantification of in vivo endogenous β-cell population [99]. Magnetic nanoprobes, especially the superparamagnetic iron oxide nanoparticles (SPIONs) are being visualized as novel biodegradable diagnostic agents for blood-glucose sensing. The design of appropriate SPION-based MRI contrast imaging agents can provide benefits in diagnosing diabetic patients [52]. Nanosensor-type Field-effect waterproof transistor films can be utilized for glucose monitoring of sweat with the help of a wearable device through pH change or change in charge. Additionally, these films can supply the required insulin through a touchable skin patch [68]. Implantation of subcutaneous electromagnetic nanosensor in free-roaming animals has shown potential in detecting variation in minimum glucose levels in a noninvasive manner [48]. Nickel-based nanomaterials are efficient nonenzymatic electrochemical glucose sensors but lose their performance and stability with time. However, their efficiency may be improved by designing ligand-free nickel nanosensors with a detection capacity of extremely low glucose concentration [13].

Electrically conductive multiwalled carbon nanotubes (MWCNTs) may generate the microphysiometer which helps in the diagnosis of diabetes by indirect monitoring of insulin levels at physiological pHs. Real-time monitoring of insulin concentration is possible. Implantable polyethylene bead nanosensors in the skin produce fluorescence signals after being displaced by glucose which is visible as a glowing tattoo. Microchip-based nanosensors have been developed recently to differentiate between two forms of diabetes by detecting antibodies through fluorescence signals in a portable manner [93]. Semiconductor quantum dots (QDs) have demonstrated high quantum yields, broad excitation ranges, specific emission characteristics, high photostability, and prolonged fluorescence lifetime leading to superior optoelectronic properties [20].

It must be emphasized that nonenzymatic nanosensors show great possibilities as glucose monitoring devices through optical or fluorescence spectroscopic methods. Table 6.1 depicts the nanostructures for glucose estimation post 2020. It is noteworthy that there is no reference to glucose nanosensors after 2023. The majority of the nanosensors are copper nanostructures, a few being coordinated with nickel ones. Nanostructures vary from nanoflowers, nanoleaves, nanoporous films, nanowire, nanofibers, and quantum dots to nanotubes. The electrochemical method is the primary process with a few studies carried out through cyclic voltammetry and amperometry. The detection limit was primarily in the micromolar range to a maximum of 7 mM and the limit of detection (LOD) was <2 mM in most cases, and it was usually in micromolar range. Two studies show glucose detection is possible with nanosensors in less than 5 s.

Nanotechnology, especially the multifunctional nanoparticle formulations, is being utilized to enhance the ease, efficacy, and safety of insulin-dependent as well as insulin-replacement therapeutic options and reduce the risks and challenges of conventional formulations such as suboptimal dosing, low therapeutic efficacy, reduced specificity, metabolism, pain, and infection at the site of injection. They have been designed to protect against their potential cargo's enzymatic and hydrolytic deterioration. Organic, inorganic, hybrid, and biomimetic nanotechnology offers various options for loading different antidiabetic formulations. They can minimize insulin resistance, reduce the varying factors to absorption, and increase insulin secretion and response, insulin being a hydrophobic molecule [43, 100].

Table 6.1 Nanostructures in glucose monitoring

Nanostructure	Detection level	Analytical method used	Reference
Gold nanoparticle-tannic acid-3-amionopropyl-triethoxysilan-activated carbon cloth flower cluster	Detection of 1 mM glucose in sweat	Cyclic voltammetry	[109]
Self-healing triboelectric biosensor with glucose adaptive hydrogel	Detection of pH sensitivity with responsivity of 12 kHz/Ph	Piezoelectric gravimetry	[41]
Coaxial size-tunable copper oxide/nickel oxide hollow nanofiber	Detection of 1–10,000 µM	Electrochemistry	[54–56]
Gold-nickel alloy nanoparticles on activated carbon support	Detection limit of 0.41 µM with sensitivity of 1955 µA mM^{-1} cm^{-2}	Electrochemistry	[11]
Hierarchical 3D-morchella-shaped nickel-copper layered double hydroxide-copper hydroxide nanowires	Detection of 0.006–1.6 µM with LOD of 1.3 µM	Electrochemistry	[106]
Cerium oxide-β-nickel hydroxide nanoarchitecture assembly on reduced graphene oxide	Detection of 13 µM and sensitivity of 33.1 µA mM^{-1} cm^{-2}	Electrochemistry	[111]
Nickel-gold bilayer nanowire arrays	Detection of 50 µM–10 mM with LOD of 10.69 µM with a rapid response time of 2.5 s	Electrochemistry	[94]
Hierarchical and engineered copper oxide nanoleaves	Detection of 0.005–5.89 mM, LOD of 0.012 µM, sensitivity of 1467.32 µA mM^{-1} cm^{-1}, response time of <3.5 s	Cyclic voltammetry and amperometry	[3]
Copper nanowire-polyaniline-reduced graphene oxide nanocomposite	Detection of 0–4 mM, LOD of 1.6 mM, sensitivity of 843.06 µA mM^{-1} cm^{-2}	Amperometry	[10]
Nickel nanostructure with fluorine-doped tin oxide electrodeposition modified with silica nanochannel electrode	Detection of 10 µM-12 mM, LOD of 0.44 µM, and sensitivity of 62.3 µA mM^{-1}/cm^2	Electrochemistry	[24]
Reduced graphene oxide and nickel oxide nanocomposite	Detection limit of 19.35 µM	Electrochemistry	[6]

(continued)

Table 6.1 (continued)

Nanostructure	Detection level	Analytical method used	Reference
Silver/silver orthophosphate/graphene oxide quenched by Pd@Au core-shell nanoflowers	Detection of insulin in a linear range of 0.0001–80 ng/mL and LOD of 0.02 pg/mL	Electrochemiluminescence resonance energy transfer (ECL-RET) immunosensor	[46]
Copper oxide nanoparticle functionalized thin film electrode	Linear detection limit of up to 2.2 mM, sensitivity of 1207 μA mM^{-1} cm^{-2} and LOD of 1.19 μM	Electrochemistry	[38]
Nickel-niobium amorphous alloy type nanoglass	Linear detection of up to 2.2 mM, sensitivity of 20 μA mM^{-1} cm^{-2} and LOD of 100 nM, reproducibility >5000 cycles	Cyclic voltammetry and chronoamperometry	[13]
3D-copper sulfide nanosheets/cuprous oxide/cupric oxide nanowire array	Detection of 0.002–4.096 mM, LOD of 0.83 μM, and sensitivity of 4262 μA mM^{-1} cm^{-2}	Amperometry	[95]
Multiwalled carbon nanotube-copper 1,3,5-benxene tricarboxylic acid epoxy composite	Detection of 0.2–1.0 mM, LOD of 10 μM	Cyclic voltammetry, linear sweep voltammetry, differential pulse voltammetry, chronoamperometry	[71]
Nanoporous copper/nickel/gold film	Detection of 0.5 μM to 3 mM and 3–7 mM, LOD of 0.1 μM	Electrochemistry	[57]
Poly(dimethyldiallylammonium chloride)-copper oxide nanocomposite-carbon nanodots	Detection of 0.5–2 mM and 2–5 mM, LOD of 0.2 mM, sensitivity of 110, and 63.3 μA mM^{-1} cm^{-2}	Amperometry	[82]

Insulin, being acid-resistant, requires protection from the gastric pH and the nanocarriers are designed to release insulin only at its appropriate pH to elicit maximum therapeutic response. Mucoadhesive nanoparticles (NPs), such as mucin-chitosan complexes and chitosan-coated cationic liposomes are formidable strategies for delivering insulin [64]. Lipid-polymer-based nanostructured cochleates have been utilized for insulin delivery [44]. Zinc oxide/silver nanocomposites have demonstrated the potential to increase insulin secretion in alloxan-induced diabetic Wistar rats, with a reduction in fasting blood glucose levels (FBG), total cholesterol (TC), and total triglycerides (TG) [67]. Dextran-poly(lactic-co-glycolic acid) (PLGA)-based amphiphilic polymersomes of insulin exhibited high encapsulation efficiency and in vitro insulin release in simulated intestinal fluid, with potential hypoglycemic effects in diabetic rats. The oral bioavailability of insulin has been improved by chitosan and zein-carboxymethylated short-chain amylose nanocomposites. Insulin absorption was markedly enhanced, and hypoglycemic effects were observed with L-valine-targeted glucose-responsive chitosan-based multifunctional nanocarriers. Glucose-responsive acetylated dextran NPs also produced glycemic control in the diabetic mouse model. However, these glucose oxidase-based glucose-responsive delivery systems delay insulin release. The micro-environment pH must attain an appropriate value for insulin release [23]. Although the insulin-based nanodelivery systems suffer from inherent disadvantages, insulin-based nanopumps with MEMS (Micro-Electromechanical Systems) or NEMS (Nano Electromechanical Systems) chips may form a suitable strategy for some patients, where insulin along with some other drug molecules can be injected at a constant rate into the patient's body [50].

For other noninsulin antidiabetic drugs, oral medication must ensure enhanced intestinal permeability bypassing mucosal absorption for which modern inorganic nanoparticles of copper (Cu), silver (Ag), zinc oxide (ZnO), selenium (Se), and the noble metals such as gold (Au) and platinum (Pt). They can be utilized for delivering the drug or for their action [29]. Another prominent strategy involves islet cell transplantation by encapsulation in immune-isolating membranes [37]. Gold nanoparticles (AuNP) have shown promise in reducing oxidative stress and ameliorating renal damage in diabetic nephropathy [7]. Similarly for diabetic neuropathy, neuron-targeted nanocarriers are potential candidates [59]. Smart PLGA-based nanocrystals have been synthesized to encapsulate gliclazide with initial burst release followed by prolonged release and significant improvement in bioavailability profile and therapeutic efficacy. Glycyrrhizin- and metformin-loaded NPs demonstrated promising antidiabetic effects at a lower dose than the conventional dosage form. Poly(catechin) NPs exhibited a better inhibitory effect on the glucosidase enzyme and carbohydrate digestion with poly(catechin) than catechin nanoparticles. Poly (rosmarinic acid) NPs could inhibit glucosidase at a lower concentration than rosmarinic acid. Thus, poly phenolic NPs can act as appropriate substitutes for noninsulin-based formulations [23]. Amphoteric NPs of silicon dioxide can be utilized as antidiabetic protein drug delivery systems [43]. Polydatin-loaded chitosan nanoparticles can act as efficient delivery systems in diabetic male Wistar albino rats [1]. Pioglitazone solid lipid nanoparticles demonstrated low particle size, high encapsulation efficiency, and prolonged release profile in streptozotocin-induced diabetic male albino Wistar rats [77].

Nanotechnology may provide benchmark therapeutic novelties in diabetes management, but the investigations have been done in animal models. No study could be found after 2022 which is alarming. They need to be standardized through human clinical trials and then only the merit exists.

3 Nanosensors and Nanotherapeutics in Obesity Management

Abnormal production of adipose tissue macrophages, hypertrophic adipocytes, adiponectin, leptin, and resistin may be responsible for obesity, alternatively termed adiposity. If unregulated and uncontrolled sedentary lifestyle persists, it may lead to metabolic syndromes, such as imbalance in glucose homeostasis, reduced insulin sensitivity, dyslipidemia, and cardiovascular diseases [49, 105].

Nanobiosensors offer promise in the detection of obesity. In situ nanosensors can monitor the underlying interactions between the biomolecules and sensing molecules and provide nutritional or metabolic status updates.

Nanometallic components exhibit superior physicochemical properties for sensing devices. Due to their high electrical conductivity, increased specific surface area, excellent electron mobility, and high interaction capability with biological molecules, bimetallic nanocomposites of gold, platinum, palladium, silver, and copper have been extensively utilized in diagnostics. In addition, they are cost-effective, highly selective, easily manufacturable, simple, and provide high sensing ability [79]. SPIONs with highly detectable signal-to-noise ratio may be suitable nanosensors in obesity detection. Magnetically responsive upconversion nanoparticles may serve as suitable diagnostic agents. Fluorimetric imaging and Raman spectroscopy can prove efficient diagnostic methods. Microfluidic nanobiosensors can potentially enhance the diagnostic potential. These methods depend on the detection of obesity-associated factors, such as leptin and adiponectin [110]

Antibody-conjugated gold nanorod (AuNR) LSPR (localized-surface plasmon resonance) nanobiosensor with adipose-tissue-on-chip platform offers multiplexed detection of cytokines and formation of adipose tissue initiation in the range of 10–10,000 pg mL^{-1} with extremely low sample volume and assay time of 30 min. This process avoids complex sample preparation procedures [108]. Engineered photoluminescent carbon nanotubes, composed of short, single-stranded DNA and a single chirality have been developed to quantify lipid content maps in live cells, without altering cell proliferation and viability [39].

Obesity is a condition that is easily preventable and very few research studies are there to detect obesity from 2020 onward as it is an entirely visible condition. Proper food intake and daily exercise can inhibit obesity.

Treatment options and strategies are more abundant than diagnostic procedures in obesity management. The nanoparticulate drug delivery system exhibits unique potential in obesity control by Notch inhibition, regulating the metabolism of adipocytes, differentiation of white adipocytes into brown tissues, and inducing adipocyte lipolysis, with minimum side effects [98]. Transcriptional factors, PPARγ and CEBPα are important in Jak-STAT3 and WNT signaling pathways, which play a pivotal role in the conversion of preadipocytes to mature adipocytes and energy homeostasis [80].

AuNP weekly dose can minimize hepatic ectopic lipid accumulation, promote weight reduction, increase muscle mass, and reduce metabolic inflammation, with a lowering in blood levels of TG in Wistar Albino rats with obesity induced by a high-fat diet (HFD). The gold nanoparticles exhibit fewer adverse effects and are tolerable in vivo. AuNPs have enormous potential to alter several hepatic gene expressions and gut microbiota profiles [74]. AuNPs loaded with *Salacia chinensis* could reduce body mass index, leptin, resistin, hepatic marker enzymes, and body weight and increase HDL-cholesterol in HFD-obese rats [31]. *Gynostemma pentaphyllum* gold nanoparticles could effectively act as an antiobesity agent in obesity-induced 3 T3-L1 cells with a simultaneous reduction in nitric oxide production in RAW 264.7 macrophage cells. The nanoparticles could reduce the PPARγ and CEBPα levels and proinflammatory cytokine levels as demonstrated by RT-PCR. Intracellular reactive oxygen species (ROS) were lowered significantly [5]. Downregulation of PPARγ and CEBPα genes demonstrated an antiadipogenic effect as evident in *Panax ginseng* leaf extract-loaded gold nanoparticles in 3 T3-L1 mature adipocytes [80].

Nutraceutical nanotechnology shows promise in obesity management with phytochemicals like quercetin, resveratrol, and capsaicin, and plants like *Garcinia cambogia*, *Curcuma longa*, *Camella Sinensis*, *Zingiber officinalis*, *Citrus aurantium*, *Aegle marmelos*, *Rosmarinus officinalis*, etc. [75]. They may act by downregulating leptin levels, inhibiting pancreatic lipase concentration, lowering adiponectin maturation, and white adipose tissue levels, ultimately reducing body weight. They may also confer energy balance and promote energy expenditure [88].

Selenium oxide NPs and zinc oxide NPs inhibited lipid fractions in the serum, reduced the total cholesterol, triglyceride, and malondialdehyde levels, increased the HDL-C, and improved the general hepatic and renal functions in adult male rats fed HFD for 10 weeks. They also elevated the beneficial hepatic enzyme levels [60]. Liposomal preparation and PLGA nanoparticles of L-carnitine could produce remarkable effects on obesity and promote a slimming effect. They also exhibited several positive effects on carbohydrate and amino acid metabolism [103]. PLGA NPs coated with chitosan and polyethylene glycol could entrap quercetin or curcumin effectively, enhance their bioavailability, and lower weights of liver, heart, gonadal, and subcutaneous adipose tissue. They could also promote the inhibition of cholesterol, glucose, and LDL-C [4].

SPION-PEG-2000 at a weekly dose and a daily dose of orlistat by intraperitoneal injection could reduce leptin, adiponectin, and nonessential fatty acid levels, lower dyslipidemia, and hyperglycemia in obese rats. Polyethylene glycol 2000 facilitated the transport and intracellular uptake of SPIONs and distribution to distant tissues [8].

Nanotherapeutics for obesity management is scarce in 2020 and ahead. Obesity is a condition evident from the outside and people can adopt suitable measures to check its progression.

4 Nanotechnology in Dyslipidemia and Cardiovascular Disease Management

4.1 Nanosensors in Dyslipidemia

Dyslipidemia refers to excessive lipid levels in the bloodstream, which poses a significant challenge to the persons possessing it. Dysregulation will cause atherosclerosis and other cardiovascular problems.

Nanotechnology in the form of nanodiagnostics can provide insights into understanding the pathophysiology of dyslipidemia and state the unprecedented risk of advancing atherosclerosis, its composition, hypertension, and coronary artery diseases. Nanosensors promise a superior understanding of the pathophysiological deterioration of atherosclerotic plaque formation and can achieve excellent targets for diagnostic imaging. Nanotechnology can augment the biocompatibility, optoelectronicity, chemical and environmental stability, and pharmacokinetic profile of the entrapped diagnostic agents in atherosclerosis. Rational nanotechnology can be employed in targeting atherosclerotic plaques and establishing the effectiveness of therapeutic interventions [21, 51]. An assessment of cholesterol levels can be of great significance in the early prevention of cardiovascular diseases, and thus, cholesterol nanobiosensors can be of great help in this context. Several instrumental analytical methods such as colorimetry, UV-vis spectrophotometry, fluorimetry, SERS, and high-performance liquid chromatography have been utilized for the purpose. Enzymatic and nonenzymatic-based sensors exhibit precise sensitivity and selectivity. The enzymatic sensors utilize the cholesterol oxidase (ChOx) or cholesterol esterase to produce fatty acids and free cholesterol as end-products of the reaction. Owing to their superior analytical possibility and high detection power, nonenzymatic cholesterol nanosensors exhibit unique potential [101, 110].

A color change in the nanomaterials upon exposure to hydrogen peroxide produced during oxidation of cholesterol by ChOx is highly preferred. Cholesterol in the linear range of 0.30–300 mM can be detected efficiently, where LOD is 0.15 mM. However, none of these methods can function without the specific oxidative reaction by the ChOx. Fluorescence imaging offers enzyme-free, highly selective, cost-effective, real-time continuous monitoring and ultrasensitive techniques in the detection of picomolar levels of cholesterol. Chemiluminescent nanomaterials have demonstrated immense potential in detecting micromolar concentrations. Electrochemical nanosensors provide a much cheaper, sensitive, and portable detection of cholesterol with a high degree of linear response, stability, and reproducibility. Plasmonic-level optical nanobiosensors show great promise in cholesterol sensing. Metal NPs, metal oxide NPs, carbonaceous materials, and metal-organic-frameworks (MOFs) have superseded nanomaterials with superior efficiency and detection potential. Platinum nanoclusters, silicon nanoparticles, polythione-modified glass carbon nanoelectrodes, β-garlic-capped silver nanoparticles, cadmium-selenium/zinc sulfide quantum dots, gold-carbon dot nanoconjugates, etc. have been immensely successful in the enzymatic and nonenzymatic sensitive quantification of cholesterol. There have been uses of nitrogen-doped carbon nanosheets, and copper-doped cerium dioxide nanoparticles which have shown high selectivity and specificity to cholesterol. Advanced techniques can be adopted for MOF-based disposable biosensors. Critical enzymes essential for cholesterol diagnosis, cholesteryl esterase, and cholesterol oxidase are detectable on cloud-enabled smartphones combined with photochemical dongles. Several modifications are also being attempted with the technology to obtain better and achievable readability at <10 mM LOD [34, 58]. Bimetallic nanorods have proven highly efficacious in identifying cholesterol via their plasmonic characteristics. Gold NP-coated ferric oxide magnetic particles, MXene-titanium carbide nanosheets combined with copper sulfide NPs, and palladium nanoparticles stabilized with indigenous gum have demonstrated remarkable potential for cholesterol biosensing [33, 70]. Harnessing miniature nanosensors with surface-immobilized enzymes and electrolyte-gated transistors have been designed for real-time sample analysis. They possess excellent shelf-life, surface roughness, sensitivity, practical applicability, performance, and reproducibility [47]. Prediction of total cholesterol level was carried out in a noninvasive manner by performing gas analysis of exhaled breath with the use of e-nose and machine learning software for the first time in human volunteers [66].

Table 6.2 summarizes the nanosensors utilized in monitoring dyslipidemia from 2020 onward. The list includes varied nanoparticles such as nanorods, MOFs, nanowires, nanosheets, nanorods, and nanoclusters which are composed of Au, Cu, Ag, Pt, etc. The detection limit is as low as 0.025–10 μM. A wide variety of analytical methods have been used ranging from colorimetry to fluorescence spectroscopy to chemiluminescence and SERS.

5 Nanosensors in Cardiovascular Diseases

Cardiovascular diseases are marked by an increasing level of apo-B-containing lipoproteins. Detection of cost-effective, efficient, highly specific, rapid, and sensitive cardiovascular nanobiomarkers is of prime importance in obtaining pathophysiological and diagnostic criteria for coronary diseases. Investigation into the biomarker concentration is crucial in diagnosis, risk stratification, and managing patients with CVD [63]. Different imaging procedures have been utilized to assess the diagnostic potential of nanosensors for atherosclerosis such as rapid immunofluorescence, electrochemiluminescence, surface Raman scattering, and electrochemical methods such as voltammetry, impedance spectroscopy, galvanometric analysis, etc. [30]. The electrochemical methods utilize the principle of direct electron transfer to the sensing nanoanalyte and can thus detect cardiac troponin. Magnetic surface-

Table 6.2 Nanosensors in dyslipidemia monitoring

Nanostructure	Detection level	Analytical method used	Reference
Bimetallic plasmonic silver-shelled gold nanorods	20-fold enhancement in the spectral peaks	Surface enhanced Raman spectroscopy	[15]
Gold nanoparticle-coated metal-organic framework (molecularly imprinted polymer)	Detection of cholesterol of 2.9–6.7 mM with LOD of 2.4 mM	Colorimetry	[19]
Cubic-shaped cuprous oxide nanoparticles	Detection of 10–1000 μM and LOD of 0.0018 nM	Electrochemistry	[102]
Silver nanowire and graphene oxide π-plasmons	>1300-fold increase in fluorescence intensity	Photonic crystal-coupled emission	[17]
Fe-Ni dual single atoms nanozyme incorporated into N-doped carbon nanosheets	Linear ranges of 0.5–5.0 mM and 0.25–5.0 mM with LOD of 0.19 mM and 0.044 mM, respectively	Colorimetry and fluorimetry	[54–56]
Garlic-capped Ag nanoparticles	Linear response in a concentration range of 0.4–5.17 mM and a LOD of 0.186 mM	Fluorescence spectroscopy	[26]
Metal-organic framework (V-MOF-700)	Detection of 2–70 μM and 70–160 μM with LOD of 0.38 μM	Fluorescence spectroscopy	[97]
Gold-titanium dioxide nanoparticles modified with multiwalled carbon nanotube	Detection of 0.05–80 μM and LOD of 0.0018 μM	Differential pulse voltammetry, cyclic voltammetry, electrochemical impedance spectroscopy	[42]
Iron oxide nanoparticles decorated zinc oxide nanorod-based electrolyte-gated transistor	Detection of 0.1–60 mM and sensitivity of 37.34 μA mM^{-1} cm^{-2}	Electrochemistry	[47]
Silicon nanoparticles	Detection of 0.025–10 μM cholesterol with LOD of 0.018 μM	Fluorescence indicator using Bandrowski's base as a quencher	[104]
Gum kondagogu-modified palladium nanoparticles	Detection in the range of 5–100 μM with LOD at 3.7 μM	Colorimetric oxidation	[70]
Copper metal-organic framework decorated with cobalt oxidenanoparticles	Nonenzymatic biosensing detection of 1–250 nM H_2O_2 with an LOD of 0.244 nM	Chemiluminescence	[9]
Silver nanoparticles with UiO-66-NH_2 and *o*-phenylenediamine	Detection limits of 10 μM with dual-emission response to H_2O_2	Fluorescence spectroscopy	[34]

(continued)

Table 6.2 (continued)

Nanostructure	Detection level	Analytical method used	Reference
Pd nanoclusters and zeolitic imidazolate framework-8 (ZIF-8)	Detection limit of 0.092 μM for cholesterol	Fluorescence spectroscopy	[53]
Cu_2O-TiO_2 hybrid nanostructure using chemical bath deposition (CBD) method	5-fold increase in the sensitivity (~6034.04 μA mM^{-1} cm^{-2}) of cholesterol with detection limit of ~0.05 μM and fast response time (3 s)	Cyclic voltammetry and amperometry	[45]
Platinum nanocluster immobilized with cholesterol oxidase and cholesterol esterase	Detection of 2 to 486 μM cholesterol with LOD of 2 μM and sensitivity of the sensor at 132 μA mM^{-1} cm^{-2}. Good specificity for ascorbic acid, uric acid, dopamine, glucose, and lactate	Electrochemistry	[28]
Polypyrrole nanoparticles with immobilized cholesterol oxidase	Linear range 10–100 μM for cholesterol in human serum with a detection limit of 3.5 μM	Colorimetry	[35]

enhanced Raman scattering (SERS) and bio-barcoded nanoparticles can facilitate screening for early CVD and potentially identify risk-prone patients [84].

Apart from these single imaging modalities, multiple imaging modalities in coronary artery disease detection may produce better diagnostic ability and synergistic benefits in a more precise and accurate manner.

Recent studies have emphasized the role of natriuretic peptides and their receptors in atherosclerosis [96]. Construction of nanosensors may provide possibilities for target-specific monitoring of atherogenic inflammation, and susceptible plaques. Nanoprobes in atherosclerotic plaque detection will broaden the scope of CVD diagnosis.

Polymeric nanomaterials with modified anti-C-C motif chemokine receptor (CCR5) peptide and ^{64}Cu have been used for PET-CT nuclear imaging of atherosclerotic plaque with the highest sensitivity and prolonged cycle time [14].

Cardiac troponins (cTnI) are complexes of 3 proteins, troponin I (cTnI), troponin T (cTnT), and troponin C (cTnC). They are crucial for diagnosis and risk stratification of patients with a high probability of heart attack in the first 2 to 6 h, but rarely for the second one, because of its profile and short half-life [63, 83, 85]. However, cTnI may serve as a potent biomarker for acute myocardial infarction, although at this stage, dyslipidemia itself is exacerbated by other causes [92].

Plasmonic gold nanoscale array structures and nanoledges with tethered DNA aptamers are selective, sensitive, and chip-based, label-free optical biosensors for detecting abnormal levels of cTnI in minute volume of blood samples. Semiconductor nanoparticles, carbon-based nanostructures, and metal chalcogenides have been utilized as field-effect transistor-based biosensors [92].

Cardiac troponin may be detected by sandwich-type electrochemical immunonanosensors with precise sensitivity, selectivity, and low background noise [25].

A brief summary is provided on the nanosensors in detecting CVD in Table 6.3. It represents the nanotechnological approaches that have been used since 2020 but few articles show studies with CVD. Gold NPs are the most common ones that have been utilized. However, compared to the methods for glucose and cholesterol detection, the limit for detection in this case is substantially low and is in the picogram level. Various analytical methods have been used.

Nanocarriers for therapeutic delivery in dyslipidemia can suppress atherosclerosis or increase HDL-C levels in blood with suitable antioxidant effects. However, there are limited studies on dyslipidemia management in 2020 and beyond. Nanotherapeutics for CVDs are not included as it is not a metabolic syndrome but its diagnosis becomes important as its extent can prove the damage caused due to obesity, dyslipidemia, and several other serious factors.

HDL mimetic nanocarriers can act as supplementary agents by restoring HDL counts, enhancing natural HDL functions by replication, and reducing lipoproteins in atherosclerosis. They can elicit anti-inflammatory responses by selectively interacting with monocytes and eliminating total cholesterol from macrophages [107]. Simvastatin nanoliposomes showed promise in fibrosis-induced hepatic microtissues with negligible burst effect. Lipid–polymer hybrid nanoparticles demonstrated better antihypercholesterolemic effects compared to polymeric nanoparticles on oral administration by promoting significant reductions in TC, TG, and LDL-C, and both showed high entrapment efficiency and controlled drug release profiles [63].

Gold nanoparticles can be used for their beneficial antioxidant effects on hepatic and cardiac tissues in young male Swiss albino mice with high cholesterol intake for 8 weeks. They could increase the hepatic superoxide dismutase levels in the liver and reduce the hypercholesterolemia-induced oxidative stress in the liver and heart [73].

Chitosan can bind to negatively charged cholesterol and can be utilized in normalizing total cholesterol level. *Pandanus tectorius* fruit extract in combination with chitosan exhibited its efficacy as an antihypercholesterolemic agent in the form of nanoparticles in Sprague Dawley rats. They could reduce TC, LDL-C, and TG in significant proportions. The nanoparticles were safe and nontoxic for HepG2 cells [65].

Phenolic compound-rich avocado peel extract, containing chlorogenic acid, catechin, and epicatechin, in bovine serum albumin NPs could lower cholesterol absorption and establish homeostasis, exhibiting an antioxidant effect, promoting healthy heart and vascular system. The nanotherapeutic system was suitable for oral administration and proved to be safe and nontoxic and can be developed as appropriate nutritional supplements [86].

Table 6.3 Nanosensors for cardiovascular diseases

Nanostructure	Detection level	Analytical method used	Reference
Gold nanoledge array sensor	Detection of 0.0001–100 ng mL^{-1} in human serum and 0.001–100 ng mL^{-1} in human plasma	Surface plasma resonance	[92]
Microfluidic biosensor	Detection of 2–25 μg L^{-1} and LOD of 0.6 μg L^{-1}	Chemiluminescence	[85]
Iron oxide nanoparticles derivatized with NOTA for ^{64}Cu labeling and MMP-2-cleavable peptide modified with PEG 2000 yielding ^{64}Cu-NOTA-IONP @MMP2c-PEG2K	Remarkable plaque uptake of 3.06 ± 0.87% ID/g and 1.83 ± 0.28% ID/g at 4 h and 12 h with excellent clearance of MMP2cNPs	PET/MRI	[91]
Paper-based analytical device	Detection of 1–1000 ng mL^{-1} C-reactive protein, 0.001–250 ng mL^{-1} cardiac troponin, 0.005–250 ng mL^{-1} procalcitonin, and LOD of 0.38 ng mL^{-1} for CRP, 0.161 pg mL^{-1} for cTnI and 0.27 pg mL^{-1} for procalcitonin	Square-wave voltammetry	[18]
Graphene quantum dots and gold nanoparticles implanted onto screen-printed gold electrode	Detection limit of 0.5 pg/mL with sensitivity of 6.81 μA cm^{-2} pg/mL	Square wave voltammetry, cyclic voltammetry, electron impedance spectroscopy, amperometry	[61]
Gold nanoparticle functionalized with malachite green isothiocyanate	Detection of 0.016–0.1 ng mL^{-1}	Surface-enhanced resonance Raman scattering	[90]
Gold nanoparticle-copper-platinum-rhodium-hollow nanobox-embedded ultrathin specialized MXene (NH_2-Ti_3-C_2)	Detection of 25 fg mL^{-1} to 100 mg mL^{-1} and LOD of 8.13 mg mL^{-1}	Electrochemical and immunosensor	[25]
Gold nanoparticle decorated cobalt oxide nanorods with single-walled carbon nanotube layer	Sensitivity of 0.5 μA mL^{-1} and LOD of 0.1 μg mL^{-1}, compatible at <14 ng mL^{-1}	Electrochemistry	[83]

Rapamycin NPs with macrophage membrane coating demonstrated excellent inhibition on the progression of atherosclerotic plaque with appropriate pharmacokinetic properties in ApoE knockout mice. Hyaluronic acid-atorvastatin NPs could be efficiently used for plaque removal by magnetic resonance imaging guided by SPION nanoworm with hyaluronic acid. Single-walled carbon nanotubes with SHP-1 loading could revive the macrophage phagocytosis which was reduced in atherosclerotic plaque [36].

The projected success of nanotherapeutics in cell lines or animals brightens up the prospects of patients with dyslipidemia but warrants clinical trials in humans [21].

6 Challenges with Nanosensor Technology and Nanomedicine in Metabolic Syndrome Management

The extent of endocytosis, biodistribution, mechanism of action, systemic circulation time, cellular interaction, biological half-life, metabolic pathways, clearance, and elimination of nanosensors are intricate and complex pathways. They may be significantly altered due to their shape, size, surface area, surface charge, and aggregation state, which may ultimately cause toxicity. Moreover, the size of nanosensors can exacerbate potential hazards. Nanosensors possess potential risks of lung inflammation, fibrosis, hepatic damage, hemorrhage, and nephrotoxicity, and may finally affect innate immunity. Thus, they have not been put to clinical applications [63]. The optimum nanostructure of the nanosensors and electrostatic interaction between surface cations of the charged nanosensors and negatively charged cell membrane may influence the spatial kinetics, pharmacokinetic, pharmacodynamic, and toxicity profile of nanosensors. However, the metabolic and toxicity reports are concerned with preclinical studies on animal models and require clinical data on humans. Safety and prolonged performance evaluation is crucial for the complete assessment of nanosensors for various metabolic symptoms, especially for those sensors that remain in the biological system and are not eliminated after their predetermined schedule. Therefore, continuous progress in nanosensor technology is essential for metabolic syndrome prediction [62, 63, 84]. Moreover, although these nanosensors demonstrate excellent specificity and selectivity, their transformation into clinical portfolio is challenging, primarily due to scalability, manufacturability, and regulatory issues [32]. Cost-effective, affordable, and durable nanosensors for metabolic syndromes are essential requirement in developing countries [87]. Therefore, the nanosensors need huge improvements and then only can mitigate the vagaries of metabolic syndrome at an affordable cost.

Enzymatic and nonenzymatic glucose nanosensors have their demerits. The shelf-life of enzymatic nanosensors is low and biofouling problems may arise. Another major problem with enzymatic nanosensors is their widespread use among diabetics, some of whom may not have proper training. Transition metal-based or carbon nanotube-based nonenzymatic nanosensors have very limited usability at physiological pH. They suffer from low selectivity, sensitivity, and clinical efficiency ([2]). The major challenge with the development of robust glucose nanosensors lies with the design of suitable assay procedures. This requires the development of appropriate responsive moieties with strong binding properties and thus enhanced sensitivity and specificity to glucose. Consistent and predictable nanosensors should minimize sensor degradation and trigger glucose monitoring with the highest sensitivity [76].

Most of the techniques used for obesity detection are invasive, painful, cause discomfort, have prolonged recovery time, and may produce sarcomas and enhanced vascular permeability [78, 89]. Therefore, future studies with human adipocytes are a prerequisite for the safety assessment of clinical translation of obesity-related nanosensors.

Precise sensing of cholesterol with MOFs provides very accurate results but with challenges owing to their need for both aqueous and organic solvent solubility, over a range of buffers with suitable pHs and temperature and thus exhaustive stability studies. Microfluidics can be a viable option but with a high cost [12, 16]. Designing continuous wearable devices for cholesterol monitoring at low concentrations seems a better option but requires miniature devices with high sensitivity and accuracy. Nowadays, newer synthetic routes with reused, recycled, and recovered materials are thought of thereby minimizing environmental concerns [81, 112].

Nanotherapeutic approaches beyond 2020 for diabetes, obesity, and dyslipidemia management have shown promise in animals and only one study showed effectiveness in human patients with machine learning programs. Thus, nanocarriers for therapeutic agents for human use seem to be a distant option for future applications.

7 Conclusion

Nanosensor and nanotherapeutics progress might have attempted great strides in revolutionizing metabolic syndrome detection and therapy. However, the ultimate challenge lies in maximizing the benefits of the clinical applications with appropriate molecular imaging techniques, biocompatibility, safety, and long-term performance profiles at the zenith and manufacturing cost at the minimum. Scale-up and regulatory approval are essential before the commercialization of nanotechnological approaches for metabolic disorders. The design of versatile nanosensor and nanotherapeutic delivery agents will depend on their studies of strengths and limitations and offer long-lasting opportunities. Endeavors in the diagnosis and therapy of diabetes, obesity, and hypercholesterolemia or dyslipidemia will reach the highest potential soon.

References

1. Abdel-Moneim, A., El-Shahawy, A., Yousef, A. I., et al. (2020). Novel polydatin-loaded chitosan nanoparticles for safe and efficient type 2 diabetes therapy: *In silico*, *in vitro* and *in vivo* approaches. *International Journal of Biological Macromolecules, 154*, 1496–1504. https://doi.org/10.1016/j.ijbiomac.2019.11.031

2. Adeel, M., Rahman, M. M., Caligiuri, I., et al. (2020). Recent advances of electrochemical and optical enzyme-free glucose sensors operating at physiological conditions. *Biosensors & Bioelectronics, 165*, 112331. https://doi.org/10.1016/j.bios.2020.112331
3. Ahmad, R., Khan, M., Mishra, P., et al. (2021). Engineered hierarchical CuO nanoleaves based electrochemical nonenzymatic biosensor for glucose detection. *Journal Of The Electrochemical Society, 168*, 017501. https://doi.org/10.1149/1945-7111/abd515
4. Ahmed, H. H., Kotob, S. E., Abd-Rabou, A. A., et al. (2021). Pre-clinical evidence for the anti-obesity potential of quercetin and curcumin loaded chitosan/PEG blended PLGA nanoparticles. *Biomedical & Pharmacology Journal, 14*(4), 1731–1759. https://doi.org/10.13005/bpj/2274
5. Akter, R., Ling, L., Rupa, E. J., et al. (2022). Binary effects of gynostemma gold nanoparticles on obesity and inflammation via downregulation of PPARγ/CEPBα and TNF-α gene expression. *Molecules, 27*(9), 2795. https://doi.org/10.3390/molecules27092795
6. Alghazzawi, W., Danish, E., Alnahdi, H., & Abdel Salam, M. (2020). Rapid microwave-assisted hydrothermal green synthesis of rGO/NiO nanocomposite for glucose detection in diabetes. *Synthetic Metals, 267*(2), 116401. https://doi.org/10.1016/j.synthmet.2020.116401
7. Alomari, G., Al-Trad, B., Hamdan, S., et al. (2020). Gold nanoparticles attenuate albuminuria by inhibiting podocyte injury in a rat model of diabetic nephropathy. *Drug Delivery and Translational Research, 10*(1), 216–226. https://doi.org/10.1007/s13346-019-00675-6
8. Alsenousy, A. H. A., El-Tahan, R. A., Ghazal, N. A., et al. (2022). The anti-obesity potential of superparamagnetic iron oxide nanoparticles against high-fat diet-induced obesity in rats: Possible involvement of mitochondrial biogenesis in the adipose tissues. *Pharmaceutics, 14*(10), 2134. https://doi.org/10.3390/pharmaceutics14102134
9. Amirzehni, M., Eskandari, H., Vahid, B., & Hassanzadeh, J. (2021). An efficient chemiluminescence system based on mimic CuMOF/Co3O4 nanoparticles composite for the measurement of glucose and cholesterol. *Sensors and Actuators B: Chemical, 348*, 130690. https://doi.org/10.1016/j.snb.2021.130690
10. Anand, V. K., Bhatt, K., Kumar, S., et al. (2021). Sensitive and enzyme-free glucose sensor based on copper nanowires/polyaniline/reduced graphene oxide nanocomposite ink. *International Journal of Nanoscience, 20*(2), 2150020. https://doi.org/10.1142/S0219581X21500204
11. Arikan, K., Burhan, H., Sahin, E., & Sen, F. (2022). A sensitive, fast, selective, and reusable enzyme-free glucose sensor based on monodisperse AuNi alloy nanoparticles on activated carbon support. *Chemosphere, 291*(3), 132718. https://doi.org/10.1016/j.chemosphere.2021.132718
12. Ayyanu, R., Arul, A., Song, N., et al. (2023). Wearable sensor platforms for real-time monitoring and early warning of metabolic disorders in humans. *Analyst, 148*(19), 4616–4636. https://doi.org/10.1039/d3an01085f
13. Bag, S., Baksi, A., Nandam, S. H., et al. (2020). Nonenzymatic glucose sensing using Ni60Nb40 nanoglass. *ACS Nano, 14*(5), 5543–5552. https://doi.org/10.1021/acsnano.9b09778
14. Bajaj, L. L., & Bajaj, B. L. (2024). Evidence-based approaches to diagnosing and treating hypertension in primary care settings: A systematic review. *Nanotechnology Perceptions, 20*(Suppl.7), 940–959. https://doi.org/10.62441/nano-ntp.vi.1305
15. Basak, M., Nemade, H. B., & Bandyopadhyay, D. (2025). Silver shelled gold nanorods for sensitive detection of cholesterol and triglycerides. *Biosensors and Bioelectronics, 268*, 116885. https://doi.org/10.1016/j.bios.2024.116885
16. Bayoumy, K., Gaber, M., Elshafeey, A., et al. (2021). Smart wearable devices in cardiovascular care: Where we are and how to move forward. *Nature Reviews. Cardiology, 18*(8), 581–599. https://doi.org/10.1038/s41569-021-00522-7
17. Bhaskar, S., Lis, S. M., Kanvah, S., et al. (2023). Single-molecule cholesterol sensing by integrating silver nanowire propagating plasmons and graphene oxide π-plasmons on a photonic crystal-coupled emission platform. *ACS Applied Optical Materials, 1*(1), 159–172. https://doi.org/10.1021/acsaom.2c00026

18. Boonkaew, S., Jang, I., Noviana, E., et al. (2021). Electrochemical paper-based analytical device for multiplexed, point-of-care detection of cardiovascular disease biomarkers. *Sensors and Actuators B: Chemical, 330*, 129336. https://doi.org/10.1016/j.snb.2020.129336
19. Cao, M., Huang, C., Zhang, Y., et al. (2024). Molecularly imprinted sensor based on cascade enzyme system supported by metal-organic framework (UiO-66-NH2) for sensitive colorimetric detection of cholesterol. *Sensors and Actuators B: Chemical, 404*, 135235. https://doi.org/10.1016/j.snb.2023.135235
20. Chen, S., Huang, D., Xu, P., et al. (2020). Semiconductor-based photocatalysts for photocatalytic and photoelectrochemical water splitting: Will we stop with photocorrosion? *Journal of Materials Chemistry A, 8*(5), 2286–2322. https://doi.org/10.1039/C9TA12799B
21. Chen, W., Schilperoort, M., Cao, Y., et al. (2022). Macrophage-targeted nanomedicine for the diagnosis and treatment of atherosclerosis. *Nature Reviews. Cardiology, 19*(4), 228–249. https://doi.org/10.1038/s41569-021-00629-x
22. Chu, D.-T., Thi, H. V., Nguyen, T. T., et al. (2023). Nanotechnology and nucleic acid nanoparticles for treatment of metabolic disorders. *OpenNano, 13*, 100181. https://doi.org/10.1016/j.onano.2023.100181
23. Debele, T. A., & Park, Y. (2022). Application of nanoparticles: Diagnosis, therapeutics, and delivery of insulin/anti-diabetic drugs to enhance the therapeutic efficacy of diabetes mellitus. *Life, 12*(12), 2078. https://doi.org/10.3390/life12122078
24. Ding, J., Li, X., Zhou, L., et al. (2020). Electrodeposition of nickel nanostructures using silica nanochannels as confinement for low-fouling enzyme-free glucose detection. *Journal of Materials Chemistry B, 8*(16), 3616–3622. https://doi.org/10.1039/C9TB02472G
25. Dong, H., Cao, L., Tan, Z., et al. (2020). A signal amplification strategy of CuPtRh CNB-embedded ammoniated Ti3C2 MXene for detecting cardiac troponin I by a sandwich-type electrochemical immunosensor. *ACS Applied Bio Materials, 3*(1), 377–384. https://doi.org/10.1021/acsabm.9b00863
26. El-Naka, M. A., El-Dissouky, A., Ali, G. Y., et al. (2023). Garlic capped silver nanoparticles for rapid detection of cholesterol. *Talanta, 253*, 123908. https://doi.org/10.1016/j.talanta.2022.123908
27. Elsaygh, J., Zaher, A., Parikh, M. A., et al. (2024). Nanotechnology: The future for diagnostic and therapeutic intervention in cardiovascular diseases is here. *Cardiology in Review*. https://doi.org/10.1097/CRD.0000000000000727
28. Eom, K. S., Lee, Y. J., Seo, H. W., et al. (2020). Sensitive and non-invasive cholesterol determination in saliva: Via optimization of enzyme loading and platinum nano-cluster composition. *Analyst, 145*(3), 908–916. https://doi.org/10.1039/C9AN01679A
29. Ferro, C., Florindo, H. F., & Santos, H. A. (2021). Selenium nanoparticles for biomedical applications: From development and characterization to therapeutics. *Advanced Healthcare Materials, 10*(16), e2100598. https://doi.org/10.1002/adhm.202100598
30. Gachpazan, M., Mohammadinejad, A., Saeidinia, A., et al. (2021). A review of biosensors for the detection of B-type natriuretic peptide as an important cardiovascular biomarker. *Analytical and Bioanalytical Chemistry, 413*, 5949–5967. https://doi.org/10.1007/s00216-021-03490-6
31. Gao, L., Hu, Y., Hu, D., et al. (2020). Anti-obesity activity of gold nanoparticles synthesized from *Salacia chinensis* modulates the biochemical alterations in high-fat diet-induced obese rat model via AMPK signaling pathway. *Arabian Journal of Chemistry, 13*(8), 6589–6597. https://doi.org/10.1016/j.arabjc.2020.06.015
32. Global Market Insight-Nanosensor market – By type (optical nanosensor, chemical nanosensor, physical nanosensor), by technology (molecular self assembly, top-down assembly, bottom-up assembly), by end user & forecast, 2024–2032 [Internet], 2024. Available from: https://www.gminsights.com/industry-analysis/nanosensor-market
33. Guan, H., Song, Y., Han, B., et al. (2020). Colorimetric detection of cholesterol based on peroxidase mimetic activity of GoldMag nanocomposites. *Spectrochimica Acta, Part A:*

Molecular and Biomolecular Spectroscopy, 241, 118675. https://doi.org/10.1016/j.saa.2020.118675
34. Guo, L., Chen, S., Yu, Y. L., et al. (2021). A smartphone optical device for point-of-care testing of glucose and cholesterol using Ag NPs/UiO-66-NH2-based ratiometric fluorescent probe. *Analytical Chemistry, 93*(48), 16240–16247. https://doi.org/10.1021/acs.analchem.1c04126
35. Hong, C., Zhang, X., Wu, C., et al. (2020). On-site colorimetric detection of cholesterol based on polypyrrole nanoparticles. *ACS Applied Materials & Interfaces, 12*(49), 54426–54432. https://doi.org/10.1021/acsami.0c15900
36. Hossaini Nasr, S., & Huang, X. (2021). Nanotechnology for targeted therapy of atherosclerosis. *Frontiers in Pharmacology, 12*, 755569. https://doi.org/10.3389/fphar.2021.755569
37. Hu, S., Kuwabara, R., Navarro Chica, C. E., et al. (2021). Toll-like receptor 2-modulating pectin-polymers in alginate-based microcapsules attenuate immune responses and support islet-xenograft survival. *Biomaterials, 266*, 120460. https://doi.org/10.1016/j.biomaterials.2020.120460
38. Inyang, D., Kibambo, G., Palmer, M., et al. (2020). One step copper oxide (CuO) thin film deposition for non-enzymatic electrochemical glucose detection. *Thin Solid Films, 709*, 138244. https://doi.org/10.1016/j.tsf.2020.138244
39. Jena, P. V., Roxbury, D., Galassi, T. V., et al. (2017). A carbon nanotube optical reporter maps endolysosomal lipid flux. *ACS Nano, 11*(11), 10689–10703. https://doi.org/10.1021/acsnano.7b04743
40. Kalambate, P. K., & Huang, Y. (2021). Nanomaterials-based electrochemical sensors and biosensors for early identification and monitoring of diseases. In S. Rajendran, M. Naushad, D. Durgalakshmi, & E. Lichtfouse (Eds.), *Metal, metal oxides and metal sulphides for biomedical applications* (Environmental Chemistry for a Sustainable World) (Vol. 58, pp. 209–226). Springer. https://doi.org/10.1007/978-3-030-56413-1_7
41. Kanokpaka, P., Chang, Y.-H., Chang, C.-C., et al. (2023). Enabling glucose adaptive self-healing hydrogel based triboelectric biosensor for tracking a human perspiration. *Nano Energy, 112*, 108513. https://doi.org/10.1016/j.nanoen.2023.108513
42. Kariz, D. G. A., Darabi, E., & Elahi, S. M. (2022). Electrochemical sensing of cholesterol based on MWCNTs loaded nanoparticles. *International Nano Letters, 12*, 113–123. https://doi.org/10.1007/s40089-021-00357-7
43. Kerry, R. G., Mahapatra, G. P., Maurya, G. K., et al. (2021). Molecular prospect of type-2 diabetes: Nanotechnology based diagnostics and therapeutic intervention. *Reviews in Endocrine & Metabolic Disorders, 22*(2), 421–451. https://doi.org/10.1007/s11154-020-09606-0
44. Khair, R., Shende, P., & Kulkarni, Y. A. (2020). Nanostructured polymer-based cochleates for effective transportation of insulin. *Journal of Molecular Liquids, 311*, 113352. https://doi.org/10.1016/j.molliq.2020.113352
45. Khaliq, N., Rasheed, M. A., Cha, G., et al. (2020). Development of non-enzymatic cholesterol bio-sensor based on TiO_2 nanotubes decorated with Cu_2O nanoparticles. *Sensors and Actuators B: Chemical, 302*, 127200. https://doi.org/10.1016/j.snb.2019.127200
46. Khan, M. S., Ameer, H., Ali, A., et al. (2020). Electrochemiluminescence behaviour of silver/silver orthophosphate/graphene oxide quenched by Pd@Au core-shell nanoflowers for ultrasensitive detection of insulin. *Biosensors & Bioelectronics, 147*, 111767. https://doi.org/10.1016/j.bios.2019.111767
47. Khan, M., Nagal, V., Masrat, S., et al. (2022). Wide-linear range cholesterol detection using Fe_2O_3 nanoparticles decorated ZnO nanorods based electrolyte-gated transistor. *Journal Of The Electrochemical Society, 169*(2), 027512. https://doi.org/10.1149/1945-7111/ac51f6
48. Kim, S., Malik, J., Seo, J. M., et al. (2022). Subcutaneously implantable electromagnetic biosensor system for continuous glucose monitoring. *Scientific Reports, 12*(1), 17395. https://doi.org/10.1038/s41598-022-22128-w
49. Kumari, R., Kumar, S., & Kant, R. (2019). An update on metabolic syndrome: Metabolic risk markers and adipokines in the development of metabolic syndrome. *Diabetes and Metabolic*

Syndrome: Clinical Research and Reviews, 13(4), 2409–2417. https://doi.org/10.1016/j.dsx.2019.06.005

50. Lagopati, N., & Pavlatou, E. A. (2021). Nanotechnology in diabetes management. *Interventions in Obesity & Diabetes, 5*(1), IOD.000603. https://doi.org/10.31031/IOD.2021.05.000603
51. Lankala, C. R., Yasir, M., Ishak, A., et al. (2023). Application of nanotechnology for diagnosis and drug delivery in atherosclerosis: A new horizon of treatment. *Current Problems in Cardiology, 48*(6), 101671. https://doi.org/10.1016/j.cpcardiol.2023.101671
52. Lemmerman, L. R., Das, D., Higuita-Castro, N., et al. (2020). Nanomedicine-based strategies for diabetes: diagnostics, monitoring, and treatment. *Trends in Endocrinology and Metabolism, 31*(6), 448–458. https://doi.org/10.1016/j.tem.2020.02.001
53. Li, Y., Li, S., Bao, M., et al. (2021). Pd nanoclusters confined in ZIF-8 matrixes for fluorescent detection of glucose and cholesterol. *ACS Applied Nano Materials, 4*(9), 9132–9142. https://doi.org/10.1016/10.1021/acsanm.1c01712
54. Li, J. Q., Mao, Y. W., Zhang, R., et al. (2023a). Fe-Ni dual-single atoms nanozyme with high peroxidase-like activity for sensitive colorimetric and fluorometric dual-mode detection of cholesterol. *Colloids and Surfaces. B, Biointerfaces, 232*, 113589. https://doi.org/10.1016/j.colsurfb.2023.113589
55. Li, M., Huan, K., Deng, D., et al. (2023b). Coaxial electrospinning synthesis of size-tunable CuO/NiO hollow heterostructured nanofibers: Towards detection of glucose level in human serum. *Colloids and Surfaces. B, Biointerfaces, 222*, 113047. https://doi.org/10.1016/j.colsurfb.2022.113047
56. Li, Y., Wang, J., & Xie, J. (2023c). Biomimetic nanoparticles targeting atherosclerosis for diagnosis and therapy. *Smart Medicine, 2*(3), e20230015. https://doi.org/10.1002/SMMD.20230015
57. Liu, G., Zhao, J., Qin, L., et al. (2020). Synthesis of an ordered nanoporous Cu/Ni/Au film for sensitive nonenzymatic glucose sensing. *RSC Advances, 10*(22), 12883–12890. https://doi.org/10.1039/d0ra01224f
58. Liu, L., Wang, J., Wang, J., et al. (2021). Colorimetric detection of cholesterol based on the peroxidase-like activity of metal-organic framework MIL-101(Cr). *ChemistrySelect, 6*(28), 7143–7149. https://doi.org/10.1002/slct.202102053
59. Lv, J., Wu, Y., Mai, Y., et al. (2020). Noncoding RNAs in diabetic nephropathy: Pathogenesis, biomarkers, and therapy. *Journal of Diabetes Research, 2020*, 3960857. https://doi.org/10.1155/2020/3960857
60. Mahmoud, A. H., AbdElMonem, H. A., & Abbas, M. M. (2024). The role of selenium and zinc oxide nanoparticles on mitigating side effects of obesity in rats. *Brazilian Journal of Biology, 84*, e264004. https://doi.org/10.1590/1519-6984.264004
61. Mansuriya, B. D., & Altintas, Z. (2021). Enzyme-free electrochemical nano-immunosensor based on graphene quantum dots and gold nanoparticles for cardiac biomarker determination. *Nanomaterials, 11*(3), 578. https://doi.org/10.3390/nano11030578
62. Mirhadi, E., Kesharwani, P., Johnston, T. P., et al. (2023). Nanomedicine-mediated therapeutic approaches for pulmonary arterial hypertension. *Drug Discovery Today, 28*(6), 103599. https://doi.org/10.1016/j.drudis.2023.103599
63. Młynarska, E., Bojdo, K., Frankenstein, H., et al. (2025). Nanotechnology and artificial intelligence in dyslipidemia management—cardiovascular disease: Advances, challenges, and future perspectives. *Journal of Clinical Medicine, 14*(3), 887. https://doi.org/10.3390/jcm14030887
64. Mumuni, M. A., Kenechukwu, F. C., Ofokansi, K. C., et al. (2020). Insulin-loaded mucoadhesive nanoparticles based on mucin-chitosan complexes for oral delivery and diabetes treatment. *Carbohydrate Polymers, 229*, 115506. https://doi.org/10.1016/j.carbpol.2019.115506
65. Oksal, E., Pangestika, I., Muhammad, T. S. T., et al. (2020). *In vitro* and *in vivo* studies of nanoparticles of chitosan-*Pandanus tectorius* fruit extract as new alternative treatment for

hypercholesterolemia via Scavenger Receptor Class B type 1 pathway. *Saudi Pharmaceutical Journal, 28*(10), 1263–1275. https://doi.org/10.1016/j.jsps.2020.08.017
66. Paleczek, A., Grochala, J., Grochala, D., et al. (2024). Noninvasive total cholesterol level measurement using an E-nose system and machine learning on exhaled breath samples. *ACS Sensors, 9*(12), 6630–6637. https://doi.org/10.1021/acssensors.4c02198
67. Pouran, S. R., Bayrami, A., Mohammadi Arvanag, F., et al. (2020). Biogenic integrated ZnO/Ag nanocomposite: Surface analysis and *in vivo* practices for the management of type 1 diabetes complications. *Colloids and Surfaces B: Biointerfaces, 189*, 110878. https://doi.org/10.1016/j.colsurfb.2020.110878
68. Puri, A. (2024). Nanosensor technologies for improved glucose monitoring [internet]. AZoNano. Available from: https://www.azonano.com/article.aspx?ArticleID=6692
69. Rajora, A., & Nagpal, K. (2021). Nanotechnology mediated diagnosis of type II diabetes mellitus. *Recent Innovations in Chemical Engineering (Formerly Recent Patents on Chemical Engineering), 14*(4), 272–298. https://doi.org/10.2174/2405520414999210111230008
70. Rastogi, L., Dash, K., & Sashidhar, R. B. (2021). Selective and sensitive detection of cholesterol using intrinsic peroxidase-like activity of biogenic palladium nanoparticles. *Current Research in Biotechnology, 3*, 42–48. https://doi.org/10.1016/j.crbiot.2021.02.001
71. Remes, A., Manea, F., Motoc, S., et al. (2020). Highly sensitive non-enzymatic detection of glucose at MWCNT-CuBTC composite electrode. *Applied Sciences, 10*, 8419. https://doi.org/10.3390/app10238419
72. Rizzo, M., & Rizvi, A. A. (2024). New advances in metabolic syndrome. *International Journal of Molecular Sciences, 25*(15), 8311. https://doi.org/10.3390/ijms25158311
73. Rodrigues, M. S., Martins, J. N., De Paula, G. C., et al. (2022). Effects of diet-induced hypercholesterolemia and gold nanoparticles treatment on peripheral tissues. *Anais da Academia Brasileira de Ciências, 94*(Suppl. 4), e20211081. https://doi.org/10.1590/0001-3765202220211081
74. Selima, R. M., Saleem, I. A., Shawki, M. M., et al. (2025). Effect of gold nanoparticles treatment on rats-induced obesity by evaluating body-composition directly and indirectly via bioelectric impedance analysis. *Scientific Reports, 15*, 4942. https://doi.org/10.1038/s41598-025-87971-z
75. Sharma, N., Behl, T., Singh, S., et al. (2022). Targeting nanotechnology and nutraceuticals in obesity: An updated approach. *Current Pharmaceutical Design, 28*(40), 3269–3288. https://doi.org/10.2174/1381612828666221003105619
76. Sharma, R., Borah, S. J., Bhawna, et al. (2023). Emerging trends in nano-based antidiabetic therapeutics: A path to effective diabetes management. *Materials Advances, 4*(15), 3091–3113. https://doi.org/10.1039/D3MA00159H
77. Shaveta, S., Singh, J., Afzal, M., et al. (2020). Development of solid lipid nanoparticle as carrier of pioglitazone for amplification of oral efficacy: Formulation design optimization, *in-vitro* characterization and *in-vivo* biological evaluation. *Journal of Drug Delivery Science and Technology, 57*, 101674. https://doi.org/10.1016/j.jddst.2020.101674
78. Sibuyi, N. R. S., Moabelo, K. L., Meyer, M., et al. (2019). Nanotechnology advances towards development of targeted-treatment for obesity. *Journal of Nanobiotechnology, 17*, 122. https://doi.org/10.1186/s12951-019-0554-3
79. Şimşek, N., Erdoğan, N. Ö., & Tığ, G. A. (2023). Metal nanoparticles-based biomarkers for clinical diagnosis. In U. P. Azad & P. Chandra (Eds.), *Handbook of nanobioelectrochemistry* (pp. 541–568). Springer. https://doi.org/10.1007/978-981-19-9437-1_25
80. Simu, S. Y., Ahn, S., Castro-Aceituno, V., et al. (2019). Gold nanoparticles synthesized with fresh *Panax ginseng* leaf extract suppress adipogenesis by downregulating PPARγ/CEBPα signaling in 3T3-L1 mature adipocytes. *Journal of Nanoscience and Nanotechnology, 19*(2), 701–708. https://doi.org/10.1166/jnn.2019.15753
81. Singh, N., Qutub, S., & Khashab, N. M. (2021). Biocompatibility and biodegradability of metal organic frameworks for biomedical applications. *Journal of Materials Chemistry B, 9*(30), 5925–5934. https://doi.org/10.1039/D1TB01044A

82. Sridara, T., Upan, J., Saianand, G., et al. (2020). Nonenzymatic amperometric glucose sensor based on carbon nanodots and copper oxide nanocomposites electrode. *Sensors, 20*(3), 808. https://doi.org/10.3390/s20030808
83. Surya, S. G., Majhi, S. M., Agarwal, D. K., et al. (2020). A label-free aptasensor FET based on Au nanoparticle decorated Co_3O_4 nanorods and a SWCNT layer for detection of cardiac troponin T protein. *Journal of Materials Chemistry B, 8*(1), 18–26. https://doi.org/10.1039/c9tb01989h
84. Tang, C., Zhou, K., Wu, D., et al. (2024). Nanoparticles as a novel platform for cardiovascular disease diagnosis and therapy. *International Journal of Nanomedicine, 19*, 8831–8846. https://doi.org/10.2147/IJN.S474888
85. Tannenberg, R., Paul, M., Röder, B., et al. (2023). Chemiluminescence biosensor for the determination of cardiac troponin I (cTnI). *Biosensors (Basel)., 13*(4), 455. https://doi.org/10.3390/bios13040455
86. Teixeira, L. M., Reis, C. P., & Pacheco, R. (2024). Potential of application of natural product nanoparticles in hypercholesterolemia. *Chemistry Proceedings, 16*(1), 84. https://doi.org/10.3390/ecsoc-28-20168
87. Thwala, L. N., Ndlovu, S. C., Mpofu, K. T., et al. (2023). Nanotechnology-based diagnostics for diseases prevalent in developing countries: Current advances in point-of-care tests. *Nanomaterials (Basel), 13*(7), 1247. https://doi.org/10.3390/nano13071247
88. Trandafir, L. M., Dodi, G., Frasinariu, O., et al. (2022). Tackling dyslipidemia in obesity from a nanotechnology perspective. *Nutrients, 14*(18), 3774. https://doi.org/10.3390/nu14183774
89. Tsou, Y. H., Wang, B., Ho, W., et al. (2019). Nanotechnology-mediated drug delivery for the treatment of obesity and its related comorbidities. *Advanced Healthcare Materials, 8*(12), e1801184. https://doi.org/10.1002/adhm.201801184
90. Tu, D., Holderby, A., & Coté, G. L. (2020). Aptamer-based surface-enhanced resonance Raman scattering assay on a paper fluidic platform for detection of cardiac troponin I. *Journal of Biomedical Optics, 25*(9), 097001. https://doi.org/10.1117/1.JBO.25.9.097001
91. Tu, Y., Ma, X., Chen, H., et al. (2022). Molecular imaging of matrix metalloproteinase-2 in atherosclerosis using a smart multifunctional PET/MRI nanoparticle. *International Journal of Nanomedicine, 17*, 6773–6789. https://doi.org/10.2147/IJN.S385679
92. Tukur, F., Mabe, T., Liu, M., et al. (2024). A plasmonic nanoledge array sensor for selective detection of cardiovascular disease biomarkers in human whole blood. *ACS Applied Nano Materials, 7*(17), 20024–20033. https://doi.org/10.1021/acsanm.4c02524
93. Villena Gonzales, W., Mobashsher, A. T., & Abbosh, A. (2019). The progress of glucose monitoring – a review of invasive to minimally and non-invasive techniques, devices and sensors. *Sensors (Basel)., 19*(4), 800. https://doi.org/10.3390/s19040800
94. Wang, M., Liu, F., Shi, M., et al. (2022). Ultra-sensitive non-enzymatic glucose sensor based on Ni/Au bilayer nanowire arrays. *Journal of Electronic Materials, 51*, 2490–2499. https://doi.org/10.1007/s11664-022-09495-w
95. Wei, C., Zou, X., Liu, Q., et al. (2020). A highly sensitive nonenzymatic glucose sensor based on CuS nanosheets modified Cu2O/CuO nanowire arrays. *Electrochimica Acta, 334*, 135630. https://doi.org/10.1016/j.electacta.2020.135630
96. Woodard, P. K., Liu, Y., Pressly, E. D., et al. (2016). Design and modular construction of a polymeric nanoparticle for targeted atherosclerosis positron emission tomography imaging: A story of 25% (64)Cu-CANF-Comb. *Pharmaceutical Research, 33*(10), 2400–2410. https://doi.org/10.1007/s11095-016-1963-8
97. Wu, Y., Fang, W., Hu, Y., et al. (2023). Optimization of the tandem enzyme activity of V-MOF and its derivatives for highly sensitive nonenzymatic detection of cholesterol in living cells. *Journal of Colloid and Interface Science, 649*, 601–615. https://doi.org/10.1016/j.jcis.2023.06.048
98. Wu, Y., Deng, S., Wei, S., et al. (2025). Adipocyte-targeted nanotechnology and cell-based therapy for obesity treatment. *ChemMedChem, 20*(2), e202400611. https://doi.org/10.1002/cmdc.202400611

99. Xin, F., Li, Y., Fu, C., et al. (2020). Multimodal nanoprobe for pancreatic beta-cell detection and amyloidosis mitigation. *Chemistry of Materials, 32*(3), 1080–1088. https://doi.org/10.1021/acs.chemmater.9b03841
100. Yadav, K. (2024). Nanotechnology in diabetes management: Revolutionizing treatment and diagnostics. *Journal of Molecular Liquids, 414A*, 126117. https://doi.org/10.1016/j.molliq.2024.126117
101. Yadav, H. M., Park, J. D., Kang, H.-C., et al. (2021). Recent development in nanomaterial-based electrochemical sensors for cholesterol detection. *Chemosensors, 9*(5), 98. https://doi.org/10.3390/chemosensors9050098
102. Yan, Q., Wu, R., Chen, H., et al. (2023). Highly sensitive cholesterol biosensor based on electron mediator thionine and cubic-shaped Cu2O nanomaterials. *Microchemical Journal, 185*, 108201. https://doi.org/10.1016/j.microc.2022.108201
103. Yaşacan, M., Erikçi, A., Eylem, C. C., et al. (2020). Polymeric nanoparticle *versus* liposome formulations: Comparative physicochemical and metabolomic studies as L-carnitine delivery systems. *AAPS PharmSciTech, 21*(8), 308. https://doi.org/10.1208/s12249-020-01852-4
104. Ye, X., Jiang, Y., Mu, X., et al. (2022). Ultrabright silicon nanoparticle fluorescence probe for sensitive detection of cholesterol in human serum. *Analytical and Bioanalytical Chemistry, 414*(13), 3827–3836. https://doi.org/10.1007/s00216-022-04024-4
105. Zafar, U., Khaliq, S., Ahmad, H. U., et al. (2018). Metabolic syndrome: An update on diagnostic criteria, pathogenesis, and genetic links. *Hormones (Athens, Greece), 17*(3), 299–313. https://doi.org/10.1007/s42000-018-0051-3
106. Zhao, Z., Huang, Y., Huang, Z., et al. (2022). Nonenzymetic glucose sensitive device based on morchella shaped nickel-copper layered double hydroxide. *Applied Surface Science, 597*, 153658. https://doi.org/10.1016/j.apsusc.2022.153658
107. Zhen, J., Li, X., Yu, H., et al. (2024). High-density lipoprotein mimetic nano-therapeutics targeting monocytes and macrophages for improved cardiovascular care: A comprehensive review. *Journal of Nanobiotechnology, 22*(1), 263. https://doi.org/10.1186/s12951-024-02529-x
108. Zhu, J., He, J., Verano, M., et al. (2018). An integrated adipose-tissue-on-chip nanoplasmonic biosensing platform for investigating obesity-associated inflammation. *Lab on a Chip, 18*(23), 3550–3560. https://doi.org/10.1039/C8LC00605A
109. Zhu, Y., Qi, Y., Xu, M., et al. (2023). Flexible biosensor based on signal amplification of gold nanoparticles-composite flower clusters for glucose detection in sweat. *Colloids and Surfaces A: Physicochemical and Engineering Aspects, 661*, 130908. https://doi.org/10.1016/j.colsurfa.2022.130908
110. Juanjuan L, Ruitao C, Huize L, et al. (2019). Nanomaterials for the theranostics of obesity. *Biomaterials*, 223, 119474. https://doi.org/10.1016/j.biomaterials.2019.119474
111. Manjushree, S. G., Adarakatti, P. S., Udayakumar, V., et al. (2022). Fabrication of cerium oxide and β-Ni(OH) nano hexagonal architectures assembled on reduced graphene oxide for non-enzymatic electrochemical detection of glucose. *Ionics, 28*, 1957–1972. https://doi.org/10.1007/s11581-022-04451-4
112. Farasati Far, B., Rabiee, N., & Iravani, S. (2023). Environmental implications of metal–organic frameworks and MXenes in biomedical applications: A perspective. *RSC Advances, 13*(49), 34562–34575. https://doi.org/10.1039/D3RA07092A

Nanomaterials in Regulating Lipid Metabolism

7

Pradeep Kumar Patel, Gitali Sen, Pooja Kathait, Soki Daeme Malang, Ankur, Muhammad Delwar Hussain, and Alakh N. Sahu

Abstract

Lipid metabolism refers to the processes by which lipids are synthesized, transported, stored, and metabolized in the body. It plays a vital role in energy homeostasis, cellular signaling, and maintain membrane structure integrity. Dysregulation of lipid metabolism contributes to metabolic disorders such as obesity, diabetes, atherosclerosis, and nonalcoholic fatty liver disease (NAFLD). Nanomaterials have gathered considerable attention in recent years for their potential applications in regulating metabolic processes, particularly lipid metabolism. Their unique physical and chemical properties enable them to be used for targeted drug delivery, biosensing, and as agents to modify metabolic pathways. Recent advancements in nanotechnology have introduced innovative strategies for the precise modulation of lipid metabolism using engineered nanomaterials. These nanoplatforms, including nanoparticles, liposomes, dendrimers, and nanostructured polymers, offer unique physicochemical properties such as high surface-to-volume ratios and controlled release capabilities. Nanomaterials can effectively deliver bioactive compounds, genetic materials, or small interfering RNA (siRNA) to specific tissues or cellular organelles, enhancing therapeutic efficacy while minimizing off-target effects. Moreover, emerging studies have established their ability to influence lipid pathways directly by interacting with enzymes, lipid transport proteins, and signaling cascades. This book chapter

P. K. Patel · G. Sen · P. Kathait · S. D. Malang · Ankur · A. N. Sahu (✉)
Phytomedicine Research Laboratory, Department of Pharmaceutical Engineering & Technology, IIT (BHU), Varanasi, India
e-mail: ansahu.phe@iitbhu.ac.in

M. D. Hussain
Department of Pharmaceutical Sciences, School of Pharmacy and Health Professions, University of Maryland Eastern Shore, Princess Anne, MD, USA

V. K. Chaturvedi et al. (eds.), *Nanotheranostics in Metabolic Syndrome*,
https://doi.org/10.1007/978-981-95-3746-4_7

unravels the potential of nanomaterials in regulating lipid metabolism, with an emphasis on their applications in targeted therapy, metabolic function, and disease management. By heading the gap between nanotechnology and lipid biology, this interdisciplinary approach holds promise for transformative advancements in metabolic health.

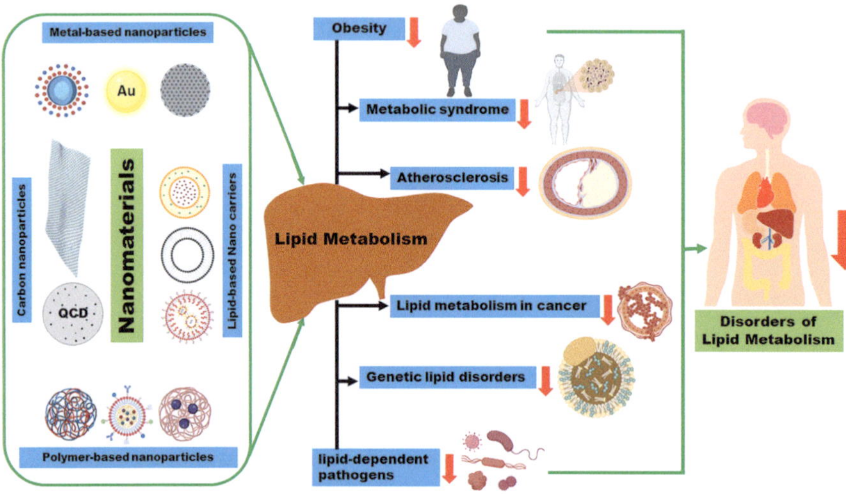

Keywords

Nanomaterials · Lipid metabolism · Lipid regulation · Drug delivery · Oxidative Stress

1 Introduction

1.1 Brief Overview of Lipid Metabolism

Nutrient, hormone, and organismal homeostasis control are all impacted by the intricate physiological process of lipid metabolism [1]. During synthesis, storage, and catabolism stages, alterations in lipids and their metabolites often change the amounts, shape, and function of plasma lipoproteins, which in turn leads to the development and advancement of associated disorders [2]. The proper operation of its several parts is necessary to keep the body healthy. Lipid metabolism problems and numerous associated human illnesses have become more common in now days due to changes in food habits, lifestyle choices, and improvements in living status [3].

1.2 Disease Associated with Lipid Metabolism

Obesity, the very common metabolic disorder of the endocrine system and mostly brought on by an excess of calories and poor lipid metabolism. Obesity also develops and occurs more frequently when lipid metabolism disorders are present. Type 2 diabetes mellitus (DM-II) is an extensive metabolic disorder that affects the endocrine system. The probability of cardiovascular disease (CVD), especially atherosclerosis and coronary heart disease, is also increased concurrently by abnormal lipid metabolism [4]. After various studies have demonstrated that deviation of lipid metabolism by directly or indirectly enhance the prevalence of cardiovascular disease and mortality [5]. It has also been found that abnormal metabolism of polyunsaturated fatty acids and lipoproteins contributes to the onset and progression of CVD. Lipid metabolic disorders are implicated in the genesis and development of various diseases, including cancer and chronic illnesses. Cancer development is characterized by disrupted lipid metabolism within cells, and abnormal lipid buildup in the tumor environment promotes tumor growth, immune evasion, and inflammation. Similarly, chronic conditions such as neurodegenerative diseases, NAFLD, Chronic Kidney Disease (CKD), and osteoarthritis are associated with lipid metabolic dysfunction.

1.3 The Role of Nanomaterials in Lipid Metabolism

The disadvantages of conventional therapies for lipid metabolic diseases are intended to be addressed using nanotechnology. Effective management of lipid metabolism is made feasible by nanotechnology, which enables better drug delivery. The production of metallic nanoparticles from a variety of metals has been studied; silver and gold nanoparticles are essential for medical purposes. Numerous ligands have been used to adorn the surface of these nanoparticles, which are readily functionalized on the surface. The first FDA-approved nanocarrier was created by lipid-based nanocarrier research, which is also making quick progress in areas including cancer and peptide delivery [6]. The phrase lipid-based nanocarrier refers to a number of systems, such as liposomes, drug delivery systems that are self-nanoemulsifying and self-microemulsifying, nanoemulsions, and nanocapsules, which are now utilized for disorders linked to lipid metabolism. Numerous techniques exist for producing polymeric nanomaterials (PNMs) for drug delivery, including the use of brush polymers, dendrimers, micelles, and various nanostructured materials. Carbon-based nanoparticles, like nanotubes and fullerenes, possess unique properties determined by their structure. Scaling up PNM production from the laboratory to industrial levels presents significant challenges, particularly in selecting appropriate excipients for pharmaceutical applications without compromising quality [7]. The multistep nanomaterial production process requires continuous monitoring of numerous properties, including size, shape, charge, and biocompatibility [8]. This discussion will focus on lipid metabolism and the use of nanoparticles in treating related diseases.

2 Lipid Metabolism

2.1 Mechanism of Lipid Metabolism

Lipid metabolism plays an essential role in various physiological functions, involving energy storage, nerve impulse transmission, hormone regulation, and the transport of fat-soluble nutrients. Lipids serve as an extremely efficient source of energy, and provide significantly more than proteins or carbohydrates 9 kcal per gram. It can store more than 100,000 kcal in the whole body, sustaining essential functions for month without food intake, provided there is enough water. Biochemical lipids are stored in cells in whole body, primarily in a specialized connective tissue also known as adipose tissue. Additionally, lipids act as a protective cushion for vital organs such as the spleen, liver, heart, and kidneys, shielding them from damage [9]. The liver acts as an additional storage site for excess fat. During prolonged energy surplus, unused energy is stored as triglycerides in both adipose tissue and hepatocytes [10]. The metabolic cycle extends to include the citric acid cycle, the urea cycle, and the citrate cycle [11]. Fatty acids undergo degradation through oxidation, generating substantial amounts of ATP and producing reactive oxygen species [12]. The three fundamental aspects of lipid biology are: (1) lipid production, (2) lipid catabolism for ATP generation, and (3) the role of lipids as signaling molecules.

- Lipids, including triacylglycerides and phospholipids, are synthesized by glycerol derivative via glucose and fatty acids produced in the mitochondria.
- Synthesis of fatty acid occurs in the cytoplasm, with mitochondrial citrate acting as a precursor for the production of palmitate, and it can then be converted into another fatty acid.
- Mitochondrial matrix is the place where fatty acid β-oxidation occurs. Fatty acids are carried out into the matrix via carnitine acyltransferase-I (CPT-I) in the mitochondrial outer membrane, along with carnitine acyltransferase-II (CPT-II) and carnitine-acylcarnitine translocase in the mitochondrial inner membrane.
- Synthesis of fatty acid is linked to the conversion of NADPH to $NADP^+$, while oxidation of fatty acid produces Acetyl-CoA, NADH, and FADH2, which drive ATP production by oxidative phosphorylation.
- Fatty acid synthesis is controlled through acetyl-CoA carboxylase (ACC), which is stimulated by citrate and inhibited by the fatty acid palmitate.
- Fatty acid β-oxidation is controlled by malonyl-CoA, which inhibits the activity of CPT-I), thereby blocking the import of fatty acids into the mitochondrial matrix for β-oxidation.
- Lipids can reshape proteins to change their function. Key changes include N-myristoylation, N- or S-palmitoylation, and S-prenylation.
- Lipids, including phosphoinositides, eicosanoids, and sphingolipids, function as key signaling molecules.

- Cholesterol biogenesis pathway begins in the cytosol and is regulated through the enzyme 3-hydroxy-3-methylglutaryl CoA reductase (HMG-CoA reductase), which is targeting statins, a division of cholesterol-lowering drugs [13].

2.2 Lipid Uptake and Storage

There is still very little known about the mechanisms that govern the uptake of certain lipid classes by macrophages and how this relates to the formation of foamy cells. It is possible that several pathways exist and mediate the uptake of different lipids. For instance, the absorption of fatty acids has been reported to happen via two ways which are passive transport and receptor-mediated endocytosis [14]. Endocytosis is also known to be involved in the uptake of lipoproteins by macrophages where extracellular material is ingested into the cell [15]. One of the types of endocytosis is clathrin-mediated endocytosis or receptor-dependent endocytosis in which lipids are endocytosed with the help of transmembrane receptors and transporters. This process involves clathrin coated pits which are membrane invaginations covered by clathrin capsules which are made of triskelia [16]. Caveolae-dependent endocytosis or raft-dependent endocytosis is a type of endocytosis that is independent of clathrin, is dependent on dynamin and is sensitive to cholesterol depletion. This pathway is also involved in lipid uptake in different kinds of cells [17]. The function of transporters in the absorption of fatty acids has been proven while certain transporters or receptors are involved in the process of uptake of fatty acids by cells including macrophages. CD36 is one of the major fatty acid transporters in macrophages and is a member of pattern recognition receptors which is a class of transporters. These receptors are important to the innate immune system since they recognize patterns of molecules found on the surface of the pathogens and lead to the stimulation of immune responses [18]. The expression of CD36 started as monocytes mature into macrophages [19]. This leads to an increase the fatty acids uptake. Expression of CD36 in macrophages is lead to macrophage colony activating factor and interleukin-4 (IL-4). These cytokines encourage the differentiation of alternatively stimulated macrophages, which help resolve proinflammatory reaction and support tissue reconditioning [20]. These findings may reveal differences in fatty acid uptake capacity based on the macrophage's function as well as phenotype properties. Especially, unsaturated fatty acids trigger gene expression of CD36 in human macrophages in a dose associated mode [21]. This suggests that fatty acids escalate the mechanisms essential for their uptake into cells. Along with CD36, the fatty acid transport proteins like FATP1 and FATP4 are also expressed in macrophages [22]. FATP1 insufficiency was found to noticeably decrease fatty acid uptake in bone marrow-originated macrophages, while overexpressing FATP1 in RAW 264.7 macrophages led to increase in fatty acid uptake up to 17% [22]. The lack of FATP4 in bone marrow-originated macrophages arises in modified levels of several lipid subfamilies, such as different phospholipids and sphingolipids, with the changes being strongly affected by the macrophage type and the particular treatment. Interestingly, FATP4 shortage did not lead to a reduction in macrophage TAG proportion [23].

2.3 Impact on Lipid Transport Proteins

Although the precise method through which each one of particular proteins contributes to the transport mechanism inside the plasma membrane is yet unknown, these proteins are frequently involved to as "fatty acid transporters" for convenience. FATPs constitute a family of six homologous proteins viz, FATP1, FATP2, FATP3, FATP4, FATP5 and FATP6. FATP1–6 [24], plasma membrane-related FABP (FABPpm) [25], CD36 [26], and from research in the tissues of animals with genetic modifications FABPpm [27], CD36, FATP1, and FATP4 [28].

Triacylglycerol production was unaffected by FABPpm's impact on fatty acid metabolism and skeletal muscle over expression, although elevated fatty acid oxidation. The concurrent mitochondrial overexpression of FABPpm/mAspAT could not change isolated mitochondrial fatty oxidation, this was because of the transfer into the muscle cell of extra fatty acids. FABPpm appears to cooperate with different fatty acid transporters, especially CD36, according to various studies. FABPpm/mAspAt ablation investigations have not yet been carried out. CD36 function in controlling fatty acid absorption and metabolism has been underappreciated during metabolic challenges as have its effects on fatty acid metabolism under baseline settings. Oxidation-enhancing agents significantly reduced the stimulation of fatty acid uptake in perfused skeletal muscles and cardiac myocytes of CD36-null mice, and insulin significantly reduced the activation of fatty acid uptake in CD36-null mice's muscle when compared to WT mice [29]. Alteration in the rates of metabolism were similarly influenced by these decreases in fatty acid absorption. For instance, AICAR-activated fatty acid oxidation has been significantly reduced in CD36-null muscle, and fatty acid oxidation in working hearts was still more than 40% lower in CD36-null animals compared to WT mice [30]. Although fatty acid oxidation rose 400% in CD36-overexpressing mice when they contracted their muscles, whereas this rise was far smaller in WT animals. In CD36-null red muscle, insulin-stimulated triacylglycerol esterification was much lower than in WT mouse red muscles [29]. Since CD36-null mice do not exhibit any changes in cellular signaling pathways or enzymatic activities [31], these studies show that during metabolic challenges CD36 involve in regulating fatty acid oxidation and esterification in the heart and skeletal muscle. Other studies have demonstrated that in 293 cells the fatty acid utilized determines the transport capacity of FATPs (oleate transport: FATP4 > FATP1 > FATP6; palmitate transport: FATP6 >> FATP1 > FATP4) [32], indicating that various FATPs have varying efficiency when it comes to transporting particular fatty acids. As of right now, the caveolin family has three known members. While most common isoform of caveolin-3 is present in muscle and the heart, caveolin-1 and -2 are distributed very widely throughout the tissues, with some exception of muscular tissues [33]. This sparked the idea that caveolin-1 might have fatty acid transporter roles in addition to its caveolar-related ones. Using cholesterol-lowering agents, caveolin-1 antisense oligonucleotides, or caveolin dominant-negative mutants. Remarkably, CD36 is also found in caveolae, indicating that CD36 plays a role in caveolae-associated fatty acid absorption [34]. The finding that caveolin-1 depletion in fibroblasts decreased the amount of CD36 in the plasma membrane concurrently with a decrease in cellular fatty acid absorption [34], further supported this theory. The overexpression of caveolin-1, alternatively, saved fatty acid absorption by redirecting CD36 to the plasma membrane [34]. Although caveolin-1 was identified as a possible fatty acid transporter in the early experiments using photoreactive fatty acid analogs [35], the more recent research suggesting caveolins work indirectly in uptake of fatty acid by providing CD36 with docking sites in the plasma membrane. These contradictory findings regarding caveolin-1 and CD36's collaboration in cellular fatty acid uptake could be the result of variations in the model systems and/or methods employed, or they could be related to tissue- or cell-specific variations. No alterations in the absorption and metabolism of cardiac fatty acids were found in the hearts of caveolin-3-mutant mice [36].

2.4 Lipid Biosynthesis Pathways

In both cellular physiology and disease, lipids are crucial. They are the structural elements of cell membranes, which are necessary for the existence of cells and life. Glycerolipids, sphingolipids, and sterols are the three kinds of lipids that make up the majority of a mammalian cell's biological membrane [37]. The Golgi and mitochondria contribute somewhat to the ER's synthesis of most lipids. In mammals, PA is the precursor of the majority of phospholipids. The routes for phosphatidylinositol (PI), phosphatidylcholine (PC), phosphatidylethanolamine (PE), and phosphatidylserine (PS) synthesis are shown in Fig. 7.1. The majority of PE synthesis occurs in the ER, with some occurring in mitochondria. Table 7.1 lists the complete names of the enzymes that are involved in the phospholipid production. Glycerol 3-phosphate acyl-transferase (GPAT) sequentially acylates glycerol-3-phosphate (G-3-P) as the initial step in the production of PA. Lyso-PA acyl transferase (LPAAT) performs the second acylation at the sn-2 position after the initial acylation at the sn-1 position to create lyso-PA (LPA) [37]. There are many isoforms of both enzymes, which are essential membrane proteins. PA phosphatases, often referred to as lipins, transform PA into (diacylglycerol) DAG for the synthesis of PC, PE, and PS. The pathway is explained in detail in Fig. 7.1.

2.5 Lipid-Associated Enzymes and Signaling Molecules

As unit structure of biological membranes (phospholipids, sphingolipids, etc.) of particular proteins (palmitoylation, myristoylation), as a basis for the synthesis of signaling molecules (long-chain fatty acids and fatty acid metabolites), and metabolic energy (long-chain fatty acids), lipids are essential for a wide range of biological processes. All lipid species are distinguished by their hydrophobic or

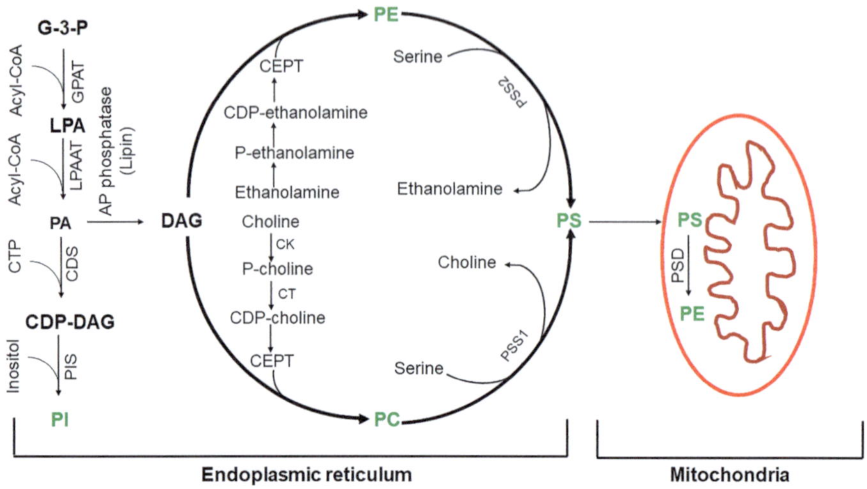

Fig. 7.1 Pathway of various lipid biosynthesis

amphiphilic characteristics, which make them virtually insoluble in aqueous solutions. This characteristic significantly impairs the movement of lipids in cytosol, including blood plasma, the interstitium, and the soluble cytoplasm of cells. However, cytosolic compartments having some proteins called "lipid binding proteins" that can bind lipids reversibly and noncovalently, significantly increasing the availability and water transport of particular lipid molecules. Cytoplasmic lipid binding proteins, and plasma albumin, like oxysterol binding protein, retinol or retinoic acid-associated binding proteins, and cytoplasmic fatty acid binding proteins (FABPc), are examples of these proteins [38]. Membrane fatty acid binding proteins are a group of proteins found in the cell plasma membrane that have the capability to bind long-chain fatty acids in a reversible and noncovalent manner [39]. The soluble and membrane-associated lipid binding proteins work together to control the bioavailability of their lipid ligands in both space and time, which has a direct impact on the substances metabolism or signaling. Cell-associated fatty acid-binding proteins are important for both health and sickness, and this brief overview focuses on their possible use as therapeutic targets. Since their discovery in 1972, the intracellular or cytoplasmic FABPs have been identified as a set of nine different proteins, each weighing 14–15 kDa and exhibiting a unique tissue distribution pattern [40]. Some examples like, heart-type FABPc is found in the kidney, brain, and (red) skeletal muscle in addition to the heart. Similarly, other organs such as small intestine, liver, and kidney all contain liver-type FABPc. On the other hand, intestinal-type FABPc is expressed only in enterocytes of intestine, while brain-type FABP is expressed exclusively in the glial cells of brain. With cellular concentrations as high as 300 mM, the FABPs are widely expressed [41]. Like plasma albumin, the FABPs each have extremely high binding affinities toward fatty acids, increasing the overall amount of fatty acids in cytoplasm by different levels of magnitude. There is a lot of

Table 7.1 Enzymes involved for lipid biosynthesis

Enzymes involved in phospholipid synthesis	Abbreviations of enzymes	Comments
Acyl-CoA:glycerol-3-phosphate acyltransferase	GPAT (α)	Integral membrane proteins; two enzymes are in the ER and two are located in mitochondria
Choline kinase	CKα and CKβ	Soluble proteins; rate-limiting enzyme
CDP-diacylglycerol synthase	CDS1 and CDS2	Integral membrane proteins located at the ER
CTP-phosphocholine cytidyltransferase	CTα and CTβ	Soluble proteins that associate with ER membranes in a regulated manner
CDP-choline:1,2-diacylglycerol choline phosphotransferase	CPT	CPT, an integral enzyme, localizes to the Golgi
CTP:phosphoethanolamine cytidyltransferase	ET	Soluble protein and does not require membrane association to be active
CDP-ethanolamine:1,2-diacylglycerol ethanolamine phosphotransferase	CEPT	CEPT, an integral enzyme localized at the ER, can use either CDP-ethanolamine or CDP-choline as substrate
Ethanolamine kinase	EK1 and EK2	Soluble proteins
Lyso-phosphatidic acid acyltransferase	LPAAT (β)	Integral membrane proteins at the ER
Phosphatidylserine synthase	PSS1 and PSS2	PSS1 and PSS2 replace the head group of PC and PE with serine, respectively. Both are integral membrane enzymes. Mice knockouts of PSS1 or PSS2 individually are viable but double knockouts are not
Phosphatidylethanolamine N-methyl transferase	PEMT	Enzyme restricted to the liver
Phosphatidylserine decarboxylase	PSD	PSD is localized in the external side of the inner mitochondrial membrane

evidence currently that membrane linked fatty acid binding proteins help to transport the long-chain fatty acids across membranes, even though they can readily enter and disperse within biological membranes [42]. Specifically, these membrane-associated proteins aid in the rate-limiting stage of transmembrane transport, which is decrease the absorption of fatty acids at the membrane [43]. There are currently at least three different kinds of membrane proteins known to aid in the fatty acids uptake by cells. First, the 43 kDa peripheral plasma membrane fatty acid-binding protein (FABPpm) protein is found in all tissues [44]. Second, there are six members of fatty acid-transport proteins (FATP; 63 kDa) family, each of which has a definite tissue distribution [39]. The only function of the trans-membrane proteins known as FATPs, which exhibit Acyl-CoA synthetase function, is to absorb fatty acids (chain length 422) and convert them straight into acyl-CoA esters [39]. Another is CD36, also known as fatty acid translocase (FAT), is a section-B scavenger receptor protein that has several roles, including acting as a gustatory lipid sensor and binding thrombospondin, anionic phospholipids, and oxidized low-density lipoprotein (oxLDL) [45]. With two transmembrane-spanning sections and a hairpin membrane shape, CD36 is highly glycosylated, increasing its 472-amino acid protein weight from 53 to 88 kDa.

2.6 Disease Associated with Lipid Metabolism

Disorders are the many health repercussions that occur in the human body as a result of rising or falling lipid levels. Typically, triglyceride, LDL, or both lipid levels rise in these kinds of diseases. The body needs HDL, a beneficial fatty acid that aids in the removal of harmful cholesterol. Triglycerides and fatty LDLs are examples of undesirable lipids that build up and harm arteries, which can have detrimental effects on cardiovascular health. According to the pathway and health consequences, Fredrickson's divided lipid metabolism-based illnesses into five categories [46]. Hyperlipidemia, or a higher accumulation of lipids in the body, is associated with more health concerns than a lower level of lipids [47]. A collection of severe lipid illnesses known as hyperlipidemia are brought through an excessively high blood level of undesirable lipids [48]. Several illnesses are recently identifiable with structural faults of apolipoproteins and lipid transfer proteins with their presence or absence respectively. The classification of distinct lipid metabolic disorders differs according to concentrations of classes of lipoproteins [49]. Nuclear fatty acid receptors, also known as peroxisome proliferator-activated receptors (PPAR), are a class of lipid that has been linked to obesity and metabolic disorders such insulin resistance, hyperlipidemia, and coronary artery disease [50]. PPAR have also been identified for use in medical treatment, and regulating pathways that control a variety of lipid diseases [51]. Furthermore, postmenopausal women face significant global health issues due to a variety of lipid metabolism disorders, including atherosclerosis, osteoporosis, and bone-related ailments [52]. The high buildup of lipid droplets on muscle fibers results in myopathy and the complication of carnitine insufficiency [53]. Numerous illnesses and disorders are brought on by lipid metabolic systems or lipid imbalances. Furthermore, metabolic disorders including Niemann–Pick disease (types A and B), methylmalonic acidemia, GM1 and GM2 gangliosidoses, Gaucher disease, xanthoma, Bassen–Kornzweig syndrome, familial lipoprotein lipase deficiency, and chylomicronemia syndrome are characterized by abnormal or too much lipid deposition within the body. Apart from this, diabetes and cardiovascular diseases, both of which were asymptomatic at the time, are the most severe outcomes.

3 Nanomaterials for Lipid Regulation

3.1 Overview of Nanomaterials

Compared to free pharmaceuticals, nanoparticles (NPs) have several advantages, including increased bioavailability, biodegradability, safety, site-specific drug delivery, and prolonged drug release [54]. The goal of nanotechnology is to overcome the drawbacks of traditional treatments for lipid metabolic disorders. Because it facilitates improved medication delivery, nanotechnology makes it possible to manage lipid metabolism more effectively. New microvessels are created in lipid metabolic diseases such as the pathophysiology of atherosclerosis, which improves permeability and enables nanoparticles to crowd around the plaques and have their therapeutic effects [55]. In addition to its application in reducing endothelial inflammation, nanotechnology has been investigated to enhance the endothelium anti-inflammatory process [56]. Collagen IV-targeted NPs containing the anti-inflammatory peptide Ac2-26 enhanced the anti-inflammatory peptide's bioavailability and decreased polymorphonuclear neutrophil mobilization [57]. High-density lipoprotein (HDL) nanoparticles have been engineered due to HDL's well-established antiatherosclerotic effects and its ability to improve endothelial function [55]. Studies indicate that HDL nanoparticles can modulate fatty acid metabolism in patients with type 2 diabetes mellitus and help reduce atherosclerotic plaque formation [58]. Additionally, the interaction between HDL and plaque macrophages makes it an excellent platform for targeted drug delivery. This chapter focuses on nanoparticles that play a role in lipid metabolism. Nanoparticles generally classified on the basis of their structure, composition, and function. They are basically classified into three categories: hybrid, inorganic, and organic, and each having specific advantages based on their application. Due to their biocompatibility, organic nanoparticles such as liposomes and polymeric nanoparticles are mostly accepted in drug delivery and biomedicine. Inorganic nanoparticles like gold, silica, or quantum dots (QDs) are mostly utilized and have potential in imaging as well as diagnostics due to their stability and optical characteristics.

3.2 Metal-Based Nanoparticles

These nanoparticles are now showing promise as medication and biosensor delivery systems. Although several metals have been investigated for the creation of metallic nanoparticles, silver and gold nanoparticles are crucial for medicinal applications. These nanoparticles are easily functionalized on the surface, and their surface has been decorated with a variety of ligands. Nanoparticles have been connected to a wide range of ligands, including sugars, peptides, proteins, and DNA. Iron oxide NPs have been used in many biological applications because of their special chemical and magnetic characteristics. They have received clinical approval as MRI contrast agents and biocompatible drug delivery systems [59]. Similar to iron oxide nanoparticles, gold nanoparticles (AuNPs) have been extensively studied and utilized across diverse fields, including biology, chemistry, and medicine. Likewise, silver nanoparticles (AgNPs) have emerged as valuable tools in various domains, particularly in healthcare and medicine. Their widespread application is primarily attributed to their unique physicochemical properties, which encompass electrical, optical, biological, and thermal features. Consequently, people are frequently exposed to AgNPs due to their extensive use in many products [60].

Table 7.2 Lipid-based nanocarriers and their characteristics

Nanocarrier type	Structure explanation
Liposomes	A vesicle with an aqueous inner core surrounded by a bilayer of phospholipids, used to encapsulate hydrophilic or lipophilic drugs.
O/W Nanoemulsions	Oil-in-water emulsions with a fluid lipid inner core, stabilized by surfactants, ideal for poorly water-soluble drugs.
O/W Microemulsions	Similar to nanoemulsions but thermodynamically stable, with smaller droplet sizes and enhanced drug solubility.
Nanocapsules	A solid or fluid lipid core surrounded by a polymer coating, providing controlled drug release and stability.
Solid lipid nanoparticles	Solid lipid core containing drugs, stabilized by surfactants, offering improved bioavailability and controlled release.
Self-nanoemulsifying systems	Systems that form nanoemulsions spontaneously upon mixing with water, with a fluid lipid core and surfactants.
Self-microemulsifying systems	Similar to self-nanoemulsifying systems but forming microemulsions, providing enhanced drug solubilization.

3.3 Lipid-Based Nanocarriers (e.g., Liposomes, Solid Lipid Nanoparticles)

When it comes to medication delivery, lipid-based nanocarriers are revolutionary. Lipid-based nanocarrier research has produced the first FDA-approved nanocarrier and is rapidly advancing in fields including peptide delivery and cancer [61]. The term lipid-based nanocarrier encompasses various systems, including liposomes, self-nanoemulsifying and self-microemulsifying drug delivery systems, nanoemulsions, and nanocapsules, among others (Table 7.2).

3.4 Polymer-Based Nanoparticles

Therapeutic disease detection and treatment have benefited greatly from the use of PNMs. The creation of PNMs as novel biomaterials has the potential to significantly raise the standard of healthcare because of their increased diagnostic precision and dependability, more efficient therapeutic agent targeting, and enhanced scaffold usability for tissue engineering and regenerative medicine. There are several techniques to create PNMs for distribution, such as brush polymers, dendrimers, micelles, nanoparticles, polymerases, nanocapsules, nanocomposites, nanogels, and nanofibers. A new path in biomedical engineering, packaging, tissue engineering, food processing, and better cure for water-insoluble and soluble drug delivery systems has resulted from the nanoparticles combination with polymer science. Nanoparticles, ranging from 1 to 100 nm in size, possess diverse properties depending on their composition (e.g., ceramic, liquid, metal, semiconductor, or carbon-based materials) [62]. These properties, such as a high surface area-to-charge ratio, ease of synthesis and characterization, stability, biocompatibility, and good absorption, make them ideal drug delivery vehicles. The transition from traditional to nanodrug delivery systems offers improved targeted drug release. Nanomedicines are broadly categorized into nanospheres (drugs encapsulated within a polymeric shell) and nanocapsules (drugs dispersed within a polymer matrix), depending on their synthesis method [63]. Their unique characteristics make nanoparticles valuable in treating chronic diseases like diabetes, cardiovascular diseases, cancer disease, and neurodegenerative complications. Natural as well as synthetic polymers play a crucial role in targeted drug delivery systems [64]. Many of these polymeric materials are readily available, biocompatible, biodegradable, and easy to modify. Their reactive groups like amino, carboxylic groups, and hydroxyl can easily interact with formulated materials, resulting in hybrid materials with enhanced physical and chemical properties [65]. Examples of some natural polymers include polysaccharides, proteins, peptides, collagen, albumin, alginate, gelatin, chitosan, etc., while synthetic polymers include polyglycolic acid (PGA) and polylactic acid (PLA). Each type of polymer has its own advantages and limitations. Biocompatible polymers are particularly favored due to their cost-effectiveness, ease of preparation, high stability in biological fluids, and superior properties such as improved proliferation, adhesion, and efficient targeting [66].

3.5 Carbon Nanomaterials

In 1991, carbon nanotubes are tubular structures made of carbon atoms [67]. These structures resemble a sheet of graphite that has been rolled up into a cylinder and has a buckyball at one or both ends. These sheets are hexagonal networks of carbon atoms that range in length from 1 to 100 nm and have a diameter of 1 nm. These carbon networks are made up of a cylinder-shaped layer of graphite. Recently, there has been a lot of interest in two carbon-linked configurations: single-walled nanotubes (SWNTs) and multiwalled nanotubes (MWNTs). In addition to these kinds, popular combinations also include C_{60} fullerenes. Carbon-based nanoparticles, such as nanotubes and fullerenes (buckyballs), are hollow, cage-like structures whose properties vary based on their cylindrical graphite arrangements. Their size, shape, and surface features make them attractive drug delivery systems. SWNTs and C_{60} fullerenes having internal diameters around 1–2 nm (approximately 1/2 diameter of a DNA helix), while MWNTs having diameters ranging from several to tens of nanometers, with interlayer spacing of 0.36 nm. Their lengths can extend from 1 micrometer to several micrometers [68]. QDs are nanocrystals typically measuring 2–10 nm in size. It is having a semiconductor inorganic core, like CdSe, and a coated organic shell, like ZnS, which enhances their optical properties and enables fluorescence when exposed to light. A capping layer on quantum dots improves their dissolving properties in aqueous buffers. Unlike bulk or atomic semiconductors, QDs emission color is determined by their core, while the outer shell allows for binding with biomolecules. This biomolecular bonding can be tailored to target specific biomarkers [69]. The unique properties of QDs, derived from their nanoscale size, include narrow emission spectra, broad UV excitation, bright fluorescence, and high photostability. These features make them ideal for applications such as long-term intracellular tracking, real-time monitoring and in vitro bioimaging.

4 Mechanisms of Nanomaterials for Lipid Metabolism

4.1 Interaction with Lipid Uptake Pathways

Because of their potential use in imaging, medication delivery, and diagnostics, nanosized materials are being extensively researched in nanomedicine [70]. Their size allows them to interact with cells similarly to other nano-sized things including viruses, proteins, and cholesterol particles. Cells use their endocytic machinery to ingest these naturally occurring nano-sized particles, which are often identified by certain cell receptors at the plasma membrane [71]. Similar to this, materials that are created at the nanoscale can take advantage of the cellular machinery and be absorbed by cells. The adherence of nanoparticles to the cell membrane and their possible interaction with cell receptors are the initial stages of nanoparticle–cell interactions that take place at the cell surface. To enhance targeted drug delivery, nanoparticles can be modified with targeting molecules (peptides, proteins, or antibodies) that bind to specific cell surface receptors [72]. However, achieving effective targeted delivery remains challenging [73]. This is due to incomplete understanding of initial nanoparticle-cell interactions, difficulties in nanoparticle design, and the impact of surface modifications. For example, chemical conjugation can reduce the binding strength of targeting ligands, and controlling ligand density and orientation, crucial for receptor recognition and cellular uptake, is difficult [74]. These challenges in surface functionalization are well-documented. Ultimately, improved control over initial cell interactions is essential for successful targeted intracellular drug delivery [75]. The surface of nanoparticles is quickly altered by the adsorption of some biomolecules, like proteins and lipids, which reduce the bare surface energy. This occurs through mechanisms like surface charge compensation, shielding of hydrophobic regions, water displacement, and others, bringing the nanoparticle's residual interactions closer to thermal energy. As a result, a tightly bound, immobile layer of proteins with strong affinities toward the particle surface forms, known as the hard corona, along with a loosely associated, mobile layer called the soft corona [76]. These strongly bound proteins significantly influence the nanoparticle interactions toward the biological system, creating a biomolecule–nanoparticle interface that can be read by living organisms.

4.2 Effect in Lipid Transport and Distribution

The liver is main organ for detoxification of body. It is capable of secretory protein synthesis, glycogen storage, and deoxidation. Through phagocytosing and separating different foreign materials, it functions like a biological barrier. Various nanoparticles, like nanometal monomers and nanometal oxides, have been found to deposited in the liver and having a wide range of harmful consequences [77]. Inflammatory cytokines alter when metal nanoparticles (MNPs) enter the body. NiO NP reduced the anti-inflammatory cytokines such as IL-4 and IL-10 levels while raising the concentrations of proinflammatory cytokines such as IL-1β and IL-6 [78]. MNP-induced liver dysfunction results in structural alterations to the liver. Inflammation brought on by MNPs may alter liver coefficients [79]. Blood serum analysis revealed liver damage in the form of increased alkaline phosphatase (ALP), decreased bilirubin and increase the aspartate aminotransferase (AST) levels [80]. This damage included structural changes in the liver, leading to metabolic dysfunction. AgNPs caused increased spleen weight and liver damage, including hepatocyte necrosis, hemorrhage, and damage to portal vein endothelium [81]. Titanium dioxide nanoparticles (TiO_2 NPs) induced liver structural changes such as inflammation, increased collagen density, fibrosis, and thickening of the Glisson's capsule [82]. AuNP have been shown to stimulate hepatic macrophages and significantly worsening the progression of practical liver injury and immune hepatitis [83]. Hepatocytes, which make up the basic working unit of the liver, the hepatic lobule, account for solid cells approx. 60% of the liver cells, while the remaining 30–35% are nonparenchymal cells, such as Kupffer cells, hepatic stellate cells, and sinusoidal endothelial cells. Nanoparticles that accumulate in hepatic tissue can disrupt normal biochemical and physiological functions by impacting both hepatic parenchymal cells and nonparenchymal cells, as well as key physiological processes in the liver.

4.3 Oxidative Stress and Lipid Peroxidation

Numerous physiological, metabolic, and pathological processes depend on reactive oxygen species (ROS). Methylene (CH_2) groups found in polyunsaturated fatty acids found in body physiological systems have the ability to react with free radicals. Bis-allylic hydrogens of unsaturated lipids combine with hydroxyl radicals, which function as catalysts for free radical chain reactions, in the first stage of lipid peroxidation. The lipid radicals ($R^\cdot$) having in carbon-centered are the end product of this process. Under physiological circumstances (concentration of around 5 nM), which is enough for oxidative eustress, hydrogen peroxide is a significant redox signaling molecule. All significant biomolecules are harmed by extremely high hydrogen peroxide concentrations (>100 nM) (oxidative distress) [84]. Scientists are exploring the use of ROS-scavenging nanotechnology to fight age-related illnesses. While small amounts of ROS are vital for processes like cell growth and the immune system, an excess of ROS overwhelms the body's natural defenses, causing damage to DNA, fats, and proteins [85]. Nanomaterials, like metals, metal oxides, and carbon-based materials, can counteract this excess ROS [86]. In particular, nanoparticles of precious metals like palladium, gold, and platinum are highly effective because of their large surface area, maximizing their interaction with ROS [87]. Noble metal nanocatalysts have been suggested as potential antioxidants.

While AuNPs are not typically recognized for their redox sensitivity, they provide an excellent framework for electrochemical biosensors. Because of their redox catalysts behaviors, facilitating electron transfer in various electrically active biological molecules without requiring electron transfer intermediates [88]. Platinum nanomaterials (Pt NMs) are also a strong candidate for addressing oxidative damage due to their significant nanozyme activities, mimicking peroxidase, catalase and superoxide dismutase functions. These activities enable the catalytic conversion of O_2 to H_2O_2 and H_2O_2 to H_2O and O_2 [89]. Carbon-based nanomaterials, including graphene, and C_{60} and its derivatives are the most commonly used ROS quenchers [90].

4.4 Effects on Lipid Catabolism and Energy Homeostasis

Even while research on the long-term impacts of nanoparticles in vivo has generated a lot of attention, the majority of these studies have concentrated on the longevity or fertility of animals exposed to nanoparticles, with little knowledge of the underlying processes [91]. Animal fertility and lifespan are significantly impacted by homeostasis in nutrition and energy metabolism, which is also closely linked to long-term illnesses such metabolic disorders. In vivo impacts of nanoparticles longer will be better understood by looking at how they affect metabolic signaling pathways and identifying the underlying mechanisms that control these processes. Numerous appealing physical and chemical characteristics of gold nanoparticles (AuNPs) have made them extremely useful in biological research [60]. Researchers investigated the long-term metabolic effects of feeding fruit fly larvae with 15-nm AuNPs. The AuNPs accumulated within the larvae's fat bodies (a key metabolic organ). Surprisingly, the AuNPs did not trigger typical stress responses like increased heat shock protein production, JNK pathway activation, or autophagy. Although the AuNPs had minimal impact on growth or stress, they slightly increased the larvae's lipid levels. Triglycerides provide greater (twice) the energy per unit mass compared to proteins and carbohydrates. Consequently, during low glucose levels, and via aerobic respiration, triglycerides are converted into acetyl-CoA to generate ATP [23]. Furthermore, NPs can induce lipid alterations in addition to causing changes in carbohydrate metabolism.

5 Safety and Biocompatibility of Nanomaterials

The safety of a drug or formulation involves detecting, assessing, understanding, and preventing side effects, enabling a deeper understanding of its risks and benefits. The development of nanomaterials has brought about changes in various fields, particularly biomedicine, due to their unique physicochemical properties and nanoscale dimensions. However, due to their small size and high surface reactivity, they can cause interactions with biological systems, raising concerns about their safety and biocompatibility. As they are commonly released into the environment, impacting soil, water, and air can cause harm to human health [92]. Biocompatibility refers to the ability of a nanomaterial to interact safely with living organisms without causing adverse interactions. It includes assessing the material's physical, mechanical, and chemical properties, as well as its toxicity, mutagenicity, and allergenicity, to ensure it poses no significant harm to cells or biological functions [93]. The size, shape, surface charge, and surface functionalization of the particles are some of the variables that affect a nanomaterial's biocompatibility [94].

To evaluate the safety of nanomaterials, standardized procedures, such as ISO/TR 10993-22:2017 (International Organization for Standardization), should be followed with detailed focus on cytotoxicity, genotoxicity, and immunotoxicity study. The ISO-supported standardized assays provide data comparability; however, assessing the toxicity of nanomaterials including liposomes and other dynamic systems has been challenging. So, for this matter, the NIST (National Institute of Standards and Technology) has created reference standards, but toxicity results are still impacted by issues like aggregation and chemical residues. Thereby there it is of great importance to crosscheck the standardized examinations which may need further validations by journals to improve comparison [95, 96].

NPs interact with biological barriers of the human body, impacting toxicity and biocompatibility. For example: in the lungs, NPs like PLGA and silica can disrupt macrophages, while silver NPs cause inflammation. On the skin, TiO_2 has limited penetration, and gold NPs show size-dependent infiltration. In the liver, quantum dots stress the need for surface modifications to prevent ion leakage, with POSS improving safety [97]. Graphene-based materials like graphene oxide (GO) and graphene quantum dots (GQDs) exhibit dose-dependent toxicity influenced by size and functionalization. High doses of GO can cause oxidative stress, cell damage, and apoptosis, while higher doses in mice may lead to chronic toxicity and fatality. PEGylation significantly reduces toxicity, ensuring biosafety for biomedical applications [98].

5.1 Toxicity of Nanoparticles on Lipid Metabolism

Although the nanomaterials have high therapeutic potential, however, their toxicity in lipid metabolism due to oxidative stress and disruption of regular metabolic processes can result in negative consequences such as altered fat storage, lipid build-up, and decreased energy output. Inflammation, diabetes, obesity, and other metabolic disorders can all be exacerbated by these disruptions. Although NPs hold great potential for use in biomedical applications, however concerns regarding their toxicity are growing, especially in relation to lipid metabolism and adipocyte differentiation. The development of adipocytes (fat cells) is impacted by the disruption of mesenchymal stem cell (MSC) differentiation caused by NPs [99]. Research has demonstrated these mechanisms with several nanomaterials below:

- Quantum dots activate the PI3K/Akt pathway, blocking fatty acid oxidation and causing lipid droplet accumulation in glial cells [100].

- Gold nanoparticles reduce fat and inflammation in the liver and adipose tissue, improving blood parameters, while Silver Nanoparticles generate reactive oxygen species, decrease SOD activity, and lead to MDA accumulation, reducing blood triglycerides and highlighting oxidative stress's role in lipid metabolism disruption [101].
- AuNPs, multiwalled carbon nanotubes (WCNTs), and hyaluronic acid nanoparticles (HA-NPs) affect mesenchymal stem cell differentiation into adipocytes by modulating the TAK, BMP, and PPAR pathways [102]. AgNPs also inhibit PGC1α, shifting energy production from fatty acid oxidation to glycolysis, impairing cellular energy efficiency, and contributing to dysfunction.
- Frequent exposure to silica nanoparticles (SiNPs) may increase the chance of atherogenesis or hepatic steatosis. Particularly, research conducted on the SiNP-treated groups shows significantly higher leptin levels. This could indicate leptin resistance in the liver, which can lead to inflammation, impaired lipid metabolism, insulin resistance, and the development of more serious diseases like steatohepatitis and atherogenesis [103].

5.2 Long-Term Effects of Nanomaterials on Cellular Lipid Dynamics

Nanomaterials have found common applications that benefit society, but there are greater concerns about their long-term safety and toxicity. When exposed to nanomaterials for a long period they can affect changes in cellular lipid dynamics and can cause alterations in the structure, composition, and function of cellular membranes. Nanomaterials are mostly uptake by the cell following two mechanisms, i.e., endocytic uptake and receptor-mediated uptake. The nanomaterial may interact with lipid bilayers, potentially disrupting membrane integrity, modulating lipid metabolism, and affecting signaling pathways related to lipids. This can lead to changes in membrane fluidity, lipid raft organization, and overall cellular lipid homeostasis, influencing processes like endocytosis, vesicle trafficking, and the regulation of cellular processes such as apoptosis and inflammation. Over time, these disturbances could contribute to cellular dysfunction and disease development [104].

5.3 Strategies to Improve Biocompatibility

Controlling nanoparticles' size and ensuring their homogeneity are essential for enhancing their biocompatibility. Smaller nanomaterials reduce their accumulation in organs and help with renal excretion, however. They can also cause toxicity by increasing their circulation times and ROS. Although very tiny particles can trigger immunological reactions, their larger surface area improves cellular absorption and tissue targeting. Larger particles might not be able to enter cells as effectively [94]. There are different methods in which we can control the size of nanomaterial to enhance biocompatibility which include using techniques like solvent evaporation, nanoprecipitation, ultrasonication, and microfluidics. It also includes optimizing experimental parameters such as solvent ratios, mixing rates, pH, and temperature. Recent research studied the influence of gradient ultracentrifugation and pH-dependent fractionation and oxidation techniques like $KMnO_4$-H_2SO_4 to reduce the size of GO. Additionally, organizing small nanoparticles into larger superstructures using DNA conjugates enables targeted accumulation in diseased areas. Maintaining a uniform size distribution through surfactant-assisted synthesis and controlled temperatures further reduces toxicity. Controlling size uniformity through surfactant-assisted synthesis and temperature regulation further reduces toxicity. Long-term impact on cellular lipid dynamics [105].

5.4 Controlling Shape to Maintain a Safe Aspect Ratio to Improve Biocompatibility

The shape of nanomaterials affects their stability and interaction with cells. Spherical particles are generally more biocompatible and less likely to cause damage, while irregularly shaped particles, like nanowires or rods, might trigger mechanical stress or toxicity due to their sharp edges [94]. Nanomaterials with higher aspect ratios are usually more toxic due to increased cellular uptake, but exceptions exist, like spherical Au NPs causing more damage than rod-shaped ones due to faster release of functional molecules. High aspect ratios typically cause oxidative stress, while spherical ones elevate inflammation. Shape control is achieved by adjusting surfactants, stabilizing agents, reaction times, ionic strength, and magnetic density. For instance, spacer length and surfactant concentration alter PbS nanoparticle shape, while ionic concentration and heating time modify ZnO and nanowire structures. Magnetic fields also influence nanowire morphology [105].

5.5 Controlling Surface Charge to Improve Biocompatibility

Surface charge plays an essential role in regulating cell responses, including adhesion, mobility, proliferation, and differentiation [106]. Nanomaterials with positive charges may enhance cellular uptake through electrostatic interactions but can cause cytotoxicity. Negatively charged and neutral particles tend to be more biocompatible, reducing membrane disruption and immune activation but potentially limiting cellular penetration [107]. Modifying the surface of the nanomaterial is crucial to enhance biocompatibility, modifications such as roughness, coatings, and charge optimization, impacting protein adsorption, wetting, and overall cell-material interactions [106].

5.6 Functionalization of Nanomaterial to Improve Biocompatibility

Functionalizing nanomaterials with specific groups like PEG or targeting ligands can improve stability, reduce immune response, and enhance targeting efficiency. It also prevents protein adsorption, boosting biocompatibility [94]. Surface modification of

NPs involves attaching organic groups (e.g., R-NH2, R-COOH) using bifunctional crosslinkers. Silica NPs are modified with aminosilanes, gold NPs with -SH or -NH_2 groups, and metal oxides through ligand exchange. Carbon-based nanomaterials are functionalized via oxidation, halogenation, or cycloaddition. These modifications enhance the performance of NPs in drug delivery, diagnostics, and therapeutics [107].

5.7 Safe Design of Nanomedicine

Nanoinformatic tools integrate data from nanomaterial studies with computational algorithms to guide nanomedicine design. These tools help manage large, often conflicting data sets by modeling drug encapsulation, delivery, release, and biological interactions. They facilitate data sharing and support the rational selection of materials for safer nanomedicines. Nano-QSAR models establish correlations between the physicochemical properties of nanomaterials and their biological effects, using high-throughput techniques to predict outcomes. These models aid in the rational design of nanomaterials, enabling safer and more effective nanomedicines by understanding toxicological mechanisms and incorporating them into the design and manufacturing process [108].

6 Challenges in Nanomaterial-Based Lipid Regulation

6.1 Limitations in Nanomaterial Synthesis and Characterization

The most difficult steps in the development of nanomaterials arise from the scale-up of pilot laboratory batch sizes to industrial scale production and choosing the suitable excipients essential for the production of pharmaceuticals without compromising the quality [109]. The production of nanomaterials is a multistep approach that necessitates continuous monitoring of the properties of nanomaterial, including shape, size, structure, charge, composition, physicochemical characteristics, pharmacokinetics, and biopharmaceutical attributes [110, 111].

The advancement of nanomaterials depends on improvements made in manufacturing technology to facilitate scalable methods that comply with the quality standards of good manufacturing procedure (GMP) [112]. However, the production and scaling up of nanomaterials face problems within the strict framework of regulated GMP guidelines [112, 113]. The GMP guidelines define minimum specifications the manufacturers should adhere to in their manufacturing protocol to guarantee consistent batch-to-batch standard quality, ensure the end product is suitable for its proposed purpose, maintain thorough documentation of all manufacturing steps, and confirm that the medicine complies with the specifications of marketing or clinical trial authorizations [112, 114].

The optimization and reproducibility of the process are easier to achieve at a small pilot scale. Nanomaterials well-characterized at this scale may be employed in preclinical studies and early-stage clinical trials that generally require a smaller quantity of drug. In contrast, scaling up presents obstacles in establishing an efficient manufacturing method to regulate the shape and size distribution of the nanoparticles, as well as ensuring the stability of the physicochemical attributes of the product with insignificant variation between the batches [113, 115]. Research and development approaches often require pilot production, and scaling up can lead to significant difficulties in manufacturing precise nanomaterials [110]. Only through extensive trial and error preparations that do not meet the standard quality can be improved. This strategy is valid for a pilot laboratory scale. However, for industrial large-scale production, the cost of raw materials must be taken into consideration, along with the additional requirement for numerous multistep manufacturing processes, which make manufacturing at an industrial scale highly expensive in both time and money [113–115].

Moreover, the nanoparticle analysis is typically carried out offline under controlled environments. The offline method provides a reliable way to maintain the uniformity and quality of the nanoparticle, but it may lead to delays and errors in sampling complicating the optimization process [114]. Despite the continuous advancement of machinery, the majority of production operations are largely furnished with analytical equipment appropriate for the analysis of conventional dosage forms. The majority of analytical tools meant for the characterization of nanomaterials are complicated and necessitate specialized and costly equipment. Additionally, owing to the complexity of the nanomaterial behavior and the expertise required in advanced techniques, a group of interdisciplinary professionals is essential for the analysis and interpretation of data, which considerably raises the already high costs of nanomaterial manufacturing [113, 116].

6.2 Complexities in Understanding Lipid–Nanomaterial Interactions

The distribution and ultimate fate of nanoparticles or nanomaterials within the body are mainly determined by how they interact with the biological barriers present at their target sites [117]. These biological barriers are vital elements of the defense mechanism in the body which restricts the invasion of foreign substances. The reticuloendothelial system (RES), which includes the liver and spleen, serves as the primary barrier to intravenously administered nanoparticles, rapidly removing particles from the bloodstream over a short duration of administration, ultimately hampering their efficient target-specific delivery to diseased areas, resulting in nonspecific distribution to healthy tissues [118, 119].

To overcome this, techniques such as grafting of standard polyethylene glycol (PEG) on the surface of nanomaterials can decrease their clearance by the RES and prolong the circulation period; however, repeated injections of PEGylated nanomaterials have led to various issues, immunogenicity, poor cellular uptake, and toxicity. Therefore, alternative strategies are being explored to prevent nanomaterial blockage by the RES, by modifying the surface of nanomaterials with peptides or proteins or using cell membranes or cells to cover the nanoparticles or nanomaterials [120, 121].

Several characteristics of NPs, such as nature, size, surface chemistry, stability, reactivity, aggregation, mobility, and storage duration, influence their toxicity [122]. The contact of nanoparticles with biological systems and the resulting harmful impact is primarily driven by their smaller size, surface properties, and large surface-to-mass ratio. The distinctive properties of nanomaterials enable them to readily penetrate cell and tissue membranes as well as intracellular compartments, leading to potential cellular damage. The large surface-to-mass ratio of nanoparticles facilitates strong chemical connections with biological macromolecules. Additionally, an enhancement in surface area for the same chemical composition improves adsorption capacity, surface interactions, and possible toxicity [123].

For instance, AgNPs of varying sizes, specifically 20 nm (Ag20Pep) and 40 nm (Ag40Pep), were studied for their cellular uptake in THP-1-derived human macrophages. The findings revealed that most silver nanoparticles were distributed throughout the cells. Oxidative stress responses, such as the generation of protein carbonyls and the activation of heme oxygenase I, were also observed. In another research, the use of charged AuNPs with a size of 15 nm was found to induce apoptosis, while neutral AuNPs caused necrosis in human epidermal keratinocyte (HaCaT) cell lines [124]. Various NPs can enter the nuclear envelope and contribute significantly to genotoxicity. AgNPs have been shown to exhibit greater toxicity compared to AuNPs. TiO_2 nanoparticles are generally regarded as biologically inert both in vivo and in vitro. However, TiO_2 nanoparticles bigger than 15 μm in size are very toxic because of their ability to generate reactive oxygen species (ROS) [125]. Another study on *Drosophila* larvae showed that dietary AuNPs of 15 nm accumulated in the larvae's fat body without causing typical stress responses. While growth and stress responses were largely unaffected, the AuNP-fed larvae had slightly increased lipid levels, associated with enhanced PI3K/Akt/mTOR signaling, higher glucose uptake, and increased fatty acid synthesis and lipid production [126].

The toxicity due to inorganic nanomaterials, resulting from their nonspecific distribution in healthy tissues and prolonged circulation in the bloodstream, can be mitigated through renal clearance. However, renal clearance remains a significant obstacle for the clinical use of many inorganic nanoparticles, especially those made from heavy or toxic metals. Therefore, inorganic nanomaterials with properties that allow for renal clearance are more likely to have successful prospects for clinical application. Therefore, such nanomaterials are to be chosen which get eliminated via the renal route, thus increasing the chances for their clinical use [121].

6.3 Barriers to Clinical Translation

Many innovative nanomaterials face challenges in obtaining regulatory approval despite positive clinical results due to government concerns, such as improper endpoints, insufficient justification for the selection of clinical comparators, and insufficient methodologies for analyzing clinical data. Therefore, pharmaceutical companies are bound to take advice from regulators during the entire developmental timeline of nanomedicine to bypass possible problems that could lead to either a lack of regulatory approval or delays in obtaining it [121]. Although the FDA recently issued guidelines on nanodrugs and liposomes, regulatory structures for nanomedicine are still developing, and further guidelines on safety and efficacy are required, especially about novel nanomaterials [127].

Also, there is a notable lack of consistency and uniformity in the global regulation of nanomedicines, with major regulatory authorities like the FDA and European Medicines Agency (EMA) adopting significantly different terminology and regulatory strategies for nanomedicines [128]. Alternative approaches in collaboration with regulatory bodies should be explored to speed up the clinical development of nanomedicines that show promising preclinical data due to the complexity of nanomaterials [127].

Regulatory bodies take preclinical toxicity testing for small-molecule products beneficial for nanomedicines, provided it is carried out using at least two animal models, with prolonged treatment duration and multiple doses [112]. Different clinical situations have distinct safety criteria, with some medical therapies allowing lower benefit-to-risk ratios. As a result, nanosafety criteria must be established for each disease use, taking into consideration factors such as the disease's stage and phase, as well as other patient-specific factors like gender, age, and past or present health conditions [129]. Proper patient selection is another criterion that can significantly impact the success of clinical trials. The poor clinical results are the outcome of choosing patients from a diverse population having multiple complications [121].

To identify the optimal formulation for further development, brief studies at subtoxic doses may permit pharmacokinetic analysis, insights on biomarkers (such as complement activation and inflammatory response), and even provide data on biodistribution and targeting (e.g., using PET-radiolabeled nanomedicines) [127].

Finally, in-depth toxicology research is crucial for the clinical translation of nanoparticles, requiring a thorough assessment of safety concerns for human use. Even though every product in clinical trials has gone through in vitro toxicological studies, in vivo toxicity-related failures in clinical trials still happen quite frequently [96].

7 Characterization Techniques for Nanomaterials

7.1 Scanning Electron Microscopy (SEM)

Scanning electron microscopy differs from light microscopy, it uses light and glass lenses to magnify and show images of specimens. Electron microscopy employs accelerated electron beams along with electromagnetic or electrostatic lenses to generate detailed images with significantly higher resolution compared to visible light, due to the shorter wavelength of electrons. Electron beam scans the sample's surface, generating signals that reveal the specimen's atomic composition and topographic features. It reveals topographic details and atomic composition by producing signals through interactions with the sample. SEM provides information about characteristics such as topography, morphology, composition, and crystallographic structure [130].

7.2 Transmission Electron Microscopy (TEM)

TEM is the most commonly employed technique in electron microscopy for analyzing nanomaterials. It offers chemical insights and atomic-level spatial resolution images of nanomaterials. When the electron beam interacts with the specimen, it generates elastically or inelastically scattered electrons, allowing the incident beam to pass through a thin foil specimen. It operates on principles analogous to optical microscopy, where photons are replaced by electrons, and glass lenses are substituted with electromagnetic lenses, and the eyepiece by a screen for viewing images [130]. The advantages of TEM include high magnification and knowledge about the structure of compounds and elements and it requires specialized maintenance and produces black-and-white images, which are costly [131].

7.3 Atomic Force Microscope (AFM)

AFM is an adaptable imaging technology for examining samples at the nanoscale. It captures a picture in three dimensions and gives a variety of surface metrics that engineers and scientists require. Images can be captured at atomic resolution with angstrom-scale height information, requiring minimal sample preparation. Surface roughness and texture may be determined and visualized on a wide range of materials using polymer nanocomposites. Furthermore, it is a nondestructive approach that provides excellent three-dimensional spatial resolution [132].

7.4 Dynamic Light Scattering (DLS)

DLS can explore a variety of nanomaterial physicochemical properties, includes hydrodynamic size, morphology, structural characteristics, aggregation state, and biomolecular conformation. To observe the temporal fluctuations in the intensity of elastically scattered light, also referred to as Rayleigh scattering, it arises from the Brownian motion of particles or molecules that are much smaller than the wavelength of incident light, observed at a fixed scattering angle [133].

7.5 X-Ray Diffraction (XRD)

XRD measures the diffraction rays as interact with a material. It is a method for the identification and quantification of distinct crystalline phases in solid materials and powders based on their structural characteristics, that can be determined, such as grain and nanomaterial sizes. When X-rays strike a regular crystal line sample, a fraction of them are diffracted and form a pattern. Crystal phases can be detected from such a pattern by comparing it to reference patterns from internationally recognized databases (such as the International Centre of Diffraction Data—ICDD). In detecting applications, XRD is commonly used to correlate a material's characteristics with its detecting capability [134].

7.6 Differential Scanning Calorimetry (DSC)

Although DSC is an established method in the development of pharmaceuticals, only a brief description of its fundamental principles and instruments will be provided. Common DSC investigations involve heating or cooling the sample at a regulated rate while quantitatively monitoring the heat transfer into or out of the sample. This technique provides valuable insights into phase transitions, such as melting or crystallization, by analyzing heat flow [135].

7.7 Thermogravimetric Analysis (TGA)

Thermogravimetric analysis is a method of analysis that measures the mass of a sample in relation to their respective temperature or time. The ligand packing density on the NP surface was calculated based on the weight fraction of the ligand and NP derived from the TGA curve, combined with the surface area of a single particle. It is a highly cost-effective method for the characterization of nanomaterials; however, it necessitates a significant sample quantity for accurate analysis [136].

7.8 Entrapment Efficiency (EE)

Entrapment efficiency is determined by the efficacy with which nanoparticles trap the drug inside them; this is to say, a measure of percentage of the actual drug that actually gets entrapped. This also goes to make an assessment regarding the encapsulation efficiency or how much amount of the drug would be available to the body in its therapeutic amount.

$$\mathrm{EE}\% = \frac{(T - C)}{T} \times 100 \tag{7.1}$$

where T is the total amount of drug used, and C is the free drug that remains unencapsulated in the supernatant.

The higher the EE value, the better the encapsulation of the drug, and thus, stability of the drug and controlled release are improved. Optimizing EE is crucial for the development of effective nanomaterial-based drug delivery systems that maximize therapeutic benefits [137].

7.9 Drug Loading Capacity (DLC)

The lyophilized nanomaterials containing entrapped drugs were weighted into a vial, and an appropriate amount of DMF was added to achieve a concentration of approximately volume. The material was agitated for 30 minutes at 350 rpm. The mixture was centrifuged, and the supernatant was then decanted, and the encapsulated drug content was subsequently quantified, add required microliters of supernatant to dimethylformamide (DMF) and evaluate the UV-Vis spectrum of the resulting solution. The drug content was estimated using the calibration curve [138].

$$\mathrm{DLC}(\%) = \frac{\text{Mass of encapsulated drug}}{\text{Mass of polymer}} \tag{7.2}$$

7.10 Pharmacokinetics

Nanomaterials offer enormous potential in biological applications, particularly drug delivery, imaging, magnetic hyperthermia, and tissue regeneration. Pharmacokinetics, the study of absorption, distribution, metabolism, and excretion (ADME), is important for understanding how medications function at the molecular level. Nanomaterials can change ADME characteristics by enabling tailored medication administration. The pharmacokinetics of these nanoparticles are dependent on protein adsorption. As a result, understanding the relationship between nanomaterial characteristics and biological activity is critical for rationally designing an efficient drug delivery system and predicting in vivo response. These studies enable the optimization of nanoformulations for a sustained, controlled, or stimuli-responsive release of drugs that ensure better treatment efficacy with reduced side effects [139].

7.11 Drug–Excipient Compatibility

The stability and efficacy of nanoformulations rely significantly on drug–excipient compatibility. Fourier Transform Infrared Spectroscopy (FTIR) detects chemical interactions by comparing the vibrational peaks of the drug, excipient, and physical mixes over a specific wavelength range. DSC evaluates thermal behavior by analyzing endothermic or exothermic transitions, which may suggest incompatibility. High-performance thin-layer chromatography (HPTLC) provides a fingerprint analysis to confirm stability using consistent retention factor (Rf) values. These investigations are critical in formulation development to avoid undesirable interactions and maintain medication stability [140].

7.12 In Vitro and In Vivo Therapeutic Activity

Nanomaterials have a high therapeutic potential due to their unique properties such as nanoscale size and vast surface area. The in vitro activity has been investigated for a variety of applications, including antibacterial, anticancer, and anti-inflammatory treatments. Metallic nanoparticles, particularly gold and silver nanoparticles, exhibit in vitro activity as a result of ROS, microbial membrane rupture, and cancer cell growth [141]. These nanocarriers increase medication solubility while shielding active compounds from degradation, resulting in continuous release. When given in vivo, liposomes and solid lipid nanoparticles boost bioavailability while posing fewer systemic toxicity. Many nanoparticles use the enhanced permeability and retention (EPR) effect for passive tumor targeting, while others are developed for active targeting using disease biomarker-specific ligands [142]. Their biocompatibility ensures minimum immunological response, and their ability to load numerous medicines or diagnostic agents concurrently broadens their therapeutic potential. Long-term toxicity, biodegradation, and regulatory difficulties are among the challenges that remain. Finally, nanoparticles are transforming in vitro and in vivo therapeutic techniques, leading to significant advances in precision medicine [143].

8 Applications of Nanomaterials for Lipid Metabolism-Based Disease Treatment

8.1 Role in Obesity Management

The effective treatment of obesity necessitates a lifetime of work since anomalies would manifest during the development of obesity. Intestinal dysbiosis, functional cell abnormalities, redox imbalance, and blood lipid metabolism disorders are among the abnormalities brought on by obesity. Improving intestinal health and lowering calorie intake, addressing and repairing anomalies in functional cells, controlling redox homeostasis, and eliminating free lipoprotein from the blood are some of the therapies for obesity [144]. Theranostics of obesity and metabolic disorders associated with obesity heavily relies on nanomaterials [145]. Serious adverse effects from antiobesity medications included headache, disorientation, and anxiety. By increasing these medications' absorption, nanomaterials might lessen their adverse effects. Digestion, occurring in the gut, significantly impacts overall health and obesity. Improving gut health and consuming fewer calories are effective obesity treatments, and nanomaterials show promise in aiding this process through various applications like food additives, gut regulation, and fat digestion inhibition. While fats enhance food flavor, overconsumption leads to obesity, high blood pressure, and high cholesterol [146]. Fat substitutes, a type of food additive, help with weight loss by reducing calorie intake, increasing feelings of fullness, and maintaining palatable taste. Nanomaterials, particularly nanocellulose, present a promising strategy for weight management owing to their distinctive physical and chemical characteristics [147]. Lipase inhibitors function by binding to the active site of lipase enzymes in the digestive system, thereby preventing fat breakdown, reducing lipid absorption, and supporting obesity treatment [148]. Additionally, nanomaterials can serve as nanocarriers for small molecule inhibitors and nanolipase inhibitors, aiding in the suppression of lipid digestion and absorption, ultimately lowering energy intake.

Table 7.3 Nanomaterials targeting diseases associated with metabolic syndrome

Disease condition	Nanoparticles used
Dyslipidemia	Lipid-based nanoparticles such as nanosponges
Type-2 diabetes mellitus	Chitosan-based, alginate-based, dextran-based, synthetic-based, polylactic-based, polyaniline-based, niosomes, polyamidoamine (PAMAM), polymeric micelles, inorganic nanoparticles, solid lipid nanoparticles, nanostructured lipid carriers, and liposomes
Endothelial dysfunction	Polymeric nanoparticles, including high-density lipoprotein nanoparticles
Nonalcoholic fatty liver disease	Chitosan-based, alginate-based, dextran-based, synthetic-based, polylactic-based, polyaniline-based, niosomes, polyamidoamine (PAMAM), polymeric micelles, inorganic nanoparticles, solid lipid nanoparticles, nanostructured lipid carriers, and liposomes
Obesity	Chitosan-based, alginate-based, dextran-based, synthetic-based, polylactic-based, polyaniline-based, niosomes, polyamidoamine (PAMAM), polymeric micelles, inorganic nanoparticles, solid lipid nanoparticles, nanostructured lipid carriers, and liposomes. Additionally, lipid nanoparticles, polymer nanoparticles, and those used in photothermal therapy or self-assembled peptides are mentioned.

8.2 Nanomaterial-Based Therapies for Metabolic Syndrome

Furthermore, the continuous use of electronic devices has made life easier and more straightforward, which has led to the generation of a number of lifestyle diseases. One of them is the metabolic syndrome, a group of conditions that predisposes people to cardiovascular diseases and includes diabetes, dyslipidemia, insulin resistance, obesity, hypertension, nonalcoholic fatty liver disease, and other conditions [149]. The involvement of nanotechnology in metabolic syndrome (MS) treatment offers the potential benefits to improve drug delivery effectiveness and reduce adverse drug reactions. Titanium oxide nanoparticles, silver nanoparticles, manganese nanoparticles, gold nanoparticles, nanocrystal cerium dioxide ($nCeO_2$), and poly (lacticglycol) acid nanoparticles (PLGANP) are among the several nanomaterials utilized to treat metabolic syndrome [150]. Numerous nanomaterials used for the diagnosis and treatment of different metabolic syndromes showing in Table 7.3.

8.3 Targeting Lipid Metabolism in Cancer Treatment

Tumor immunosuppressive microenvironment development and tumor immune evasion are mostly caused by changes in lipid metabolism inside the tumor microenvironment (TME). One extremely promising approach to cancer treatment is to improve tumor immunity by controlling metabolism of lipid in the TME. Due to the barriers to effective drug delivery in the complex and heterogeneous TME to reach the target cells, it is challenging to precisely regulate lipid metabolism in the TME. The use of nanotechnology to redesign TME has garnered a lot of attention lately [151]. Drug delivery based on nanotechnology can greatly increase bioavailability and lengthen the drug's circulation period. Controlling lipid metabolism within the tumor microenvironment (TME) is a promising approach to boost the immune system's antitumor response. Nanomaterials can influence cellular lipid metabolism and function [152]. Metabolic changes in the TME impair CD8+ T cells, hindering immune function and promoting tumor evasion [153]. Tumor hypoxia shifts metabolism from oxidative phosphorylation to glycolysis, favoring carbohydrate use over lipids. A targeted nanoparticle system designed to regulate lipid metabolism and enhance the anticancer effects of T cells has been developed. Similarly, manipulating the lipid metabolism of CAR-T cells within the TME can strengthen their antitumor activity. Cholesterol plays crucial role for T-cell function and aids in the clustering of T-cell receptors and immune synapses formation. Enhancing cholesterol levels in the cell membrane can boost the antitumor activity of T cells. Some drugs like avasimibe, an inhibitor of the cholesterol-esterification enzyme acetyl coenzyme A acetyltransferase 1 (ACAT1), have been shown to elevate plasma membrane cholesterol levels, facilitate TCR aggregation, and thereby enhance T-cell antitumor efficacy [154]. This suggests that metabolism of cholesterol could be a promising target for optimizing the antitumor performance of CAR-T cells. By enable efficient drug delivery, Hao et al. designed a liposome like NPs to deliver avasimibe directly to the T-cell membrane through lipid insertion without disrupting T-cell functionality [155]. The encapsulated avasimibe liposome effectively inhibited ACAT1 activity in CAR-T cells via autocrine and paracrine mechanisms as well. This elevated membrane cholesterol level, accelerated T-cell receptor (TCR) aggregation, and sustained T-cell activation, significantly enhancing the therapeutic potential of adoptive T cells against solid tumors.

8.4 Nanomaterials for Atherosclerosis

The accumulation of lipids, mostly cholesterol, in the arterial wall is the hallmark of atherosclerosis [156]. In the previous century, atherosclerosis was thought to be caused only by cholesterol buildup and deposition in the artery wall, which led to the blood vessel lumen narrowing. One Since the majority of biological processes, like atherogenesis, take place at the nanoscale, molecular imaging and targeted atherosclerosis therapy are made possible by nanotechnology [157]. By the use of nanoparticles various parameters such as aqueous solubility, stability, and absorption of diagnostic or therapeutic properties of compounds can be improved. They can also increase the duration of their circulation, allow for high interaction and cellular uptake efficiency in the target cells compared to other cells, shield them from enzyme degradation in tissues as well as physiological fluids, and lessen their toxicity and unwanted effects [158]. Targeting the immune cells (macrophages and foam cells) within the artery wall is a key strategy for diagnosing and treating atherosclerosis. Iron oxide nanoparticles, readily absorbed by macrophages, are frequently used for MRI detection of these cells [159, 160]. Endothelial-targeted nanoparticles, combined with various imaging techniques (MRI, PET, MDCT), allow for visualization of atherosclerotic artery wall structure and function [161]. These nanoparticles also offer a means for delivering therapeutic agents directly to damaged endothelial cells, thus preventing or treating atherosclerosis [162]. Studies have shown the ability of nanoparticles to detect MMP9 activity, and protease-sensitive polymeric nanoparticles have been developed as biosensors. After administering the nanoparticles to apoE$^{-/-}$ mice, the researchers used conjugated fluorescence molecular tomography (FMT) as well as CT imaging. The outcomes showed that these nanoparticles successfully visualized protease behavior in atherogenesis lesions and effectively distinguish the therapeutic impact of an anti-inflammatory drug [163]. A secretory serine protease, Proprotein convertase subtilisin/kexin type 9 (PCSK9), binds to LDL receptors and prevents their reprocess to the cell surface, stimulating their degradation in cells, particularly hepatocytes [164]. Intravenous delivery of PCSK9 siRNA-loaded nanoparticles to various animal experimental models, such as mice, rats, and nonhuman primates, reduced PCSK9 transcript levels in the hepatocytes of the liver [165]. These NPs further decreased plasma PCSK9 protein as well as LDL-cholesterol levels, while minimally affecting HDL cholesterol and triglyceride concentrations [165].

8.5 Drug Delivery for Lipid-Dependent Pathogens

Although inflammatory bowel disease (IBD) etiology has been thoroughly investigated, the underlying causes are still unclear. According to our present understanding of complicated disorders, several factors frequently contribute to their development [166]. It is believed that IBD develops as a result of a combination of endogenous triggers (genetic susceptibility and immune-regulatory dysfunction), external environmental triggers, and microbial exposure [167]. The etiology of this significant illness is further complicated by early research on the various human microbiome, which has appear to generate dysbiosis, a change in the normal makeup of the microbiota and also involve a significant role in the pathogenesis of IBD3 [168]. Traditional methods for delivering anti-inflammatory and immunosuppressive drugs for IBD include capsules, tablets, and solutions. In contrast, targeted biological therapies are usually administered intravenously or subcutaneously [169]. Nanoparticle-based drug delivery systems (DDS) are attracting attention for their advantages in size, drug capacity, and surface modification properties making them well-suited for cell-specific targeting. Because most gut bacteria reside in the intestines, where the pH is much higher (5.5–7.5) than in the stomach (1.5–2.0), delivering antibiotics to the intestine can disrupt the gut microbiota. By using antibiotics to target the microbiota, numerous nanodelivery systems utilize pH-sensitive polymeric materials [170]. Synthetic polymers, such as poly acrylic acid, poly acrylamide, poly diethyl-aminoethyl methacrylate, poly dimethyl-aminoethyl methacrylate, and poly methacrylic acid, have been extensively used for designing of pH-sensitive antibiotic nanodrugs that enhance antimicrobial efficacy via sustained drug release

[171]. Additionally, natural polymers like gelatin and albumin have been explored for creating well-organized pH-responsive nanosystems, which enhance the antimicrobial activity of antibiotics by enabling targeted and controlled drug release [172].

8.6 Genetic Lipid Disorders and Enzyme Replacement Therapy

The substrate that builds up in the impacted cells, as indicated in the table below, can be used to categorize lysosomal storage diseases. Cell and organ failure results from the buildup of undergraded material caused by a lysosomal malfunction. As the years passed, several studies have been conducted in a huge range of tissues and organs to try to understand how substrate buildup leads to illness. These investigations have revealed that the pathophysiology of lysosomal storage diseases is influenced by a wide range of variables. In Gaucher disease, the buildup of glucosylceramide in macrophages triggers the release of inflammatory cytokines and chemokines [173]. Early research led to the development of enzyme replacement therapy (ERT), which involves delivering the missing lysosomal enzyme to treat lysosomal storage disorders. In recent years, a variety of nanotechnology-based drug delivery vehicles, including metal nanoparticles, carbon nanotubes, dendrimers, liposomes, polymeric nanoparticles, nanocapsules, and nanovesicles, have been created and tested for various purposes [174]. Up to recently, the most promising of these for lysosomal storage disorders (LSDs) were polymeric nanoparticles, liposomes, nanocapsules, and nanovesicles. The first generation of nanovectors to be considered for the delivery of enzymes in LSDs were liposomes. Both of them are composed of lipid vesicles with an aqueous core and one or more phospholipid bilayers. While amphiphilic and lipophilic medicines can dissolve in the phospholipid bilayer, the latter can be helpful for encapsulating hydrophilic ones. Because the reticuloendothelial system eliminates conventional liposomes, they have short plasma circulation periods. To extend the lifespan of liposomes in the bloodstream, researchers reduce their size (to under 100 nm) or coat them with PEG, creating stealth liposomes. Liposomes were first explored for ERT of LSDs in 1970s, initially for Gaucher disease and Pompe disease [175]. Recent studies using liposomes to deliver the enzyme velaglucerase alfa (VPRIV) showed a significant (68%) increase in enzyme delivery to macrophages and fibroblasts compared to free enzyme. In addition to liposomes, polymeric nanoparticles (NPs), especially poly (lactic-co-glycolic acid) (PLGA)—FDA-approved for intravenous use—have been extensively studied for enzyme delivery. Research has shown that PLGA nanoparticles carrying the enzyme GAA are taken up by cells more efficiently than free GAA.

9 Future Directions

Development of targeted drug delivery through nanomaterials to specifically address various aspects of lipid metabolism such as triglycerides to the associated tissue. Through enhancing their pharmacokinetics properties and exploring the use of biocompatible materials for improved safety. Further need developed nanocarriers that interact naturally with the circulatory system, and enhance delivery to target sites. Also required to developed nanoparticles that are able to deliver small molecules associated with lipid metabolism like enzymes as well as developed imaging nanoparticles for diagnostic purpose. As the application point of view such as management of obesity, hypercholesterolemia, and metabolic syndrome need to developed nanoparticles that can help to improve and potentiate the nanomaterials for better treatment. Some challenges associated with nanomaterials regulating lipid metabolism such as toxicity concern, biodistribution and clearance mechanism are challenging aspect. All of them further need to developed such kind of nanomaterials that can overcome the challenges associated with nanomaterials for the lipid metabolism.

10 Summary and Conclusion

- Numerous physiological processes, including hormone control, energy storage, nerve impulse transmission, and the movement of fat-soluble nutrients, are influenced by lipid metabolism. Lipid metabolism is a complex physiological process that also affects hormone, nutrient, and organismal homeostasis management.
- When external material is swallowed into the cell, endocytosis is also known to have a role in the absorption of lipoproteins by macrophages. A mammalian cell's biological membrane is mostly composed of three types of lipids: glycerolipids, sphingolipids, and sterols.
- Type 2 diabetes mellitus, obesity, cancer, cardiovascular disease, and metabolic syndrome are all highly prevalent endocrine system metabolic disorders that are primarily caused by inadequate lipid metabolism and an excess of calories.
- Nanoparticles provide a number of benefits over free medicines, such as enhanced bioavailability, biodegradability, safety, site-specific drug delivery, and extended drug release. Nanotechnology makes it possible to handle lipid metabolism effectively, which improves medication delivery. Nanomaterials' size enables them to interact with cells in a manner comparable to that of other nanoscale objects, including as viruses, proteins, and cholesterol particles. These naturally occurring nanoparticles, which are frequently recognized by specific cell receptors at the plasma membrane, are taken up by cells through their endocytic machinery.
- Standardized protocols should be followed in order to assess the safety of nanomaterials, and a thorough investigation of cytotoxicity, genotoxicity, and immunotoxicity should be conducted. Selecting the appropriate excipients necessary for pharmaceutical manufacture without sacrificing quality and scaling up pilot laboratory batch sizes to industrial scale production are the most challenging phases in the development of nanomaterials. The process of creating nanomaterials involves several steps and requires constant observation of the material's form, size, structure, charge, composition, physicochemical properties, pharmacokinetics, and biopharmaceutical qualities.

Authors' Contributions Pradeep Kumar Patel, Gitali Sen, Pooja Kathait, Soki Daeme Malang, Ankur, and Muhammad Delwar Hussain were responsible for the conceptualization, thought process, and writing the original draft. Alakh N. Sahu provided the scientific input, reviewed the final version, and incorporated appropriate content in the final version of the chapter.

References

1. Li, Y., et al. (2017). A global perspective on FOXO1 in lipid metabolism and lipid-related diseases. *Progress in Lipid Research, 66*, 42–49.
2. Hoy, A. J., Nagarajan, S. R., & Butler, L. M. (2021). Tumour fatty acid metabolism in the context of therapy resistance and obesity. *Nature Reviews Cancer, 21*(12), 753–766.
3. Li, Z., et al. (2020). Sulforaphane prevents type 2 diabetes-induced nephropathy via AMPK-mediated activation of lipid metabolic pathways and Nrf2 antioxidative function. *Clinical Science, 134*(18), 2469–2487.
4. Balling, M., et al. (2023). Elevated LDL triglycerides and atherosclerotic risk. *Journal of the American College of Cardiology, 81*(2), 136–152.
5. Liu, L., et al. (2022). Adipokines, adiposity, and atherosclerosis. *Cellular and Molecular Life Sciences, 79*(5), 272.
6. Obeid, M. A., et al. (2018). Lipid-based nanoparticles for cancer treatment. In *Lipid nanocarriers for drug targeting* (pp. 313–359). Elsevier.
7. Tong, X., et al. (2020). Recent advances in natural polymer-based drug delivery systems. *Reactive and Functional Polymers, 148*, 104501.
8. Sur, S., et al. (2019). Recent developments in functionalized polymer nanoparticles for efficient drug delivery system. *Nano-Structures & Nano-Objects, 20*, 100397.
9. Church, C., Horowitz, M., & Rodeheffer, M. (2012). WAT is a functional adipocyte? *Adipocyte, 1*(1), 38–45.
10. Huang, J., Borensztajn, J., & Reddy, J. K. (2011). Hepatic lipid metabolism. In *Molecular pathology of liver diseases* (pp. 133–146). Springer.
11. Ramakrishnan, A., & Vijayakumar, N. (2017). Urea cycle pathway targeted therapeutic action of naringin against ammonium chloride induced hyperammonemic rats. *Biomedicine & Pharmacotherapy, 94*, 1028–1037.
12. Rosca, M. G., et al. (2012). Oxidation of fatty acids is the source of increased mitochondrial reactive oxygen species production in kidney cortical tubules in early diabetes. *Diabetes, 61*(8), 2074–2083.
13. Chandel, N. S. (2021). Lipid metabolism. *Cold Spring Harbor Perspectives in Biology, 13*(9), a040576.
14. Abumrad, N., Harmon, C., & Ibrahimi, A. (1998). Membrane transport of long-chain fatty acids: Evidence for a facilitated process. *Journal of Lipid Research, 39*(12), 2309–2318.
15. Deng, L., et al. (2022). Macrophages take up VLDL-sized emulsion particles through caveolae-mediated endocytosis and excrete part of the internalized triglycerides as fatty acids. *PLoS Biology, 20*(8), e3001516.
16. Mettlen, M., et al. (2018). Regulation of clathrin-mediated endocytosis. *Annual Review of Biochemistry, 87*(1), 871–896.
17. Pilch, P. F., & Liu, L. (2011). Fat caves: Caveolae, lipid trafficking and lipid metabolism in adipocytes. *Trends in Endocrinology & Metabolism, 22*(8), 318–324.
18. Pohl, J., et al. (2005). FAT/CD36-mediated long-chain fatty acid uptake in adipocytes requires plasma membrane rafts. *Molecular Biology of the Cell, 16*(1), 24–31.
19. Nicholson, A. C. (2004). Expression of CD36 in macrophages and atherosclerosis the role of lipid regulation of PPARγ signaling. *Trends in Cardiovascular Medicine, 14*(1), 8–12.
20. Porcheray, F., et al. (2005). Macrophage activation switching: An asset for the resolution of inflammation. *Clinical and Experimental Immunology, 142*(3), 481–489.

21. Vallvé, J.-C., et al. (2002). Unsaturated fatty acids and their oxidation products stimulate CD36 gene expression in human macrophages. *Atherosclerosis, 164*(1), 45–56.
22. Johnson, A. R., et al. (2016). Metabolic reprogramming through fatty acid transport protein 1 (FATP1) regulates macrophage inflammatory potential and adipose inflammation. *Molecular Metabolism, 5*(7), 506–526.
23. Zhang, Y., et al. (2021). FATP4 inactivation in cultured macrophages attenuates M1-and ER stress-induced cytokine release via a metabolic shift towards triacylglycerides. *Biochemical Journal, 478*(10), 1861–1877.
24. DiRusso, C. C., et al. (2005). Comparative biochemical studies of the murine fatty acid transport proteins (FATP) expressed in yeast. *Journal of Biological Chemistry, 280*(17), 16829–16837.
25. Isola, L., et al. (1995). 3T3 fibroblasts transfected with a cDNA for mitochondrial aspartate aminotransferase express plasma membrane fatty acid-binding protein and saturable fatty acid uptake. *Proceedings of the National Academy of Sciences of the United States of America, 92*(21), 9866–9870.
26. Ibrahimi, A., et al. (1996). Expression of the CD36 homolog (FAT) in fibroblast cells: Effects on fatty acid transport. *Proceedings of the National Academy of Sciences, 93*(7), 2646–2651.
27. Holloway, G. P., et al. (2007). Fatty acid binding protein facilitates sarcolemmal fatty acid transport but not mitochondrial oxidation in rat and human skeletal muscle. *The Journal of Physiology, 582*(1), 393–405.
28. Nickerson, J. G., et al. (2009). Greater transport efficiencies of the membrane fatty acid transporters FAT/CD36 and FATP4 compared with FABPpm and FATP1 and differential effects on fatty acid esterification and oxidation in rat skeletal muscle. *Journal of Biological Chemistry, 284*(24), 16522–16530.
29. Bonen, A., et al. (2007). A null mutation in skeletal muscle FAT/CD36 reveals its essential role in insulin-and AICAR-stimulated fatty acid metabolism. *American Journal of Physiology-Endocrinology and Metabolism, 292*(6), E1740–E1749.
30. Koonen, D. P., et al. (2007). CD36 expression contributes to age-induced cardiomyopathy in mice. *Circulation, 116*(19), 2139–2147.
31. Habets, D. D., et al. (2007). AMPK-mediated increase in myocardial long-chain fatty acid uptake critically depends on sarcolemmal CD36. *Biochemical and Biophysical Research Communications, 355*(1), 204–210.
32. Gimeno, R. E., et al. (2003). Characterization of a heart-specific fatty acid transport protein. *Journal of Biological Chemistry, 278*(18), 16039–16044.
33. Cohen, A. W., et al. (2004). Role of caveolae and caveolins in health and disease. *Physiological Reviews, 84*(4), 1341–1379.
34. Pohl, J., et al. (2004). Long-chain fatty acid uptake into adipocytes depends on lipid raft function. *Biochemistry, 43*(14), 4179–4187.
35. Trigatti, B. L., Anderson, R. G., & Gerber, G. E. (1999). Identification of caveolin-1 as a fatty acid binding protein. *Biochemical and Biophysical Research Communications, 255*(1), 34–39.
36. Augustus, A. S., et al. (2008). Substrate uptake and metabolism are preserved in hypertrophic caveolin-3 knockout hearts. *American Journal of Physiology-Heart and Circulatory Physiology, 295*(2), H657–H666.
37. Cockcroft, S. (2021). Mammalian lipids: Structure, synthesis and function. *Essays in Biochemistry, 65*(5), 813–845.
38. Storch, J., & Thumser, A. E. (2010). Tissue-specific functions in the fatty acid-binding protein family. *Journal of Biological Chemistry, 285*(43), 32679–32683.
39. Kazantzis, M., & Stahl, A. (2012). Fatty acid transport proteins, implications in physiology and disease. *Biochimica et Biophysica Acta (BBA)-Molecular and Cell Biology of Lipids, 1821*(5), 852–857.
40. Smathers, R. L., & Petersen, D. R. (2011). The human fatty acid-binding protein family: Evolutionary divergences and functions. *Human Genomics, 5*, 1–22.

41. Glatz, J. F., & van der Vusse, G. J. (1996). Cellular fatty acid-binding proteins: Their function and physiological significance. *Progress in Lipid Research, 35*(3), 243–282.
42. Bonen, A., et al. (2007). Mechanisms and regulation of protein-mediated cellular fatty acid uptake: Molecular, biochemical, and physiological evidence. *Physiology, 22*, 15.
43. Hamilton, J. A. (2007). New insights into the roles of proteins and lipids in membrane transport of fatty acids. *Prostaglandins, Leukotrienes and Essential Fatty Acids, 77*(5–6), 355–361.
44. Duttaroy, A. K. (2009). Transport of fatty acids across the human placenta: A review. *Progress in Lipid Research, 48*(1), 52–61.
45. Degrace-Passilly, P., & Besnard, P. (2012). CD36 and taste of fat. *Current Opinion in Clinical Nutrition & Metabolic Care, 15*(2), 107–111.
46. Quispe, R., et al. (2019). Characterization of lipoprotein profiles in patients with hypertriglyceridemic Fredrickson-Levy and Lees dyslipidemia phenotypes: The Very Large Database of Lipids Studies 6 and 7. *Archives of Medical Science, 15*(5), 1195–1202.
47. Natesan, V., & Kim, S.-J. (2021). Diabetic nephropathy–a review of risk factors, progression, mechanism, and dietary management. *Biomolecules & Therapeutics, 29*(4), 365.
48. Verma, N. (2016). Introduction to hyperlipidemia and its treatment: A review. *International Journal of Current Pharmaceutical Research, 9*(1), 6–14.
49. Schonfeld, G. (1990). Inherited disorders of lipid transport. *Endocrinology and Metabolism Clinics of North America, 19*(2), 229–257.
50. Azhar, S. (2010). Peroxisome proliferator-activated receptors, metabolic syndrome and cardiovascular disease. *Future Cardiology, 6*(5), 657–691.
51. Lee, C.-H., Olson, P., & Evans, R. M. (2003). Minireview: Lipid metabolism, metabolic diseases, and peroxisome proliferator-activated receptors. *Endocrinology, 144*(6), 2201–2207.
52. Bagger, Y., et al. (2006). Radiographic measure of aorta calcification is a site-specific predictor of bone loss and fracture risk at the hip. *Journal of Internal Medicine, 259*(6), 598–605.
53. Di Mauro, S., Trevisan, C., & Hays, A. (1980). Disorders of lipid metabolism in muscle. *Muscle & Nerve: Official Journal of the American Association of Electrodiagnostic Medicine, 3*(5), 369–388.
54. Sharma, K., Kumar, K., & Mishra, N. (2016). Nanoparticulate carrier system: a novel treatment approach for hyperlipidemia. *Drug Delivery, 23*(3), 684–699.
55. Adetunji, C. O., et al. (2022). "Potentialities of nanomaterials for the management and treatment of metabolic syndrome: a new insight." *Materials Today Advances, 13*, 100198.
56. Klingenberg, R., & Hansson, G. K. (2009). Treating inflammation in atherosclerotic cardiovascular disease: Emerging therapies. *European Heart Journal, 30*(23), 2838–2844.
57. Kamaly, N., et al. (2013). Development and in vivo efficacy of targeted polymeric inflammation-resolving nanoparticles. *Proceedings of the National Academy of Sciences, 110*(16), 6506–6511.
58. Shaw, J. A., et al. (2008). Infusion of reconstituted high-density lipoprotein leads to acute changes in human atherosclerotic plaque. *Circulation Research, 103*(10), 1084–1091.
59. Katsuki, S., et al. (2017). Anti-inflammatory nanomedicine for cardiovascular disease. *Frontiers in Cardiovascular Medicine, 4*, 87.
60. Zhang, X.-F., et al. (2016). Silver nanoparticles: Synthesis, characterization, properties, applications, and therapeutic approaches. *International Journal of Molecular Sciences, 17*(9), 1534.
61. Niu, Z., et al. (2016). Lipid-based nanocarriers for oral peptide delivery. *Advanced Drug Delivery Reviews, 106*, 337–354.
62. Pugazhendhi, A., et al. (2018). Synthesis and characterization of silver nanoparticles using Gelidium amansii and its antimicrobial property against various pathogenic bacteria. *Microbial Pathogenesis, 114*, 41–45.
63. Nejati-Koshki, K., et al. (2014). Synthesis and in vitro study of cisplatin-loaded Fe3O4 nanoparticles modified with PLGA-PEG6000 copolymers in treatment of lung cancer. *Journal of Microencapsulation, 31*(8), 815–823.

64. Dragan, E. S., & Bucatariu, F. (2016). Design and characterization of anionic hydrogels confined in Daisogel silica composites microspheres and their application in sustained release of proteins. *Colloids and Surfaces A: Physicochemical and Engineering Aspects, 489*, 46–56.
65. Qi, X., et al. (2017). Salecan-based pH-sensitive hydrogels for insulin delivery. *Molecular Pharmaceutics, 14*(2), 431–440.
66. Pande, V. (2015). Studies on the characteristics of zaltoprofen loaded gelatin nanoparticles by nanoprecipitation. *Inventi Rapid NDDS, 3*, 1–7.
67. Iijima, S. (1991). Helical microtubules of graphitic carbon. *Nature, 354*(6348), 56–58.
68. Reilly, R. M. (2007). Carbon nanotubes: Potential benefits and risks of nanotechnology in nuclear medicine. *Journal of Nuclear Medicine, 48*(7), 1039–1042.
69. Iga, A. M., et al. (2007). Clinical potential of quantum dots. *BioMed Research International, 2007*(1), 076087.
70. Shi, J., et al. (2017). Cancer nanomedicine: Progress, challenges and opportunities. *Nature Reviews Cancer, 17*(1), 20–37.
71. Sandvig, K., & Van Deurs, B. (2005). Delivery into cells: Lessons learned from plant and bacterial toxins. *Gene Therapy, 12*(11), 865–872.
72. Bertrand, N., et al. (2014). Cancer nanotechnology: The impact of passive and active targeting in the era of modern cancer biology. *Advanced Drug Delivery Reviews, 66*, 2–25.
73. Danhier, F. (2016). To exploit the tumor microenvironment: Since the EPR effect fails in the clinic, what is the future of nanomedicine? *Journal of Controlled Release, 244*, 108–121.
74. Olivier, V., et al. (2003). Influence of targeting ligand flexibility on receptor binding of particulate drug delivery systems. *Bioconjugate Chemistry, 14*(6), 1203–1208.
75. Kogot, J. M., et al. (2009). Analysis of the dynamics of assembly and structural impact for a histidine tagged FGF1– 1.5 nm Au nanoparticle bioconjugate. *Bioconjugate Chemistry, 20*(11), 2106–2113.
76. Casals, E., et al. (2010). Time evolution of the nanoparticle protein corona. *ACS Nano, 4*(7), 3623–3632.
77. Li, T., et al. (2016). Acute toxicity and bio-distribution of silver nitrate and nano-silver with different particle diameters in rats. *Zhonghua shao shang za zhi= Zhonghua shaoshang zazhi = Chinese Journal of Burns, 32*(10), 606–612.
78. Liu, F., et al. (2017). Nano NiO induced liver toxicity via activating the NF-κB signaling pathway in rats. *Toxicology Research, 6*(2), 242–250.
79. Tang, H., et al. (2018). Effects and mechanism of nano-copper exposure on hepatic cytochrome P450 enzymes in rats. *International Journal of Molecular Sciences, 19*(7), 2140.
80. Recordati, C., et al. (2015). Tissue distribution and acute toxicity of silver after single intravenous administration in mice: Nano-specific and size-dependent effects. *Particle and Fibre Toxicology, 13*, 1–17.
81. Bartneck, M., et al. (2012). Peptide-functionalized gold nanorods increase liver injury in hepatitis. *ACS Nano, 6*(10), 8767–8777.
82. Barandeh, F., et al. (2012). Organically modified silica nanoparticles are biocompatible and can be targeted to neurons in vivo. *PLoS One, 7*(1), e29424.
83. Chen, P., et al. (2011). Gold nanoparticles for high-throughput genotyping of long-range haplotypes. *Nature Nanotechnology, 6*(10), 639–644.
84. Jomova, K., et al. (2024). Several lines of antioxidant defense against oxidative stress: Antioxidant enzymes, nanomaterials with multiple enzyme-mimicking activities, and low-molecular-weight antioxidants. *Archives of Toxicology, 98*(5), 1323–1367.
85. Sies, H., et al. (2022). Defining roles of specific reactive oxygen species (ROS) in cell biology and physiology. *Nature Reviews Molecular Cell Biology, 23*(7), 499–515.
86. Dai, D.-F., et al. (2014). Mitochondrial oxidative stress in aging and healthspan. *Longevity & healthspan, 3*, 1–22.
87. Shibuya, S., et al. (2014). Palladium and platinum nanoparticles attenuate aging-like skin atrophy via antioxidant activity in mice. *PLoS One, 9*(10), e109288.

88. Sims, C. M., et al. (2017). Redox-active nanomaterials for nanomedicine applications. *Nanoscale, 9*(40), 15226–15251.
89. Yoshihisa, Y., et al. (2011). SOD/catalase mimetic platinum nanoparticles inhibit heat-induced apoptosis in human lymphoma U937 and HH cells. *Free Radical Research, 45*(3), 326–335.
90. Qiu, Y., et al. (2014). Antioxidant chemistry of graphene-based materials and its role in oxidation protection technology. *Nanoscale, 6*(20), 11744–11755.
91. Yeo, W. W. Y., et al. (2014). Safety profile of dextran-spermine gene delivery vector in mouse lungs. *Journal of Nanoscience and Nanotechnology, 14*(5), 3328–3336.
92. Hu, X., et al. (2016). Knowledge gaps between nanotoxicological research and nanomaterial safety. *Environment International, 94*, 8–23.
93. John, A., et al. (2015). Physico-chemical modification as a versatile strategy for the biocompatibility enhancement of biomaterials. *RSC Advances, 5*(49), 39232–39244.
94. Thakur, A., & Kumar, A. (2024). Strategies for enhancing biocompatibility of nanoparticles. *Nanoparticle Toxicity and Compatibility, 161*, 182–224.
95. Kyriakides, T. R., et al. (2021). Biocompatibility of nanomaterials and their immunological properties. *Biomedical Materials, 16*(4), 042005.
96. Nyström, A. M., & Fadeel, B. (2012). Safety assessment of nanomaterials: Implications for nanomedicine. *Journal of Controlled Release, 161*(2), 403–408.
97. Li, X., et al. (2012). Biocompatibility and toxicity of nanoparticles and nanotubes. *Journal of Nanomaterials, 2012*(1), 548389.
98. Liao, C., Li, Y., & Tjong, S. C. (2018). Graphene nanomaterials: Synthesis, biocompatibility, and cytotoxicity. *International Journal of Molecular Sciences, 19*(11), 3564.
99. Martín-Pardillos, A., & Martin-Duque, P. (2023). Cellular alterations in carbohydrate and lipid metabolism due to interactions with nanomaterials. *Journal of Functional Biomaterials, 14*(5), 274.
100. Przybytkowski, E., et al. (2009). Nanoparticles can induce changes in the intracellular metabolism of lipids without compromising cellular viability. *The FEBS Journal, 276*(21), 6204–6217.
101. Ansari, S., et al. (2019). Gold nanoparticles synthesized with Smilax glabra rhizome modulates the anti-obesity parameters in high-fat diet and streptozotocin induced obese diabetes rat model. *Journal of Photochemistry and Photobiology B: Biology, 201*, 111643.
102. Liu, D., et al. (2010). Inhibition of proliferation and differentiation of mesenchymal stem cells by carboxylated carbon nanotubes. *ACS Nano, 4*(4), 2185–2195.
103. Duan, J., et al. (2018). Silica nanoparticles trigger hepatic lipid-metabolism disorder in vivo and in vitro. *International Journal of Nanomedicine, 13*, 7303–7318.
104. Li, B., Zhang, T., & Tang, M. (2022). Toxicity mechanism of nanomaterials: Focus on endoplasmic reticulum stress. *Science of the Total Environment, 834*, 155417.
105. Hu, X., et al. (2017). Strategies and knowledge gaps for improving nanomaterial biocompatibility. *Environment International, 102*, 177–189.
106. Metwally, S., & Stachewicz, U. (2019). Surface potential and charges impact on cell responses on biomaterials interfaces for medical applications. *Materials Science and Engineering: C, 104*, 109883.
107. Sanità, G., Carrese, B., & Lamberti, A. (2020). Nanoparticle surface functionalization: How to improve biocompatibility and cellular internalization. *Frontiers in Molecular Biosciences, 7*, 587012.
108. Yan, L., et al. (2019). A safe-by-design strategy towards safer nanomaterials in nanomedicines. *Advanced Materials, 31*(45), 1805391.
109. Eaton, M. A., Levy, L., & Fontaine, O. M. (2015). Delivering nanomedicines to patients: a practical guide. *Nanomedicine: Nanotechnology, Biology and Medicine, 11*(4), 983–992.
110. Paliwal, R., Babu, R. J., & Palakurthi, S. (2014). Nanomedicine scale-up technologies: Feasibilities and challenges. *AAPS PharmSciTech, 15*(6), 1527–1534.

111. Dahlgren, D., Sjöblom, M., & Lennernäs, H. (2019). Intestinal absorption-modifying excipients: A current update on preclinical in vivo evaluations. *European Journal of Pharmaceutics and Biopharmaceutics, 142*, 411–420.
112. Đorđević, S., et al. (2022). Current hurdles to the translation of nanomedicines from bench to the clinic. *Drug Delivery and Translational Research*, 1–26.
113. Agrahari, V., & Hiremath, P. (2017). Challenges associated and approaches for successful translation of nanomedicines into commercial products. *Nanomedicine, 12*(8), 819–823.
114. Operti, M. C., et al. (2021). PLGA-based nanomedicines manufacturing: Technologies overview and challenges in industrial scale-up. *International Journal of Pharmaceutics, 605*, 120807.
115. Wu, L.-P., Wang, D., & Li, Z. (2020). Grand challenges in nanomedicine. *Materials Science and Engineering: C, 106*, 110302.
116. Landesman-Milo, D., & Peer, D. (2016). Transforming nanomedicines from lab scale production to novel clinical modality. *Bioconjugate Chemistry, 27*(4), 855–862.
117. Barua, S., & Mitragotri, S. (2014). Challenges associated with penetration of nanoparticles across cell and tissue barriers: a review of current status and future prospects. *Nano Today, 9*(2), 223–243.
118. Yoo, J.-W., Chambers, E., & Mitragotri, S. (2010). Factors that control the circulation time of nanoparticles in blood: Challenges, solutions and future prospects. *Current Pharmaceutical Design, 16*(21), 2298–2307.
119. Rodriguez, P. L., et al. (2013). Minimal "Self" peptides that inhibit phagocytic clearance and enhance delivery of nanoparticles. *Science, 339*(6122), 971–975.
120. Parodi, A., et al. (2013). Synthetic nanoparticles functionalized with biomimetic leukocyte membranes possess cell-like functions. *Nature Nanotechnology, 8*(1), 61–68.
121. Huang, H., et al. (2020). Inorganic nanoparticles in clinical trials and translations. *Nano Today, 35*, 100972.
122. Pandey, R. K., & Prajapati, V. K. (2018). Molecular and immunological toxic effects of nanoparticles. *International Journal of Biological Macromolecules, 107*, 1278–1293.
123. Bakand, S., & Hayes, A. (2016). Toxicological considerations, toxicity assessment, and risk management of inhaled nanoparticles. *International Journal of Molecular Sciences, 17*(6), 929.
124. Haase, A., et al. (2011). Toxicity of silver nanoparticles in human macrophages: Uptake, intracellular distribution and cellular responses. In *Journal of Physics: Conference Series*. IOP Publishing.
125. Baranowska-Wójcik, E., et al. (2020). Effects of titanium dioxide nanoparticles exposure on human health—a review. *Biological Trace Element Research, 193*, 118–129.
126. Wang, B., et al. (2012). Akt signaling-associated metabolic effects of dietary gold nanoparticles in Drosophila. *Scientific Reports, 2*(1), 563.
127. Anchordoquy, T., et al. (2024). Mechanisms and barriers in nanomedicine: Progress in the field and future directions. *ACS Nano, 18*, 13983.
128. Hertig, J. B., et al. (2021). Tackling the challenges of nanomedicines: Are we ready? *American Journal of Health-System Pharmacy, 78*(12), 1047–1056.
129. Poley, M., et al. (2022). Nanoparticles accumulate in the female reproductive system during ovulation affecting cancer treatment and fertility. *ACS Nano, 16*(4), 5246–5257.
130. Kumar, P. S., Pavithra, K. G., & Naushad, M. (2019). Characterization techniques for nanomaterials. In *Nanomaterials for solar cell applications* (pp. 97–124). Elsevier.
131. Barhoum, A., & García-Betancourt, M. L. (2018). Physicochemical characterization of nanomaterials: Size, morphology, optical, magnetic, and electrical properties. In *Emerging applications of nanoparticles and architecture nanostructures* (pp. 279–304). Elsevier.
132. Lin, P.-C., et al. (2014). Techniques for physicochemical characterization of nanomaterials. *Biotechnology Advances, 32*(4), 711–726.
133. Rodriguez-Loya, J., Lerma, M., & Gardea-Torresdey, J. L. (2023). Dynamic light scattering and its application to control nanoparticle aggregation in colloidal systems: a review. *Micromachines, 15*(1), 24.

134. Salame, P. H., Pawade, V. B., & Bhanvase, B. A. (2018). Characterization tools and techniques for nanomaterials. In *Nanomaterials for green energy* (pp. 83–111). Elsevier.
135. Bunjes, H., & Unruh, T. (2007). Characterization of lipid nanoparticles by differential scanning calorimetry, X-ray and neutron scattering. *Advanced Drug Delivery Reviews, 59*(6), 379–402.
136. Mudalige, T., et al. (2019). Characterization of nanomaterials: Tools and challenges. In *Nanomaterials for food applications* (pp. 313–353). Elsevier.
137. Song, X., et al. (2008). Dual agents loaded PLGA nanoparticles: Systematic study of particle size and drug entrapment efficiency. *European Journal of Pharmaceutics and Biopharmaceutics, 69*(2), 445–453.
138. Zatorska-Płachta, M., et al. (2021). Encapsulation of curcumin in polystyrene-based nanoparticles—Drug loading capacity and cytotoxicity. *ACS Omega, 6*(18), 12168–12178.
139. Khalbas, A. H., et al. (2024). Drug loading methods and kinetic release models using of mesoporous silica nanoparticles as a drug delivery system: A review. *South African Journal of Chemical Engineering, 50*, 261.
140. Chadha, R., & Bhandari, S. (2014). Drug–excipient compatibility screening—Role of thermoanalytical and spectroscopic techniques. *Journal of Pharmaceutical and Biomedical Analysis, 87*, 82–97.
141. Kuppusamy, P., et al. (2016). Biosynthesis of metallic nanoparticles using plant derivatives and their new avenues in pharmacological applications–An updated report. *Saudi Pharmaceutical Journal, 24*(4), 473–484.
142. Deivayanai, V., et al. (2024). A comprehensive review on advances in nanoparticle-mediated cancer therapeutics: Current research and future perspectives. *Cancer Pathogenesis and Therapy, 2*, E01–E16.
143. Al-Thani, A. N., et al. (2024). Nanoparticles in cancer theragnostic and drug delivery: A comprehensive review. *Life Sciences, 352*, 122899.
144. Veiseh, O., et al. (2015). Managing diabetes with nanomedicine: Challenges and opportunities. *Nature Reviews Drug Discovery, 14*(1), 45–57.
145. Liu, Y., et al. (2017). pH switchable nanoassembly for imaging a broad range of malignant tumors. *ACS Nano, 11*(12), 12446–12452.
146. Liu, L., et al. (2018). Influence of nano-fibrillated cellulose (NFC) on starch digestion and glucose absorption. *Carbohydrate Polymers, 196*, 146–153.
147. DeLoid, G. M., et al. (2018). Reducing intestinal digestion and absorption of fat using a nature derived biopolymer: Interference of triglyceride hydrolysis by nanocellulose. *ACS Nano, 12*(7), 6469–6479.
148. Sánchez, J., et al. (2008). Oral supplementation with physiological doses of leptin during lactation in rats improves insulin sensitivity and affects food preferences later in life. *Endocrinology, 149*(2), 733–740.
149. Kassi, E., et al. (2011). Metabolic syndrome: Definitions and controversies. *BMC Medicine, 9*, 1–13.
150. Bahadori, E., et al. (2019). Recent advances in nanotechnology for the treatment of metabolic syndrome. *Diabetes & Metabolic Syndrome: Clinical Research & Reviews, 13*(2), 1561–1568.
151. Fernandes, C., Suares, D., & Yergeri, M. C. (2018). Tumor microenvironment targeted nanotherapy. *Frontiers in Pharmacology, 9*, 1230.
152. Kim, D., et al. (2021). Nanoparticle-mediated lipid metabolic reprogramming of T cells in tumor microenvironments for immunometabolic therapy. *Nano-Micro Letters, 13*, 1–27.
153. Daneshmandi, S., Wegiel, B., & Seth, P. (2019). Blockade of lactate dehydrogenase-A (LDH-A) improves efficacy of anti-programmed cell death-1 (PD-1) therapy in melanoma. *Cancers, 11*(4), 450.
154. Dempsey, L. A. (2016). Good cholesterol. *Nature Immunology, 17*(5), 480–480.
155. Hao, M., et al. (2020). Combination of metabolic intervention and T cell therapy enhances solid tumor immunotherapy. *Science Translational Medicine, 12*(571), eaaz6667.

156. Quillard, T., & Croce, K. J. (2014). Pathobiology and mechanisms of atherosclerosis. In *Cardiovascular imaging: Arterial and aortic valve inflammation and calcification* (pp. 3–38). Springer.
157. Patel, S. K., & Janjic, J. M. (2015). Macrophage targeted theranostics as personalized nanomedicine strategies for inflammatory diseases. *Theranostics, 5*(2), 150.
158. Peer, D., et al. (2020). Nanocarriers as an emerging platform for cancer therapy. In *Nano-enabled medical applications* (pp. 61–91). Jenny Stanford Publishing.
159. Tang, T. Y., et al. (2009). The ATHEROMA (Atorvastatin Therapy: Effects on Reduction of Macrophage Activity) Study: Evaluation using ultrasmall superparamagnetic iron oxide-enhanced magnetic resonance imaging in carotid disease. *Journal of the American College of Cardiology, 53*(22), 2039–2050.
160. Kooi, M. E., et al. (2003). Accumulation of ultrasmall superparamagnetic particles of iron oxide in human atherosclerotic plaques can be detected by in vivo magnetic resonance imaging. *Circulation, 107*(19), 2453–2458.
161. Burtea, C., et al. (2012). Development of a magnetic resonance imaging protocol for the characterization of atherosclerotic plaque by using vascular cell adhesion molecule-1 and apoptosis-targeted ultrasmall superparamagnetic iron oxide derivatives. *Arteriosclerosis, Thrombosis, and Vascular Biology, 32*(6), e36–e48.
162. Jayagopal, A., et al. (2010). Insights into atherosclerosis using nanotechnology. *Current Atherosclerosis Reports, 12*, 209–215.
163. Nahrendorf, M., et al. (2009). Hybrid in vivo FMT-CT imaging of protease activity in atherosclerosis with customized nanosensors. *Arteriosclerosis, Thrombosis, and Vascular Biology, 29*(10), 1444–1451.
164. Shimada, Y. J., & Cannon, C. P. (2015). PCSK9 (Proprotein convertase subtilisin/kexin type 9) inhibitors: Past, present, and the future. *European Heart Journal, 36*(36), 2415–2424.
165. Frank-Kamenetsky, M., et al. (2008). Therapeutic RNAi targeting PCSK9 acutely lowers plasma cholesterol in rodents and LDL cholesterol in nonhuman primates. *Proceedings of the National Academy of Sciences, 105*(33), 11915–11920.
166. Venkataraman, G. R., & Rivas, M. A. (2019). Rare and common variant discovery in complex disease: The IBD case study. *Human Molecular Genetics, 28*(R2), R162–R169.
167. Hammer, T., et al. (2019). Dietary risk factors for inflammatory bowel diseases in a high-risk population: Results from the Faroese IBD study. *United European Gastroenterology Journal, 7*(7), 924–932.
168. Putignani, L., et al. (2016). Gut microbiota dysbiosis as risk and premorbid factors of IBD and IBS along the childhood–adulthood transition. *Inflammatory Bowel Diseases, 22*(2), 487–504.
169. Sabino, J., et al. (2019). New biologics and small molecules in inflammatory bowel disease: An update. *Therapeutic Advances in Gastroenterology, 12*, 1756284819853208.
170. Holota, Y., et al. (2019). The long-term consequences of antibiotic therapy: Role of colonic short-chain fatty acids (SCFA) system and intestinal barrier integrity. *PLoS One, 14*(8), e0220642.
171. Anderson, E. M., et al. (2009). Sustained release of antibiotic from poly (2-hydroxyethyl methacrylate) to prevent blinding infections after cataract surgery. *Biomaterials, 30*(29), 5675–5681.
172. Gao, P., et al. (2011). Recent advances in materials for extended-release antibiotic delivery system. *The Journal of Antibiotics, 64*(9), 625–634.
173. Hollak, C. E., et al. (1997). Elevated levels of M-CSF, sCD14 and IL8 in type 1 Gaucher disease. *Blood Cells, Molecules, and Diseases, 23*(2), 201–212.
174. Mitchell, M. J., et al. (2021). Engineering precision nanoparticles for drug delivery. *Nature Reviews Drug Discovery, 20*(2), 101–124.
175. Braidman, I., & Gregoriadis, G. (1976). *Preparation of glucocerebroside β-glucosidase for entrapment in liposomes and treatment of patients with adult Gaucher's disease*. Portland Press Ltd.

Cancer Nanotheranostics: Emerging Strategies for Diagnosis and Therapeutics in Cancer Metabolism

8

Singh Shreya, Debadatta Mohapatra, and Alakh N. Sahu

Abstract

Cancer is characterized by uncontrolled proliferation of cells and remains a major health burden across the globe. Altered metabolism/abnormal metabolic signaling are the key features of cancer progression. Conventional therapy for cancer possesses various limitations, such as drug resistance, low targeting ability, limited tissue distribution, off-targeted effects, high doses, and systemic toxicity. Current advancements have harnessed the exclusive properties of nanomaterials for diagnosis and treatment of cancer via targeted drug delivery, improved imaging, and precise therapeutic action with negligible toxicity. Nanotheranostics integrate bioimaging and therapeutic functionalities. Exploring cancer metabolic abnormalities via nanotheranostics provides a novel strategy for early diagnosis, real-time monitoring, and therapy for cancer progression. This book chapter highlights various theranostic nanoformulations designed to modulate cancer cell metabolism such as glucose metabolism, lactate metabolism, amino acid metabolism, nucleotide metabolism and lipid metabolism. These

S. Shreya (✉)
SBS College of Pharmacy, Fatehpur, India

Phytomedicine Research Laboratory, Department of Pharmaceutical Engineering & Technology, IIT (BHU), Varanasi, India
e-mail: sshreya.rs.phe16@iitbhu.ac.in

D. Mohapatra
Phytomedicine Research Laboratory, Department of Pharmaceutical Engineering & Technology, IIT (BHU), Varanasi, India

Leslie Dan Faculty of Pharmacy, University of Toronto, Toronto, ON, Canada

A. N. Sahu
Phytomedicine Research Laboratory, Department of Pharmaceutical Engineering & Technology, IIT (BHU), Varanasi, India

V. K. Chaturvedi et al. (eds.), *Nanotheranostics in Metabolic Syndrome*,
https://doi.org/10.1007/978-981-95-3746-4_8

nanotheranostics exploits tumor-specific metabolic vulnerabilities, offering promising strategy to enhance diagnostic accuracy and therapeutic efficacy in cancer treatment.

Keywords

Cancer · Nanotheranostics · Nanomedicine · Drug delivery · Diagnosis · Treatment · Metabolism

1 Introduction

Metabolism, the basic process of life, represents a cascade of biochemical pathways that drive biomolecule synthesis and energy production [1, 2]. Unlike normal cells, cancerous cells reprogram their metabolism to sustain their bioenergetics demands, promote biosynthesis, and facilitate immune evasion. Besides oncogenes and tumor suppressors, metabolic reprogramming is regarded as a hallmark of cancer [3, 4]. Notably, cancerous cells undergo metabolic reprogramming by rewriting anabolic and catabolic processes such as glucose metabolism, lactate metabolism, amino acid metabolism, lipid metabolism, and nucleotide metabolism [5]. Cancerous cells enhanced their glucose uptake to meet their energy needs and produced more lactate irrespective of oxygen availability, a process called the Warburg effect or aerobic glycolysis, identified by positron emission tomography of ^{18}F-flurodeoxy glucose in different cancers [6, 7]. Even without glucose and oxygen, cancerous cells rewrite their catabolism by upregulating glutamine metabolism for energy sources [8]. Meanwhile, reprogramming anabolism involves aberrant production of lactate, protein, and nucleotide not only to support rapid proliferation but also to create an immune-suppressive microenvironment [9]. Nanotechnology has various advantages in addressing the limitations of cancer chemotherapy, including low solubility, poor bioavailability, off-target, rapid clearance, and tumor microenvironment barrier. Nanotechnology improves drug solubility, targeting, retention, and intracellular delivery while lowering toxicity, overcoming resistance, and optimizing combination therapy [10–12] (Fig. 8.1).

The explored nanotechnology approach for targeting cancer metabolism includes metal-based nanoparticles (gold-based nanoparticles, silver-based nanoparticles, manganese-based nanoparticles, ferrous-based nanoparticles, platinum-palladium-based nanoparticles, lithium-based nanoparticles, and tin-selenium-based nanoparticles), organic-based nanoparticles (Albumin-based nanoparticles, liposomes, and polymeric-based nanoparticles (poly(lactic-co-glycolic acid) (PLGA), and polyethylene glycol-poly(lactic-co-glycolic acid) (PEG-PLGA)), and hybrid-based nanoparticles: hallow mesoporous nanoparticles (silica-based, carbon-based, and metal doped), and metal-organic framework (zeolitic imidazolate framework (ZIF-8), and Materials of Institute Lavoisier (MIL-101, and MIL-100)). Different types of nanoparticles excel in different types of properties, like, lipid-based nanoparticles for enhanced solubility, polymeric-based for controlled release,

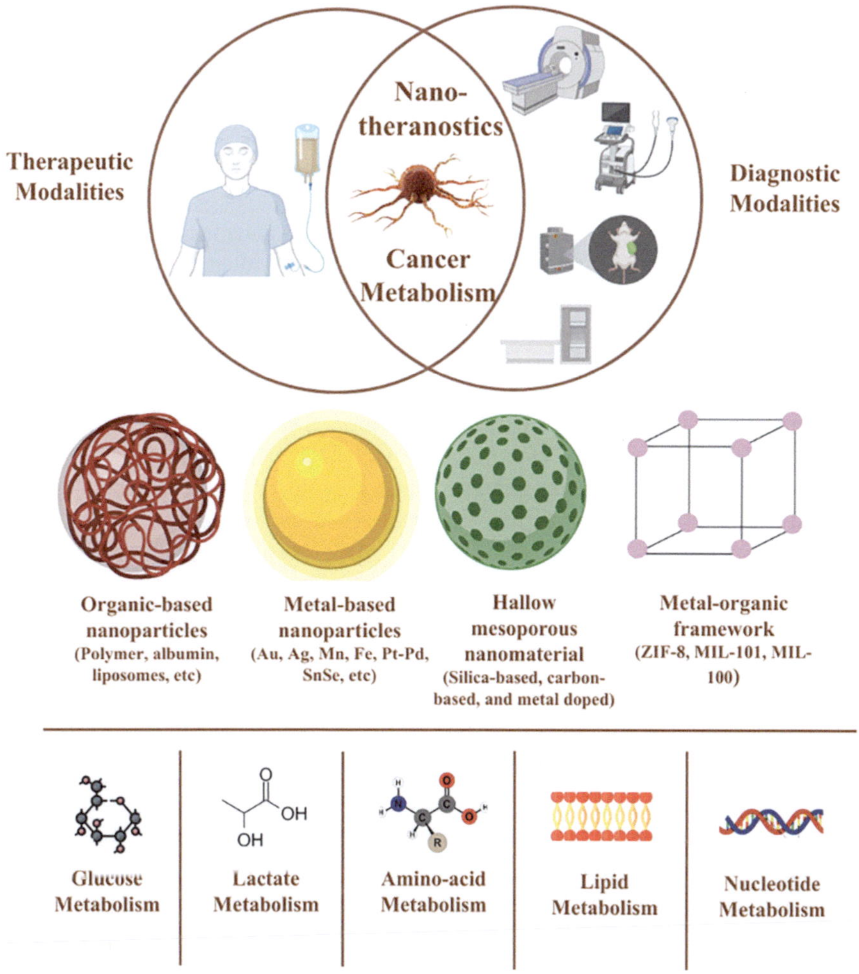

Fig. 8.1 Nanotheranostics targeting cancer metabolism

manganese-based, and iron-based for catalyzing Fenton reaction, and gold-silver-based for high stability and photothermal effect. Whereas, hybrid-based nanoparticles such as hallow mesoporous nanoparticles and metal-organic framework possess high loading capacity due to high surface area and mesoporous structure, along with a controlled release approach involving pH-controlled and/or stimuli-responsive release properties. Thus, these exceptional properties, such as high loading capacity, high stability, high biocompatibility, and controlled release make them ideal nano-carriers for starvation therapy, that targets cancer metabolism by depriving nutrients from cancerous cells. Besides this, nanotechnology approaches camouflaged nanoparticles surface with cell membrane biomimetics such as cancerous cell membranes, red blood cell (RBC) membranes, platelet membranes, and hyaluronic acid. This biomimetic modification facilitates homotypic targeting, preferential accumulation in cancerous cells, prevents immune evasion and prolongs their circulation time. To achieve a synergic effect against cancerous cells, a nanotechnology approach targeting cancer metabolism is usually employed in combination with other treatment interventions such as photothermal therapy, photodynamic therapy, chemodyanmic therapy, sonodynamic therapy, and chemotherapy [13–15].

Over and above exploiting therapeutic responses, nanotechnology also enables real-time monitoring of therapeutic responses by using imaging modalities like magnetic resonance imaging (MRI), computed tomography (CT), fluorescence, surface enhanced raman scattering (SERS), and photoacoustic imaging. Fluorescence imaging offers high sensitivity and enables visualization of nanoparticle distribution at the cellular and molecular levels. On other hand, MRI provides excellent spatial resolution and deep tissue penetration, especially when nanoparticles are engineered with magnetic elements such as iron or manganese. SERS employs plasmonic nanoparticles such as gold or silver to amplify weak Raman signals, enabling real-time, non-invasive detection of nanoparticles and monitoring of therapeutic responses at cellular and tissue levels. While, CT provides high-resolution three-dimensional anatomical imaging, allowing precise visualization of nanoparticle accumulation and tissue architecture. Photoacoustic imaging integrates optical and ultrasonic techniques, where nanoparticles act as photoabsorbers to generate acoustic signals, enabling real-time visualization of biodistribution and tumor microenvironment changes. Collectively, these multimodal imaging [16]. Nanotheranostics integrates bioimaging and therapeutic functionalities [17, 18]. Nanotheranostics emerges as a powerful strategy for targeting cancer metabolism, that offers a combination of diagnostic imaging and therapeutic intervention within a single nanosystem. Exploring metabolic abnormalities via nanotheranostics is an excellent tactic for early diagnosis, real-time monitoring, and therapy for cancer progression. This book chapter includes a nanotheranostics approach for targeting metabolic pathways, such as glucose metabolism, lactate metabolism, amino acid metabolism, nucleotide metabolism, and lipid metabolism.

2 Nanotheranostics Targeting Cancer Metabolism

2.1 Glucose Metabolism

Glycolysis is distinguished as a mutual pathway of aerobic respiration (oxidative phosphorylation) and anaerobic respiration wherein glucose molecule is broken down to pyruvate in the cytoplasm. Unlike normal cells, even in the presence of a sufficient amount of oxygen, cancerous cells reprogram their further glycolysis and decompose glucose to lactate (anaerobic respiration), rather than relying on oxidative phosphorylation. Cancerous cells prefer this aberrant glycolysis to meet their high energy and biosynthetic needs for proliferation, invasion, and metastasis.

Understanding this metabolic shift and addressing this aberrant glycolysis by lowering glucose uptake, impairing glycolysis, and depleting glucose are potential treatment strategies. Moreover, targeting glucose metabolism always remains a hot spot for cancer researchers. Cancerous cell overexpresses glucose transporter (GLUT1/GLUT3) to increase glucose uptake [19]. Conjugating GLUT inhibitors or antagonists (BAY-876, WZB117, and Glutipyran) [20–22] in nanocarrier can effectively inhibit GLUT, reduce the influx of glucose into the cancerous cell, and impair glycolysis, thereby disrupting energy and biosynthetic needs for proliferation, invasion, and metastasis of cancerous cells [23–28]. This observed metabolic reprogramming of glycolysis, is also associated with upregulation of key rate-limiting enzymes regulating glycolysis such as hexokinase, phosphofructokinases, and pyruvate kinase [29, 30]. Nanotechnology approaches can disrupt glycolysis at various stages, by inhibiting these rate-limiting enzymes, leading to metabolic starvation and enhanced therapeutic efficacy [31–38]. Glucose oxidase (GOx) in the presence of oxygen catalyzes the conversion of glucose to gluconic acid and hydrogen peroxide (H_2O_2). GOx depletes intracellular glucose sources from cancerous cells and starve them to death. Despite depriving energy sources, GOx can also elevate acidic microenvironment, and H_2O_2 levels and build up oxidative stress, demonstrating a dual approach to cancer cell targeting: starvation therapy and oxidative stress [39–41]. However, despite its effectiveness as a starvation therapy, GOx possesses a few limitations, which involve poor stability, inactivation on encapsulation, decrease in enzyme bioactivity on chemical conjugation, degradation by proteases enzyme in physiological conditions resulting in short half-life, and off-target effects (can affect normal cell). Nanotechnology has been utilized to overcome this limitation of GOx, along with offering imaging assistance. Moreover, GOx has been used in conjunction with other combination treatment strategies for synergetic anti-cancer therapy [42–65] (Table 8.1).

2.2 Lactate Metabolism

Majorly, lactate produced from pyruvate contributes to an acidic tumor microenvironment. High production of lactate meets the energy and biosynthetic needs of cancerous cells, playing a vital role in proliferation, invasion, and metastasis. Owing to the increased production need of lactate in progressive and aggressive cancerous cells, lactate transporter (monocarboxylate transporter (MCT1, MCT2, and MCT4)) and rate-limiting enzymes (pyruvate dehydrogenase kinase, and lactate dehydrogenase A) are upregulated [66–69]. Hence interfering with lactate metabolism is likely to combat cancerous cell invasion and metastasis. Nanotechnology explores various strategies interfering with lactate metabolism including inhibiting lactate uptake, impairing lactate anabolism, and depleting lactate [70–78] (Table 8.2).

Table 8.1 Nanotheranostics targeting glucose metabolism

Nanotechnology	Type of nanoparticle	Combination therapy	Therapeutic modality	Diagnostic modality	In-vitro and in-vivo tumor model	Reference
Paclitaxel and BAY-876 conjugated in human serum albumin nanoparticles, and surface disguised with *Escherichia coli*	Human serum albumin-based nanoparticle	Chemotherapy: paclitaxel	Inhibiting GLUT *E. coli* promotes homotypic targeting of cancerous cells. Once accumulated in a cancerous cell, consumes intracellular tumor glucose, thereby depriving the cancerous cell glucose and facilitating the release of paclitaxel and BAY-876 from bounded human serum albumin. Released BAY-876 inhibits GLUT, causing reduced glucose levels that further activate AMPK. Activated AMPK promotes pinocytosis	Fluorescence imaging	CT26 tumor-bearing mice	[24]
BAY-876/V-9302 incorporated in human serum albumin nanoparticles	Human serum albumin-based nanoparticle		Inhibiting GLUT and Alanine, Serine, Cysteine Transporter (ASCT2) Impaired glucose uptake causes cancerous cells to reprogram their metabolism and depend on glutamine as an energy source. Released BAY-876 and V-9302 block glucose and glutamine uptake respectively. Inhibition of	Fluorescence imaging	CFPAC-1-luc tumor-bearing mice	[26]

			glutamine inhibits glutathione resulting in enhanced oxidative stress. Enhanced oxidative stress induces pyroptosis			
siRNA GLUT3 in poly(lactic-co-glycolic acid)-block-poly (ethylene glycol) (PLGA-b-PEG) nanoparticles	Polymeric-based nanoparticles		Inhibit GLUT Nanoparticles efficiently silenced GLUT3 and reduced the expression of GLUT3 by delivering siRNA GLUT3 to cancerous cells	Fluorescence imaging	U87MG tumor-bearing mice	[23]
GLUT1 mRNA-cleaving DNAzyme entrapped in ZIF and camouflaged with hyaluronic acid surface modification	Metal-organic framework		Inhibit GLUT Hyaluronic acid surface modification triggers targeting cancerous cells overexpressed with CD44, followed by degradation of nanoparticles framework by hyaluronidase. Released zinc ion blocked NAD^{+}synthesis. This promotes catalytic cleavage of GLUT1 mRNA by DNAzyme, thereby degrading GLUT1 mRNA	Fluorescence imaging	B16-F10 tumor-bearing mice	[25]
BAY-876 and doxorubicin-duplex entrapped into polymer nanoparticles containing disulfide bond	Polymer-based nanoparticles	Chemotherapy: Doxorubicin	Inhibit GLUT Polymeric nanoparticles disulfide bond cleaved by intracellular glutathione. Released BAY-876 inhibits GLUT1. While, doxorubicin-	Fluorescence imaging	4T1 tumor-bearing mice	[28]

(continued)

Table 8.1 (continued)

Nanotechnology	Type of nanoparticle	Combination therapy	Therapeutic modality	Diagnostic modality	In-vitro and in-vivo tumor model	Reference
			duplex interacts with ATP, reducing intracellular ATP and releasing activated doxorubicin to ablate the cancerous cell			
Diclofenac encapsulated in gold nanoparticles, disguised with a hyaluronic acid surface alteration	Gold-based-nanoparticles	Photothermal therapy: Near-Infrared (NIR) 808 nm	Inhibit GLUT Hyaluronic acid directs nanoparticles preferential accumulation in cancerous cells overexpressed with CD44 and subsequently cleaved nanoparticles by its hyaluronidase activity. Released diclofenac, inhibits GLUT1, and deprives cancerous cell glucose	Fluorescence imaging	MCF-7 cancerous cell	[27]
Lonidamine and NLG919 dimer connected with a disulfide bond, further coated with F-127 polymer	Polymer-based nanoparticles		Inhibit hexokinase-II Intracellular glutathione cleaved disulfide bond between lonidamine and NLG919. Released lonidamine inhibits hexokinase-II and NLG919 reduces kynurenine levels and Treg	Fluorescence imaging	4T1 tumor-bearing mice	[38]

2-deoxy-D-glucose (glucose analog) encapsulated in PLGA nanoparticle	Polymer-based nanoparticles		Inhibit hexokinase Decreased glucose and lactate production in cancerous cells	Fluorescence imaging	B16F10 tumor-bearing mice	[37]
2-deoxy-D-glucose and doxorubicin-based nanoliposome	Liposome	Chemotherapy: doxorubicin	Inhibit hexokinase Inhibit glycolysis of cancerous cells. Doxorubicin working synergistically causes DNA damage, subsequently leading to apoptosis	Fluorescence imaging	HeLa tumor-bearing mice	[31]
Ethylenediaminetetraacetic acid (EDTA) and rotenone coloaded in thiolated poly (acrylic acid) mesoporous silica nanoparticles	Hollow mesoporous nanoparticles		Inhibit glycolytic enzyme EDTA inhibits glycolytic enzymes by chelating magnesium ions. While rotenone inhibits mitochondrial oxidative phosphorylation	Fluorescence imaging	4T1 tumor-bearing mice	[32]
2-deoxy-d-glucose, BAY-876, and chloroquine enveloped in ZIF-8	Metal-organic framework	Chemotherapy: chloroquine	Inhibit GLUT1 and hexokinase BAY-876 inhibits GLUT1, parallelly 2-deoxy-d-glucose inhibits hexokinase, leading to impaired glycolysis. Meanwhile, chloroquine exerts autophagy	Fluorescence imaging	4T1 tumor-bearing mice	[33]
PFK15 enveloped in silver-based metal-organic framework, camouflaged with RBC cell membrane surface modification	Metal-organic framework		Inhibit 6-phosphofructo-2-kinase/fructose-2,6-bisphosphatase 3 (PFKFB3) enzyme Constructed nanoparticles target CD20 aptamer and	Fluorescence imaging	OCI-LY8, OCI-LY10, Raji, and Daudi cells	[34]

(continued)

Table 8.1 (continued)

Nanotechnology	Type of nanoparticle	Combination therapy	Therapeutic modality	Diagnostic modality	In-vitro and in-vivo tumor model	Reference
			release PFK15 inhibiting PFKFB3, thus depriving cancerous cells glucose			
Paclitaxel and PFK15 are enclosed in solid lipid nanoparticles, disguised by the hybrid cancerous cell membrane and activated fibroblast membrane	Solid lipid nanoparticle	Chemotherapy: paclitaxel	Inhibit PFKFB3 enzyme PFK15 inhibits PFKFB3 and deprives glucose from cancerous cells. Simultaneously, paclitaxel, being an anti-microtubule, halts the growth of cancerous cells	Fluorescence imaging	4T1 tumor-bearing mice	[35]
siRNA programmed death-ligand 1 (PD-L1) and resveratrol co-loaded in polymer-based nanoparticles	Polymer-based nanoparticles		Inhibit pyruvate kinase M2 Codelivered siRNA PD-L1 silenced PD-L1 and resveratrol inhibited pyruvate kinase M2 and interfered with glycolysis	Fluorescence imaging	B16F10 and CT26 tumor-bearing mice	[36]
GOx entrapped in gold-silver hallow nanotriangles (Au-Ag HTN)	Gold-silver nanoparticles	Photothermal therapy: NIR 1064 nm	Deplete glucose GOx converts glucose to gluconic acid and H_2O_2. Laser irradiation induce hot electrons and trigger the activation of peroxidase activity of Au-Ag HTNs nanozymes to produce highly toxic hydroxyl (•OH) radicals, that further cause apoptosis in the cancer cell	Photothermal imaging Multi-spectral optoacoustic tomography CT	4T1 tumor-bearing mice	[57]

GOx and ultrasmall iron(II, III) oxide (Fe_3O_4) nanoparticles incorporated into mesoporous silica nanoparticles	Hollow mesoporous nanoparticles		Deplete glucose Firstly, GOx depletes glucose and produces H_2O_2. H_2O_2 subsequently utilized for a Fenton-like reaction, catalyzed by Fe_3O_4 nanoparticles, and produced a highly toxic •OH radical. This further leads to apoptosis in the cancer cell	Fluorescence imaging	4T1 tumor-bearing mice	[58]
GOx loaded on manganese dioxide (MnO_2) nanoparticles overlayed with hyaluronic acid	Manganese based nanoparticles		Deplete glucose Nanoparticle targets CD44 overexpressed tumor cells and later, cause depletion of glucose by GOx. Furthermore, MnO_2 utilized generated H_2O_2 to produce more oxygen, which is constantly used for glucose depletion. The whole reaction resembles a loop	Fluorescence imaging	CT-26 tumor-bearing mice	[55]
GOx conjugated in PLGA nanoparticle	Polymeric nanoparticles	Photothermal therapy: NIR 780 nm	Deplete glucose Assisted by NIR780, the nanosphere projected into deep inside the tumor region, particularly mitochondria. Upon irradiation, mitochondria get susceptible to reactive oxygen species (ROS) and get damaged. This further accelerates the starving therapy of GOx	Photoacoustic imaging	4T1 tumor-bearing mice	[56]

(continued)

Table 8.1 (continued)

Nanotechnology	Type of nanoparticle	Combination therapy	Therapeutic modality	Diagnostic modality	In-vitro and in-vivo tumor model	Reference
GOx incorporated in Prussian blue (PB)-gellan hydrogel	Hollow mesoporous nanoparticles	Photothermal therapy: NIR 720 nm	Deplete glucose Intratumoral injection leads to the conversion of light to heat by PB particles for NIR irradiation and also acts as a nanozyme with catalase activity decomposing H_2O_2 to produce oxygen. Concurrently, GOx in the presence of oxygen depletes glucose to form H_2O_2, which is further catalyzed by the catalase activity of PB nanoparticles to produce more oxygen	Photothermal imaging	4T1 tumor-bearing mice	[51]
GOx fabricated in metal-organic framework MIL, along with sonosensitizer	Metal-organic frameworks	Sonothermal therapy: ultrasound irradiation	Deplete glucose Intratumoral injection produces toxic •OH radical. Being a nanozyme, the catalase activity of the formed nanoparticle converts H_2O_2 to oxygen, which is further utilized in the depletion of glucose by GOx and forms gluconic acid and H_2O_2. Generated H_2O_2 can further accelerate Fenton's reaction. Additionally, sonosensitizer aids in producing $^1O_2$	Fluorescence imaging	4T1 tumor-bearing mice	[50]

Cupric ion (Cu^{2+}) incorporated into hollow mesoporous silica nanoparticles, decorated with GOx	Hollow mesoporous nanoparticles		Deplete glucose GOx depletes glucose. Reduced pH causes the release of Cu^{2+}. Cu^{2+} induces glutathione (GSH) depletion and catalyzes a Fenton-like reaction with generated H_2O_2 to produce highly toxic •OH radicals, that further hinder tumorigenesis	Fluorescence imaging	MCF-7 tumor-bearing mice	[52]
Biomimetic tumor cell membrane and GOx decorated onto ZIF-8, loaded with prodrug H_2O_2 sensitive doxorubicin	Metal-organic frameworks	Chemotherapy: doxorubicin	Deplete glucose Immune evasion, along with preferential uptake in tumors via the biomimetic tumor cell membrane. Following this GOx depletes glucose to form gluconic acid and H_2O_2. Increased intracellular acid environment and increased H_2O_2 production promote controlled release of doxorubicin from the porous framework	Fluorescence imaging	4T1 tumor-bearing mice	[53]
GOx and dexamethasone encapsulated in ZIF-8 further conjugated with tannic acid	Metal-organic frameworks	Chemotherapy: dexamethasone	Deplete glucose GOx converts glucose to gluconic acid and H_2O_2, resulting in reduced pH. This leads to the degradation of ZIF-8 and the controlled release of dexamethasone. Tannic acid reduces toxicity	Fluorescence imaging	4T1 tumor-bearing mice	[49]

(continued)

Table 8.1 (continued)

Nanotechnology	Type of nanoparticle	Combination therapy	Therapeutic modality	Diagnostic modality	In-vitro and in-vivo tumor model	Reference
			by forming coordinates with metal ions and improves biocompatibility			
Gold nanoparticles incorporated in MIL-101, decorated with GOx and lactate oxidase (LOx)	Metal-organic frameworks	SERS	Deplete glucose Oxidases catalyzed the conversion of glucose or lactate to H_2O_2. Peroxidase activity of gold nanoparticle utilized generated H_2O_2 to catalyze the conversion of inactive caged Raman reporters to active Raman reporters thereby enhancing Raman signal	SERS	CT-26 tumor-bearing mice	[48]
Gold nanoparticles assimilated in the amino-functionalized mesoporous sphere mobil composition of matter no. 41 (MCM-41)	Hollow mesoporous nanoparticles		GOx mimicking strategy resulting in starvation therapy			[47]
Self-assembly and crosslinking of haemoglobin and GOx into multimeric protein superstructure encapsulated with RBC cell membrane	Polymer-based nanoparticles		Deplete glucose RBC cell membrane increased systemic retention, reduced reticuloendothelial system (RES) uptake, and reduced immune evasion. Degradation of Hb releases	Fluorescence imaging	U87MG tumor-bearing mice	[46]

			free ferric ions (Fe^{3+}), catalyzing the Fenton reaction utilizing H_2O_2 to form highly toxic •OH. Meanwhile, GOx consumes glucose to form H_2O_2			
Biotin, cell-penetrating peptide R8, and polyethylene glycol (PEG) were assembled to form nanocarrier, GOx, and paclitaxel loaded in nanocarrier	Polymeric nanoparticles	Chemotherapy: paclitaxel	Deplete glucose GOx causes the degradation of tumoral glucose. This increase in acidity causes degradation of acid-sensitive hydrazone bond, thus causing enhanced penetration of nanocarrier in the tumor, thereby increasing the sensitivity of paclitaxel	Fluorescence imaging	4T1 tumor-bearing mice	[43]
GOx decorated on palladium-core platinum-shell (Pd@Pt) nanosheet, enveloped with hyaluronic acid	Pd@Pt based nanoparticles		Deplete glucose Hyaluronic acid targets tumor cells overexpressed with CD44. Hyaluronidase enzyme in the tumor cell degrades hyaluronic acid. GOx carries out oxygen-dependent depletion of glucose forming H_2O_2. The catalase-like activity of the nanoparticle converts generated H_2O_2 to oxygen which is later used in the GOx-based glucose depletion. Besides this, the	Fluorescence imaging	4T1 tumor-bearing mice	[42]

(continued)

Table 8.1 (continued)

Nanotechnology	Type of nanoparticle	Combination therapy	Therapeutic modality	Diagnostic modality	In-vitro and in-vivo tumor model	Reference
			peroxidase-like activity of nanoparticles converts H_2O_2 to highly toxic •OH radicals that further cause apoptosis			
GOx decorated on N-doped carbon nanoparticles	Carbon nanomaterials	Photothermal therapy: NIR 808 nm	Deplete glucose GOx depletes glucose. The peroxidase-like activity of carbon nanoparticles catalyzes H_2O_2 to highly toxic •OH radical Along with NIR-II irradiation, nanoparticles generate more heat and increase the sensitivity of irradiation toward tumor cells	Fluorescence imaging	4T1 tumor-bearing mice	[54]
GOx incorporated with bovine serum albumin, further coated with tannic acid	Albumin-based nanoparticles		Deplete glucose Deprivation of glucose was assessed by GOx. Tannic acid aids the conversion of Fe^{3+} to ferrous ion (Fe^{2+}), which further catalyzes Fenton reaction utilizing H_2O_2 and forming •OH radical	Fluorescence imaging	4T1 tumor-bearing mice	[59]
GOx and doxorubicin incorporated in hallow iron-tannic acid nanocapsules	Iron-based nanoparticles	Chemotherapy: doxorubicin	Deplete glucose Intracellular ATP-driven release of GOx and doxorubicin from hallow	Fluorescence imaging	4T1 tumor-bearing mice	[60]

			nanocapsules. Meanwhile, GOx depletes glucose forming H_2O_2 and decreases intracellular ATP production. Decreased ATP production inhibits p-glycoprotein efflux of doxorubicin. Further, iron ion catalyzes Fenton reaction with H_2O_2 and forms a highly toxic •OH radical			
Rattle-shaped nanoparticle constructed with polydopamine core and mesoporous silica shell. Chloroquine was incorporated into the shell and GOx was decorated on nanoparticle	Hollow mesoporous nanoparticles	Chemotherapy: chloroquine Photothermal therapy: NIR 808 nm	Deplete glucose Formulated nanoparticles cause deprivation of glucose and form gluconic acid and toxic H_2O_2. Glucose depletion suppresses heat shock protein expression thereby enhancing photothermal therapy of NIR laser irradiation. Along with NIR laser irradiation, chloroquine synchronously inhibits autophagy	Photoacoustic imaging Photothermal imaging	HepG-2 tumor-bearing mice	[63]
GOx coated on organosilica nanoparticles and further conjugated with MnO2 nanoparticles carrying photosensitizer chlorin e6	Hollow mesoporous nanoparticles	Photodynamic therapy: photosensitiser chlorin e6 and NIR 606 nm	Deplete glucose GOx consumes glucose-forming gluconic acid and toxic H_2O_2 and accelerates the decomposition of MnO_2 nanosheet, thus releasing photosensitizer chlorin e6. Besides being a nanocarrier,	Fluorescence imaging	HeLa cell	[62]

(continued)

Table 8.1 (continued)

Nanotechnology	Type of nanoparticle	Combination therapy	Therapeutic modality	Diagnostic modality	In-vitro and in-vivo tumor model	Reference
			MnO_2 parallelly utilizes H_2O_2-producing oxygen, which relieves the hypoxia tumor microenvironment and enhances starvation and photodynamic therapy			
Photosensitizer chlorin e6) and GOx conjugated with PEG functionalized cerium dioxide nanoparticle	Polymeric nanoparticles	Photodynamic therapy: photosensitiser chlorin e6 and NIR 660 nm	Deplete glucose GOx catalyzes glucose to H_2O_2. Being an enzyme mimic, cerium dioxide acts as superoxide dismutase (SOD) and catalase-like enzyme and generates a superoxide anion, which is further converted to oxygen in the presence of H_2O_2. Generated oxygen alleviates hypoxia tumor microenvironment and promotes starvation and photodynamic therapy	Fluorescence imaging	HeLa tumor-bearing mice	[61]
A ultra-small zero valent iron ion (Fe^{0}) nanoparticle incorporated with GOx further enveloped with RBC membrane	Iron-based nanoparticles	Photothermal therapy: NIR 808 nm	Deplete glucose RBC cell membrane endows tumor target site preferential accumulation. Following laser irradiation, the erythrocyte membrane ruptures to release GOx-Fe^{0}.	Fluorescence imaging	C6 tumor-bearing mice	[64]

			GOx depletes glucose to form gluconic acid and H_2O_2. Parallelly, Fe^0 ionizes to Fe^{2+}catalyzing Fenton reaction converting H_2O_2 into a highly toxic •OH radical			
ZIF-8 backbone enveloped with hyaluronic acid, conjugating with GOx, photosensitizer indocyanine green, and prodrug tilazamine	Metal-organic frameworks	Photodynamic therapy: photosensitiser indocyanine green and NIR 660 nm Chemotherapy: tilazamine	Deplete glucose GOx consumes glucose, forming gluconic acid and H_2O_2. Fe^{3+} ions catalyze the Fenton reaction decomposing H_2O_2 and generating highly toxic •OH radical. Laser irradiation along with photosensitizer indocyanine green works synergistically and enhances the production of singlet oxygen ($^1O_2$). Based on a series of reactions, the hypoxic favors the conversion of prodrug tilazamine into an active state	Photothermal imaging	4T1 tumor-bearing mice	[45]
GOx and prodrug tilazamine incorporated with polydopamine, then decorated on silver nanocubes, camouflaged with tumor cytomembrane	Silver nanoparticles	Chemotherapy: tilazamine	Deplete glucose Following tumor target accumulation, GOx catalyzes glucose conversion to gluconate and H_2O_2. Resulting in hypoxic microenvironment that promotes the conversion of the prodrug to the active form	Fluorescence imaging	U14 tumor-bearing mice	[44]

(continued)

Table 8.1 (continued)

Nanotechnology	Type of nanoparticle	Combination therapy	Therapeutic modality	Diagnostic modality	In-vitro and in-vivo tumor model	Reference
GOx, soya lecithin, paclitaxel, and MnO_2 conjugated and disguised with hyaluronic acid	Manganese-based nanoparticle	Chemotherapy: paclitaxel	Deplete glucose Nanoreactor accumulates in tumor targeting CD44, MnO_2 consumes GSH and GOx depletes glucose to H_2O_2. MnO_2 ionizes, catalase Fenton-like reaction forming highly toxic •OH radical, all this reaction promotes drug release	T2-weighted MRI	U87mg tumor-bearing mice	[65]

Table 8.2 Nanotheranostics targeting lactate metabolism

Nanotechnology	Type of nanoparticle	Combination therapy	Therapeutic modality	Diagnostic modality	In-vitro and in-vivo tumor model	Reference
Ferrocene and fluvastatin loaded in nanostructured lipid carriers, with surface modification with protein	Nanostructured lipid carrier		Inhibit lactate efflux Fluvastatin inhibits MCT-4, inhibiting lactate efflux. Meanwhile, ferrocene stimulates ROS production	Fluorescence imaging	ADC tumor-bearing mice	[77]
Hydroxycamptothecin and siRNA MCT-4 encapsulated in GSH-responsive organosilica mesoporous nanomaterials with PEG surface modification	Hollow mesoporous nanomaterial	Chemotherapy: hydroxycamptothecin	Inhibit lactate efflux High intracellular GSH and acidic levels cleave nanoparticles and release cargo. siRNA MCT-4 silenced MCT-4, and impaired lactate efflux, whereas hydroxycamptothecin induces apoptosis. Reduced lactate level polarized M2 to M1 tumor-associated macrophages and increased CD8+ T-cell expression	Fluorescence imaging	B16F10 and 4T1 tumor-bearing mice	[76]
GSK2837808A entrapped in the metal phenolic network, enveloped with hyaluronic acid	Metal phenolic network		Inhibit lactate dehydrogenase A Upon preferential accumulation directed by hyaluronic acid, degradation of the network occurs Released GSK2837808A inhibits lactate dehydrogenase	Fluorescence imaging	4T1 tumor-bearing mice	[75]

(continued)

Table 8.2 (continued)

Nanotechnology	Type of nanoparticle	Combination therapy	Therapeutic modality	Diagnostic modality	In-vitro and in-vivo tumor model	Reference
			A. Reduced lactate levels activate CD8+ T-cells and destabilize Treg			
Conjugating prodrug dichloroacetate in mitochondria targeting triphenylphosphonium-poly (lactic-co-glycolic acid) (TPP-PLGA) nanoparticle	Polymeric-based nanoparticle		Inhibit pyruvate dehydrogenase kinase 1 Upon mitochondrial targeting, the prodrug gets activated to mito-dichloroacetate and inhibited pyruvate dehydrogenase kinase As a result, reduced lactate level, stimulated expression of $CD8^+$ T cell, destabilized Treg, and downregulated PD-L1, thus elicits immune reaction	Fluorescence imaging	CT-26 tumor-bearing mice	[78]
GOx, LOx, and catalase in fluorinated polymer	Fluorinated-based nanoparticles		Deplete glucose and lactate LOx in the presence of oxygen catalyzed the conversion of lactate to pyruvate and H_2O_2. Thus, GOx and LOx deplete glucose and lactate. Further catalase convert generated H_2O_2 into oxygen, which further catalyzes GOx and LOx activity	Photoacoustic imaging	4T1 tumor-bearing mice	[74]

LOx and catalase were loaded into Fe_3O_4 nanoparticles and encapsulated with indocyanine green	Iron-based nanoparticles	Photothermal Therapy: NIR 808 nm	Deplete lactate LOx depletes lactate to pyruvate and H_2O_2, and catalase further converts H_2O_2 to oxygen Further, the Fe^{3+} ion carried out a Fenton reaction with H_2O_2 and produced highly toxic •OH radicals Meanwhile, laser irradiation produced $^1O_2$ that causes cancerous cell death	Fluorescence imaging	SMMC-7721 tumor-bearing mice	[73]
Doxorubicin encapsulated in MIL-101, along with surface medication with hyaluronic acid and live *Shewanella oneidensis*	Metal-organic framework	Chemotherapy: doxorubicin	Deplete lactate *S. oneidensis* oxidizes lactate, reduces iron ions, and leads to subsequent degradation of MIL-101, thereby releasing doxorubicin	Fluorescence imaging	4T1 tumor-bearing mice	[72]
Carbonic anhydrase IX inhibitor loaded in tin-selenide (SnSe) nanosheets	SnSe-based nanoparticle		Lactate dehydrogenase A mimicking-like activity SnSe nanosheet adsorbs lactate, and hydrogen atoms from lactate are transferred to Sn and Se, thereby converting pyruvate Carbonic anhydrase inhibition creates an acidic environment, that facilitates this reaction	Fluorescence imaging	4T1 tumor-bearing mice	[71]

(continued)

Table 8.2 (continued)

Nanotechnology	Type of nanoparticle	Combination therapy	Therapeutic modality	Diagnostic modality	In-vitro and in-vivo tumor model	Reference
SnSe nanoparticle	SnSe-based nanoparticle		Lactate dehydrogenase A mimicking-like activity Lactate adsorbs the SnSe nanosheet, and the H_2 atom from lactate is transferred to Sn and Se, thereby converting to pyruvate	Fluorescence imaging	CT-26 tumor-bearing mice	[70]

2.3 Amino Acid Metabolism

Amino acid, being a carbon and nitrogen source, plays a pivotal role in carcinogenesis. In a cancerous cell, the requirement rate exceeds the supply rate, metabolic enzymes are mutated and metabolic pathways are dysregulated, Hence, unlike normal cells, cancerous cell reprogram their amino acid metabolism and rely on nonessential amino acids rather than the essential amino acid, which are referred to as conditionally essential amino acid [79, 80]. Cancerous cells use conditionally essential amino acids (glutamine, and arginine), essential amino acids (branched-chain amino acids, and tryptophan), and non-essential amino acids (asparagine, and aspartate) to meet their energy and biosynthetic needs. Cancerous cells rely heavily on glutamine for energy and biosynthetic needs due to increasing demand; this dependence is known as glutamine addiction. ASCT2, a glutamine transporter, plays a crucial role in glutamine uptake; thus, its inhibition can effectively hinder glutamine uptake in malignant cells [81, 82]. Glutaminolysis converts glutamine to glutamic acid, which is then converted to α-ketoglutarate by glutamine dehydrogenase (GDH), resulting in ATP production. Thus, hindering the uptake of glutamine, metabolizing or degrading amino acid and inhibiting glutaminolysis can majorly impact the proliferation, invasion, and metastasis of cancerous cells. Specific malignant cells rely on specific amino acids like leukemia cell rely on asparagine, hence targeting asparagine metabolism with asparaginase could potentially treat leukemia. Nanoscale technology can limit the uptake of these amino acids and metabolize or decompose amino acids, thereby ensuring targeted delivery, minimizing off-target effects, and overcoming potential resistance [83–88] (Table 8.3).

2.4 Nucleotide Metabolism

Cancerous cells undergoing fast cell division require a significant supply of nucleotides for nuclei acid production. Nucleotides, the building blocks of nucleic acids, can be synthesized by salvage processes, which reuse pre-existing nucleosides and nucleobases, or through de novo synthesis pathways that use amino acids and small molecules. Cancerous cells, unlike normal cells, rewrite nucleotide metabolism, enhancing the salvage pathway and de novo nucleotide synthesis pathway to fulfill the increasing demand for nucleotides [89–91]. Antimetabolites, a purine and pyrimidine analog (methotrexate and 5-fluorouracil) inhibit nucleotide synthesis via a competitive mechanism, thereby halting nucleic acid synthesis. Encapsulating antimetabolites, within nanoparticles enhances their delivery to cancerous cells, effectively reducing their nucleotide synthesis and thus hindering tumor growth [92, 93]. Oligonucleotides including small interfering RNAs (siRNAs), antisense oligonucleotides (ASOs), and microRNAs (miRNAs), modulate target genes involved in nucleotide metabolism and inhibit cancer cell proliferation. However, short half-live, poor cellular uptake, endosome trap, clearance by the RES, and off-target outweigh their promising strategy for silencing oncogenes and modulating cancer-related pathways. Nanotechnology-based carriers can protect

Table 8.3 Nanotheranostics targeting amino acid metabolism

Nanotechnology	Type of nanotechnology	Combination therapy	Therapeutic modality	Diagnostic modality	In vitro and in vivo tumor model	Reference
Poly(ethylene glycol) monomethyl ether-b-polyglutamate-g-spermine carrying siRNA	Polymeric-based nanoparticles		Utilizing ASCT2 A549/DPP cells (cisplatin-resistant human lung cancerous cell) overexpress ASCT2. Polyglutamate-g-spermine, a glutamine analog competitively uses ASCT2 transporter to effectively deliver siRNA, thus reducing intracellular glutamine levels in cancerous cells and inhibiting tumorigenesis. Besides this, polyglutamate-g-spermine upregulated ASCT2	Micro-CT	A549/DPP cells	[83]
Hollow Fe^{3+}/tannic acid nanocapsules delivering amino acid oxidase, disguised with the cancerous cell membrane	Iron-based nanoparticles		Causing degradation of amino acid and chemodynamic therapy The cancerous cell membrane causes preferential accumulation of nanocapsules in the cancerous cell. Upon internalization, amino acid oxidases get delivered and catalyze oxidative deamination of amino acid, thereby reducing the intracellular level of amino acid. Increased intracellular acid and H_2O_2 facilitate the Fenton reaction and produce highly reactive •OH	Fluorescence imaging	4T1 tumor-bearing mice	[85]

Pt-Pd nanocarrier containing 6-diazo-5-oxo-L-norleucine	Pt-Pd-based nanoparticles	Electron dynamic therapy	Glutamine antagonist: 6-diazo-5-oxo-L-norleucine Besides reducing intracellular glutamine levels, upregulated $CD8^+$ T cells, thereby causing immune-based cell death	Fluorescence imaging	4T1 tumor-bearing mice	[84]
Ferrocene, and purpurin encapsulated in bovine serum albumin nanoparticles	Bovine serum albumin-based nanoparticles		Inhibit glutaminolysis and chemodynamic therapy Purpurin inhibits the GDH1 enzyme, leading to the inhibition of glutaminolysis and depriving energy sources from cancerous cells. Ferrocene promotes Fenton reaction via iron ion, producing highly reactive •OH	T2-weighted MRI	MCF-7 tumor-bearing mice	[88]
2-Deoxyglucose (2-DG), and V9302 coloaded in polymeric-based nanoparticles	Polymeric-based nanoparticles		Utilizing ASCT2 Synergistic effect via inhibition of glycolysis by 2-DG and V9302 mediated reduced intracellular glutamine level by competitively targeting ASCT2	Fluorescence imaging	4T1 tumor-bearing mice	[87]
Asparaginase shielded by urchin-like lysine conjugated polypeptide-based nanoparticle	Polymeric-based nanoparticles		Asparaginase Metabolizes asparagine and reduces intracellular asparagine in cancerous cells, thereby causing energy deprivation	Fluorescence imaging	NKYS tumor-bearing mice	[86]

oligonucleotides and enable targeted delivery to tumor cells [94]. Adenosine, a nucleoside produced from extracellular ATP, through the enzymatic action of ectonucleotidases such as CD39 or CD73, binds to four G protein-coupled receptors and promotes nucleotide synthesis. Hence it is required for rapid mitosis and even for creating an immunosuppressive tumor microenvironment by suppressing cytotoxic T-cell stimulation [95–97]. Nanotechnology-carrying cargo such as ectonucleotidases inhibitors, adenosine metabolizing enzymes, and adenosine receptor antagonists can downregulate adenosine production, resulting in decreased nucleic acid synthesis and stimulating cytotoxic T-cells [98–101] (Table 8.4).

2.5 Lipid Metabolism

Lipids including fatty acids, cholesterol, triglycerides, and phospholipids are integral parts of the cell membrane and rely on the intracellular signals as secondary messengers [102–106]. Moreover, under a limited nutrient supply, lipids are recognized as the source of energy [107, 108]. In normal cells, once the level of lipids reaches a certain level, the lipid synthesis process is inactivated by the inhibition of key regulators such as sterol regulatory element binding proteins (SREBP) in the endoplasmic reticulum [109–111]. Unlike normal cells, cancerous cells substantially reprogram lipid metabolism, activate SREBP, and stimulate lipid metabolism to meet their demand, thus supporting biosynthesis, membrane formation, and signaling pathways [112, 113]. Lipid synthesis occurs through key substrate acetyl coenzyme A (acetyl- CoA). Nanotechnology revolutionizes lipid metabolism-targeted cancer therapy and delivers targets key enzymes involved in acetyl-CoA formation, fatty acid biosynthesis, cholesterol biosynthesis, and fatty acid oxidation that can inhibit lipid metabolism, modulate lipid uptake, or induce ferroptosis for enhanced anti-cancer activity [114–123] (Table 8.5).

3 Challenges and Future Outlook

Understanding cancer metabolism is critical and complex, as targeting single-multiple metabolic molecules will determine an individual's life or death. However, the FDA has approved treatment for acute myeloid leukemia: Oncaspar® (pegaspargase). Oncaspar® is L-asparaginase covalently conjugated with monomethoxypolyethylene glycol (mPEG) that targets asparagine. This standard has piqued researchers' attention in addressing malignant metabolism. Nonetheless, tremendous progress in nanotechnology has been made, but there are some limitations and challenges that must be addressed.

- The first key challenges lie is monitoring specificity. Despite exploring metabolic variabilities in the cancerous, this explored targets also plays crucial role in regulating normal metabolism. As a result, the question arises as to what degree

Table 8.4 Nanotheranostics targeting nucleotide metabolism

Nanotechnology	Type of nanotechnology	Combination therapy	Therapeutic modality	Diagnostic modality	In vitro and in vivo tumor model	Reference
5-fluorouracil and zinc oxide formulated in nanocarrier ZIF-90	Metal-organic frameworks		Inhibit the synthesis of nucleotide Zinc oxide prevents premature leakage of 5-fluorouracil in a physiological environment. On the degradation of the metal-organic framework, produced Zn^{2+} also overcomes cancer resistance to 5-fluorouracil and works synergistically targeting cancerous cells	Fluorescence imaging	H22 tumor-bearing mice	[93]
Methotrexate decorated on the cerium(III)-doped lithium yttrium fluoride ($LiYF_4$: Ce^{3+}) nanoparticles surface through a photocleavable linker molecule	Lithium-based nanoparticles	Photodynamic therapy (X-ray: 0.1 Gy)	Inhibit the synthesis of nucleotide X-rays stimulate formulated nanoparticles, causing cleavage of the linker to release methotrexate	Fluorescence imaging	AsPC-1 and PANC-1 tumor-bearing mice	[92]
Sequential enveloped mesoporous silica nanoparticles with siRNA, followed by polydopamine, then removing sacrificial mesoporous silica nanoparticles core forming a	Hallow mesoporous nanomaterial		Oligonucleotide therapy Polydopamine recruits serum albumin, facilitating transendothelial transport, cellular uptake in cancerous cells, and effectively silencing gene expression of PD-L1	Fluorescence imaging	CT-26 tumor-bearing mice	[94]

(continued)

Table 8.4 (continued)

Nanotechnology	Type of nanotechnology	Combination therapy	Therapeutic modality	Diagnostic modality	In vitro and in vivo tumor model	Reference
noncationic hallow nanocapsule						
Phenyl boronic ester nanoparticles carrying an ectonucleotidase inhibitor ARL67156	Phenyl boronic ester-based nanoparticles	Photothermal therapy NIR 700 nm	Inhibition adenosine synthesis Upon near-infrared irradiation, ROS produced, causing degradation of phenyl boronic ester nanoparticles and releasing ectonucleotidase inhibitor ARL67156	Fluorescence imaging	MOC1 tumor-bearing mice	[100]
Anti-PD-L1 antibody, adenosine metabolizing enzyme: adenosine deaminase, sonosenstizer (hematoporphyrin), and bovine serum albumin, covalently immobilized into one nanoparticle	Bovine serum albumin-based nanoparticle	Sonodynamic therapy Ultrasound irradiation	Metabolizing adenosine In the physiological environment, the constructed nano-immunocomplex is inactivated Upon sono-irradiation in the acidic tumor environment, the nano-immuno complex gets activated Adenosine deaminase causes the conversion of adenosine to inosine, depletes intracellular adenosine levels, inhibits adenosine receptors, and reduces the	Fluorescence imaging	4T1 and CT26 tumor-bearing mice	[101]

			immunosuppressive environment of the tumor			
8-Cyclopentyl-1, 3-dipropylxanthine loaded in the PLGA core and anti-PD-L1 DNAzyme encapsulated with Fe^{3+}/Mn^{2+}-tannic acid coordination metal-organic frameworks shell	Hybrid nanostructure: polymeric-based nanoparticles metal-organic frameworks		Inhibiting adenosine receptors ADORA1 Through synergetic modes, Released 8-Cyclopentyl-1, 3-dipropylxanthine inhibit adenosine receptors, and released Mn^{2+} causes self-activation of anti-PD-L1 DNAzymes, and reduced expression of PD-L1 in the cancerous cell	Fluorescence imaging	B16F10 tumor-bearing mice	[99]
SCH442416 and aptamer AS1411 conjugated on phosphorous nanoparticle	Phosphorous-based nanoparticles	Photothermal therapy: NIR 808 nm	Inhibiting adenosine receptors A_{2A} Apatmer AS1411 promotes preferential accumulation in the cancerous cell overexpressed with nucleolin. Photothermal therapy ablates the tumor cell and also causes degradation of crosslinked collagen and cancer-associated fibroblast. Inhibiting adenosine receptors by SCH442416 causes maturation of dendritic cells, inhibition of Treg activity, and stimulation of the activity of cytotoxic T-cells in cancerous cells	Fluorescence imaging	B16F10 tumor-bearing mice	[98]

Table 8.5 Nanotheranostics targeting lipid metabolism

Nanotechnology	Type of nanotechnology	Combination therapy	Therapeutic modality	Diagnostic modality	In-vitro and in-vivo tumor model	Reference
Fenofibrate encapsulated in PLGA with surface modification anti-CD3 fragment	Polymeric-based nanoparticles		Fatty acid metabolism Fenofibrate activates peroxisome proliferator-activated receptor and gene of fatty acid metabolism	Fluorescence imaging	B16F10 tumor-bearing mice	[123]
Avasimibe entrapped in human serum albumin-based nanoparticles	Human serum albumin-based nanoparticles		Cholesterol metabolism Avasimibe inhibits acetyl-CoA acetyltransferase 1 (ACAT-1) and blocks cholesterol esterification, thus elevating free cholesterol and inducing apoptosis	Fluorescence imaging	PC3 and HCT116 tumor-bearing mice	[120]
Methyl-β-cyclodextrin conjugated to hyaluronic acid-ceramide nano-assembly	Ceramide-based nanoparticles		Disrupting lipid raft Methyl-β-cyclodextrin depletes cholesterol in lipid rafts in cancerous cells	Fluorescence imaging	MDA-MB-231 tumor-bearing mice	[121]
Simvastatin entrapped into polymer-coated magnetic nanoparticles	Polymeric iron-based nanoparticles		Ferroptosis Polymer-coated magnetic nanoparticles prolong the circulation duration of nanoparticles and increase preferential accumulation in cancerous cells. Simvastatin downregulates 3-hydroxy-3-methyl-glutaryl-coenzyme A reductase (HMGCR) and glutathione peroxide X4 (GPX4) expression, leading to excessive lipid peroxidation and induces ferroptosis-induced apoptosis	Fluorescence imaging	MDA-MB-231 tumor-bearing mice	[122]

Monoacylglycerol lipase (MGLL) siRNA and cannabinoid receptor (CB-2) siRNA conjugated with polydisulfide amide-based nanoplatforms	Amide-based nanoparticles		Catalysing lipid Released siRNA simultaneously silenced MGLL and CB-2, thus promoting lipid metabolism and activating repolarization of tumor-associated macrophages	Fluorescence imaging	LTPA tumor-bearing mice	[119]
Cholesterol oxidases (COD) adsorbed on a metal-organic framework containing photosensitizer	Metal-organic framework	Photothermal therapy NIR 660 nm	Cholesterol metabolism COD depletes cholesterol, and increases membrane tension and osmotic fragility, making it susceptible to oxidative stress. This enhanced the photodynamic effect	Fluorescence imaging	4T1 and MC38 tumor-bearing mice	[118]
COD and Doxorubicin conjugated with Cu^{2+}-modified zirconium-based metal-organic framework and disguised with cancerous cell membrane	Metal-organic framework	Chemotherapy: doxorubicin	Cholesterol metabolism Preferential accumulation in cancerous cells is achieved by cancerous cell membrane surface coating Further, COD depletes cholesterol from cancerous cell membranes, as a consequence the rigidity of resistant cancerous cell membranes decreases. This restores the activity and sensitivity of doxorubicin to cancerous cells	Fluorescence imaging	MCF-7/ADR tumor-bearing mice	[117]
COD and Liver X agonist entrapped in hydrogen-bonded organic framework, followed by rabies virus glycoprotein surface modification	Hydrogen-bonded organic framework		Cholesterol metabolism Followed by preferential accumulation in cancerous cell COD depletes cholesterol from cancerous cell membranes and helps in immune invasion	T2-weighted MRI	GL261 tumor-bearing mice	[114]

(continued)

Table 8.5 (continued)

Nanotechnology	Type of nanotechnology	Combination therapy	Therapeutic modality	Diagnostic modality	In-vitro and in-vivo tumor model	Reference
Folate, cisplatin, and siRNA GPX-4 entrapped in iron oxide-based nanoparticles	Iron-based nanoparticles	Chemotherapy: cisplatin	Ferroptosis On internalization in cancerous cells, formulated nanoparticles released Fe^{2+} and Fe^{3+} ions. While co-delivery of siRNA GPX-4 silenced GPX-4 and cisplatin caused nuclear and mitochondrial DNA damage. Iron oxide nanoparticles increased H_2O_2 levels by activating NADPH oxidase. Fe^{2+} and Fe^{3+} ions react with increased levels of H_2O_2 and generate highly toxic ROS	Fluorescence imaging	U87MG tumor-bearing mice	[116]
Prodrug cisplatin loaded in manganese-deposited iron oxide nanoparticles	Iron-based nanoparticles	Chemotherapy: cisplatin	Ferroptosis Formed nanoparticles release manganese ions, iron ions, and prodrug cisplatin. Activated cisplatin triggered DNA damage and increased H_2O_2 levels. Released manganese ion catalyzed Fenton reaction with this increased H_2O_2, thus favoring ferroptosis	Effect contrast ability Prominent T2-weighted MRI	HeLa tumor-bearing mice	[115]

of therapeutic index should be used, and whether it should be extended or narrowed based on the severity of their targeting of a normal metabolism. Parallelly, cancerous metabolism is directly linked to the tumor microenvironment. In this perspective, researchers should investigate characteristics of the tumor microenvironment. One goal should be to target cancer metabolism along with simultaneously reducing tumor proliferating cells and replenishing tumor suppressor cells. Meanwhile, cancer metabolism and nutrition metabolism are inextricably linked, thus metabolic disorders like hyperglycemia and hyperlipidemia might have an impact on cancer metabolism, making it more complex. This impact highlights the need for personalised therapies and the use of theranostic strategies to assess real-time monitoring and biodistribution.

- The second major challenge is heterogeneity. The phenotype of tumor microenvironments is continually evolving, emphasizing their metabolic heterogeneity. Aside from that, tumors acquired from various sources exhibit varying metabolic expression. Even subtypes of the same tumor exhibit varying metabolism. All of the foregoing, including various regions of malignant cells, exhibit metabolic variability. Above all, malignant cells will modify their metabolic process based on the macromolecule's concentration. As a result, nanotechnology, which regulates multitude metabolic variables, or stimuli-responsive nanostructures can address this issue. In accordance to this viewpoint, adopting effective combination therapies such as chemotherapy, photothermal therapy, chemodynamic therapy, and sonodynamic therapy can tackle this problem.
- The third challenge is the nanoparticles' biosafety and long-term safety. Nonetheless, nanoparticles employed in nanotechnology have been shown to be safe in preclinical trials, although the translation to clinical studies remains unclear. To hasten this translation, biosafety of nanoparticles should be investigated in larger animals. The majority of the nanoparticles employed are inorganic (Au, Ag, Mn, Fe, Si), and their biodegradation and bioaccumulation is still a challenge. Thus, the safety of nanoparticles should not be overlooked. Optimizing material composition, structural, physical, and chemical parameters, offers minimal toxicity with targeted drug delivery. Moreover, despite promising preclinical results, regulatory approval also remains a barrier to widespread clinical application.

4 Summary and Conclusion

Recent innovations and studies demonstrating the efficacy of cancer metabolism-targeted nanotheranostics are highlighted, underscoring their potential to overcome the limitations of conventional treatments. This emerging approach holds significant potential for transforming cancer management through simultaneous diagnosis and targeted metabolic intervention. The chapter concludes that exploiting metabolic vulnerabilities through nanotheranostics holds immense potential in revolutionizing cancer therapeutics offers a promising path for improved therapeutic outcomes and reduced adverse effects. However, overcoming biological, regulatory, and translational challenges is critical for its successful integration into clinical oncology.

Continued advancements in nanotechnology, imaging, and metabolic research will pave the way for more effective, personalized, and minimally invasive cancer treatments in the future.

References

1. Finley, L. W. S. (2023). What is cancer metabolism? *Cell, 186*(8), 1670–1688. https://doi.org/10.1016/j.cell.2023.01.038
2. Pavlova, N. N., Zhu, J., & Thompson, C. B. (2022). The hallmarks of cancer metabolism: Still emerging. *Cell Metabolism, 34*(3), 355–377. https://doi.org/10.1016/j.cmet.2022.01.007
3. DeBerardinis, R. J., & Chandel, N. S. (2016). Fundamentals of cancer metabolism. *Science Advances, 2*(5), e1600200. https://doi.org/10.1126/sciadv.1600200
4. Vander Heiden, M. G. (2011). Targeting cancer metabolism: A therapeutic window opens. *Nature Reviews. Drug Discovery, 10*(9), 671–684. https://doi.org/10.1038/nrd3504
5. Hanahan, D. (2022). Hallmarks of cancer: New dimensions. *Cancer Discovery, 12*(1), 31–46. https://doi.org/10.1158/2159-8290.CD-21-1059
6. Liberti, M. V., & Locasale, J. W. (2016). The Warburg effect: How does it benefit cancer cells? *Trends in Biochemical Sciences, 41*(3), 211–218. https://doi.org/10.1016/j.tibs.2015.12.001
7. Groves, A. M., Win, T., Haim, S. B., & Ell, P. J. (2007). Non-[18F]FDG PET in clinical oncology. *The Lancet. Oncology, 8*(9), 822–830. https://doi.org/10.1016/S1470-2045(07)70274-7
8. Le, A., Lane, A. N., Hamaker, M., Bose, S., Gouw, A., Barbi, J., & Dang, C. V. (2012). Glucose-independent glutamine metabolism via TCA cycling for proliferation and survival in B cells. *Cell Metabolism, 15*(1), 110–121. https://doi.org/10.1016/j.cmet.2011.12.009
9. Elia, I., & Haigis, M. C. (2021). Metabolites and the tumour microenvironment: from cellular mechanisms to systemic metabolism. *Nature Metabolism, 3*(1), 21–32. https://doi.org/10.1038/s42255-020-00317-z
10. Brannon-Peppas, L., & Blanchette, J. O. (2004). Nanoparticle and targeted systems for cancer therapy. *Advanced Drug Delivery Reviews, 56*(11), 1649–1659. https://doi.org/10.1016/j.addr.2004.02.014
11. Alexis, F., Pridgen, E. M., Langer, R., & Farokhzad, O. C. (2010). Nanoparticle technologies for cancer therapy. In *Handbook of experimental pharmacology* (Vol. 197, pp. 55–86). https://doi.org/10.1007/978-3-642-00477-3_2
12. Pérez-Herrero, E., & Fernández-Medarde, A. (2015). Advanced targeted therapies in cancer: Drug nanocarriers, the future of chemotherapy. *European Journal of Pharmaceutics and Biopharmaceutics: Official Journal of Arbeitsgemeinschaft für Pharmazeutische Verfahrenstechnik E.V., 93*, 52–79. https://doi.org/10.1016/j.ejpb.2015.03.018
13. Ren, M., Zheng, X., Gao, H., Jiang, A., Yao, Y., & He, W. (2022). Nanomedicines targeting metabolism in the tumor microenvironment. *Frontiers in Bioengineering and Biotechnology, 10*. https://doi.org/10.3389/fbioe.2022.943906
14. Yang, J., Zhao, Y., Zhou, Y., Wei, X., Wang, H., Si, N., & Zhao, H. (2022). Advanced nanomedicines for the regulation of cancer metabolism. *Biomaterials, 286*, 121565. https://doi.org/10.1016/j.biomaterials.2022.121565
15. Dong, X., Xia, S., Du, S., Zhu, M.-H., Lai, X., Yao, S. Q., & Fang, C. (2023). Tumor metabolism-rewriting nanomedicines for cancer immunotherapy. *ACS Central Science, 9*(10), 1864–1893. https://doi.org/10.1021/acscentsci.3c00702
16. Wang, B., Hu, S., Teng, Y., Chen, J., Wang, H., Xu, Y., & Gao, X. (2024). Current advance of nanotechnology in diagnosis and treatment for malignant tumors. *Signal Transduction and Targeted Therapy, 9*(1), 200. https://doi.org/10.1038/s41392-024-01889-y

17. Lan, H., Jamil, M., Ke, G., & Dong, N. (2023). The role of nanoparticles and nanomaterials in cancer diagnosis and treatment: A comprehensive review. *American Journal of Cancer Research, 13*(12), 5751–5784.
18. Jin, C., Wang, K., Oppong-Gyebi, A., & Hu, J. (2020). Application of nanotechnology in cancer diagnosis and therapy – A mini-review. *International Journal of Medical Sciences, 17*(18), 2964–2973. https://doi.org/10.7150/ijms.49801
19. Zhang, Z., Li, X., Yang, F., Chen, C., Liu, P., Ren, Y., & Li, X. (2021). DHHC9-mediated GLUT1 S-palmitoylation promotes glioblastoma glycolysis and tumorigenesis. *Nature Communications, 12*(1), 5872. https://doi.org/10.1038/s41467-021-26180-4
20. Zhao, F., Ming, J., Zhou, Y., & Fan, L. (2016). Inhibition of Glut1 by WZB117 sensitizes radioresistant breast cancer cells to irradiation. *Cancer Chemotherapy and Pharmacology, 77*(5), 963–972. https://doi.org/10.1007/s00280-016-3007-9
21. Siebeneicher, H., Cleve, A., Rehwinkel, H., Neuhaus, R., Heisler, I., Müller, T., & Buchmann, B. (2016). Identification and optimization of the first highly selective GLUT1 inhibitor BAY-876. *ChemMedChem, 11*(20), 2261–2271. https://doi.org/10.1002/cmdc.201600276
22. Kawatani, M., Aono, H., Hiranuma, S., Shimizu, T., Muroi, M., Ogawa, N., & Osada, H. (2021). Identification of a small-molecule glucose transporter inhibitor, Glutipyran, that inhibits cancer cell growth. *ACS Chemical Biology, 16*(8), 1576–1586. https://doi.org/10.1021/acschembio.1c00480
23. Xu, C.-F., Liu, Y., Shen, S., Zhu, Y.-H., & Wang, J. (2015). Targeting glucose uptake with siRNA-based nanomedicine for cancer therapy. *Biomaterials, 51*, 1–11. https://doi.org/10.1016/j.biomaterials.2015.01.068
24. Wang, J.-W., Chen, Q.-W., Luo, G.-F., Ji, P., Han, Z.-Y., Song, W.-F., & Zhang, X.-Z. (2022). Interference of glucose bioavailability of tumor by engineered biohybrids for potentiating targeting and uptake of antitumor nanodrugs. *Nano Letters, 22*(21), 8735–8743. https://doi.org/10.1021/acs.nanolett.2c03608
25. Wu, S., Zhang, K., Liang, Y., Wei, Y., An, J., Wang, Y., & Shi, J. (2022). Nano-enabled tumor systematic energy exhaustion via Zinc (II) interference mediated glycolysis inhibition and specific GLUT1 depletion. *Advanced Science, 9*(7), 2103534. https://doi.org/10.1002/advs.202103534
26. Wang, X., Ding, B., Liu, W., Qi, L., Li, J., Zheng, X., & Ma, P. (2024). Dual starvations induce pyroptosis for orthotopic pancreatic cancer therapy through simultaneous deprivation of glucose and glutamine. *Journal Of The American Chemical Society, 146*(26), 17854–17865. https://doi.org/10.1021/jacs.4c03478
27. Chen, W.-H., Luo, G.-F., Lei, Q., Hong, S., Qiu, W.-X., Liu, L.-H., & Zhang, X.-Z. (2017). Overcoming the heat endurance of tumor cells by interfering with the anaerobic glycolysis metabolism for improved photothermal therapy. *ACS Nano, 11*(2), 1419–1431. https://doi.org/10.1021/acsnano.6b06658
28. Jiang, W., Luo, X., Wei, L., Yuan, S., Cai, J., Jiang, X., & Hu, Y. (2021). The sustainability of energy conversion inhibition for tumor ferroptosis therapy and chemotherapy. *Small, 17*(38), 2102695. https://doi.org/10.1002/smll.202102695
29. DiTullio, D., & Dell'Angelica, E. C. (2019). Glucose metabolism. In *Fundamentals of biochemistry: Medical course. Step 1 review*. McGraw-Hill Education. Retrieved from accesspharmacy.mhmedical.com/content.aspx?aid=1159163353
30. Al Hasawi, N., Alkandari, M. F., & Luqmani, Y. A. (2014). Phosphofructokinase: A mediator of glycolytic flux in cancer progression. *Critical Reviews in Oncology/Hematology, 92*(3), 312–321. https://doi.org/10.1016/j.critrevonc.2014.05.007
31. Yang, B., Chen, Y., & Shi, J. (2020). Tumor-specific chemotherapy by nanomedicine-enabled differential stress sensitization. *Angewandte Chemie International Edition, 59*(24), 9693–9701. https://doi.org/10.1002/anie.202002306
32. Xu, Y., Guo, Y., Chen, L., Ni, D., Hu, P., & Shi, J. (2021). Tumor chemical suffocation therapy by dual respiratory inhibitions. *Chemical Science, 12*(22), 7763–7769. https://doi.org/10.1039/D1SC00929J

33. Luo, Y., Li, Y., Huang, Z., Li, X., Wang, Y., Hou, J., & Zhou, S. (2022). A nanounit strategy disrupts energy metabolism and alleviates immunosuppression for cancer therapy. *Nano Letters, 22*(15), 6418–6427. https://doi.org/10.1021/acs.nanolett.2c02475
34. Zhao, Q., Li, J., Wu, B., Shang, Y., Huang, X., Dong, H., & Nie, X. (2020). Smart biomimetic nanocomposites mediate mitochondrial outcome through aerobic glycolysis reprogramming: A promising treatment for lymphoma. *ACS Applied Materials & Interfaces, 12*(20), 22687–22701. https://doi.org/10.1021/acsami.0c05763
35. Zang, S., Huang, K., Li, J., Ren, K., Li, T., He, X., & He, Q. (2022). Metabolic reprogramming by dual-targeting biomimetic nanoparticles for enhanced tumor chemo-immunotherapy. *Acta Biomaterialia, 148*, 181–193. https://doi.org/10.1016/j.actbio.2022.05.045
36. Jia, L., Gao, Y., Zhou, T., Zhao, X.-L., Hu, H.-Y., Chen, D.-W., & Qiao, M.-X. (2021). Enhanced response to PD-L1 silencing by modulation of TME via balancing glucose metabolism and robust co-delivery of siRNA/resveratrol with dual-responsive polyplexes. *Biomaterials, 271*, 120711. https://doi.org/10.1016/j.biomaterials.2021.120711
37. Sasaki, K., Nishina, S., Yamauchi, A., Fukuda, K., Hara, Y., Yamamura, M., & Hino, K. (2021). Nanoparticle-mediated delivery of 2-deoxy-D-glucose induces antitumor immunity and cytotoxicity in liver tumors in mice. *Cellular and Molecular Gastroenterology and Hepatology, 11*(3), 739–762. https://doi.org/10.1016/j.jcmgh.2020.10.010
38. Liu, X., Li, Y., Wang, K., Chen, Y., Shi, M., Zhang, X., & Tang, B. (2021). GSH-responsive nanoprodrug to inhibit glycolysis and alleviate immunosuppression for cancer therapy. *Nano Letters, 21*(18), 7862–7869. https://doi.org/10.1021/acs.nanolett.1c03089
39. Li, S., Wang, Q., Jia, Z., Da, M., Zhao, J., Yang, R., & Chen, D. (2023). Recent advances in glucose oxidase-based nanocarriers for tumor targeting therapy. *Heliyon, 9*(10), e20407. https://doi.org/10.1016/j.heliyon.2023.e20407
40. Fu, Y., Sun, J., Wang, Y., & Li, W. (2023). Glucose oxidase and metal catalysts combined tumor synergistic therapy: Mechanism, advance and nanodelivery system. *Journal of Nanobiotechnology, 21*(1), 400. https://doi.org/10.1186/s12951-023-02158-w
41. Fu, L.-H., Qi, C., Lin, J., & Huang, P. (2018). Catalytic chemistry of glucose oxidase in cancer diagnosis and treatment. *Chemical Society Reviews, 47*(17), 6454–6472. https://doi.org/10.1039/c7cs00891k
42. Ming, J., Zhu, T., Yang, W., Shi, Y., Huang, D., Li, J., & Zheng, N. (2020). Pd@Pt-GOx/HA as a novel enzymatic Cascade nanoreactor for high-efficiency starving-enhanced chemodynamic cancer therapy. *ACS Applied Materials & Interfaces, 12*(46), 51249–51262. https://doi.org/10.1021/acsami.0c15211
43. Lu, R., Zhou, L., Liu, Q., Wang, S., Yang, C., Hai, L., & Wu, Y. (2021). Skillfully collaborating chemosynthesis with GOx-enabled tumor survival microenvironment deteriorating strategy for amplified chemotherapy and enhanced tumor ablation. *Biomaterials Science, 9*(5), 1855–1871. https://doi.org/10.1039/D0BM01950J
44. Hao, Z., Cheng, X., Cong, C., Zhang, X., Zhang, W., Zhao, Q., & Gao, D. (2021). Nanoreactor of "butterfly effect" inciting a triple interlocked combination of starvation/chemo/metal ion therapy by remodeling tumor microenvironment. *Chemical Engineering Journal, 405*, 126571. https://doi.org/10.1016/j.cej.2020.126571
45. Li, G., Zhang, J., Zhang, S., Teng, L., & Sun, F. (2023). Multifunctional nanoadjuvant-driven microenvironment modulation for enhanced photothermal immunotherapy in breast cancer. *Journal of Controlled Release, 362*, 309–324. https://doi.org/10.1016/j.jconrel.2023.08.046
46. Zheng, T., Wang, W., Ashley, J., Zhang, M., Feng, X., Shen, J., & Sun, Y. (2020). Self-assembly protein superstructures as a powerful chemodynamic therapy nanoagent for glioblastoma treatment. *Nano-Micro Letters, 12*(1), 151. https://doi.org/10.1007/s40820-020-00490-6
47. Ortega-Liebana, M. C., Bonet-Aleta, J., Hueso, J. L., & Santamaria, J. (2020). Gold-based nanoparticles on amino-functionalized mesoporous silica supports as nanozymes for glucose oxidation. *Catalysts, 10*(3). https://doi.org/10.3390/catal10030333

48. Hu, Y., Cheng, H., Zhao, X., Wu, J., Muhammad, F., Lin, S., & Wei, H. (2017). Surface-enhanced Raman scattering active gold nanoparticles with enzyme-mimicking activities for measuring glucose and lactate in living tissues. *ACS Nano, 11*(6), 5558–5566. https://doi.org/10.1021/acsnano.7b00905
49. Geng, S., Lou, R., Yin, Q., Li, S., Yang, R., & Zhou, J. (2021). Reshaping the tumor microenvironment for increasing the distribution of glucose oxidase in tumor and inhibiting metastasis. *Journal of Materials Chemistry B, 9*(5), 1424–1431. https://doi.org/10.1039/D0TB02468F
50. Hu, C., Wang, J., Liu, S., Cai, L., Zhou, Y., Liu, X., & Pang, M. (2021). Urchin-shaped metal organic/hydrogen-bonded framework nanocomposite as a multifunctional nanoreactor for catalysis-enhanced synergetic therapy. *ACS Applied Materials & Interfaces, 13*(4), 4825–4834. https://doi.org/10.1021/acsami.0c19584
51. Hao, Y., Liu, Y., Wu, Y., Tao, N., Lou, D., Li, J., & Liu, Y.-N. (2020). A robust hybrid nanozyme@hydrogel platform as a biomimetic cascade bioreactor for combination antitumor therapy. *Biomaterials Science, 8*(7), 1830–1839. https://doi.org/10.1039/C9BM01837A
52. Wang, Y., & Song, M. (2020). pH-responsive cascaded nanocatalyst for synergistic like-starvation and chemodynamic therapy. *Colloids and Surfaces B: Biointerfaces, 192*, 111029. https://doi.org/10.1016/j.colsurfb.2020.111029
53. Cheng, H., Jiang, X.-Y., Zheng, R.-R., Zuo, S.-J., Zhao, L.-P., Fan, G.-L., & Zhang, X.-Z. (2019). A biomimetic cascade nanoreactor for tumor targeted starvation therapy-amplified chemotherapy. *Biomaterials, 195*, 75–85. https://doi.org/10.1016/j.biomaterials.2019.01.003
54. Xu, K., Wu, X., Cheng, Y., Yan, J., Feng, Y., Chen, R., & Zhang, H. (2020). A biomimetic nanoenzyme for starvation therapy enhanced photothermal and chemodynamic tumor therapy. *Nanoscale, 12*(45), 23159–23165. https://doi.org/10.1039/D0NR05097K
55. Zhang, Y.-H., Qiu, W.-X., Zhang, M., Zhang, L., & Zhang, X.-Z. (2018). MnO2 motor: A prospective cancer-starving therapy promoter. *ACS Applied Materials & Interfaces, 10*(17), 15030–15039. https://doi.org/10.1021/acsami.8b01818
56. Wang, Y., Wang, B., Zhang, L., Huang, J., Li, P., Zhao, Y., & He, J. (2020). Mitochondria-targeted nanospheres with deep tumor penetration for photo/starvation therapy. *Journal of Materials Chemistry B, 8*(34), 7740–7754. https://doi.org/10.1039/D0TB00001A
57. Xu, M., Lu, Q., Song, Y., Yang, L., Ren, C., Li, W., & Li, N. (2020). NIR-II driven plasmon-enhanced cascade reaction for tumor microenvironment-regulated catalytic therapy based on bio-breakable Au–Ag nanozyme. *Nano Research, 13*(8), 2118–2129. https://doi.org/10.1007/s12274-020-2818-5
58. Huo, M., Wang, L., Chen, Y., & Shi, J. (2017). Tumor-selective catalytic nanomedicine by nanocatalyst delivery. *Nature Communications, 8*(1), 357. https://doi.org/10.1038/s41467-017-00424-8
59. Deng, F., Fan, G., Yuan, P., Liu, Y., Huang, C., Zhao, L., & Li, S. (2020). A self-accelerated biocatalyst for glucose-initiated tumor starvation and chemodynamic therapy. *Chemical Communications, 56*(93), 14633–14636. https://doi.org/10.1039/D0CC06483A
60. Zhu, H., Cao, G., Fu, Y., Fang, C., Chu, Q., Li, X., & Han, G. (2021). ATP-responsive hollow nanocapsules for DOX/GOx delivery to enable tumor inhibition with suppressed P-glycoprotein. *Nano Research, 14*(1), 222–231. https://doi.org/10.1007/s12274-020-3071-7
61. Liu, X., Liu, J., Chen, S., Xie, Y., Fan, Q., Zhou, J., & Dai, Z. (2020). Dual-path modulation of hydrogen peroxide to ameliorate hypoxia for enhancing photodynamic/starvation synergistic therapy. *Journal of Materials Chemistry B, 8*(43), 9933–9942. https://doi.org/10.1039/D0TB01556C
62. Yang, C., Liu, Y., Su, S., Gao, N., Jing, J., & Zhang, X. (2020). A multifunctional oxygen-producing MnO2-based nanoplatform for tumor microenvironment-activated imaging and combination therapy in vitro. *Journal of Materials Chemistry B, 8*(43), 9943–9950. https://doi.org/10.1039/D0TB00529K
63. Shao, L., Li, Y., Huang, F., Wang, X., Lu, J., Jia, F., & Wu, Y. (2020). Complementary autophagy inhibition and glucose metabolism with rattle-structured

polydopamine@mesoporous silica nanoparticles for augmented low-temperature photothermal therapy and in vivo photoacoustic imaging. *Theranostics, 10*(16), 7273–7286. https://doi.org/10.7150/thno.44668
64. Liu, W., Ruan, M.-L., Liu, L., Ji, X., Ma, Y., Yuan, P., & Xue, W. (2020). Self-activated *in vivo* therapeutic cascade of erythrocyte membrane-cloaked iron-mineralized enzymes. *Theranostics, 10*, 2201–2214. https://doi.org/10.7150/thno.39621
65. Li, X., Zhang, X., Zhang, W., Li, L., Gao, W., Zhang, X., & Gao, D. (2020). Biocatalysis of MnO2-mediated nanosystem for enhanced multimodal therapy and real-time tracking. *ACS Sustainable Chemistry & Engineering, 8*(35), 13206–13214. https://doi.org/10.1021/acssuschemeng.0c02367
66. Payen, V. L., Mina, E., Van Hée, V. F., Porporato, P. E., & Sonveaux, P. (2020). Monocarboxylate transporters in cancer. *Molecular Metabolism, 33*, 48–66. https://doi.org/10.1016/j.molmet.2019.07.006
67. Singh, M., Afonso, J., Sharma, D., Gupta, R., Kumar, V., Rani, R., & Kumar, V. (2023). Targeting monocarboxylate transporters (MCTs) in cancer: How close are we to the clinics? *Seminars in Cancer Biology, 90*, 1–14. https://doi.org/10.1016/j.semcancer.2023.01.007
68. Sharma, D., Singh, M., & Rani, R. (2022). Role of LDH in tumor glycolysis: Regulation of LDHA by small molecules for cancer therapeutics. *Seminars in Cancer Biology, 87*, 184–195. https://doi.org/10.1016/j.semcancer.2022.11.007
69. Feng, Y., Xiong, Y., Qiao, T., Li, X., Jia, L., & Han, Y. (2018). Lactate dehydrogenase A: A key player in carcinogenesis and potential target in cancer therapy. *Cancer Medicine, 7*(12), 6124–6136. https://doi.org/10.1002/cam4.1820
70. Wang, H., Wang, B., Jiang, J., Wu, Y., Song, A., Wang, X., & Wang, C. (2022). SnSe nanosheets mimic lactate dehydrogenase to reverse tumor acid microenvironment metabolism for enhancement of tumor therapy. *Molecules (Basel, Switzerland), 27*(23). https://doi.org/10.3390/molecules27238552
71. Ling, J., Chang, Y., Yuan, Z., Chen, Q., He, L., & Chen, T. (2022). Designing lactate dehydrogenase-mimicking SnSe nanosheets to reprogram tumor-associated macrophages for potentiation of photothermal immunotherapy. *ACS Applied Materials & Interfaces, 14*(24), 27651–27665. https://doi.org/10.1021/acsami.2c05533
72. Wang, J.-W., Chen, Q.-W., Luo, G.-F., Han, Z.-Y., Song, W.-F., Yang, J., & Zhang, X.-Z. (2021). A self-driven bioreactor based on bacterium–metal–organic framework biohybrids for boosting chemotherapy via cyclic lactate catabolism. *ACS Nano, 15*(11), 17870–17884. https://doi.org/10.1021/acsnano.1c06123
73. Qin, X., Wu, C., Niu, D., Qin, L., Wang, X., Wang, Q., & Li, Y. (2021). Peroxisome inspired hybrid enzyme nanogels for chemodynamic and photodynamic therapy. *Nature Communications, 12*(1), 5243. https://doi.org/10.1038/s41467-021-25561-z
74. Xiao, Y.-P., Chen, P.-H., Lei, S., Bai, F., Fu, L.-H., Lin, J., & Huang, P. (2022). Biocatalytic depletion of tumorigenic energy sources driven by Cascade reactions for efficient antitumor therapy. *Angewandte Chemie (International ed. in English), 61*(42), e202204584. https://doi.org/10.1002/anie.202204584
75. Yan, J., Li, W., Tian, H., Li, B., Yu, X., Wang, G., & Dai, Y. (2023). Metal-phenolic nanomedicines regulate T-cell antitumor function for sono-metabolic cancer therapy. *ACS Nano, 17*(15), 14667–14677. https://doi.org/10.1021/acsnano.3c02428
76. Li, K., Lin, C., He, Y., Lu, L., Xu, K., Tao, B., & Cai, K. (2020). Engineering of Cascade-responsive nanoplatform to inhibit lactate efflux for enhanced tumor chemo-immunotherapy. *ACS Nano, 14*(10), 14164–14180. https://doi.org/10.1021/acsnano.0c07071
77. Wang, W., Fu, F., Huang, Z., Wang, W., Chen, M., Yue, X., & Pan, X. (2022). Inhalable biomimetic protein Corona-mediated nanoreactor for self-amplified lung adenocarcinoma ferroptosis therapy. *ACS Nano, 16*(5), 8370–8387. https://doi.org/10.1021/acsnano.2c02634
78. Kolb, D., Kolishetti, N., Surnar, B., Sarkar, S., Guin, S., Shah, A. S., & Dhar, S. (2020). Metabolic modulation of the tumor microenvironment leads to multiple checkpoint inhibition

and immune cell infiltration. *ACS Nano, 14*(9), 11055–11066. https://doi.org/10.1021/acsnano.9b10037

79. Butler, M., van der Meer, L. T., & van Leeuwen, F. N. (2021). Amino acid depletion therapies: Starving cancer cells to death. *Trends in Endocrinology & Metabolism, 32*(6), 367–381. https://doi.org/10.1016/j.tem.2021.03.003
80. Wei, Z., Liu, X., Cheng, C., Yu, W., & Yi, P. (2020). Metabolism of amino acids in cancer. *Frontiers in Cell and Developmental Biology, 8*, 603837. https://doi.org/10.3389/fcell.2020.603837
81. Morita, M., Kudo, K., Shima, H., & Tanuma, N. (2021). Dietary intervention as a therapeutic for cancer. *Cancer Science, 112*(2), 498–504. https://doi.org/10.1111/cas.14777
82. Chen, J., Cui, L., Lu, S., & Xu, S. (2024). Amino acid metabolism in tumor biology and therapy. *Cell Death & Disease, 15*(1), 42. https://doi.org/10.1038/s41419-024-06435-w
83. Wang, C., Wu, J., Wang, Z., Yang, Z., Li, Z., Deng, H., & Feng, M. (2018). Glutamine addiction activates polyglutamine-based nanocarriers delivering therapeutic siRNAs to orthotopic lung tumor mediated by glutamine transporter SLC1A5. *Biomaterials, 183*, 77–92. https://doi.org/10.1016/j.biomaterials.2018.08.035
84. Chen, G., Xu, Q., Feng, Z., Xu, Q., Zhang, X., Yang, Y., & Yu, M. (2022). Glutamine antagonist synergizes with electrodynamic therapy to induce tumor regression and systemic antitumor immunity. *ACS Nano, 16*(1), 951–962. https://doi.org/10.1021/acsnano.1c08544
85. Chu, Q., Zhu, H., Liu, B., Cao, G., Fang, C., Wu, Y., & Han, G. (2020). Delivery of amino acid oxidase via catalytic nanocapsules to enable effective tumor inhibition. *Journal of Materials Chemistry B, 8*(37), 8546–8557. https://doi.org/10.1039/D0TB01425G
86. Hu, Y., Wang, D., Wang, H., Zhao, R., Wang, Y., Shi, Y., & Lu, H. (2021). An urchin-like helical polypeptide-asparaginase conjugate with mitigated immunogenicity. *Biomaterials, 268*, 120606. https://doi.org/10.1016/j.biomaterials.2020.120606
87. Luo, Z., Xu, J., Sun, J., Huang, H., Zhang, Z., Ma, W., & Li, S. (2020). Co-delivery of 2-Deoxyglucose and a glutamine metabolism inhibitor V9302 via a prodrug micellar formulation for synergistic targeting of metabolism in cancer. *Acta Biomaterialia, 105*, 239–252. https://doi.org/10.1016/j.actbio.2020.01.019
88. Xu, L., Xu, R., Saw, P. E., Wu, J., Cheng, S.-X., & Xu, X. (2021). Nanoparticle-mediated inhibition of mitochondrial glutaminolysis to amplify oxidative stress for combination cancer therapy. *Nano Letters, 21*(18), 7569–7578. https://doi.org/10.1021/acs.nanolett.1c02073
89. Vander Heiden, M. G., & DeBerardinis, R. J. (2017). Understanding the intersections between metabolism and cancer biology. *Cell, 168*(4), 657–669.
90. Villa, E., Ali, E. S., Sahu, U., & Ben-Sahra, I. (2019). Cancer cells tune the signaling pathways to empower de novo synthesis of nucleotides. *Cancers, 11*(5). https://doi.org/10.3390/cancers11050688
91. Tong, X., Zhao, F., & Thompson, C. B. (2009). The molecular determinants of de novo nucleotide biosynthesis in cancer cells. *Current Opinion in Genetics & Development, 19*(1), 32–37.
92. Manoharan, D., Chang, L.-C., Wang, L.-C., Shan, Y.-S., Lin, F.-C., Wu, L.-C., & Yeh, C.-S. (2021). Synchronization of nanoparticle sensitization and radiosensitizing chemotherapy through cell cycle arrest achieving ultralow X-ray dose delivery to pancreatic tumors. *ACS Nano, 15*(5), 9084–9100. https://doi.org/10.1021/acsnano.1c02283
93. Xiao, X., Liang, S., Zhao, Y., Huang, D., Xing, B., Cheng, Z., & Lin, J. (2020). Core-shell structured 5-FU@ZIF-90@ZnO as a biodegradable nanoplatform for synergistic cancer therapy. *Nanoscale, 12*(6), 3846–3854. https://doi.org/10.1039/c9nr09869k
94. Kim, H., Yuk, S. A., Dieterly, A. M., Kwon, S., Park, J., Meng, F., & Yeo, Y. (2021). Nanosac, a noncationic and soft polyphenol nanocapsule, enables systemic delivery of siRNA to solid tumors. *ACS Nano, 15*(3), 4576–4593. https://doi.org/10.1021/acsnano.0c08694
95. Bours, M. J. L., Swennen, E. L. R., Di Virgilio, F., Cronstein, B. N., & Dagnelie, P. C. (2006). Adenosine 5′-triphosphate and adenosine as endogenous signaling molecules in immunity and

inflammation. *Pharmacology & Therapeutics, 112*(2), 358–404. https://doi.org/10.1016/j.pharmthera.2005.04.013

96. Boison, D., & Yegutkin, G. G. (2019). Adenosine metabolism: Emerging concepts for cancer therapy. *Cancer Cell, 36*(6), 582–596. https://doi.org/10.1016/j.ccell.2019.10.007
97. Chiarella, A. M., Ryu, Y. K., Manji, G. A., & Rustgi, A. K. (2021). Extracellular ATP and adenosine in cancer pathogenesis and treatment. *Trends in Cancer, 7*(8), 731–750. https://doi.org/10.1016/j.trecan.2021.04.008
98. Zhao, Y., Xie, Z., Deng, Y., Huang, A., He, Y., Wen, B., & Liu, H. (2022). Photothermal nanobomb blocking metabolic adenosine-A2AR potentiates infiltration and activity of T cells for robust antitumor immunotherapy. *Chemical Engineering Journal, 450*, 138139. https://doi.org/10.1016/j.cej.2022.138139
99. Guo, J., Liu, P., Wei, B., Peng, Y., Ding, J., Zhang, H., & Chen, X. (2023). Reversing the negative effect of adenosine A1 receptor-targeted immunometabolism modulation on melanoma by a co-delivery nanomedicine for self-activation of anti-PD-L1 DNAzyme. *Nano Today, 48*, 101722. https://doi.org/10.1016/j.nantod.2022.101722
100. Mao, C., Yeh, S., Fu, J., Porosnicu, M., Thomas, A., Kucera, G. L., & Ming, X. (2025). Delivery of an ectonucleotidase inhibitor with ROS-responsive nanoparticles overcomes adenosine-mediated cancer immunosuppression. *Science Translational Medicine, 14*(648), eabh1261. https://doi.org/10.1126/scitranslmed.abh1261
101. Zhang, C., Huang, J., Zeng, Z., He, S., Cheng, P., Li, J., & Pu, K. (2022). Catalytical nano-immunocomplexes for remote-controlled sono-metabolic checkpoint trimodal cancer therapy. *Nature Communications, 13*(1), 3468. https://doi.org/10.1038/s41467-022-31044-6
102. Maxfield, F. R. (2002). Plasma membrane microdomains. *Current Opinion in Cell Biology, 14*(4), 483–487. https://doi.org/10.1016/s0955-0674(02)00351-4
103. Pomorski, T., Hrafnsdóttir, S., Devaux, P. F., & van Meer, G. (2001). Lipid distribution and transport across cellular membranes. *Seminars in Cell & Developmental Biology, 12*(2), 139–148. https://doi.org/10.1006/scdb.2000.0231
104. van Meer, G., Voelker, D. R., & Feigenson, G. W. (2008). Membrane lipids: Where they are and how they behave. *Nature Reviews. Molecular Cell Biology, 9*(2), 112–124. https://doi.org/10.1038/nrm2330
105. Mukherjee, S., & Maxfield, F. R. (2004). Membrane domains. *Annual Review of Cell and Developmental Biology, 20*, 839–866. https://doi.org/10.1146/annurev.cellbio.20.010403.095451
106. Holthuis, J. C. M., & Menon, A. K. (2014). Lipid landscapes and pipelines in membrane homeostasis. *Nature, 510*(7503), 48–57. https://doi.org/10.1038/nature13474
107. Zechner, R., Strauss, J. G., Haemmerle, G., Lass, A., & Zimmermann, R. (2005). Lipolysis: pathway under construction. *Current Opinion in Lipidology, 16*(3), 333–340. https://doi.org/10.1097/01.mol.0000169354.20395.1c
108. Efeyan, A., Comb, W. C., & Sabatini, D. M. (2015). Nutrient-sensing mechanisms and pathways. *Nature, 517*(7534), 302–310. https://doi.org/10.1038/nature14190
109. Shao, W., & Espenshade, P. J. (2012). Expanding roles for SREBP in metabolism. *Cell Metabolism, 16*(4), 414–419. https://doi.org/10.1016/j.cmet.2012.09.002
110. Jeon, T.-I., & Osborne, T. F. (2012). SREBPs: metabolic integrators in physiology and metabolism. *Trends in Endocrinology and Metabolism: TEM, 23*(2), 65–72. https://doi.org/10.1016/j.tem.2011.10.004
111. Guo, D., Prins, R. M., Dang, J., Kuga, D., Iwanami, A., Soto, H., & Mischel, P. S. (2009). EGFR signaling through an Akt-SREBP-1-dependent, rapamycin-resistant pathway sensitizes glioblastomas to antilipogenic therapy. *Science Signaling, 2*(101), ra82. https://doi.org/10.1126/scisignal.2000446
112. Menendez, J. A., & Lupu, R. (2007). Fatty acid synthase and the lipogenic phenotype in cancer pathogenesis. *Nature Reviews. Cancer, 7*(10), 763–777. https://doi.org/10.1038/nrc2222

113. Guo, D., Bell, E. H., & Chakravarti, A. (2013). Lipid metabolism emerges as a promising target for malignant glioma therapy. *CNS Oncology, 2*(3), 289–299. https://doi.org/10.2217/cns.13.20
114. Yin, N., Wang, Y., Liu, Y., Niu, R., Zhang, S., Cao, Y., & Zhang, H. (2023). A cholesterol metabolic regulated hydrogen-bonded organic framework (HOF)-based biotuner for antibody non-dependent immunotherapy tailored for glioblastoma. *Advanced Materials, 35*(44), 2303567. https://doi.org/10.1002/adma.202303567
115. Cheng, J., Zhu, Y., Xing, X., Xiao, J., Chen, H., Zhang, H., & Liu, Y. (2021). Manganese-deposited iron oxide promotes tumor-responsive ferroptosis that synergizes the apoptosis of cisplatin. *Theranostics, 11*(11), 5418–5429. https://doi.org/10.7150/thno.53346
116. Zhang, Y., Fu, X., Jia, J., Wikerholmen, T., Xi, K., Kong, Y., & Ni, S. (2020). Glioblastoma therapy using codelivery of cisplatin and glutathione peroxidase targeting siRNA from iron oxide nanoparticles. *ACS Applied Materials & Interfaces, 12*(39), 43408–43421. https://doi.org/10.1021/acsami.0c12042
117. Guo, J., Du, X., Huang, J., Liu, C., Zhou, Y., Li, Y., & Du, B. (2022). Robust dual enzyme Cascade-catalytic cholesterol depletion for reverse tumor multidrug resistance. *Advanced Healthcare Materials, 11*(19), 2200859. https://doi.org/10.1002/adhm.202200859
118. Zhen, W., Luo, T., Wang, Z., Jiang, X., Yuan, E., Weichselbaum, R. R., & Lin, W. (2023). Mechanoregulatory cholesterol oxidase-functionalized nanoscale metal–organic framework stimulates pyroptosis and reinvigorates T cells. *Small, 19*(52), 2305440. https://doi.org/10.1002/smll.202305440
119. Cao, S., Saw, P. E., Shen, Q., Li, R., Liu, Y., & Xu, X. (2022). Reduction-responsive RNAi nanoplatform to reprogram tumor lipid metabolism and repolarize macrophage for combination pancreatic cancer therapy. *Biomaterials, 280*, 121264. https://doi.org/10.1016/j.biomaterials.2021.121264
120. Lee, S. S.-Y., Li, J., Tai, J. N., Ratliff, T. L., Park, K., & Cheng, J.-X. (2015). Avasimibe encapsulated in human serum albumin blocks cholesterol esterification for selective cancer treatment. *ACS Nano, 9*(3), 2420–2432. https://doi.org/10.1021/nn504025a
121. Lee, S. Y., Ko, S.-H., Shim, J.-S., Kim, D.-D., & Cho, H.-J. (2018). Tumor targeting and lipid rafts disrupting hyaluronic acid-cyclodextrin-based nanoassembled structure for cancer therapy. *ACS Applied Materials & Interfaces, 10*(43), 36628–36640. https://doi.org/10.1021/acsami.8b08243
122. Yao, X., Xie, R., Cao, Y., Tang, J., Men, Y., Peng, H., & Yang, W. (2021). Simvastatin induced ferroptosis for triple-negative breast cancer therapy. *Journal of Nanobiotechnology, 19*(1), 311. https://doi.org/10.1186/s12951-021-01058-1
123. Kim, D., Wu, Y., Li, Q., & Oh, Y.-K. (2021). Nanoparticle-mediated lipid metabolic reprogramming of T cells in tumor microenvironments for immunometabolic therapy. *Nano-Micro Letters, 13*(1), 31. https://doi.org/10.1007/s40820-020-00555-6

Polymer-Based Hybrid Nanomaterials for the Treatment of Metabolic Syndrome

9

Amitabh Aharwar and D. K. Parihar

Abstract

A variety of metabolic irregularities define the complex condition known as metabolic syndrome. Innovative solutions to these problems can be found in nanotechnology, particularly in the form of hybrid nanomaterials based on polymers. To improve the effectiveness of medication delivery, these systems combine polymers with either organic or inorganic nanocomponents. At least two distinct nanoparticles combine to form hybrid nanoparticles. Improvements in circulation time, structural disintegration, high stability, premature release, low encapsulation rate, and unspecific release kinetics are just a few of the benefits that come from combining these characteristics into a single hybrid system as opposed to non-hybrid platforms. Over the past three decades, nanomaterials have attracted a lot of scientific attention due to their special qualities. Hybrid nanomaterials based on polymers can be used to treat metabolic syndrome and have several modes of action such as gene therapy, insulin sensitivity modification, targeted medication delivery, anti-inflammatory effects, and antioxidant activities. The potential of hybrid nanomaterials based on polymers in the treatment of metabolic syndrome is covered in this chapter.

Keywords

Metabolic syndrome · Hybrid nanomaterial · Polymer · Diabetes · Obesity

A. Aharwar (✉)
Government College Harrai, Chhindwara, Madhya Pradesh, India

D. K. Parihar
Department of Biotechnology, Guru Ghasidas University, Bilaspur, Chhattisgarh, India

V. K. Chaturvedi et al. (eds.), *Nanotheranostics in Metabolic Syndrome*,
https://doi.org/10.1007/978-981-95-3746-4_9

1 Introduction

Insulin resistance, dyslipidemia, hypertension, type 2 diabetes, inflammation, non-alcoholic fatty liver disease, obesity, and other interconnected risk factors are all part of the metabolic syndrome. Due in large part to poor dietary practices, genetic predispositions, and sedentary lifestyles, the frequency of metabolic syndrome has increased significantly in recent years. According to some estimates, metabolic syndrome affects almost 30% of the world's population, making it a serious world-wide public health issue. A combination of metabolic, genetic, and environmental variables plays a crucial role in its pathogenesis. Furthermore, studies have indicated that people with metabolic syndrome have a significant risk of myocardial infarction and stroke. There is growing evidence that some illnesses are caused by the overexpression or deficiency of certain genes in the affected organ. As a result, sophisticated solutions are needed to regulate them at the molecular level [8]. Metabolic disorders are linked to many health problems, including ischemia, cardiovascular disease, hypertension, several forms of cancer, and abnormalities in reproduction. In the USA, treating these illnesses is predicted to cost $66 billion by 2030 [7, 8].

Nanomaterials have garnered a lot of research interest over the past three decades because of their unique properties primarily the quantum confinement effect and extremely large surface-to-volume ratios. To address the drawbacks of single-component nanoparticles, enhance their characteristics, acquire new features not possible for single nanoparticles, and accomplish numerous functions for single nanoparticles, hybrid nanoparticles are composed of at least two different nanoparticles. Heterodimers, Janus, core-shell, yolk-shell, dot-in-nanotube, dot-on-nanorod, nanobranches, and other hybrid nanostructures have all been reported. To further their prospective applications, careful synthesis, logical design, and a thorough comprehension of structure-property interactions are all crucial [20, 28].

Polymer-based hybrid nanomaterials have many mechanisms of action and can be utilized to treat metabolic syndrome. The primary elements of these procedures include gene therapy, insulin sensitivity modulation, targeted drug delivery, anti-inflammatory effects, and antioxidant activities [26, 28]. Nanotechnology has a wide range of applications in biomedicine, particularly for the management of metabolic syndrome and its associated disorders like diabetes and obesity. Innovative solutions to metabolic syndrome can be found in nanotechnology, particularly in the form of hybrid nanomaterials based on polymers. Nanotechnology improves drug delivery systems and early detection of disease; moreover it also improves bioavailability, permeability compatibility and therapeutic efficacy and also minimizes side effects [1, 8]. To improve the effectiveness of medication delivery, these systems combine polymers with either organic or inorganic nanocomponents. For the treatment of metabolic syndrome, their capacity to deliver many therapeutic drugs with regulated release kinetics and target specificity has drawn interest [30]. This chapter discusses the potential of polymer-based hybrid nanomaterials in the management of metabolic syndrome.

2 Polymer-Based Hybrid Nanomaterials

Composites of two or more materials produced by combining both components in a way that works well at the nanoscale are known as hybrid nanoparticles. Mixing materials with varied chemical compositions can result in improved properties. The two types of hybrid materials are inorganic-in-organic (inorganic components used to change organic materials or matrices) and organic-in-inorganic (organic moieties used to modify inorganic materials) [27, 30]. Polymer-hybrid nanoparticles may be produced by mixing inorganic materials, including metal oxide nanoparticles, carbon nanotubes, graphene, and polymers. However, integrating natural or synthetic polymers with organic components (phospholipids, proteins, and lipids) can create new hybrid nanosystems that combine the benefits of these biomacromolecules with those of specially designed synthetic polymers. A number of benefits over non-hybrid platforms, including enhanced bloodstream circulation time, early leakage, low encapsulation rate, and unspecific release kinetics, can be made possible by the combination of several features in hybrid systems [15, 16, 28]. The creation of hybrid nanosystems that may combine various organic-inorganic and organic-organic materials. Combinations of nanosized components with different compositions and categories, such as nanoplates, nanoparticles, nanotubes, and nanofibers, are known as hybrid nanoparticles. These nanoparticles are usually in the nanometer size range and frequently have nanostructures [15, 16]. To create hybrid materials, many synthetic routes can be employed. Advances in polymer synthesis, primarily in reversible deactivation radical polymerization (RDRP) techniques have been made possible by the need to synthesize macromolecules with controlled sequences and configurations. These methods make it possible to precisely and consistently adjust chain length, monomer content, and, to a lesser extent, the monomer sequence [24]. Various hybrid nanostructures have been conceived and created, including core-shell, yolk-shell, heterodimer, Janus, dot-in-nanotube, dot-on-nanorod, and nanobranches [20]. Natural products such liposomes, core-shell nanoparticles, hydrogels, nano-capsules, nanostructured lipid carriers, nano-emulsions, solid lipid nanoparticles, and micelles have been used to create a variety of nano-formulations [4]. A new class of materials called polymer-based hybrid nanomaterials has the potential to completely transform the way metabolic syndrome is treated. These nanomaterials, which offer better drug delivery, increased bioactivity, and multifunctional qualities, are made by mixing polymers with inorganic nanoparticles, biomolecules, or medicinal substances. Polymer-based hybrid nanomaterials have great potential for treating the complex nature of metabolic syndrome because of their special qualities, which include biocompatibility, biodegradability, and the capacity to be customized for particular therapeutic requirements [1, 8]. Since 2009, there have been advancements in the production of hybrid nanoparticles in miniemulsions; examples of these hybrid nanoparticles include organic–inorganic, polymeric, and natural macromolecule/synthetic polymer hybrid nanoparticles [22].

3 Polymer-Based Hybrid Nanomaterials for the Treatment of Metabolic Syndrome

In medicine, there is a constant demand for treatments with high therapeutic efficacy, low toxicity, and patient compliance. The use of nanoparticles for a range of therapeutic purposes, or "nanomedicine," has emerged as a result of developments in nanotechnology. One of the most popular new materials in cancer treatment and diagnosis is multifunctional hybrid nanosystems [19]. They tackle issues including low water solubility, nonspecific targeting, and less than ideal therapeutic results that plague conventional nanoparticle delivery methods. Additionally, HNPs ensure accuracy in the diagnosis of lung cancer by facilitating the switch from anatomical to molecular imaging. Selecting nanoplatforms with enhanced functional and structural properties, cost-effectiveness, ease of fabrication, and exceptional repeatability is crucial in therapeutic settings [5].

Hybrid silver nanoparticles are one of example of hybrid nanoparticles. The controlled production of Ag nanoparticles (AgNPs) has attracted a lot of attention lately Because of its effective uses in biological and chemical sensors, surface enhanced Raman spectroscopy (SERS), catalysis, and many other domains such as metabolic syndromes [27]. Curcumin-loaded hybrid nanoparticles of vanillin-chitosan coated with super paramagnetic calcium ferrite were used in cancer. L929 fibroblast cells were used to examine the hybrid material's biocompatibility. To evaluate the hybrid nano carrier's anticancer properties with the curcumin medication, a cytotoxicity test was conducted against the MCF-7 breast cancer cell line [12].

Hydrogel-based nanocarriers that transport calcium channel blockers or long-acting angiotensin-converting enzyme inhibitors can be used to treat hypertension in people with metabolic syndrome. These systems reduce the frequency of dosage and offer prolonged release. The creation of an injectable hydrogel system for controlled release of medications or therapeutic compounds is offered as an alluring alternative for treating metabolic syndrome. Hydrogel-based controlled release systems improve therapeutic efficacy in various metabolic syndrome components. Hydrogel systems achieve better plasma levels of the loaded compound, higher bioavailability, and low cytotoxicity in comparison to conventional systems [3, 25].

A composite material with additional features such as drug delivery, biosensing, photothermal, magnetic response, and antibacterial qualities is created by combining nanoparticles with nanofibers. Using the electrospinning technology, nanofiber and nanoparticle hybrids (NNHs) were produced in a single step. An alternative method for producing nanoparticles in fibers is to employ a self-assembly process. It integrates new research findings to concentrate on the use of NNHs in tissue engineering, medication release, and antimicrobial applications [34]. Polymer dots have been created and used for complex hybrid nanomaterials and applications. Conducting and nonconducting polymers are combined with the right processing methods to create polymer dots. Polymer dots have been used to create sophisticated hybrid nanomaterials by combining them with other quantum dots and nanoparticles. Nanomaterials based on polymer dots possess distinct morphology, conductivity, electrochemical, luminescence, and sensing properties. There are now numerous uses for polymer dot nanomaterials, including drug delivery, gas sensors, biosensors, electronics, probes, solar cells, supercapacitors, and bioimaging [13].

Research on therapeutics based on nucleic acid nanoparticles for the treatment of different illnesses (such as metabolic illness) has grown rapidly in recent years. Clinical trials and extensive research have been conducted on a number of therapies based on nucleic acid nanoparticles. Nanoscale nucleic acid particles aid in the absorption of nucleic acids into the target host tissues and cells while shielding them from immunological reactions and premature enzymatic destruction. They are broken down cytosolically in the cells to release DNAs and RNAs that either stimulate or inhibit the creation of encoded proteins [8]. MicroRNAs (miRNAs) are a significant class of endocrine factors that are essential for preserving metabolic homeostasis and energy balance. Clinical use is still hampered by its undesirable characteristics, such as its off-target action and quick blood breakdown. The efficacy of gene therapy is aided by the robust protection, high bioavailability, and controlled release rate of nanosystem-based administration. According to the study, miRNA-29b and miRNA-122 combination treatment may be useful in enhancing liver function and lowering hepatic fibrosis. For other metabolic problems, a more thorough and comparable strategy ought to be employed [17].

Lipid polymer hybrid nanoparticles (LPHNPs) are sophisticated core-shell nanostructures and a combination of polymeric nanounits and liposomes. They have a polymeric core area surrounded by a lipidic layer. They show a more powerful and limitless effect in delivering medication to the specified area without endangering the body. There are three parts to the LPHNPs assembly. (i) The polymer core, which contains the drug; (ii) The lipid monolayer, which envelops the polymeric core and prevents drug release while shielding the core by preventing water transport; and (iii) The lipid PEG layer, which allows for the conjugation of specific targeting moieties. LPHNPs are reliable, practical, and highly efficient drug delivery vehicles for cancer medication treatment [9]. Polymer-Lipid Hybrid Nanoparticles (PLHNPs) are perfect for increasing bioavailability and delivering hydrophobic medications orally [6]. Lipid-polymer hybrid nanoparticles (LPHNPs) are used in administration technologies for regulated medication administration at tumor locations. The single-step ionic gelation process, which relies on the ionic interaction between positively charged chitosan and negatively charged lipid, was used to create cisplatin-loaded lipid-chitosan hybrid nanoparticles. The A2780 ovarian cancer cell line was used in in vitro cell-based tests to examine the cytotoxicity and cellular interaction of cisplatin-loaded LPHNP [14]. In models of nonalcoholic fatty liver disease, which frequently co-occurs with Metabolic syndrome, chitosan-coated lipid-polymer nanoparticles loaded with silymarin showed strong lipid-lowering efficacy [18]. For improved intestinal absorption and hepatic targeting, lipid-modifying medications such as simvastatin and fenofibrate can be encapsulated in hybrid nanoparticles [33].

Excellent properties including nontoxicity, nonirritability, superior water solubility, and biocompatibility make polymer–protein complexes particularly attractive as cancer treatment agents. Their ability to deliver medications with targeted and prolonged release is impressive, as it can significantly enhance the effectiveness of cancer treatment and minimize adverse effects. But there are still a lot of obstacles to be solved, such lowering the toxicity of materials, safeguarding protein functions, and managing protein release. Polymers increase the variety of medications that are available, including nucleic acid, peptide, and protein medications. FDA-approved protein conjugates covalently bound to PEG. PEGylated protein medicines, which are delivered less often and have a longer half-life in the blood, demonstrating substantial advantages for patients. Polymer–protein systems has application in gene therapy, phototherapy, immunotherapy, targeted treatment, and vaccinations [2, 29].

By boosting the pharmacokinetics and tumor deposition of the payloads through the increased permeability and retention (EPR) effect, nanoparticle drug delivery may be a game-changing tool for increasing treatment efficacies and decreasing adverse effects [10]. mPEG-b-PMSPEP nanoparticles were used to coencapsulate the photosensitizer chlorin e6 (Ce6) with the chemotherapeutic drug paclitaxel (PTX). PTX and Ce6 coencapsulated nanoparticles showed combination cell-killing impact in MDA-MB-231 tumor cells, according to in vitro cytotoxicity studies and also showed tremendous potential for drug delivery systems that realize synergistic chemo-photodynamic therapy for the treatment of cancer [31]. To achieve targeted specificity in computed tomography (CT) imaging and cancer cell treatment, a multifunctional dendrimer-based theranostic nanosystem has been developed. In order to create dendrimer-entrapped Au NPs (Au DENPs), doxorubicin (DOX), a model anticancer drug, was first covalently attached to the partially acetylated poly (amidoamine) dendrimers of generation 5 (G5) prefunctionalized with folic acid (FA) using acid-sensitive cis-aconityl linkage. This resulted in G5·NHAc-FA-DOX conjugates [35].

One of the main causes of metabolic syndrome, obesity, is linked to dysregulated lipid metabolism and chronic inflammation. In order to stabilize and release antiobesity medicines in a spatiotemporally regulated manner, emerging drug delivery systems employ polymeric materials and chemical techniques to increase therapeutic effectiveness and specificity. Human health is seriously threatened by the growth in obesity and related metabolic syndrome prevalence worldwide. In the therapeutic environment, a variety of methods have been employed to manage obesity, such as bariatric surgery and medication [11]. Insulin sensitizers with low oral absorption in free form, such as pioglitazone, berberine, and curcumin, have their bioavailability increased by hybrid nanocarriers. By specifically targeting and modifying Bcl2, the Curcuma-derived diferuloylmethane chemical CUR, loaded on Poly (lactide-co-glycolic) acid (PLGA) nanoparticles, was used to counteract Diabetic Nephropathy-induced kidney apoptosis [26]. Polymetformin and Penetratin-Based Hybrid Nanoparticles were used to treat nonalcoholic fatty liver disease by delivering the targeted interleukin-22 gene to the liver [32]. Gum rosin, a biocompatible polymer, was used in the nanoprecipitation process to create polymeric nanocapsules (NCs) containing metformin and thymoquinone. In type-2 diabetic rats, thymoquinone nanocapsules—which actually contained half of the dosages of thymoquinone—produced a greater antihyperglycemic effect than thymoquinone alone [23].

4 Challenges and Future Perspectives

Although polymer-based hybrid nanomaterials show promise in treating metabolic syndrome, a number of issues need to be resolved before they can be widely used in clinical settings. While many polymeric nanomaterials are friendly, certain inorganic nanoparticles' toxicity is still a cause for worry. Preclinical and clinical research must carefully evaluate the long-term impacts of nanoparticle exposure to guarantee safety. Large-scale manufacture of hybrid nanomaterials based on polymers is still difficult. Standardization and improvement of synthesis processes are essential for the cost-effective manufacturing of these materials in clinical settings. Before being authorized for clinical usage, nanomaterials must pass a rigorous regulatory review process. To address each patient's distinct metabolic profile, individualized treatment plans are required. Maximizing treatment effects will require customizing the functioning and composition of nanomaterials for certain patient populations [31]. Clinical trials for formulations based on nanoparticles; the creation of scalable and repeatable production methods; and integration with biosensors for theranostic applications must be the main areas of future study. The creation of sophisticated hybrid nanoparticles for determined, as needed illness diagnosis and treatment has quickly emerged as a popular area of nanomedicine research. The goal of the research is to create new classes of programmable hybrid nanoparticles that are specifically designed to increase medication concentrations in sick cells. This would improve the effectiveness of chemotherapy and lessen its negative effects. Specifically, the hybrid nanoparticle platforms have the ability to concurrently target sick cells, allow for optical imaging of the area, and deliver therapeutic medications to the diseased cells upon command [21].

5 Conclusion

Advanced characterizations, controlled synthesis, logical design, and a thorough comprehension of structure-property correlations are all crucial for advancing the creation of highly functional hybrid nanomaterials and their practical uses in everyday life. To quickly progress this field of study, which has the potential to transform many facets of our lives and society, including health, renewable energy, and the environment, a combined experimental and theoretical approach is greatly desired. By facilitating the targeted, regulated, and efficient distribution of medicines, polymer-based hybrid nanomaterials provide a revolutionary approach to the treatment of metabolic syndrome. These technologies have the potential to greatly enhance patient outcomes by overcoming the drawbacks of traditional medication. Successful translation from bench to bedside requires ongoing multidisciplinary research in clinical medicine, pharmacology, and materials science.

References

1. Adetunji, C. O., Michael, O. S., Rathee, S., Singh, K. R., Ajayi, O. O., Adetunji, J. B., Ojha, A., Singh, J., & Singh, R. P. (2022). Potentialities of nanomaterials for the management and treatment of metabolic syndrome: A new insight. *Materials Today Advances, 13*, 100198.
2. Alconcel, S. N. S., Baas, A. S., & Maynard, H. D. (2011). FDA-approved poly(ethylene glycol)-protein conjugate drugs. *Polymer Chemistry, 2*, 1442–1448.
3. Arauna, D., Vijayakumar, S., & Durán-Lara, E. (2021). Latest advances in hydrogel-based drug delivery systems for optimization of metabolic syndrome treatment. *Current Medicinal Chemistry, 28*(30), 6274–6286. https://doi.org/10.2174/0929867328666210223141555
4. Aswathanarayan, J. B., & Vittal, R. R. (2019). Nanoemulsions and their potential applications in food industry. *Frontiers in Sustainable Food Systems, 3*, 95. https://doi.org/10.3389/FSUFS.2019.00095
5. Bhardwaj, H., Khute, S., Sahu, R. K., & Jangde, R. K. (2024). Emerging trends in hybrid nanoparticles: Revolutionary advances and promising biomedical applications. *Current Drug Metabolism, 25*(4), 248–265.
6. Bose, R. J., Ravikumar, R., Karuppagounder, V., Bennet, D., Rangasamy, S., & Thandavarayan, R. A. (2017). Lipid–polymer hybrid nanoparticle-mediated therapeutics delivery: advances and challenges. *Drug Discovery Today, 22*(8), 1258–1265.
7. Chu, D. T., Nguyet, N. T. M., Dinh, T. C., Lien, N. V. T., Nguyen, K. H., Ngoc, V. T. N., Tao, Y., Le, D. H., Nga, V. B., Jurgoński, A., & Tran, Q. H. (2018). An update on physical health and economic consequences of overweight and obesity. *Diabetes & Metabolic Syndrome: Clinical Research & Reviews, 12*(6), 1095–1100. https://doi.org/10.1016/j.dsx.2018.05.004
8. Chu, D. T., Thi, H. V., Nguyen, T. T., Vu, T. D., Thi, Y. V. N., Mani, I., Gohil, N., Bhattacharjee, G., Ramakrishna, S., & Singh, V. (2023). Nanotechnology and nucleic acid nanoparticles for treatment of metabolic disorders. *OpenNano, 13*, 100181.
9. Gajbhiye, K. R., Salve, R., Narwade, M., Sheikh, A., Kesharwani, P., & Gajbhiye, V. (2023). Lipid polymer hybrid nanoparticles: a custom-tailored next-generation approach for cancer therapeutics. *Molecular Cancer, 22*(1), 160.
10. He, C., Lu, J., & Lin, W. (2015). Hybrid nanoparticles for combination therapy of cancer. *Journal of Controlled Release, 219*, 224–236.
11. Huang, D., Deng, M., & Kuang, S. (2019). Polymeric carriers for controlled drug delivery in obesity treatment. *Trends in Endocrinology & Metabolism, 30*(12), 974–989.
12. Kamaraj, S., Palanisamy, U. M., Mohamed, M. S. B. K., Gangasalam, A., Maria, G. A., & Kandasamy, R. (2018). Curcumin drug delivery by vanillin-chitosan coated with calcium ferrite hybrid nanoparticles as carrier. *European Journal of Pharmaceutical Sciences, 116*, 48–60.
13. Kausar, A. (2021). Polymer dots and derived hybrid nanomaterials: A review. *Journal of Plastic Film & Sheeting, 37*(4), 510–528.
14. Khan, M. M., Madni, A., Torchilin, V., Filipczak, N., Pan, J., Tahir, N., & Shah, H. (2019). Lipid-chitosan hybrid nanoparticles for controlled delivery of cisplatin. *Drug Delivery, 26*(1), 765–772.
15. Koseva, N. S., Rydz, J., Stoyanova, E. V., & Mitova, V. A. (2015). Hybrid protein–synthetic polymer nanoparticles for drug delivery. *Advances in Protein Chemistry and Structural Biology, 98*, 93–119.
16. Krishnamurthy, S., Vaiyapuri, R., Zhang, L., & Chan, J. M. (2015). Lipid-coated polymeric nanoparticles for cancer drug delivery. *Biomaterials Science, 3*(7), 923–936.
17. Li, T., Zhu, L., Zhu, L., Wang, P., Xu, W., & Huang, J. (2021). Recent developments in delivery of MicroRNAs utilizing nanosystems for metabolic syndrome therapy. *International Journal of Molecular Sciences, 22*(15), 7855. https://doi.org/10.3390/ijms22157855
18. Liang, J., Liu, Y., Liu, J., Li, Z., Fan, Q., Jiang, Z., Yan, F., Wang, Z., Huang, P., & Feng, N. (2018). Chitosan-functionalized lipid-polymer hybrid nanoparticles for oral delivery of silymarin and enhanced lipid-lowering effect in NAFLD. *Journal of Nanobiotechnology, 16*, 1–12.

19. Lotey, N. K., Pednekar, S., & Chaughule, R. (2022). Hybrid nanoparticles in biomedical applications. In *Nanomaterials for cancer detection using imaging techniques and their clinical applications* (pp. 365–400). Springer International Publishing.
20. Ma, D. (2019). Hybrid nanoparticles: an introduction. In *Noble metal-metal oxide hybrid nanoparticles* (pp. 3–6). Woodhead Publishing.
21. Nguyen, K. T., & Zhao, Y. (2015). Engineered hybrid nanoparticles for on-demand diagnostics and therapeutics. *Accounts of Chemical Research, 48*(12), 3016–3025.
22. Qi, D., Cao, Z., & Ziener, U. (2014). Recent advances in the preparation of hybrid nanoparticles in miniemulsions. *Advances in Colloid and Interface Science, 211*, 47–62.
23. Rani, R., Dahiya, S., Dhingra, D., Dilbaghi, N., Kim, K. H., & Kumar, S. (2018). Improvement of antihyperglycemic activity of nano-thymoquinone in rat model of type-2 diabetes. *Chemico-Biological Interactions, 295*, 119–132. https://doi.org/10.1016/j.cbi.2018.02.006. Epub 2018 Feb 5. PMID: 29421519.
24. Russell, A. J., Baker, S. L., Colina, C. M., Figg, C. A., Kaar, J. L., Matyjaszewski, K., Simakova, A., & Sumerlin, B. S. (2018). Next generation protein-polymer conjugates. *AICHE Journal, 64*(9), 3230.
25. Saharan, R., Kaur, J., Dhankhar, S., Garg, N., Chauhan, S., Beniwal, S., & Sharma, H. (2024). Hydrogel-based drug delivery system in diabetes management. *Pharmaceutical Nanotechnology, 12*(4), 289–299.
26. Samadder, A., Bhattacharjee, B., Dey, S., Chakrovorty, A., Dey, R., Sow, P., Tarafdar, D., Biswas, M., & Nandi, S. (2024). Enhanced drug carriage efficiency of curcumin-loaded PLGA nanoparticles in combating diabetic nephropathy via mitigation of renal apoptosis. *Journal of Pharmacopuncture, 27*(1), 1.
27. Singh, P., Singh, S., Maddiboyina, B., Kandalam, S., Walski, T., & Bohara, R. A. (2024). Hybrid silver nanoparticles: Modes of synthesis and various biomedical applications. *Electron, 2*(2), e22.
28. Soares, D. C. F., Domingues, S. C., Viana, D. B., & Tebaldi, M. L. (2020). Polymer-hybrid nanoparticles: Current advances in biomedical applications. *Biomedicine & Pharmacotherapy, 131*, 110695.
29. Sun, Q., Yang, Z., & Qi, X. (2023). Design and application of hybrid polymer-protein systems in cancer therapy. *Polymers, 15*(9), 2219. https://doi.org/10.3390/polym15092219
30. Vargas-Bernal, R. (2020). Introductory chapter: Hybrid. In *Hybrid nanomaterials: flexible electronics materials* (p. 3). IntechOpen.
31. Wang, J., Li, D., Tao, W., Lu, Y., Yang, X., & Wang, J. (2019). Synthesis of an oxidation-sensitive polyphosphoester bearing thioether group for triggered drug release. *Biomacromolecules, 20*(4), 1740–1747. https://doi.org/10.1021/acs.biomac.9b00101
32. Zai, W., Chen, W., Wu, Z., Jin, X., Fan, J., Zhang, X., Luan, J., Tang, S., Mei, X., Hao, Q., & Liu, H. (2019). Targeted interleukin-22 gene delivery in the liver by polymetformin and penetratin-based hybrid nanoparticles to treat nonalcoholic fatty liver disease. *ACS Applied Materials & Interfaces, 11*(5), 4842–4857.
33. Zhang, L. I., & Zhang, L. (2010). Lipid–polymer hybrid nanoparticles: synthesis, characterization and applications. *Nano Life, 1*(01n02), 163–173.
34. Zhang, M., Song, W., Tang, Y., Xu, X., Huang, Y., & Yu, D. (2022). Polymer-based nanofiber–nanoparticle hybrids and their medical applications. *Polymers, 14*(2), 351.
35. Zhu, J., Wang, G., Alves, C. S., Tomás, H., Xiong, Z., Shen, M., Rodrigues, J., & Shi, X. (2018). Multifunctional dendrimer-entrapped gold nanoparticles conjugated with doxorubicin for pH-responsive drug delivery and targeted computed tomography imaging. *Langmuir, 34*(41), 12428–12435.

Carbon-Based Hybrid Nanomaterials for Treatment of Metabolic Syndrome

10

Debadatta Mohapatra, Singh Shreya, Alakh N. Sahu, Sudarsan Biswal, Gyanada Sankar Dash, and Prakash Ch. Senapati

Abstract

Metabolic syndrome comprises a cluster of medical conditions, including hypertension, diabetes, excessive abdominal fat, decreased high-density lipoprotein (HDL) levels, raised serum triglycerides, chronic inflammation, and nonalcoholic fatty liver disease (NAFLD). It is a prime public health associated with obesity, physical inactivity, and poor dietary habits. Metabolic syndrome is a principal component of the global burden of cardiovascular diseases and type 2 diabetes mellitus. The growing incidence of metabolic syndrome has urged interest in developing innovative therapeutics. Despite an improved understanding of the

D. Mohapatra (✉)
Phytomedicine Research Laboratory, Department of Pharmaceutical Engineering & Technology, IIT (BHU), Varanasi, India

Leslie Dan Faculty of Pharmacy, University of Toronto, Toronto, ON, Canada
e-mail: debadattamohapatra.rs.phe19@itbhu.ac.in

S. Shreya
SBS College of Pharmacy, Malwan, Fatehpur, India

A. N. Sahu
Phytomedicine Research Laboratory, Department of Pharmaceutical Engineering & Technology, IIT (BHU), Varanasi, India

S. Biswal
Directorate of Drugs Control Administration, Health and Family Welfare Department, Bhubaneswar, Government of Odisha, Bhubaneswar, Odisha, India

G. S. Dash
KIIT School of Pharmacy, KIMS, KIIT Deemed to University, Patia, Bhubaneswar, Odisha, India

P. C. Senapati (✉)
School of Pharmaceutical Sciences-II, Phulnakhara Campus, Siksha 'O' Anusandhan (Deemed to be University), Bhubaneswar, Odisha, India

V. K. Chaturvedi et al. (eds.), *Nanotheranostics in Metabolic Syndrome*,
https://doi.org/10.1007/978-981-95-3746-4_10

mechanisms underlying the pathogenesis of metabolic syndrome, current treatment remains ineffective. Nanomaterials offer a broad spectrum of therapeutic utility, especially for treating metabolic syndrome with enhanced targeted delivery and early diagnosis using nano-imaging techniques. Among various nanomaterials, carbon-based hybrid nanomaterials (CBHNs) have been demonstrated as promising multifunctional smart theranostic nanomaterials due to their unique combination of carbon-based nanostructures (e.g., carbon nanotubes, carbon dots, graphene, nanodiamonds, and fullerenes) and other functional components, such as polymers, or biomolecules or inorganic nanoparticles. Because of their high surface area, biocompatibility, chemical stability, excellent photostability, optical/fluorescence properties, surface functionalization capacity, tunable properties, biosensing, and the ability to control biological events at the molecular level, these materials provide diverse advantages in the diagnosis and treatment of metabolic syndrome. The CBHNs can improve the delivery of therapeutics, improve antioxidant activity, reduce inflammation, and regulate metabolic pathways involved in metabolic syndrome. In this chapter, the potential of CBHNs for the treatment of metabolic syndrome is discussed.

Keywords

Carbon-based hybrid nanomaterials · Metabolic syndrome · Nanomedicine · Drug delivery · Theranostics

Learning Objectives

- Brief introduction to metabolic syndrome, etiology, risk factors, pathophysiology, and management/prevention/treatment of metabolic syndrome.
- Carbon-based nanomaterials, carbon-based hybrid nanomaterials, their advantages, fabrication methodology, and their applications in metabolic syndrome.
- Characterizations of carbon-based hybrid nanomaterials.
- Other applications of carbon-based hybrid nanomaterials.
- Challenges for carbon-based hybrid nanomaterials, summary, and future outlook.

1 Introduction

Metabolic syndrome is a noninfective syndrome that is prevalent across the globe. It adversely impacts public health and the global economy [1–3]. It is defined by a collection of cardiovascular diseases, insulin resistance, hypertension, disrupted glucose metabolism, abdominal obesity, and abnormal lipid levels [1]. Metabolic syndrome results in multiple clinical outcomes, including cardiovascular diseases (CVD), chronic kidney disease, nonalcoholic steatohepatitis (NASH), other liver dysfunctions, and various cancer and neurodegenerative disorders [1, 2]. Nanomaterials that possess a size below 100 nm are currently used widely in the field of drug delivery because of their excellent physicochemical possessions. Based on the shape, nanomaterials are classified into zero dimensional (0D), one dimensional (1D), two dimensional (2D), and three dimensional (3D) materials [4]. Their physicochemical characteristics, including shape, porosity, size chemical composition, functionalization, charge, stability, hydrophilicity/hydrophobicity, and aggregation, predominantly govern their interactions with biological molecules within the body [4]. Different nanomaterials (e.g., carbon nanomaterials, nanostructured lipid carriers, liposomes, niosomes, dendrimers, micelles, chitosan nanoparticles, solid lipid nanoparticles, alginate nanoparticles, and inorganic nanoparticles) have been well-documented for the treatment of metabolic syndrome [3]. Nanomaterials enable effective drug delivery while minimizing side effects in the treatment of metabolic syndrome. This book chapter primarily explores carbon-based nanomaterials, such as fullerenes, graphene, carbon dots, reduced graphene oxide, graphene oxide, and carbon nanotubes, along with their hybrid forms incorporating polymers or metals or ceramics or biomolecules. These materials have garnered significant attention among nanomaterials because of their outstanding theranostic attributes, offering both therapeutic and diagnostic potential. Carbon-based hybrid nanomaterials for metabolic disorders have been comprehensively described in this chapter. Various types of carbon-based nanohybrids and their therapeutic/theranostic role in metabolic syndrome, characterizations, safety, toxicity, challenges, and future prospects have been demonstrated.

2 Metabolic Syndrome

Metabolic syndrome is a multifaceted disorder having a group of risk factors that accelerate serious conditions like cardiac diseases and type 2 diabetes. A person with metabolic syndrome possesses a group of medical conditions, such as abdominal obesity, impaired metabolism of glucose, insulin resistance, hypertension, dyslipidemia, pro-inflammatory along with pro-thrombotic states [1, 5, 6]. Various medical conditions such as diabetes mellitus (type 2), polycystic ovary syndrome (PCOS), frank hypertension, and Cushing's syndrome are regarded as intense examples of the building blocks of metabolic syndrome, and all increase the risk of CVDs [5]. Several alternative terms, including Syndrome X, plurimetabolic syndrome, Reaven's syndrome, and insulin resistance syndrome, are occasionally used but less interchangeably to describe metabolic syndrome [7].

The World Health Organization (WHO) first described metabolic syndrome in 1998. As per the definition by WHO, insulin resistance is the major pathophysiology of metabolic syndrome, so it is considered a fundamental need for inclusion in the definition. In addition to insulin resistance, two additional conditions, such as obesity, hypertension, microalbuminuria, and dyslipidemia. As per the definition, a patient cannot be diagnosed with metabolic syndrome if insulin resistance is not present, even if all other criteria are fulfilled. However, under the WHO definition, a patient with type 2 diabetes who meets the additional criteria may also be diagnosed with metabolic syndrome [8, 9].

In 1999, the European Group for the Study of Insulin Resistance (EGIR) recommended an alternative explanation to that of the WHO. Like the WHO definition, the EGIR also identified insulin resistance as a major pathophysiological characteristic of metabolic syndrome and incorporated it into its criteria. Like the definition by WHO, the EGIR considered two supplementary criteria, including obesity, hypertension, and dyslipidemia, along with insulin resistance. In this definition, obesity was determined based on waist circumference, while the WHO definition allowed for either waist-to-hip ratio or body mass index. According to EGIR, microalbuminuria was excluded as a diagnostic criterion [8, 9].

The National Cholesterol Education Program (NCEP) Adult Treatment Panel III (ATP III) proposed a definition for metabolic syndrome in 2001, which was revised later by the American Heart Association and the National Heart, Lung, and Blood Institute in 2005.

According to the NCEP ATP III definition, metabolic syndrome is diagnosed when at least three of the following five criteria are met: increased waist circumference (>40 inches in men or >35 inches in women), increased blood pressure (≥130/85 mmHg), fasting serum triglyceride (TG) levels >150 mg/dL, declined fasting HDL cholesterol levels (<40 mg/dL in men or <50 mg/dL in women), and raised fasting blood glucose levels >100 mg/dL [3, 8]. This definition is widely used for metabolic syndrome. It includes insulin resistance/ hyperglycemia, abdominal adiposity, high blood pressure as well as atherogenic dyslipidemia. This definition is straightforward to recall, as it does not mandate the presence of any specific condition; rather, a diagnosis of metabolic syndrome requires at least three out of five criteria to be met [8, 9].

The International Diabetes Foundation (IDF) provided another definition for metabolic syndrome in 2005. Like other definitions, it includes the same pathological conditions. However, it requires obesity to be present as a major criterion, but not necessarily insulin resistance. Body weight and waist circumference vary across populations due to differences in nationality and ethnicity. It has been recognized that the association between these values and the risk of CVD or Type 2 diabetes differs across various populations. South Asian individuals exhibit a higher risk of CVD and type 2 diabetes at lower waist circumferences, which may fall below the threshold used to define metabolic risk in Western populations. Despite the fact that visceral obesity is now acknowledged as a major component of metabolic disorders, the IDF definition is criticized for emphasizing obesity in the pathophysiology rather than insulin resistance [8, 9].

The progression and severity of the associated components are primarily influenced by age and adiposity (weight gain) [5]. Most importantly, the increased accumulation of central fat (around large waist circumference) is a stronger contributor than total body fat in both thin and overweight individuals [5]. Metabolic syndrome is characterized by a range of endocrine and biochemical abnormalities, including glucose intolerance, hyperinsulinemia, insulin resistance, hypercortisolism, hypertriglyceridemia, reduced HDL, and elevated low-density lipoprotein (LDL). Additionally, it is associated with various pathophysiological conditions such as type 2 diabetes mellitus, PCOS, central fat accumulation, stress-related disorders, morbid obesity, depression, hypertension, coronary heart disease and NASH [7]. Various other metabolic abnormalities may coexist, including inflammatory markers such as elevated serum uric acid, C-reactive protein (CRP), and cytokines, as well as a prothrombotic state characterized by elevated levels of plasminogen activator inhibitor-1 [7]. Genome-wide association studies have offered perspectives into the genetic underpinnings of obesity and metabolic syndrome [7]. Impaired growth of the fetus, weight gain in adolescence, genetic factors, and pro-inflammatory mediators develop manifestations of the metabolic syndrome [7]. Early identification of metabolic syndrome prevents the underlying causes of associated diseases, for example, cardiovascular diseases and coronary heart disease.

The most extensively used criteria for the identification of metabolic syndrome is described by the ATP III of the NCEP and the IDF [5]. The NCEP criteria include specific thresholds for HDL cholesterol, fasting glucose level, blood pressure, and fasting TG. In contrast, the IDF criteria focus on diabetes and CVD prevention, incorporating lower cut-off values for waist circumference and fasting blood glucose [5]. The primary objective of these diagnostic criteria is to facilitate early proactive measures for weight management, aiming to lower the risk of CVD and diabetes [7].

2.1 Abdominal Obesity

Abdominal obesity is also known as central obesity. Deposition of excess fat in the waist and abdomen is a prime contributing factor for metabolic syndrome. Visceral fat, i.e., the fat surrounding internal organs, is metabolically active, which can lead to insulin resistance and inflammation. Visceral obesity causes a reduction of insulin-mediated glucose uptake and develops insulin resistance [8]. It is primarily driven by adipokines secreted by adipose tissue, which regulate the interplay between metabolism and vascular function. Notably, pro-inflammatory markers including interleukin-6 (IL-6) and tumor necrosis factor-alpha (TNF-α) contribute to insulin resistance and vascular dysfunction. Adipose tissue also activates the renin-angiotensin system, which results in insulin resistance and hypertension. In contrast, adiponectin (protective adipokine) couples insulin sensitivity and energy metabolism. Studies have shown that adiponectin levels are reduced in individuals with obesity, type 2 diabetes, and metabolic syndrome [8].

2.2 Insulin Resistance

Diabetes mellitus (DM) is a key component of metabolic syndrome, resulting from impaired insulin function and secretion, leading to excessive glucose production [10]. Various types of diabetes include: (i) Type 1 (insulin-dependent), (ii) Type 2 (insulin-independent), and (iii) gestational diabetes. Type 1 diabetes influences approximately 5–10% of the population and emerges from the destruction of pancreatic β-cells. Type 2 diabetes occurs in about 90–95% of patients, which is noninsulin-dependent diabetes or adult-onset diabetes. In this case, the β-cells are not destroyed, and the patients are not dependent on the insulin. The insulin does not work properly, due to the development of insulin resistance in such individuals. Gestational diabetes mellitus (GDMs) is also a critical public health challenge, which is further categorized into GDM1 and GDM2. The GDMs in pregnancy lead to short-term and long-term complications for both mother as well as newborn [10]. Insulin is secreted by the pancreas in response to high blood glucose levels which utilize glucose in various tissues [8]. Tissues, such as skeletal muscle, adipose tissue, and liver mostly utilize glucose and remove it from the circulation [8]. In adipose tissue and skeletal muscle, insulin facilitates glucose uptake by promoting the movement of the GLUT4 transporter to the cell membrane. In the liver and skeletal muscle, it inhibits glycogenolysis and causes the synthesis of glycogen from glucose. In the hepatocyte, insulin reduces hepatic gluconeogenesis, preventing the influx of glucose into the bloodstream. Insulin causes lipolysis and stimulates the uptake of glucose in adipose tissue [8]. Overall, insulin enhances glucose uptake, decreases the levels of circulating glucose, and causes the conversion of glucose into storage molecules, fat, or glycogen [8]. Insulin resistance is a condition when the body's cells become less responsive to insulin, which results in type 2 diabetes [7]. In such conditions, the adipose tissue, liver cells, and muscle do not respond properly to insulin and leads to hyperglycemic conditions. This situation is made worse by the deregulation of feedback systems [8]. Insulin resistance develops type-2 diabetes. Hyperinsulinemia is a potential contributor to insulin resistance [8].

2.3 Hypertension

Hypertension is characterized by a blood pressure higher than 140/80 mmHg that considerably enhances the risk of various CVD, including damage to blood vessels, stroke, and kidney damage.

2.4 Dyslipidemia/Atherogenic Dyslipidemia

Dyslipidemia is the irregular lipid level in the blood, such as high TG (>150 mg/dL), reduced level of good cholesterol/HDL (<40 mg/dL in men or <50 mg/dL in women), and elevated level of bad cholesterol/LDL [8]. Visceral obesity and insulin resistance are accompanied by atherogenic dyslipidemia [8]. Insulin generally suppresses the lipolysis in adipose tissues. So, the impaired insulin secretion enhances lipolysis, causing increased levels of free fatty acid. The free fatty acids in the liver act as a substrate for the production of TG [8]. The free fatty acids also control the synthesis of apoB, a major lipoprotein of very low density lipoproteins (VLDL), resulting in more production of VLDL [8].

Insulin denaturates apoB via PI3K-dependent molecular pathways. Hence, the insulin resistance directly enhances VLDL synthesis. The insulin also controls the activity of the major rate-limiting enzyme, "lipoprotein lipase" that causes VLDL clearance [8]. Insulin resistance leads to increased VLDL synthesis and reduced

VLDL clearance, resulting in hypertriglyceridemia. VLDL is then metabolized into trace lipoproteins and small dense LDL, both of which contribute to atheroma formation [8].

2.5 Endothelial Dysfunction

Endothelial cells primarily reside in the innermost layer of blood vessels, where they perform diverse cellular functions. The endothelium detects and responds to various physiological and pathological stimuli, producing vasoactive molecules such as nitric oxide (NO), endothelins, and prostacyclin. The expression of endothelial cell adhesion molecules regulates interactions with circulating monocytes as well as leukocytes, promoting inflammation, as well as with platelets, contributing to hemostasis and thrombosis. Additionally, the endothelium modulates vascular smooth muscle responses, potentially facilitating atherosclerotic plaque formation. While normal endothelial function provides protection against these systems, endothelial dysfunction is a key driver of atherosclerosis and associated cardiovascular risk factors. Endothelial dysfunction primarily arises when the endothelial layer is impaired or absent, leading to a failure in its normal physiological and protective functions [8]. The normal endothelial responses are affected by free fatty acids, oxidative stress hyperglycemia, advanced glycation products, and inflammatory cytokines or adipokines [8]. Endothelial dysfunction also causes a reduction in NO bioavailability in vasculature [1].

2.6 Pro-inflammatory and Pro-thrombotic Conditions

It is defined by raised concentrations of inflammatory markers, such as CRP, along with clotting factors, including fibrinogen. The body experiences increased inflammation and a heightened tendency to form blood clots, primarily due to visceral fat accumulation, which releases inflammatory chemicals and disrupts normal blood clotting mechanisms; this significantly elevates the risk of CVD.

3 Etiology and Risk Factors of Metabolic Syndrome

Metabolic syndrome is mainly correlated with a "Westernized" lifestyle, which is characterized by low physical activity and eating high-fat diets. Obesity in childhood increases the likelihood of metabolic syndrome in adults. Psychosocial stress is also an etiology of metabolic syndrome. In addition to that, genetic factors also influence the constituents of the metabolic syndrome [7]. Lifestyle factors, such as smoking, a diet high in saturated fat, weight gain, more alcohol intake, and physical inactivity, lead to increased intra-abdominal fat and metabolic risk factors [7] (Fig. 10.1).

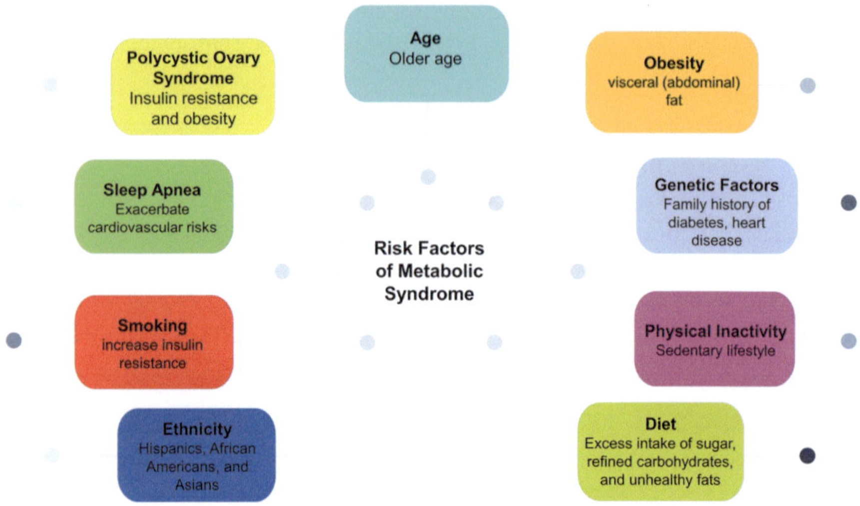

Fig. 10.1 Risk factors of metabolic syndrome

4 Pathophysiology

The pathophysiology of metabolic syndrome involves: (i) elevated glucose levels and insulin resistance, (ii) increased blood pressure associated with insulin resistance, (iii) abnormal lipid profiles (dyslipidemia), (iv) PCOS and depression, (v) excessive waist circumference with intra-abdominal fat accumulation, (vi) partial lipodystrophy, (vii) skeletal muscle atrophy, (viii) sleep disorders and hypoxia, and (x) inflammation [7].

4.1 High Glucose and Insulin Resistance

Insulin is a pleiotropic molecule that has effects on protein synthesis, amino acid uptake, proteolysis, lipoprotein lipase activity, adipose tissue triglyceride lipolysis, VLDL secretion, glucose uptake in muscle and adipose tissue, endogenous glucose production, and hepatocyte and muscle glycogen synthesis [2].

Insulin resistance is represented by the failure of plasma glucose suppression even at a high plasma insulin concentration. Cellular insulin resistance is attributed to modifications in receptor binding or impairments in post-receptor signaling pathways [2, 7, 11]. Insulin resistance contributes to hypertension by diminishing vasodilation while enhancing the vasoconstrictive effects of free fatty acids. Additionally, it increases serum viscosity, stimulates sympathetic neuron activation, and promotes renal sodium reabsorption [3]. Insulin resistance results in reduced glucose tolerance, the development of diabetes along with an increased risk of CVD.

4.2 Increased Blood Pressure and Insulin Resistance

It has been identified that numerous hypertensive patients exhibit hyperinsulinemia and impaired glucose tolerance. There is an indirect relationship between hypertension and insulin resistance. Both hypertension and hyperinsulinemia are exacerbated by obesity, and these conditions are typically improved by losing weight. By increasing catecholamine activity, insulin resistance, and hyperinsulinemia may directly contribute to hypertension [7]. Insulin resistance induces hypertension by loss of insulin's vasodilatory activity and free fatty acid (FFA)-induced vasoconstriction because of the generation of Reactive oxygen species (ROS) and subsequent scavenging of NO [11].

4.3 Dyslipidemia

It is characterized by elevated levels of TG, reduced HDL, and elevated LDL levels in the plasma. Irregularities in the level of these factors are coupled with an elevated risk of CVD [7].

4.4 Polycystic Ovarian Syndrome

Obesity and insulin resistance are commonly observed in women with PCOS, significantly increasing the risk of cardiovascular disorder and metabolic disorders [2]. The metabolic syndrome may be partially described by abnormalities in insulin, sex hormones, and cortisol, which increases the risk of depression and PCOS in women [7]. Women with PCOS suffer fertility problems and the consequences of insulin resistance, including the development of type-2 diabetes and CVD risk factors [2]. Metabolic syndrome is prevalent in case of women, especially those with obesity and PCOS [2]. The hypothalamic-pituitary axis, ovary, and insulin resistance impact the metabolic syndrome.

4.5 Excessive Waist Circumference and Intra-abdominal Deposition of Fat

The buildup of abdominal fat, particularly in the intra-abdominal region, is visibly recognized by an "apple-shaped" body and may play a direct role in the onset of metabolic syndrome. Visceral intra-abdominal fat, which is highly metabolically active, releases substantial amounts of free fatty acids into the liver via the portal circulation, potentially impairing hepatic insulin clearance. Additionally, intra-abdominal fat functions as an endocrine organ, secreting several cytokines, such as leptin, interleukins (IL-1 & IL-6), resistin, adiponectin, as well as TNF-α. An excess of these cytokines is connected to an enhanced risk of metabolic disorders [7].

4.6 Lipodystrophy

Lipodystrophy is an acquired or inherited disorder represented by the loss of selective adipose tissue depots. Patients with partial or generalized lipodystrophy exhibit severe insulin resistance and characteristic features of metabolic syndrome, placing them at an increased risk for diabetes (Type 2), dyslipidemia, NAFLD, and CVD [2]. Individuals with lipodystrophy exhibit a functionless mutation in the peroxisome proliferator-activated receptor gamma (PPARγ) gene, leading to subcutaneous adipose tissue atrophy. This condition is related to severe insulin resistance, diabetes, hepatic steatosis, dyslipidemia, and PCOS in the case of female [7].

4.7 Skeletal Muscle Atrophy

Muscle atrophy occurs gradually with age and inactivity, which adversely affects myofiber degeneration or regeneration. It occurs when muscle fibers shrink in size due to various causes, resulting in weakened muscles and reduced function. The reduced muscle mass is important in metabolic disorders. Loss of muscle mass may precipitate type 2 diabetes due to low glucose oxidation capacity. Muscles play a major role in insulin action and the oxidation of glucose and fat. A relative lacking of type I (oxidative) muscle fibers are linked with both type 2 diabetes and metabolic syndrome. Studies indicate that metabolic disorders are characterized by a reduction in oxidative type I red muscle fibers and a raised proportion of type IIB (glycolytic) white muscle fibers [7].

4.8 Sleep Disorders, Hypoxia, and Metabolic Syndrome

Obstructive sleep apnea (OSA) originates primarily from obesity. Insulin resistance is associated with OSA [2]. OSA is linked to heightened inflammation and reduced adiponectin levels [2]. People with OSA are more likely to show metabolic syndrome and are especially vulnerable to hypertension because of their elevated sympathetic nervous system, which is triggered by hypoxia. Oxidative stress is induced by sleep disorder and hypoxia, which results in inflammation, neurohumoral alterations, and metabolic abnormalities [7].

4.9 Inflammation

Inflammation is considered a key pathway in the development of metabolic syndrome, as individuals with the condition exhibit elevated levels of inflammatory markers [3]. Pro-inflammatory markers, including TNF-α, IL-6, and CRP play a significant role in the pathogenesis. TNF-α contributes to insulin receptor inactivation in adipose tissue, promotes lipolysis and fatty acid production, and suppresses adiponectin release [3]. The immune cells and adipocytes develop IL-6, which stimulates the liver to synthesize CRP, whose increased level has been implicated in the pathogenesis of metabolic syndrome [3]. CRP is a mediator of inflammation. Chronic mild elevation of CRP can cause cardiovascular consequences. Elevated CRP can cause metabolic syndrome, including low HDL cholesterol, hyperinsulinemia, adiposity, hypertriglyceridemia, and type 2 diabetes [1, 7]. Inflammatory signals and other stimuli lead to the activation of pattern recognition receptor following the activation of nuclear factor κβ, increasing the expression of interleukin-1β as well as interleukin-18. Together with the elevated ROS levels, the inflammatory process leads to the development of CVD [1, 2].

4.10 Neurohormonal Stimulation

The activation of the renin-angiotensin system has been implicated in the development of metabolic syndrome. Adipose tissue produces angiotensin II, which stimulates nicotinamide adenine dinucleotide phosphate oxidase (NADPH oxidase), leading to ROS generation. This process contributes to endothelial damage, platelet aggregation, LDL oxidation, and the expression of lectin-like oxidized low-density lipoprotein receptor-1 (LOX-1) on endothelial and vascular smooth muscle cells. These events further trigger inflammation and fibroblast proliferation, playing a crucial role in the pathogenesis of type 2 diabetes mellitus, hypertension, and dyslipidemia [3].

4.11 Hypogonadism

Similar to the increased risk of metabolic syndrome in women with PCOS, a correlation exists between metabolic syndrome and male gonadal as well as erectile dysfunction [2]. Men with metabolic syndrome possess a greater prevalence of hypogonadism [2]. Hypogonadism is a possible contributing factor to the onset of type 2 diabetes and metabolic syndrome [2].

4.12 Health Implications of Metabolic Syndrome

If left unmanaged, metabolic syndrome greatly elevates the risk of CVD (including heart attacks, coronary artery disease, and strokes), NAFLD (including NASH and liver cirrhosis), type 2 diabetes, chronic kidney disease (CKD), various cancers (such as colorectal, breast, and liver cancer), and sleep apnea.

4.13 Management, Prevention, and Treatment of Metabolic Syndrome

It includes the following components:

4.13.1 Lifestyle Changes

Diet: Eating a Mediterranean diet/whole foods (fruits, vegetables, lean proteins (e.g., fish, poultry), whole grains and healthy fats (e.g., nuts, olive oil) and reducing processed foods (trans fats, refined sugars, and high-fructose corn syrup) [2, 7, 11].

Exercise: Aerobic exercise (e.g., cycling, brisk walking, swimming, climbing stairs, dancing), and strength training improve muscle mass and insulin sensitivity. Body weight loss significantly improves insulin sensitivity and reduces the risk of developing heart disease and diabetes [2, 7].

Stress management: Chronic stress can contribute to insulin resistance, obesity, and hypertension. Meditation, Yoga, and relaxation techniques are beneficial for stress management.

Avoid smoking: Avoid smoking and drugs [7].

4.13.2 Medications

Medications for control of hypertension (antihypertensive drugs: diuretics, angiotensin-converting enzyme (ACE) inhibitor, beta-blockers, calcium channel blocker), to lower LDL cholesterol (statins), diabetes (antihyperglycemic drugs: metformin, dipeptidyl peptidase inhibitors: sitagliptin, vildagliptin, PPAR-γ agonists/thiazolidinediones: pioglitazone, dopamine D2 receptor agonist (bromocriptine mesylate), glucagon-like peptide-1 receptor agonists (GLP-1): liraglutide and exenatide, antithrombotics (aspirin), obesity (tetrahydrolipstatin: orlistat), and drugs to lower triglycerides or improve HDL cholesterol are commonly used [2, 7].

4.13.3 Monitoring

Regular checkups of blood pressure, blood sugar levels, and lipid profiles are essential for individuals with metabolic syndrome. Early diagnosis and follow-up are essential for preventing complications.

5 Carbon-Based Nanomaterials

Nanomaterials offer a diverse array of therapeutic utility because of their unique properties. Among various nanomaterials, carbon-based nanomaterials, including quantum dots, carbon dots, carbon nanotubes, graphene, nanodiamonds, and fullerenes, have been widely exploited as theranostics in pharmaceutical and biomedical fields for various diseases. A range of carbon-based nanomaterials is shown in Fig. 10.2. Currently, the carbon-based hybrid nanomaterials (CBHNs) have been demonstrated as promising multifunctional smart theranostic nanomaterials due to their unique combination of carbon-based nanomaterials (e.g., carbon nanotubes, carbon dots, graphene) and other functional components, such as polymers, or biomolecules or inorganic nanoparticles [12, 13].

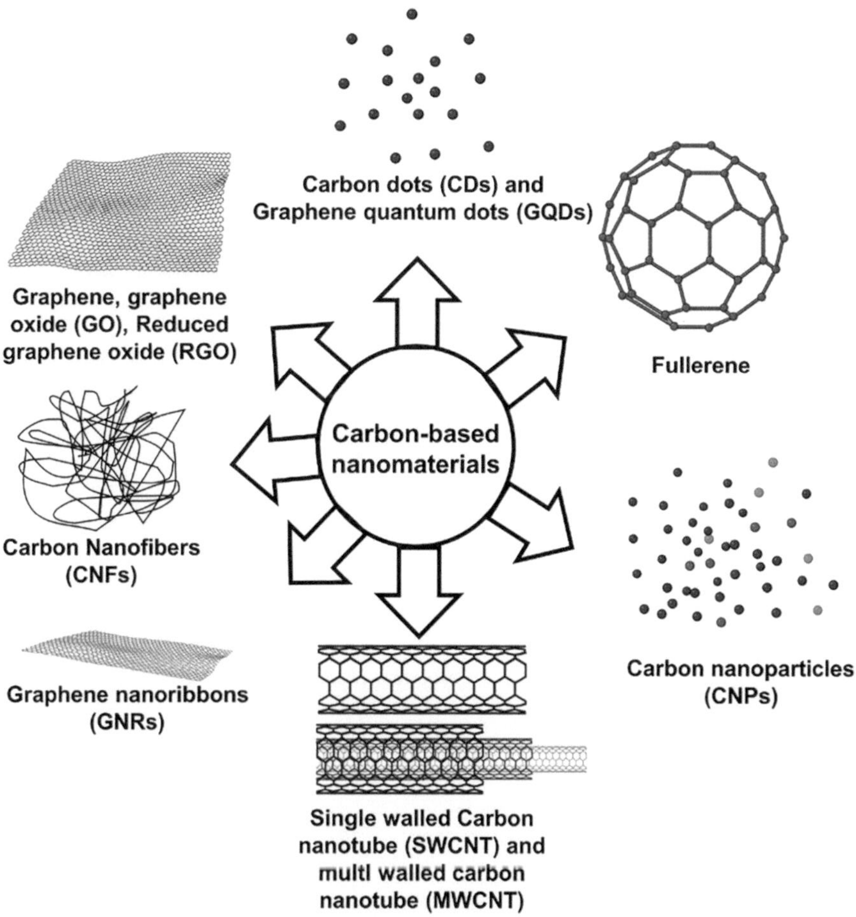

Fig. 10.2 Various carbon-based nanomaterials

5.1 Carbon-Based Zero-Dimensional (0D) Nanomaterials

0D carbon materials possess all three dimensions in the nanometric range. Such nanomaterials demonstrate unique optical, electronic, and mechanical properties due to their ultra-low dimension. Major 0D carbon nanomaterials include carbon nanodots/carbon dots (CDs), fullerenes, graphene quantum dots (GQDs), and carbon nanoparticles (CNPs).

5.1.1 Carbon Dots (CDs)/Carbon Quantum Dots (CQDs)

CDs are widely explored 0D (<10 nm) quasi-spherical fluorescent nanomaterials similar to semiconductor-based quantum dots (QDs) [13–16]. CDs are comprised of C, O, and H; however, heteroatoms, such as S, B, and N are doped onto their surface to produce CDs with improved fluorescence/quantum yield, stability, sensing, and biological activity. In contrast, the QDs are fabricated from semiconductors, such as CdS (cadmium sulfide), CdSe (cadmium selenide), PbSe (lead selenide), or ZnS (zinc sulfide). CDs possess bioimaging, sensing, optoelectronic devices, cancer theranostics, drug delivery, disease diagnostics, gene therapy, photodynamic therapy, catalysis, and broad therapeutic activity [15]. These are mostly synthesized by top-down (such as chemical oxidation, laser ablation, and electrochemical oxidation) and bottom-up (such as ultrasonic, hydrothermal, and microwave) approaches using various carbon sources (e.g., plants, biomass, sugar) [15]. CDs possess excellent fluorescence, electro-chemiluminescence, phosphorescence, surface functionalization capacity, aqueous dispersibility, tunable emission properties, excellent safety/biocompatibility, chemical stability, photobleaching resistance, nontoxic, chemically inertness, highly water solubility, excellent photochemical and physicochemical stability [17]. They have amorphous or nanocrystalline cores with sp^2 hybridized carbon; however, sp^3 carbon sometimes exists [18–21]. The integration of nitrogen-rich PEI as a passivating agent in CDs increases the fluorescence, photoluminescence, DNA loading, and cellular uptake [17]. Noble metals (Au, Ag, Ru, Pt) are often integrated with CDs to fabricate nanohybrids, which can integrate their individual strengths after hybridization for diverse applications. The electronic property, localized surface plasmon resonance, and up-conversion ability are enhanced. Techniques such as hydrothermal method, reductive chemical method, and assembled method are mostly used to prepare noble metal-CDs nanohybrid [22].

5.1.2 Graphene Quantum Dots (GQDs)

GQDs are the counterparts of semiconductor inorganic quantum dots that possess low cost and are free from toxic metals. They possess 0D and are obtained from 2D graphene sheets. Such materials possess various layers with lateral dimensions ranging from less than 10 nm to 100 nm and possess a graphene-like framework within the dot [23]. They possess excellent resistance to thermal and light-induced degradation due to their robust carbon framework with strong aromatic stabilization and π-electrons delocalization across the planar structures. The GQDs are often fused with other small molecule organic semiconductors (SMOSs) to produce a variety of optoelectronic possessions. In addition to these, they possess excellent optical properties (absorption and emission), chemical stability, surface functionalization capacity, good dispersibility in solvents, negligible photobleaching, and can be fabricated with cost-effective and environmental impact via top-down and bottom-up approaches. The bottom-up approach permits the functionalization of the GQDs with heteroatoms, for example, nitrogen, oxygen, sulfur, and boron, with varied optical properties. Nanocomposite of GQD offers excellent prospects in pharmaceutical and healthcare applications owing to its photoluminescence, higher surface area, and excellent mechanical strength [17]. The surface modification of GQD with PLA and PEG offers excellent fluorescence properties over a wide pH range and improves physiological stability, which facilitates its use in biological systems for cell imaging and other biological applications [17].

5.1.3 Fullerenes

Fullerenes are carbon-based 0D spherical nanomaterials that typically exist in the form of C_{60} extensively applied in the area of nanomedicine, drug delivery, and solar cells [4, 14]. The carbon atoms in fullerenes are aligned like a soccer ball comprising hexagons and pentagons with a size close to 1 nm [4]. The carbon atoms in fullerenes are linked by single (sigma) and double (pi) bonds [17]. Fullerenes exist in 3 different forms: spherical, tubular, and ellipsoidal. The chemical formula for spherical fullerenes is represented as C_n, where "n" denotes the overall number of carbon atoms [4]. They occur in numerous states and dimensions, containing 30 to 3000 carbon atoms. The fullerenes having <300 carbon atoms are known as buckyballs whereas the fullerenes having >300 carbon atoms are known as giant fullerenes [4]. They possess unique electrical conductivity, chemical reactivity, and high hydrophobicity. Fullerenes exhibit high resistance to pressure, high temperature, and radiation [4]. The low solubility of fullerenes limited its biological utility [17]. Covalent surface modifications are done to enhance its solubility. The introduction of amine, hydroxyl, and carboxyl groups into fullerene produces water-soluble derivatives that have utility in the biological environment [17].

5.1.4 Carbon Nanoparticles (CNPs)

CNPs are ultra-low spherical/quasi-spherical nanomaterials having a size of 1 10 nm. They are mostly similar to CDs but possess more irregular structures. They have been studied in sensing, drug delivery, and as catalysts in many chemical reactions.

5.1.5 Small Molecule Organic Semiconductors (SMOSs)

SMOSs are one of the 0D carbon nanomaterials. These carbon materials have well-defined chemical structures, high purity, and are often easier to synthesize and modify. They exhibit tunable electrical, optical, and mechanical properties.

5.2 Carbon-Based One-Dimensional (1D) Nanomaterials

1D carbon materials possess 2D in the nanometric range. 1D carbon nanomaterials include carbon nanotubes (CNTs), carbon nanofibers (CNFs), and graphene nanoribbons (GNRs). They are generally used as drug delivery systems, sensors, catalysts, and optoelectronic devices.

5.2.1 Carbon Nanotubes (CNTs)

CNTs are seamless, hollow, cylindrical nanostructures composed of graphite and formed via a hexagonal lattice of carbon atoms [14, 24]. Based on the structure, CNTs are broadly regarded as single-walled carbon nanotubes (SWCNTs) and multiwalled carbon nanotubes (MWCNTs) [4, 24, 25]. The SWCNTs and MWCNTs possess a diameter of 0.4–2.5 nm and 10–100 nm, respectively [14, 17]. SWCNTs comprised of a single graphene sheet rolled into a cylindrical structure. The diameter of the tube and the chirality influence the properties of SWCNTs. SWCNTs are the 1D carbon-based nanomaterial that has been reported for wide theranostic utilities in the biomedical field. Integrating 1D SWCNTs with 0D carbon materials, such as SMOs, produces a hybrid nanostructure with altered optical properties. The SWCNTs possess a wide scope for surface passivation/functionalization with varieties of chemicals. The SWCNTs possess improved cellular internalization compared to MWCNTs [17]. MWCNTs comprise several layers of concentric graphene cylinders, each nested within the other. Mechanically, the MWCNTs are more robust but less flexible compared to SWCNTs. Surface modification of CNTs via non-covalent or covalent interactions enhances its solubility and biocompatibility [17]. MWCNTs are widely used for various applications as they are cheaper than SWCNTs [24]. The exceptional properties of CNT, for example, high surface area, well-defined structure, cell penetrability, and carrier for biomolecules (enzyme, nucleic acid, and protein) and drugs, represent its suitability for therapeutic applications [17]. The coating of CNTs with polymers or biopolymers (poly (amidoamine) dendrimers (Poly(amidoamine) or polyethyleneimine) enables them to serve as efficient carriers for gene delivery [17].

5.2.2 Graphene Nanoribbons (GNRs)

GNRs are thin strips of graphene [17]. When graphene is cut as a strip with <10 nm in a particular direction, the nanostructure is called GNRs. The GNRs possess altered optical properties and electronic structure compared to graphene. The surface passivation with heteroatoms is done with GNRs to improve the optical and therapeutic properties. They exhibit semiconducting or semi-metallic behavior based on their width and edge arrangement (zigzag, armchair, or chiral).

5.2.3 Carbon Nanofibers (CNFs)

CNFs are filamentous nanomaterials with excellent electrical and mechanical properties comparable to CNTs, with diameters ranging from 5 to 200 nm [24]. However, compared to CNTs, the CNFs are not hollow [24]. CNFs are similar to CNTs but typically possess less defined, more disordered structures. In CNFs, the graphene layers are oriented either perpendicularly or at an angle relative to the fiber axis [24]. Very common structures of CNFs are "stack of coins" (or stacked or plane-parallel), "Christmas-tree structure" (or stack of cones, coaxial-conical, fishbone), and "stack of cups" (or lampshades or bamboo) [24]. They are fabricated via electrospinning and chemical vapor deposition. They possess high surface area, can be either hollow or solid, and possess excellent mechanical strength. CNFs are explored for usage in several fields, including materials science and the chemical industry, owing to their larger surface area, nanoscale diameter, superior electrical conductivity, mechanical properties, and excellent chemical stability [25]. They are fabricated by electrospinning, drawing, phase separation, templating, and chemical vapor deposition methods [13, 25]. CNFs are classified into three categories, such as herringbone, platelet, and strip, according to the arrangement of the graphene layers. Herringbone refers to the layers of graphene positioned diagonally to the fiber axis, the platelet is perpendicular to the fiber axis, and the strip is parallel to the growth axis [25]. CNFs possess excellent mechanical properties and high thermal and electrical conductivity making them highly appropriate for incorporation into various matrices, such as thermoplastic, thermoset, and ceramic metal [25]. Their unique properties allow surface functionalization and modifications for various applications [25].

5.3 Carbon-Based Two-Dimensional (2D) Nanomaterials

2D nanomaterials possess one of the dimensions in the nanoscale range.

5.3.1 Graphene

Graphene is a nontoxic, less expensive carbon material synthesized from graphite. Graphene comprises of a single layer of sp^2-bonded carbon atoms arranged in a hexagonal framework like a honeycomb [4, 14, 25, 26]. They possess high surface area, optical transparency, extremely high electrical conductivity, thermal conductivity, optical properties, catalytic properties, high Young's modulus, mechanical strength, and flexibility [4, 25, 26]. They offer sufficient electroactive centers by improving the surface area for biomolecule immobilization [25]. Graphene-based materials (GBMs) possess a distinctive structure with excellent physicochemical properties. The structure of such carbon materials is mostly layered and stacked upon one another by van der Waals inter-sheet forces. Their planar structure makes them a good acceptor of electrons. The bonding of 2D nanomaterials with 0D nanomaterials greatly improves the physicochemical attributes and stability. The GBMs possess a high surface area that offers a larger scope for therapeutics (drug/gene) delivery and allows better interaction with biomolecules/chemicals [17]. The presence of delocalized pi (π) electrons offers electrophilic interactions of graphene with drugs, biomolecules, and organic molecules containing aromatic structures [17]. The hydrophobic properties of graphene allow the adsorption of hydrophobic polymers or molecules onto its surface through van der Waals bonding that enables easy penetration through lipophilic cell membranes during therapy [17]. Graphene is often functionalized with pnictogens (P, Sb, N, As, Bi), halogens (Br, F, I, Cl), chalcogens (Se, O, S), silicon, boron, and other functional groups (aryl and alkyl hydrocarbons) to modify its functionality [26]. The synthesis of graphene is carried out by top-down and bottom-up approaches [26]. The top-down approach includes the separation of stacked graphite layers to obtain single graphene (e.g., exfoliation: solid phase, liquid phase, electrochemical, reduction: chemical, thermal, electrochemical), whereas the bottom-up approach includes synthesis of graphene sheet using various carbon sources (e.g., chemical vapor deposition, epitaxial growth, chemical and thermal synthesis) [26].

5.3.2 Graphene Oxide (GO)

GO is the oxidized derivative of graphene, consisting of a single layer of carbon atoms organized in a hexagonal lattice [4]. They possess exceptional surface attributes, improved surface area, and layered structure [25]. Graphite, after oxidation, produces graphite oxide, which, after exfoliation, produces graphene oxide [26]. These are very similar to graphene but contain oxygen-containing functional groups, such as hydroxyls, epoxides, carbonyls, and carboxyls, that are adhered to the edges or basal plane of the graphene [4]. These functional groups increase the interlayer distance, leading to the formation of a more hydrophilic layer [25]. Pure graphene is hydrophobic and aggregates in water. In contrast, the existence of the oxygenated functional groups in GO renders them more hydrophilic and highly reactive, allowing excellent solubility with water or other solvents for various applications. They possess surface functionalities, greater surface area, hydrophilicity, electrical conductivity, and excellent mechanical properties. They are mostly used for drug delivery, energy storage and supercapacitors, filter beds, sensors, catalysts, electronics, and various therapeutic purposes [25]. The integration of cationic polymers increases the hydrophilicity and dispersibility of GO.

5.3.3 Reduced Graphene Oxide (RGO)

The chemical reduction of GO produces RGO [14, 26]. The reduction usually requires thermal treatment and hazardous reducing agents, for example, hydrazine or sodium borohydride. Replacing hazardous reducing agents with green reducing agents has gained significant attention for the green synthesis of RGO [4, 25, 26]. RGO is also known as chemically modified graphene, functionalized graphene, or chemically converted graphene [4]. Various criteria, such as visual properties, carbon-to-oxygen atomic ratio (C/O ratio), and electrical conductivity, are evaluated to check the changes that occurred in GO before and after the reduction [4].

6 Carbon-Based Hybrid Nanomaterial

Carbon-based hybrid nanomaterials are the nanomaterials constituted by carbon atoms, often integrated with other elements or materials (polymers, semiconductors, metals, or inorganic compounds) or carbon-based nanomaterials or other nanomaterials to improve their properties or achieve new functionalities [27]. Such hybrid nanomaterials are fabricated in such a way that their size falls within the nanometric range, i.e., 1–100 nm. The bonding of carbon-based nanomaterials with various materials produces hybrid nanomaterials, which can possess synergistic properties that neither of the components would possess on their own. Various carbon-based nanomaterials, such as semiconductor QDs carbon dots, can integrated with organic polymers, and the resultant hybrid nanomaterials can be used for

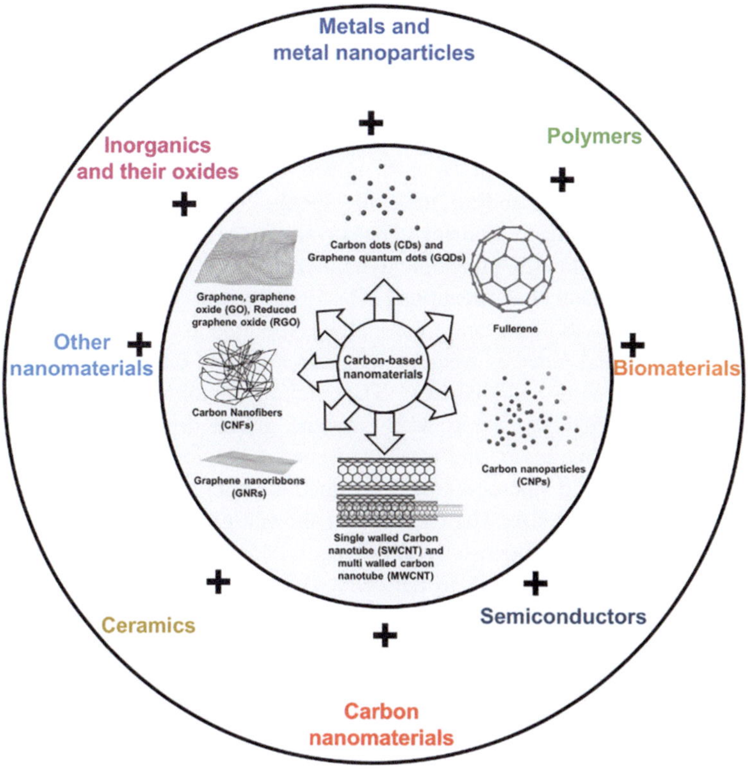

Fig. 10.3 Various carbon-based hybrid nanomaterials

theranostics use [28]. Various nanomaterials, including carbon nanotubes, QDs, graphene, metal nanoparticles, and biomolecules, have been integrated into metal-organic frameworks to produce hybrid nanomaterials [28, 29] (Fig. 10.3).

6.1 Carbon-Based Hybrid Nanomaterials and Their Applications for Metabolic Syndrome

Carbon-based hybrid nanomaterials include:

6.1.1 Carbon Quantum Dots (CQDs)-Based Hybrids

CQDs are 0D carbon nanomaterials that possess fluorescence properties and can be surface functionalized with various organic or inorganic materials to modulate their physicochemical and therapeutic properties [15, 16]. They are used in sensing, imaging, drug delivery, and electronics. CQDs-based hybrids or carbon dots-based hybrids (CDHs) are nanocomposites obtained via the incorporation of CDs with various other nanomaterials (e.g., semiconductor materials, metal nanoparticles, biomaterials, inorganic oxides) in several forms [16]. The hybridization occurs by physical and chemical methods [27]. The fabrication of CDHs is done in two key approaches: (a) dispersion in a matrix of another material or (b) doping [27]. Due to the broad advantages of CDs, scientists have broadened their attention outside CDs alone to integration with various materials with the aim of enhancing their potential applicability. By integrating CDs with various nanomaterials, the newly formed CDHs inherit the benefits of CDs with added functionalities of the hybridized material, thus further expanding their therapeutic utility [27]. The CDHs are exploited for bioimaging, biosensing, photodynamic therapy (PDT), drug & gene delivery, cancer theranostics, ultrasonic dynamics, phototherapy, sonodynamic therapy, and many biomedical applications [27].

CDs-loaded phenylboronic acid microgel was reported as a smart nanohybrid for glucose detection and glucose-responsive insulin secretion under physiological pH for effective treatment of diabetes. The carbon nanohybrid was fabricated utilizing N, S-doped CDs, along with ethylene glycol dimethacrylate which is used as a cross-linking agent, and acrylamide and 4-vinylbenzene boronic acid as monomers. The hybrid microgels demonstrated a spherical shape with an even surface with nanometric size. The nanohybrid showed glucose-responsive swelling at a physiological pH and glucose-triggered insulin release. The in vitro cytotoxicity displayed good biocompatibility to the cells. Overall, the nanohybrid displayed a novel platform for theranostic use [30].

Cholesterol is recognized as the potential cause for the formation of blockage in the coronary artery and, subsequently, heart attacks/strokes. CDs-hemoglobin nanohybrid-based biosensors have been stated for cholesterol monitoring in blood. The cholesterol sensing was achieved via improved fluorescence of CDs that initially existed in a quenched state as a CDs-hemoglobin complex. With the existence of cholesterol, the CDs detached from the CDs-hemoglobin complex by hydrophobic interactions between cholesterol and Hb, which enhances the intensity of the fluorescent dye. The biosensor was reported to sense rapidly, with selective detection of cholesterol within 0 to 800 μM, with a LOD of 56 μM in human plasma [31].

In another work, Au-CDs nanohybrid was reported as a dual-mode enzyme-free sensor for cholesterol sensing. The gold particles showed a dual role in colorimetric detection and fluorescence quenching in the presence of cholesterol, wherein the CDs serve as the fluorescent components. The nanohybrid showed high specificity towards cholesterol with a LOD of 0.12 ppm within a linear range of 50–300 ppm [32].

6.1.2 Graphene Quantum Dots (GQDs)-Based Hybrid

The integration of CDs and GQDs facilitates the physical attachment of biomolecules. They allow fast electron transfer dynamics and have specificity and flexible surface properties owing to their distinctive plasmonic structure. The CDs-GQDs-based nanohybrid was reported as a dual immune-monitoring platform to sense a variety of circulating biomarkers, comprising CRP. The CRP serves as a potential marker for assessing CVD risk in healthy individuals and individuals with a medical history of acute ischemia and myocardial infarction (MI). The CRP has an affinity to LDL that mainly triggers atherosclerotic plaque [33].

The hydrothermally fabricated GQDs loaded with antibodies have been reported for myoglobin sensing, a potential cardiac biomarker, having a LOD of 0.01 $ngmL^{-1}$. It also demonstrated myoglobin selectivity in the presence of other proteins [34]. Acetic acid-conjugated GQDs were reported for the detection of cardiac troponin I using cyclic voltammetry and amperometry with a LOD of 0.02 $ngmL^{-1}$. The carbodiimide chemistry among the N-H group of cardiac troponin I and the -COOH of GQDs permitted the biomarker detection [35]. In another work, PEG-conjugated GQDs demonstrated its ROS scavenging potential that often contributes to CVDs [36]. Amine-functionalized GQDs integrated with troponin antibodies were reported as cardiac immunosensors for sensing troponin I antigens in blood. The anti-cardiac troponin I was covalently linked to amine-functionalized GQDs through carbodiimide chemistry. The developed biosensor was effective in early and selective sensing with linearity within 0.001 to 1000 ng mL^{-1} and a LOD of 0.19 pg mL^{-1} through a fluorescence resonance energy transfer mechanism [37].

The risk of CVD increases with the advancement of age. MicroRNA-34a is a significant marker associated with cellular senescence and aging, which can be beneficial for early diagnosis. Gold nanoflower integrated with GQDs to produce hybrid nanomaterial, which has been reported as a fluorescent probe for sensing and imaging microRNA-34a in vitro as well as in vivo. The developed nanoprobe demonstrated no cytotoxicity (H9C2 cells) and showed excellent in vitro and in vivo signal imaging. By applying the concept of Förster resonance energy transfer (FRET) the hybrid probe is reported to measure target miRNA-34a at a low concentration (0.1×10^{-15} M). The nanohybrid used for bioimaging heart in C57BL/6 J mice. The hybrid nanoprobe has been reported for early diagnosis of CVD and proposing therapeutic protocols [38].

6.1.3 Graphene, Graphene Oxide (GO), and Reduced Graphene Oxide (RGO)-Based Nanohybrids

Graphene is often combined with various materials like polymers (polyetheretherketone, polycaprolactone), metals (for example, silver, gold, platinum), metal oxides (for example, ZnO, TiO_2,), or ceramics (hydroxyapatite), and other carbon nanomaterials (e.g., CNTs). Such nanohybrids are widely used for drug delivery, biomedical applications, sensors, optics, conductive materials, photonics, electronics, actuators, and storing of energy [26]. In graphene-based nanohybrids, metallic/polymeric/ceramic nanoparticles act as stabilizers that avoid the accumulation of graphene sheets via van der Waals forces among graphene layers [26]. Techniques such as the solvothermal/hydrothermal carbonization method, sol-gel method, and in situ growth are mostly utilized for the fabrication of graphene-based nanohybrids [26]. Silver nanoparticle-graphene orgraphene oxide (GO) nanocomposites have been considered as potential hybrid materials for broad purposes [26]. GO can be hybridized with polymers, metallic nanoparticles, or other nanomaterials. Such nanohybrids are used for drug delivery, biomedical purposes, electronics, and water purification. Silver nanoparticle-graphene hybrid nanomaterials are synthesized by the linked-polymer method, electron beam irradiation, laser ablation, modified photochemical, and green synthesis methods

[26]. CNT-graphene nanohybrid materials also achieved significant consideration due to their excellent physical attributes, for instance, thermal stability, improved electronic conductivity, and mechanical strength. 1D CNTs can integrate with 2D graphene horizontally, perpendicularly, or in a wrapped manner to produce hybridized nanomaterials in a variety of topological architectures [26]. Cellulose nanocrystals/nanofibrils often integrate with graphene to develop hybrid nanomaterials with improved physical and mechanical properties [26]. Ferrite nanoparticles possess excellent biocompatibility and nontoxicity, are easily synthesized, and are often hybridized with GO to produce hybrid nanomaterials [26]. Titanium dioxide (TiO_2) is one of the attractive commercially available photocatalysts. Because of its nontoxicity, chemical stability, and relatively low cost, TiO_2 (ultrafine TiO2 nanocrystals, titania peroxo complex) is widely used in composite preparation with graphene [26]. Recently, zinc oxide (ZnO) has attracted significant attention because of its cost-effective, nontoxicity, accessibility, abundant, and electrochemical activity for the fabrication of novel nanohybrids with graphene [26]. Liquid phase exfoliation and solvothermal techniques are mostly employed for the fabrication of ZnO-graphene or ZnO-reduced graphene oxide (RGO) nanohybrids [26].

Blood glucose sensors are important in clinical diagnostics because of their significance in diabetes management [39]. Regular monitoring of blood glucose levels is required to manage diabetes. Electrochemical sensor-based porous tertiary graphitic carbon nanohybrid doped with polymelamine/cobalt/nitrogen was studied for monitoring glucose in human blood. The fabrication of nanocomposite was accomplished via a single-step electrodeposition technique under cyclic voltammetry. The fabricated carbon-based nanosensor demonstrated high sensitivity, excellent stability, and reproducibility. It demonstrated excellent electrochemical sensing for monitoring glucose with a LOD of 7.8 μM [40].

In another work, electrochemical sensing of glucose was carried out using palladium and silver nanoparticles fixed on RGO/MWCNT carbon nanohybrids. The fabricated sensors displayed sensitivities of 13.16 ± 0.4 and 5.22 ± 0.07 $\mu A\ mM^{-1}\ cm^{-2}$ across the concentrations of 0.05 to 4 and 4 to 42 mM, respectively [41].

Type-2 diabetes mellitus is now recognized as an inflammatory disease because of the involvement of meta-inflammation. The adipose tissue inflammation triggered by obesity is one of the initial pathologies in the case of meta-inflammation, causing the initiation of insulin resistance and Type-2 diabetes mellitus. Hyaluronic acid-graphene oxide quantum dots nanocomposite was developed and reported as a dual-acting agent in terms of therapeutic agent and drug delivery for meta-inflammation. Metformin-loaded nanohybrid effectively downregulated the proinflammatory cytokine levels and at a low dose reestablished antioxidant status compared to free metformin in diet-induced obese mice as well as in RAW264.7 cells. The carbon-based nanohybrid improved the effectiveness of metformin by primarily functioning as both a therapeutic agent and a drug-delivery platform [42].

6.1.4 Fullerene-Based Hybrids

Fullerene can be integrated with polymers, metals, or biomolecules to fabricate nanomaterials with specialized electronic, optical, or catalytic properties. Such nanohybrids are mostly used as carriers for drug delivery, solar cells, and catalysis. Fullerene or C60 derivatives act as a category of carbon nanomaterials with antioxidant properties. The electron-lacking atoms on the fullerene cage make them suitable for ROS adherence and move the electron density to the fullerene, leading to the reduction of ROS-induced oxidative stress. The fullerene-integrated alginate hydrogel was explored as an injectable therapeutic delivery platform with antioxidant activities. The hybrid nanomaterial showed outstanding injectable and mechanical strength with scavenging of the superoxide and hydroxyl free radicals. The formulation showed an improved survival rate of stem cells derived from brown adipose tissue and caused cardiomyogenic differentiation even under ROS conditions. The nanohybrid showed reduced ROS level, improved survival rate, triggered angiogenesis, and recovered cardiac functionality in the myocardial infarct rat model [43].

Atherosclerosis involves the buildup of plaque in the wall of the blood vessel and the narrowing of the lumen. Plaque can dislodge and enter the bloodstream, leading to acute myocardial infarction and stroke. Gadolinium metallofullerenes have been reported for detecting atherosclerotic plaque lesions using cardiovascular magnetic resonance. The atherosclerotic-targeting contrast agent identifies foam cells expressing the scavenger receptor (CD36)- in vitro and is selectively internalized through the CD36 receptor present on the atherosclerotic plaque, enabling enhanced visualization of lesions in a physiological environment. The novel nanomaterial demonstrated time-dependent buildup in atherosclerotic plaque lesions of ApoE $^{-/-}$ mice and can detect atherosclerotic disease [44].

6.1.5 Carbon Nanotube (CNT)-Based Hybrids

CNTs are cylindrical molecules made up of carbon atoms, which can be functionalized or hybridized with materials like metals (e.g., gold, silver), polymers (polyurethane, polyhydroxybutyrate, chitosan/sodium alginate), carbohydrates (starch) or semiconductors [16]. These hybrids can be employed in areas such as sensors, energy storage (batteries and supercapacitors), and catalysis. People with diabetes require frequent monitoring of glucose levels. Self-evaluation of glucose in blood via finger-prick blood or venepuncture is regularly carried out for diabetes management. Such blood sampling techniques are painful and often lead to finger calluses and are also associated with mental and physical stress to the patient. The development of noninvasive, painless, low-cost glucose-sensing using other body fluids, such as saliva, sweat, and tears, would be more patient compliance for regular glucose monitoring [39]. A cobalt phthalocyanines/SWCNTs with RGO nanohybrid was successfully prepared and used for nonenzymatic detection of saliva glucose content. The developed sensor displayed excellent, reproducibility, selectivity and storage stability. The synergistic effect among the three materials showed improved analytical performance. The results showed excellent glucose detection in saliva with a sensitivity of 992.4 $\mu A{\cdot}mM^{-1}{\cdot}cm^{-2}$, a LOD of 0.12 μM, and a fast response time of 1.2 s [39].

CNTs-chitosan nanohybrid was prepared and evaluated as an immunosensor for myoglobin detection, a potential marker for CVD, diabetes, metabolic syndrome, and hypertension. The CNTs-chitosan nanohybrid immunosensor showed a LOD of 4.2 ng/mL within a linear range of 1 to 4000 ng/mL for myoglobin with high reproducibility. The nanohybrid of carboxylic acid-engineered CNTs with N-succinylated chitosan significantly increased the aqueous solubility of CNTs. The developed novel immunosensor showed excellent detection of myoglobin, unaffected by the presence of nonspecific antigens. The myoglobin-spiked serum sample reflected the highest recovery (100.262–118.55%), demonstrating great promise for myoglobin detection in actual clinical samples [45].

In another work, highly efficient CNTs hybridized with molybdenum disulfide and nickel nanoparticles were developed by hydrothermal reaction and evaluated as nonenzymatic electrochemical glucose sensors. The electrocatalytic property of the fabricated carbon nanocomposite towards glucose oxidation was analyzed using amperometry and cyclic voltammetry in alkaline media. It demonstrated an excellent sensitivity of 1212 $\mu A \cdot mM^{-1} \cdot cm^{-2}$ with linearity within 0.05–0.65 mM and LOD of 0.197 μM within 3 s response time, reflecting promising applications for direct sensing of glucose [46].

Elastomeric polyester-CNT scaffolds were reported for cardiac tissue engineering. The polymer is employed to fabricate scaffolds for tissue engineering to provide mechanical support, organization, and maturation of tissues. The cardiac tissue possesses flexibility, contractility, and electrical conductance, so the polymeric scaffolds for cardiac tissue regeneration should show both elasticity and high electrical conductivity. The CNTs were introduced into poly(octamethylene maleate (anhydride) 1,2,4-butanetricarboxylate) to develop an elastomeric scaffold for cardiac tissue engineering. The fabricated scaffold exhibited improved conductivity, surface moduli, increased swelling, increased structural support in an aqueous environment, and greater tissue maturity. The overall result demonstrated the scaffold as a versatile biomaterial for cardiac tissue engineering [47].

CNTs embedded in embryoid bodies were reported for cardiac differentiation. The CNTs improved the mechanical integrity and electrical conductivity of the embryoid bodies. The cardiac differentiation due to embryoid-CNTs was found to be significantly higher than that of the embryoid. The embryoid-CNTs specifically enhanced the differentiation of cardiac tissue and the beating activity of the embryoids [48].

In another work, 3D printed polycaprolactone CNTs nanocomposite scaffolds were prepared using a 3D printer for cardiac tissue engineering. The scaffold showed enhancement of hardness, elastic modulus, and greater peak load. The cytotoxicity assay and fluorescence imaging results revealed the attaching of myoblast cells to the scaffolds while maintaining a healthy state [49].

CNTs-gelatin-methacrylate anhydride nanocomposite hydrogels demonstrated a promising scaffold for the construction of cardiac tissue. It is demonstrated to support the growth and functionality of cardiomyocytes. The upregulation of the electrical and mechanical junction proteins was observed in the study. It showed intercalated disc formation and improved cardiac tissue regeneration [50].

6.1.6 Carbon Nanofibers (CNFs)-Based Hybrids

CNFs are thin and strong fibers primarily made up of carbon atoms. CNF-reinforced composites are mainly employed in orthopaedic implants, for example, screws, bone plates, and artificial joints, because of their excellent strength-to-weight ratio [16]. These hybrid nanomaterials provide exceptional durability and strength along with lightweight and biocompatible properties. These properties make them applicable for load-bearing biomedical uses, for example, bridge support and dental crowns [16]. Materials such as polyether ether ketone, ZnO nanoparticles, $NiCo_2O_4$ nanoneedle, and polylactic acid are integrated with CNFs to fabricate CNFs-based hybrids [16].

$NiCo_2O_4$ nanoneedle-decorated electrospun CNFs nanohybrids were reported for sensitive nonenzymatic detection of glucose. The nanohybrid was prepared using a one-pot hydrothermal process. The fabricated nanohybrid demonstrated a three-dimensional hierarchical structure with excellent glucose sensitivity within 0.005–19.175 mM, LOD of 1.5 μM, and a sensitivity of 1947.2 μA mM^{-1} cm^{-2}. The developed biosensor showed long-term stability and excellent selectivity for the detection of glucose in real serum samples [51].

CNFs loaded with chitosan scaffold were prepared using the precipitation method. The CNFs are reported for homogeneous dispersion in the chitosan matrix. The scaffold demonstrated elastic modulus comparable to that of rat myocardium and conductivity with electrical properties. The carbon-based nano scaffold was reported to enhance the characteristics of neonatal rat cardiomyocytes, promoting cardiac gene expression related to muscle relaxation and improving the electrical conductivity of cardiac tissue. The integration of carbon CNFs into porous chitosan scaffolds enhanced the properties of cardiac tissue via improved transmission of electrical signals between the cells, presenting a promising stage for cardiac tissue engineering [52].

7 Advantages of Carbon-Based Hybrid Nanomaterials

The carbon-based hybrid nanomaterials possess the following advantages:

- *Improved/synergetic properties*: The hybridization of carbon with other materials produces materials with improved therapeutic activity, mechanical strength, thermal stability, chemical reactivity, and electrical conductivity for a wide range of utilities.
- *Tailored functions*: By choosing hybridizing suitable materials, the properties of the carbon-based nanomaterials, like targeting ability, optical behavior, or catalytic activity can be tailored.
- *Broad applicability*: Carbon-based nanohybrids possess wide utility in diverse fields, including drug delivery, sensing/biosensing, bioimaging, energy, electronics, and environmental applications.
- *Safety, low toxicity, and biocompatibility*: They possess excellent biocompatibility, negligible toxicity, and a good safety profile.

- *Economic and sustainability*: Carbon-based nanohybrids are abundant, relatively low-cost, and eco-friendly compared to other nanomaterials.

8 Synthesis of Hybrid Carbon-Based Hybrid Nanomaterials

These are mostly synthesized by in-situ and ex-situ methods. An organic or inorganic component is prepared using the in situ approach while functionalized or nonfunctionalized carbon nanomaterials (CNTs, CNFs) are present. The component grows on the CNTs as particles, thin films, or nanowires [24]. Alternatively, the ex situ method concern the separate synthesis of the nanoparticles of the desired size and morphology followed by attachment to the carbon exterior surface via covalent, non-covalent bonds or electrostatic interactions [24]. The synthesis of hybrid nanomaterial depends on the type of method used, purity of carbon nanomaterial and surface functionalization [24].

8.1 Sol-Gel Method

The sol-gel method is a novel procedure for the fabrication of carbon-based nanohybrids. The surface of carbon nanomaterials can have a homogeneous distribution of nanoparticles by this technique. The process is diffusion regulated, meaning that precursors polymerize and create nanoparticles as the pH changes. After fabrication of nanohybrids, various coatings can be added on the surface of carbon nanomaterials based on precursors [24].

8.2 Hydrothermal/Solvothermal Method

The hydrothermal method involves the synthesis of nanohybrids with the use of a special autoclave under fixed high temperature and pressure. When solvents other than water (ethanol, acetone) are used, the technique is called the solvothermal method. Precursor or reagents are kept into the autoclave and allowed to be heat under a thermostatic oven and left a certain time without any supervision. The nanomaterials are synthesized via carbonization under the provided temperature and induced pressure in the specialized autoclave [24].

8.3 Chemical Vapor Deposition on Catalyst Nanoparticles

The chemical vapor deposition process creates the required carbon nanomaterial by immersing a catalyst (substrate) in a precursor vapor. By eliminating them from the surface of the nanoparticle-catalyst (substrate), chemical vapor deposition produces pure carbon nanomaterials [24], or to create a hybrid nanomaterial, nanomaterials and catalyst particles can be combined without separation. The catalyst particles serve as both an arming dopant and a substrate [24]. Uniformly thin coatings applied to surfaces improve biocompatibility and reduce implant rejection risk [16]. This method is widely used for the development of drug delivery systems, where a thin film is deposited onto a substrate and loaded with therapeutics [16]. This provides sustained and regulated medication delivery. Poly (ethylene glycol dimethacrylate) has been applied to the surface of GNFs using this method [16].

8.4 Plasma-Enhanced Chemical Vapor Deposition (PECVD)

A more sophisticated kind of chemical vapor deposition is PECVD. This method was created to prevent the problem of high coating temperatures in chemical vapor deposition, which might change some materials' integrity [16]. PECVD can create thin films at lower temperatures than chemical vapor deposition, which makes it appropriate for a wider variety of materials, both organic and inorganic [16]. The process involves generating a plasma with electrical energy that ionizes natural gas and produces free radicals, which then polymerize to form the deposition layer [16]. The PECVD is widely used for coating biomedical implants using precursor gases, such as methane, silane, and phosphine [16]. The PECVD coatings have been reported to be non-cytotoxic and biocompatible across multiple biomedical uses [16].

8.5 Electrostatic Self-Assembly (ESA)

This technique allows robust bond formation between charged particles and a charged substrate on the surface [24]. ESA is a simple and effective method for fabricating carbon-based hybrid nanomaterials through the attraction between oppositely charged components. In this technique, carbon nanomaterials such as GO or CNT are first functionalized to carry a specific surface charge, while their counterparts, such as metal nanoparticles, metal oxides, or polymers are oppositely charged. When mixed or alternately deposited onto a substrate, these species spontaneously assemble into uniform layers via Coulombic interactions. The process, often performed in a layer-by-layer manner, allows precise control over film thickness and composition.

8.6 Ion-Beam (IB)

The IB deposition is seen to be a potential method for enhancing biomedical coatings, primarily through precise surface changes that increase surface selectivity and biocompatibility. The IB, in addition to plasma treatment, improves biomaterials by offering superior surface selectivity [16]. In this method, a focused beam of high-energy ions is directed onto a target material, such as graphite, causing the ejection (sputtering) of atoms or clusters. These sputtered species are then deposited onto a substrate, forming a thin carbon-based film or hybrid layer.

9 Characterizations of Carbon-Based Hybrid Nanomaterials

9.1 Organoleptic Properties

The odor, color, transparency, agglomeration, precipitation, or sedimentation are mainly evaluated for asserting the organoleptic properties [19].

9.2 Elemental Characterization

X-ray photoelectron spectroscopy (XPS) and energy dispersive analysis of X-rays (EDAX) are often used for the analysis of elements, chemical states, and functional groups for surface chemistry analysis [19, 22, 26, 41]. Apart from that, XPS is also used to determine the oxidation states of metal atoms and the nitrogen bonding in nanohybrids [26]. The efficiency of the surface functionalization and confirmation of effective synthesis of nanohybrids are also obtained from XPS analysis [26].

9.3 Chemical Integrity

The chemical integrity of carbon-based nanomaterials and nanohybrids is characterized by Fourier transform infrared spectroscopy (FTIR) and EDAX. The FTIR provides information on surface functional groups in the nanomaterials [19, 26, 46]. The interactions among the individual components in the nanohybrid are studied by observing the changes in the spectral vibrational frequencies using FTIR [26]. Analyzing the FTIR spectra of individual components and nanohybrid and observing the shift/disappearing of specific vibrational peaks specific to certain functional groups provides information on the effective formation of nanohybrid or possible chemical interactions [46] (Fig. 10.4). EDAX is used to identify the elemental composition and distribution of carbon-based nanomaterials.

9.4 Raman Spectroscopy

Raman spectroscopy is used to characterize and validate the effective synthesis of nanocomposite [46]. It is a promising technique for investigating the effective synthesis of CNT nanohybrids [46], and the order of graphene particle layers before and after carbon-based nanohybrids are fabricated [26]. Information, such as edge, disorder, grain boundaries, thickness, thermal conductivity, doping, and strain of graphene, can be acquired from Raman spectroscopy [26]. The disorder bands and tangential bands are represented by the D and G bands in the Raman spectra, respectively [46]. The spectral intensity (D-to-G band ratio) or the spectral shifting of pure carbon nanostructures and the hybrid nanomaterials are compared to assure the effective formation of nanohybrid [26].

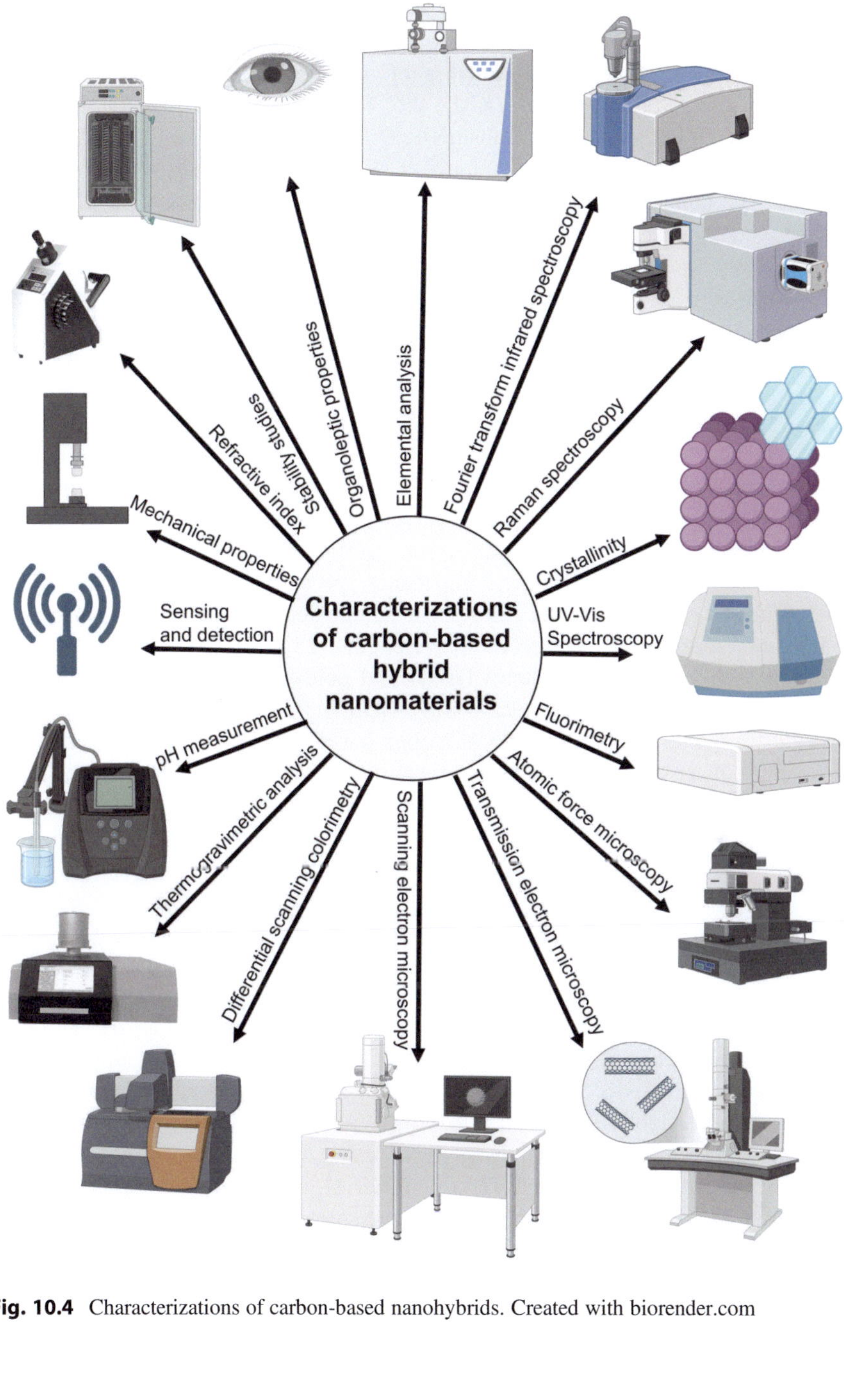

Fig. 10.4 Characterizations of carbon-based nanohybrids. Created with biorender.com

9.5 Crystallinity

The crystalline or amorphous structure of carbon nanohybrids has often been investigated by X-ray diffraction (XRD) [26, 41] and selected area electron diffraction (SAED) in high-resolution transmission electron microscopy (HRTEM). XRD is used to identify a polymorph phase, analyze the abundance of a phase, and compare the compositional efficiency of composites [26]. A reduction or shifting or appearance or disappearance of diffraction peaks provides the idea of effective synthesis/functionalization/hybridization/ anchoring of components in hybrid nanomaterials [26]. The diffused and dotted diffraction pattern of the SAED represents the amorphous or crystalline nature of the nanohybrid.

9.6 Optical and Photochemical Properties

The most often employed optical and photochemical features of carbon-based nanohybrids are fluorescence/photoluminescence (PL), diffuse reflectance spectroscopy (DRS), transient photocurrent response, UV-visible-near-infrared (UV-vis-NIR), and electrochemical impedance spectroscopy (EIS) [22]. The effective synthesis of carbon-based fluorescent nanomaterials can be ascertained from UV-visible spectroscopy at specific wavelengths [18]. Further, the appearance of new spectral peaks of components (silver and gold nanoparticles) due to surface plasmon resonance (SPR) at a certain wavelength in the carbon-based nanohybrid is an indication of effective synthesis of carbon-based nanohybrid [26]. UV-vis spectroscopy is also used for quantitative estimation of various analytes (transition metal ions, such as Pt, Pb, Cu, Au, Fe, Ag, Zn, and conjugated organic compounds) in hybrid nanomaterials [26].

9.7 Morphology

The most common methods for assessing the 2D and 3D surface morphology of carbon-based nanohybrids are transmission electron microscopy (TEM), HRTEM, high-resolution scanning electron microscopy (HRSEM), and atomic force microscopy (AFM) [22, 41, 46]. Morphology, size, thickness, size distribution, roughness, voids or gaps, fracture, and state of dispersion of particles (homogeneity or heterogenicity) in nanohybrids are analyzed using these microscopies [26]. Some of these microscopy, such as HRTEM or HRSEM, contain additional instruments, such as an elemental analyzer (EDAX) or electron diffraction system (SAED), to identify the elemental composition and crystallinity of the nanohybrid, respectively [26, 41, 46]. These techniques are also useful for confirming the effective synthesis of nanohybrid by comparing the morphology of individual components and nanohybrid.

9.8 Quantum Yield (QY)

QY is mainly evaluated for the fluorescent carbon hybrid nanomaterials. The absolute QY or relative QY is mostly evaluated to ascertain the fluorescence behavior of nanomaterials. The relative QY is evaluated with respect to standard fluorescent materials, such as rhodamine-B and quinine sulfate.

9.9 Thermal Properties Analysis

Thermogravimetric analysis (TGA), derivative thermogravimetry (DTG), and differential scanning calorimetry (DSC) are primarily employed to examine the thermal stability and characteristics of carbon-based hybrid nanoparticles. They measure weight loss, heat transfer, and mechanical properties as a function of changes in temperature [26]. In DSC, the heat flow in or out of a material is estimated as a function of temperature or time [26]. It is mainly used to study heat of fusion (ΔHf), degree of crystallinity (Xc), melting temperature (Tm), glass transition temperature (Tg), and compatibility study of nanohybrids [26]. The change in weight/ % weight loss with respect to temperature is studied using TGA/DTG [26]. TGA is also used to study the thermal stability, moisture content, and compatibility among the components of the synthesized hybrid nanomaterials.

9.10 Percent Transmittance (%T) and Refractive Index (RI)

The %T and RI is mostly used to evaluate the transparency of the liquid nanocomposite dispersed in suitable media. UV-visible spectrophotometry is used as a function of light wavelength for estimating the %light transmittance through hybrid nanocomposites [26]. RI is evaluated using a refractometer. The evaluation of RI at various locations of nanocomposite also gives information on isotropicity.

9.11 pH

The pH of liquid nanocomposites is evaluated to confirm their suitability in body physiological fluid for biomedical applications.

9.12 Sensing Study (Metal Ion, Drug, Biomolecule)

Carbon-based nanohybrids are reported as sensors for sensing wide chemicals, therapeutics, and biomolecules. Based on the quenching or enhancement of fluorescent intensity, they mostly sense various moieties. The hybrid nanomaterial utilizes the sensing property of each component for effective, selective detection of target molecules at a very low concentration.

9.13 Mechanical Strength Analysis

It is evaluated by analyzing the tensile strength and dynamic mechanical analysis (DMA) [26]. The mechanical properties of nanohybrids rely on their structure [26]. Good dispersion and a strong interfacial interaction between carbon nanoparticles and the selected polymeric/metallic matrix produce hybrid nanomaterial with superior mechanical properties [26]. The mechanical strength of the nanocomposite and its constituent parts is studied by comparing their tensile stress-strain curves, tensile strength, Young's modulus, compressive strength, toughness of fracture, strain at break, and breaking elongation [26]. Nanocomposite/nanohybrid with improved mechanical strength could be produced by homogenous dispersion or alignment of the nanomaterials in a matrix of another material [26]. DMA is one of the most useful techniques for analyzing the viscoelastic properties of carbon-based nanomaterials and their composites [26]. The dynamic mechanical behavior of the nanocomposite is represented as a change in storage modulus and loss modulus as a function of temperature [26]. The results of DMA vary with changes in concentration, type of dispersion (homogeneous or heterogeneous), crystallinity, crosslinking, adhesion, etc. [26].

9.14 Stability

Photostability: It provides the influence of light (UV radiation) on carbon-based nanohybrids during their use in in vitro or in vivo bioimaging. The change in optical possessions (fluorescence or absorbance) is observed after a prolonged exposure to the light.

Chemical stability: The chemical integrity of hybrid nanomaterials is evaluated by analyzing the changes in functional groups or elemental composition over the lapse of time using FTIR, XPS, UV-visible, Raman, and fluorescence spectroscopy intermittently.

Physical Stability: The physical stability of the prepared nanocomposite is evaluated by analyzing the changes in appearance, color, homogeneity, refractive index, precipitation, percent transmittance, signs of agglomeration, and sedimentation.

Thermal stability: It is mainly evaluated using TGA. The stability of carbon-based nanohybrids relative to temperature is studied using TGA analysis.

Heating-cooling cycle and freeze-thaw stability: It is mainly evaluated to identify the best storage temperature and stability of the synthesized nanocomposite with changes in temperature during exposure to various environmental conditions. The sample is exposed to 3 cycles. In the heating-cooling cycle (40 °C for 24 h and 4 °C for 24 h) and 3 cycles of freeze-thaw (−20 °C for 24 h and 25 °C for 24 h).

9.15 Therapeutic Activity

After carbon-based nanohybrids are characterized, they are evaluated for therapeutic/theranostic utilities through in vitro and in vivo models.

10 Other Applications of Carbon-Based Hybrid Nanomaterials

- Pharmaceutical and biomedical applications: Carbon-based nanohybrids are widely exploited for drug delivery, targeted drug/gene delivery, bioimaging-guided therapy, bioimaging, bioelectronic medicine, personalized medicine, regenerative medicine, and various therapeutic purposes, including cancer, metabolic disorder, microbial infections, and many complications [23, 53].
- Environmental Utilities: They are mostly used for environmental pollution remediation [13]. These are mostly used in water and air purification systems to remove heavy metals and organic pollutants. They are also used for the detection of environmental pollutants (heavy metals: Pb^{2+}, Hg^{2+}, toxic organic compounds: methyl blue) [22, 26]. TiO_2/Graphene or TiO_2/CNT-based nanocomposites have been studied for their photocatalyst activity for the photodegradation of toxic methyl blue for environmental utility [26]. Graphene and its hybrid derivatives demonstrated excellent activity for the detection of vapors (NO, SO_2, Cl_2, CO_2, H_2, CO, NH_3, NO_2) and a few organic vapors (acetone, toluene, and benzene) [26].
- Sensing/detection: These nanohybrids possess sensing and complexing abilities with drugs, heavy metals (Pb^{2+}, Hg^{2+}), biomolecules, pollutants, chemicals, and food adulterants. During sensing, they change their optical possessions (quenching of fluorescence or absorbance). The sensing can be based on fluorescent detection, colorimetric detection, electrochemical detection, and electrochemiluminescence detection [22]. The carbon nanohybrid shows improved detection performance compared to free carbon-based nanomaterials [22]. Graphene-based nanocomposites are widely exploited for detecting inorganic ions, microorganisms, and biomolecules and eliminating hazardous elements in the atmosphere [26]. Electrochemical sensors based on carbon nanomaterials have been developed to diagnose metabolic disease markers (glucose) [29, 53].
- Catalysis and co-catalyst: Hybrid carbon nanomaterials act as excellent catalysts in various chemical reactions with improved catalytic performance [22]. Graphene-based nanocomposites demonstrated wide applications in the electrocatalysts of fuel cells, oxygen evolution reaction, methanol oxidation fuel cells, hydrogen evolution reaction, and oxygen reduction reaction [26]. TiO_2/graphene-based nanohybrids are used as co-catalysts in inorganic semiconductors-based photochemical splitting of water [26].

- Energy Storage: Carbon-based hybrid nanomaterials are used in electrochemical energy devices, photovoltaic devices, sediment batteries, solar energy conversion, lithium-ion batteries, renewable energy, and supercapacitors for storage or energy [13, 26].
- Tissue engineering: It is possible to introduce carbon nanomaterials and their hybrids into scaffolds for tissue engineering purposes. Cell attachment, development, and tissue repair are made possible by their physicochemical characteristics and minuscule size [53].

11 Challenges for Carbon-Based Nanohybrids

- Synthesis: Reproducibility in the physicochemical and therapeutic performance of the hybrid nanomaterials is important for achieving uniform dosing and therapy. Stable and rapid synthetic methods should be adopted to guarantee the physicochemical attributes within less time. Each top-down and bottom-up synthetic method produces hybrids of different sizes, surface functionality, and therapeutic properties, which cannot always guarantee reproducibility. Hence, a comprehensive understanding of the synthetic methods mechanism and figuring out the impact synthesis independent variables (functional groups on carbon nanomaterials, nature of hybridizing material, synthetic period, reaction temperature, synthetic process, etc.) on nanohybrid can guide the synthesis of excellent nanohybrid with consistent properties and batch-to-batch uniformity [22].
- Reproducibility, Scale-up, and Commercialization: It is still difficult to produce carbon-based hybrid nanomaterials in large quantities with reliable quality and characteristics. Reproducibility, purity, and scalability require better manufacturing techniques and optimizations [53].
- Cost of production: The extensive use of such nanoparticles in biological applications may be constrained by their functionalization and manufacturing costs. It is critical for clinical implementation to keep production costs down while maintaining quality requirements [53].
- Stability: The durability of carbon-based nanomaterials in biological contexts is hampered by their ultra-small size (flocculation, aggregation, coalescence, creaming, Ostwald ripening, etc.). Thus, in order to ensure their long-term safety and efficacy, temperature is essential for ensuring stability under a range of physiological conditions, such as pH and ionic strength [53].
- Biocompatibility, safety, and toxicity: Although carbon-based nanomaterials are generally regarded as biocompatible, their possible toxicity remains a concern [13, 53]. The safety profile and toxicity of the carbon-based nanomaterials should be evaluated by in vitro (viability, hemolysis, metabolic activity, apoptosis, necrosis, DNA damage, gene expressions, immunogenicity, oxidative stress etc.,) and in vivo parameters evaluation (lethality, carcinogenicity, biodistribution, clearance, hematology, histopathology, hepatotoxicity, nephrotoxicity, cardiotoxicity, dermatotoxicity, immunotoxicity, genotoxicity,

epigenetic toxicity, etc.) [4, 54]. The hybridization of carbon nanomaterials with biocompatible polymers reduces toxicity and increases biocompatibility for pharmaceutical applications [17].

- Lack of knowledge on side effects: The use of nanomaterials to treat metabolic syndrome is severely limited by the lack of knowledge regarding potential adverse effects on the central nervous system and peripheral organs [3].
- Long-term in vivo effects: In biological systems, the long-term therapeutic/theranostic properties have not been very well investigated. Given the novelty of these hybrid nanomaterials, thorough research is necessary to evaluate their long-term biocompatibility and effectiveness [53].
- Targeting specificity: Achieving a high degree of targeting specificity is crucial for effective biomedical applications. Since functionalization is required for active targeting, the exact mechanism of target-ligand interaction has to be fully understood to achieve site-specific targeted delivery [53].
- Photobleaching: Photobleaching can occur in fluorescent carbon-based nanohybrids, gradually decreasing their fluorescence intensity. Long-term imaging requires the creation of photostable carbon nanohybrids or techniques to reduce photobleaching [53].
- Impact on the environment: The special features of carbon-based nanomaterials make them useful for environmental applications. Because of their tiny size and capacity to interact with biological systems, they have the potential to pollute ecosystems and cause harm to living species through bioaccumulation, which may have a drastic detrimental effect on the environment if not appropriately handled. To comprehend their environmental impact and create mitigating solutions, research is still being conducted.
- Regulatory challenges: When it comes to regulatory approval for medicinal uses, carbon-based hybrid nanomaterials are a unique nanomaterial. To receive regulatory approval, comprehensive preclinical and clinical research is necessary to prove their efficacy and safety [3, 53].

12 Summary and Future Outlook

- Carbon-based hybrid nanomaterials possess unique properties and are exploited for multiple applications. They are used as potential carriers for pharmaceutical and biomedical applications [16, 17, 22].
- The integration of various agents with carbon-based nanomaterials produces hybrid nanomaterials that offer a versatile platform for the delivery of therapeutics [17].
- In addition to promising therapeutic effects, they also offer diagnostic properties and, hence, can be used as theranostics.
- The carbon-nanohybrids possess the synergetic possessions of individual components instead of the sum of each component [24].

- Hybrid carbon-based nanomaterials (hybrid metal/polymer/ceramic/biomolecule) have been reported as multifunctional materials that offer versatile solutions in diagnostics and therapeutics [16].
- The nanohybrids possess high surface areas, adaptable layer thickness, high hydrophilicity, modifiable structures, and a diverse range of compositions [16].
- Such hybrid nanomaterials provide improved optical, conductivity, mechanical, catalytic, sorption, and magnetic properties [24]. The unique physicochemical possessions of the hybridized carbon nanomaterial also offer improved biocompatibility and targeted delivery [27].
- A very high surface area-to-volume ratio facilitates effective drug delivery, with improving therapeutic efficacy [16].
- Their superior electrical conductivity makes it possible to create bioelectrodes and biosensors, which allow for precise physiological parameter monitoring. Furthermore, carbon-based nanohybrids' adjustable mechanical characteristics make them ideal for scaffold reinforcement, artificial joints, bone creation, and implants [16].
- A better understanding of the desired characteristics of carbon-based nanohybrids and synthetic feasibility can develop more affordable strategies for the fabrication of advanced hybrid materials.
- Several studies reported excellent safety, biocompatibility, negligible toxicity, and favorable in vivo therapeutic effects of the carbon-based nanohybrids. The safety and efficacy profile of the carbon nanohybrids should be thoroughly investigated to guarantee their fate during real-time clinical applications. In-depth research via preclinical animal models with a high degree of homology to humans and clinical investigations should be performed to prove their safety and efficacy [16, 27].
- The formulation strategy should be optimized to fabricate carbon nanohybrids with batch-to-batch consistency, constant physicochemical attributes, and consistent therapeutic activity. Addressing the aforementioned challenges through research, synthesis optimization, and thorough preclinical studies will pave the way for harnessing their potential application in the biomedical field. The synthetic method with controllable size and surface properties should be considered during the development of carbon-based nanohybrids.
- It further necessitates comprehensive long-term stability studies to ensure their stability and safety [16].
- It is still necessary to conduct a thorough examination into the characterization of surface characteristics in carbon-based nanomaterials, as well as the ways in which surface properties affect the stability of carbon-based nanomaterials [16]. The development of theoretical models could be helpful in order to gain a greater grasp of their properties. To improve comprehension of their surface condition, these models ought to be somewhat similar to the real dimensions [27]. With the use of artificial Intelligence (AI), valuable insights into their physicochemical properties, self-assembly mechanisms, photoluminescence, and interaction with complex body environments can be established [27].

- Currently, there are very few clinical trials on carbon-based nanohybrids for metabolic disorders. There is a bright area for developing such novel nanomaterials for treating metabolic syndrome. The book chapter summarized the applications of carbon-based nanohybrids as nanomedicine platforms for metabolic disorders. The formulation of novel carbon-based nanohybrid systems with desirable functionalization, biocompatibility, stability, and multimodal applicability could open new insight into metabolic disorder.
- Only a few carbon-based nanohybrids are reported for their therapeutic activity against metabolic disorders. Most of the research has been carried out on carbon-nanohybrid-based diagnostics for metabolic disorders, such as CVD and diabetes. A lot of opportunities for researcher is available in this domain for carrying out innovative research for the development of translational theranostic products for the effective management of metabolic disorders.
- Although a lot of work has been carried out for carbon-based nanomaterials, their safety, toxicity, and pharmacokinetic studies should be comprehensively conducted towards their clinical applications.

References

1. Silveira Rossi, J. L., et al. (2022). Metabolic syndrome and cardiovascular diseases: Going beyond traditional risk factors. *Diabetes/Metabolism Research Reviews, 38*(3), e3502.
2. Cornier, M.-A., et al. (2008). The metabolic syndrome. *Endocrine Reviews, 29*(7), 777–822.
3. Adetunji, C. O., et al. (2022). Potentialities of nanomaterials for the management and treatment of metabolic syndrome: A new insight. *Materials Today Advances, 13*, 100198.
4. Madannejad, R., et al. (2019). Toxicity of carbon-based nanomaterials: Reviewing recent reports in medical and biological systems. *Chemico-Biological Interactions, 307*, 206–222.
5. Rotariu, D., et al. (2022). Oxidative stress–complex pathological issues concerning the hallmark of cardiovascular and metabolic disorders. *Biomedicine and Pharmacotherapy, 152*, 113238.
6. Lemieux, I., & Després, J.-P. (2020). Metabolic syndrome: past, present and future. *Nutrients, MDPI, 12*, 3501.
7. Han, T. S., & Lean, M. E. (2015). Metabolic syndrome. *Medicine, 43*(2), 80–87.
8. Huang, P. L. (2009). A comprehensive definition for metabolic syndrome. *Disease Models & Mechanisms, 2*(5–6), 231–237.
9. Ahima, R. S. (2024). Overview of metabolic syndrome. In *Metabolic syndrome: a comprehensive textbook* (pp. 3–14). Springer.
10. Jiang, H., et al. (2023). Carbon nanomaterials: A growing tool for the diagnosis and treatment of diabetes mellitus. *Environmental Research, 221*, 115250.
11. Fahed, G., et al. (2022). Metabolic syndrome: updates on pathophysiology and management in 2021. *International Journal of Molecular Sciences, 23*(2), 786.
12. Semchuk, O. Y., Gatti, T., & Osella, S. (2023). Carbon based hybrid nanomaterials: overview and challenges ahead. *Surface, 14*, 78–94.
13. Ayanda, O. S., et al. (2024). Recent progress in carbon-based nanomaterials: critical review. *Journal of Nanoparticle Research, 26*(5), 1–42.
14. Patel, K. D., Singh, R. K., & Kim, H.-W. (2019). Carbon-based nanomaterials as an emerging platform for theranostics. *Materials Horizons, 6*(3), 434–469.
15. Liu, J., Li, R., & Yang, B. (2020). Carbon dots: a new type of carbon-based nanomaterial with wide applications. *ACS Central Science, 6*(12), 2179–2195.

16. Kim, C.-H., et al. (2024). Carbon-based composites in biomedical applications: a comprehensive review of properties, applications, and future directions. *Advanced Composites and Hybrid Materials, 7*(2), 55.
17. Taghavi, S., et al. (2020). Hybrid carbon-based materials for gene delivery in cancer therapy. *Journal of Controlled Release, 318*, 158–175.
18. Mohapatra, D., et al. (2021). Carbon dots from an immunomodulatory plant for cancer cell imaging, free radical scavenging and metal sensing applications. *Nanomedicine, 16*(23), 2039–2059.
19. Mohapatra, D., et al. (2023). Bioengineered dual fluorescent carbon nano dots from Indian long pepper leaves for multifaceted environmental and health utilities. *Environmental Science and Pollution Research, 30*(18), 52182–52208.
20. Mohapatra, D., et al. (2023). In vitro cancer cell imaging, free radical scavenging, and Fe3+ sensing activity of green synthesized carbon dots from leaves of Piper longum. *Journal of Cluster Science, 34*(3), 1269–1290.
21. Mohapatra, D., et al. (2022). Tinospora cordifolia leaves derived carbon dots for cancer cell bioimaging, free radical scavenging, and Fe 3+ sensing applications. *Journal of Fluorescence*, 1–18.
22. Fu, Y., et al. (2020). Hybrid architectures based on noble metals and carbon-based dots nanomaterials: A review of recent progress in synthesis and applications. *Chemical Engineering Journal, 399*, 125743.
23. Veerubhotla, K., & Lee, C. H. (2020). Emerging trends in Nanocarbon-based cardiovascular applications. *Advanced Therapeutics, 3*(7), 1900208.
24. Navrotskaya, A. G., et al. (2020). Hybrid materials based on carbon nanotubes and nanofibers for environmental applications. *Frontiers in Chemistry, 8*, 546.
25. Cetinkaya, A., et al. (2024). Carbon nanomaterials-based novel hybrid platforms for electrochemical sensor applications in drug analysis. *Critical Reviews in Analytical Chemistry, 54*(5), 1227–1242.
26. Bouhfid, R., & Jawaid, M. (2021). *Graphene and nanoparticles hybrid nanocomposites: From preparation to applications*. Springer Nature.
27. Chen, H., et al. (2024). Nanotechnology of carbon dots with their hybrids for biomedical applications: A review. *Chemical Engineering Journal, 496*, 153915.
28. Park, W., et al. (2020). Advanced hybrid nanomaterials for biomedical applications. *Progress in Materials Science, 114*, 100686.
29. Kulyk, B., et al. (2019). A short review on carbon-based nanomaterials and their hybrids. *Boletín del Grupo Español del Carbón, 54*, 44–54.
30. Zhu, J., et al. (2022). Carbon dots embedded hybrid microgel with phenylboronic acid as monomer for fluorescent glucose sensing and glucose-triggered insulin release at physiological pH. *Nanomaterials, 12*(17), 3065.
31. Bui, T. T., & Park, S.-Y. (2016). A carbon dot–hemoglobin complex-based biosensor for cholesterol detection. *Green Chemistry, 18*(15), 4245–4253.
32. Priyadarshini, E., & Rawat, K. (2017). Au@ carbon dot nanoconjugates as a dual mode enzyme-free sensing platform for cholesterol. *Journal of Materials Chemistry B, 5*(27), 5425–5432.
33. Campuzano, S., Yáñez-Sedeño, P., & Pingarrón, J. M. (2019). Carbon dots and graphene quantum dots in electrochemical biosensing. *Nanomaterials, 9*(4), 634.
34. Tuteja, S. K., et al. (2016). A label-free electrochemical immunosensor for the detection of cardiac marker using graphene quantum dots (GQDs). *Biosensors Bioelectronics, 86*, 548–556.
35. Lakshmanakumar, M., et al. (2019). Functionalized graphene quantum dot interfaced electrochemical detection of cardiac Troponin I: An antibody free approach. *Scientific Reports, 9*(1), 17348.
36. Nilewski, L., et al. (2019). Highly oxidized graphene quantum dots from coal as efficient antioxidants. *ACS Applied Materials Interfaces, 11*(18), 16815–16821.

37. Bhatnagar, D., et al. (2016). Graphene quantum dots FRET based sensor for early detection of heart attack in human. *Biosensors Bioelectronics, 79*, 495–499.
38. Sun, J., et al. (2018). Comet-like heterodimers "gold nanoflower@ graphene quantum dots" probe with FRET "off" to DNA circuit signal "on" for sensing and imaging microRNA in vitro and in vivo. *Analytical Chemistry, 90*(19), 11538–11547.
39. Adeniyi, O., et al. (2021). Nanohybrid electrocatalyst based on cobalt phthalocyanine-carbon nanotube-reduced graphene oxide for ultrasensitive detection of glucose in human saliva. *Sensors and Actuators B: Chemical, 348*, 130723.
40. Elancheziyan, M., et al. (2023). Eco-friendly fabrication of nonenzymatic electrochemical sensor based on cobalt/polymelamine/nitrogen-doped graphitic-porous carbon nanohybrid material for glucose monitoring in human blood. *Environmental Research, 223*, 115403.
41. Uzunoglu, A., et al. (2020). Electrochemical glucose detection using PdAg nanoparticles anchored on rGO/MWCNT nanohybrids. *Journal of Cluster Science, 31*(1), 231–239.
42. Sarkar, K., et al. (2023). Hyaluronic acid-graphene oxide quantum dots nanoconjugate as dual purpose drug delivery and therapeutic agent in meta-inflammation. *Journal of Nanobiotechnology, 21*(1), 246.
43. Hao, T., et al. (2017). Injectable fullerenol/alginate hydrogel for suppression of oxidative stress damage in brown adipose-derived stem cells and cardiac repair. *ACS Nano, 11*(6), 5474–5488.
44. Dellinger, A., et al. (2013). Functionalization of gadolinium metallofullerenes for detecting atherosclerotic plaque lesions by cardiovascular magnetic resonance. *Journal of Cardiovascular Magnetic Resonance, 15*(1), 7.
45. Rabbani, G., et al. (2024). Synthesis of carbon nanotubes-chitosan nanocomposite and immunosensor fabrication for myoglobin detection: An acute myocardial infarction biomarker. *International Journal of Biological Macromolecules, 265*, 130616.
46. Fall, B., et al. (2023). Highly efficient non-enzymatic electrochemical glucose sensor based on carbon nanotubes functionalized by molybdenum disulfide and decorated with nickel nanoparticles (GCE/CNT/MoS2/NiNPs). *Sensors and Actuators Reports, 5*, 100136.
47. Ahadian, S., et al. (2017). Moldable elastomeric polyester-carbon nanotube scaffolds for cardiac tissue engineering. *Acta Biomaterialia, 52*, 81–91.
48. Ahadian, S., et al. (2017). Carbon nanotubes embedded in embryoid bodies direct cardiac differentiation. *Biomedical Microdevices, 19*, 1–12.
49. Ho, C. M. B., et al. (2017). 3D printed polycaprolactone carbon nanotube composite scaffolds for cardiac tissue engineering. *Macromolecular Bioscience, 17*(4), 1600250.
50. Sun, H., et al. (2017). Carbon nanotube-composite hydrogels promote intercalated disc assembly in engineered cardiac tissues through β1-integrin mediated FAK and RhoA pathway. *Acta Biomaterialia, 48*, 88–99.
51. Liu, L., et al. (2018). NiCo2O4 nanoneedle-decorated electrospun carbon nanofiber nanohybrids for sensitive non-enzymatic glucose sensors. *Sensors and Actuators B: Chemical, 258*, 920–928.
52. Martins, A. M., et al. (2014). Electrically conductive chitosan/carbon scaffolds for cardiac tissue engineering. *Biomacromolecules, 15*(2), 635–643.
53. Parvin, N., et al. (2024). Emerging trends in nanomedicine: carbon-based nanomaterials for healthcare. *Nanomaterials, 14*(13), 1085.
54. Sundararaman, G., et al. (2024). Safety matters pertaining to carbon-based nanomaterials in drug designing for targeted drug delivery in biomedical sciences. In *Carbon-based nanomaterials in biosystems* (pp. 573–589). Elsevier.

Debadatta Mohapatra is a researcher who has made significant contributions to pharmaceutical engineering and technology. Currently, he is working as a postdoctoral researcher at Leslie Dan Faculty of Pharmacy, University of Toronto, Canada. He has completed his Ph.D. from the Department of Pharmaceutical Engineering & Technology, IIT (BHU), Varanasi, India. His research primarily focuses on Quality-by-Design (QbD)-based formulation development and characterization of various novel drug delivery systems (NDDS), especially solid dispersion, transferosome, nanoemulsion, carbon dots, hydrogel, and phytoformulations for therapeutical benefits. He has a core competency in analytical and bioanalytical method development and validation, quality control and standardization of herbal drugs, pharmacokinetics, cancer therapeutics, and in vitro/ in vivo pharmacological activity evaluations. His acumen for pharmaceutical research is also reflected in his publications. Dr. Mohapatra has published numerous peer-reviewed research articles in various high-impact pharmaceutical/biomedical/environmental journals and has presented his work at national and international conferences. He has demonstrated leadership in managing and coordinating multidisciplinary research teams. In addition to his academic and research endeavors, Dr. Mohapatra actively collaborates with industries, academia, and research institutions to bridge the gap between academic research and practical applications.

Prakash Ch. Senapati is an illustrious professional known for his expertise in the field of drug delivery, pharmacokinetics, and nanoformulations for various health modalities. He has made significant contributions to the pharmaceutical field by developing and characterizing innovative therapeutics, which is reflected in various peer-reviewed international and national publications. Throughout his last 20 years of career, Dr. Senapati has been involved in industrial activity, teaching, research, consulting, and various professional activities and has gained recognition for his innovations, leadership, and mentoring. He is an active member of various professional bodies and a reviewer of various peer-reviewed international journals. His work in pharmaceutics has been influential, impacting both the academic community and industry practices. Dr. Senapati is currently working as an Associate Professor at the School of Pharmaceutical Sciences-II, Phulnakhara Campus, Siksha 'O' Anusandhan (Deemed to be University), Bhubaneswar, Odisha, India.

Safety and Biocompatibility of Nanomaterials for Metabolic Syndrome 11

Anjali Yadav, Bhaskar Sharma, Renu Negi, and Vivek K. Chaturvedi

Abstract

Nanotechnology is rapidly transforming the landscape of healthcare, particularly in the management of complex chronic conditions like metabolic syndrome. While nanomaterials offer immense promise for targeted drug delivery, biosensing, and personalized interventions, their clinical application demands thorough investigation into their safety and biocompatibility. This chapter explores the biological interactions of commonly used nanomaterials—such as gold, silver, lipid-based carriers, and polymeric systems—within the unique physiological context of metabolic syndrome. Emphasis is placed on understanding how factors like oxidative stress, chronic inflammation, and altered metabolic environments influence nanoparticle behaviour and toxicity in affected individuals. Real-world case studies and in vivo models are discussed to assess hemocompatibility, immunogenicity, and organ-specific accumulation. The chapter also reviews regulatory concerns and presents practical strategies to design safer nanotherapeutics, including surface functionalization, green synthesis, and responsive delivery systems. By bridging laboratory science with clinical relevance, this chapter aims to guide researchers and clinicians toward responsible and effective integration of nanotechnology into metabolic health care.

Anjali Yadav and Bhaskar Sharma contributed equally with all other contributors.

A. Yadav · B. Sharma (✉)
Department of Anatomy, All India Institute of Medical Sciences, New Delhi, India

R. Negi
Department of Pharmacology and Experimental Therapeutics, Boston University School of Medicine, Boston, MA, USA

V. K. Chaturvedi
Department of Gastroenterology, Banaras Hindu University, Varanasi, Uttar Pradesh, India

V. K. Chaturvedi et al. (eds.), *Nanotheranostics in Metabolic Syndrome*,
https://doi.org/10.1007/978-981-95-3746-4_11

Keywords

Nanomedicine · Metabolic syndrome · Biocompatibility · Targeted therapy · Nanotoxicity · Clinical translation

1 Introduction

Metabolic syndrome (MS) is characterized by several risk factors that include obesity, hypertension, insulin resistance, dyslipidemia, and non-alcoholic fatty liver disease leading to conditions like type 2 diabetes, heart disease or even strokes. It is a multifactorial condition (Fig. 11.1). Its prevalence has drastically increased in

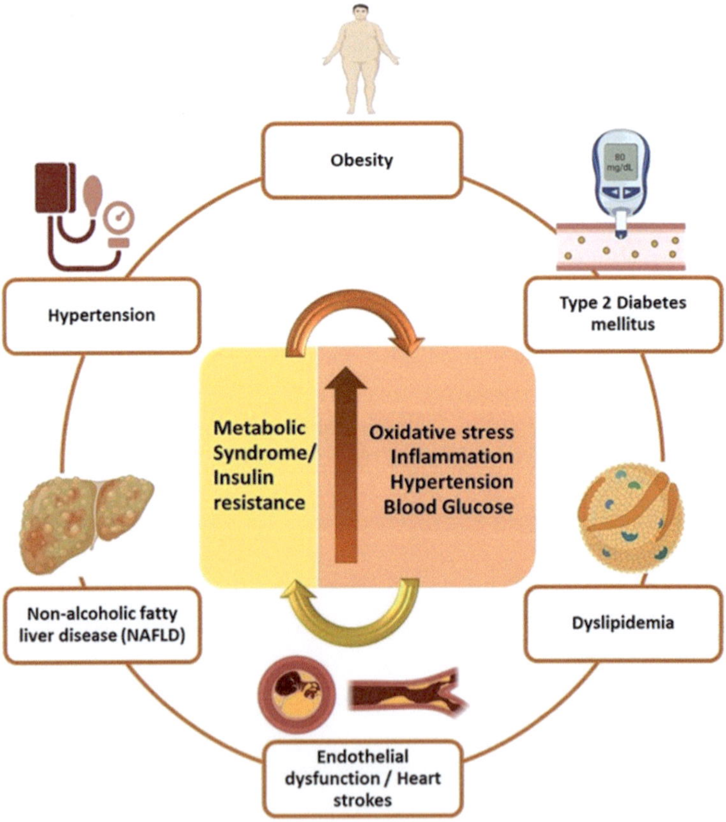

Fig. 11.1 Components and consequences of metabolic syndrome. The figure summarizes major risk factors—obesity, insulin resistance, dyslipidemia, hypertension, and NAFLD—and their progression to diabetes, cardiovascular disease, and stroke

the last decade demanding the urgent need for novel treatment approaches to reduce the risks of type 2 diabetes and cardiovascular disease.

Nanomaterials are essential to biomedical applications because they provide advanced methods for biosensing, targeted drug delivery, and better diagnostics. Numerous studies have demonstrated that medication targeting a specific region can be achieved using nano-drug delivery systems such as hydrogel-based nanocomposites, liposomes, and nano-micelles [56]. Numerous nanomaterials, including poly (lactic-glycol) acid nanoparticles (PLGANP), manganese nanoparticles (MnNP), titanium oxide nanoparticles (TiO_2 NP), silver nanoparticles (AgNP), nanocrystal cerium dioxide (nCeO2), and gold nanoparticles (AuNP), have been utilized to treat metabolic syndrome, according to a recent study. The development of customized medicine and the improvement of outcomes in complicated diseases including cancer, diabetes, and metabolic syndrome are supported by their distinct physicochemical features, which also improve bioavailability, cellular absorption, and therapeutic precision. Size, shape, surface characteristics, charge, solubility, coating, doping, the number of atoms in a nanoparticle, and exposure pathways can all affect a nanomaterial's toxicity [5].

Evaluating the safety and biocompatibility of nanomaterials is essential, as they interact with biological systems at the molecular and cellular levels. A comprehensive assessment of their toxicity, immune response, and biodistribution is critical to ensure their safe integration into clinical practice and to minimize potential adverse health effects (Table 11.1).

2 Safety Assessment Parameters

2.1 In Vitro Cytotoxicity Studies

In vitro cytotoxicity studies serve as the primary step in evaluating the biological safety of nanomaterials. These assays help to determine the adverse effect of nanoparticles on cellular viability, morphology, or function. Commonly used assays include the MTT assay, which assesses mitochondrial activity as an indicator of cell viability; the LDH (lactate dehydrogenase) release assay, which measures membrane integrity [37]; and ROS (reactive oxygen species) generation tests, which evaluate oxidative stress, which is a key mechanism of nanoparticle-induced cytotoxicity.

In vitro toxicity tests for nano-safety are primarily conducted using cell cultures. These include various models such as upright nanomaterial exposure to cells, surface-based presentation, inverted cultures for buoyant materials, and air–liquid interfaces for respiratory cells. Several studies focus on minimizing animal testing by employing in vitro models that are applicable to humans and ensuring safety prior to industrial use [21].

3D models offer more complexity, including co-cultures with multiple cell types, spheroids or organoids mimicking tissue structure, Organ-on-Chip platforms for dynamic conditions, and precision-cut tissue slices retaining native architecture. These methods aim to assess nanomaterial toxicity in a faster and less expensive

Table 11.1 Summary of key nanomaterials commonly used in the management of metabolic syndrome, detailing their biomedical applications, safety optimization strategies, and associated toxicity concerns

Nanomaterial	Application	Safety measures	Toxicity/ concerns	References
Gold nanoparticles (AuNPs)	Drug delivery for insulin resistance	PEGylation, surface modification	Oxidative stress, liver accumulation	[1, 73]
Silver nanoparticles (AgNPs)	Antimicrobial for diabetic wounds	Surface functionalization	Genotoxicity, mitochondrial toxicity	[4, 28]
Silica nanoparticles	Oral delivery of metformin	Mesoporous coating, pH-sensitive layers	Inflammation, chronic exposure risk	[15, 75]
Lipid Nanocarriers (SLNs, NLCs)	Delivery of curcumin, omega-3	Physiological lipids used	Lipid peroxidation	[7, 58]
Carbon nanotubes (CNTs)	Biosensors, glucose monitors	Oxidation, length reduction	Fibrosis, low clearance	[60, 89]
Zinc oxide nanoparticles	Modulate insulin signaling	Green synthesis, encapsulation	Beta-cell toxicity	[43, 68]
Iron oxide nanoparticles (SPIONs)	Imaging, drug delivery	Dextran/PEG coatings	Iron overload, ROS generation	[33, 72]
PLGA nanoparticles	GLP-1 analog delivery	FDA-approved polymer	Minimal toxicity (size-dependent)	[17, 46]
Chitosan nanoparticles	Controlled insulin release	Mucoadhesive, biodegradable	Charge-dependent toxicity	[76]
Titanium dioxide nanoparticles	Antioxidant therapy	Surface passivation	ROS, genotoxic effects	[70]
Cerium oxide nanoparticles	Anti-inflammatory, antioxidant	Redox cycling, surface modification	Liver/spleen accumulation	[10]
Quantum dots	Glucose sensing	PEGylation, silica coating	Heavy metal leaching	[19]
Graphene oxide	Smart biosensors	Surface functionalization	Membrane damage	[90]
Copper oxide nanoparticles	Anti-obesity formulations	Stabilized surfaces	Hepatotoxicity	[41]
Polycaprolactone (PCL)	Long-term drug delivery	Biodegradable polyester	Slow degradation	
Dendrimers	Gene/drug co-delivery	Functionalized surfaces	Dose-dependent cytotoxicity	[40]
Albumin nanoparticles	Insulin and GLP-1 analog carriers	Protein-based, biocompatible	Denaturation risk	[24]
Selenium nanoparticles	Antioxidant therapy	Controlled release, green synthesis	Narrow therapeutic index	[61]

(continued)

Table 11.1 (continued)

Nanomaterial	Application	Safety measures	Toxicity/ concerns	References
Nanogels	Oral insulin delivery	pH-responsive hydrogel systems	Swelling ratio variability	[59]
Magnesium oxide nanoparticles	Anti-diabetic agents	Stabilized via green synthesis	GI irritation at high doses	[67]
Poly(alkyl cyanoacrylate) NPs	Oral peptide delivery	PEG or polysorbate 80 coatings	Brain accumulation risk	[14]
Calcium phosphate nanoparticles	Encapsulation of bioactives	Bioabsorbable, bone-compatible	Poor solubility in acidic environments	[69]

The references cited provide supporting evidence from experimental, clinical, or review studies

way than in vivo tests, reducing animal demand [93]. Also, they are inexpensive and high-throughput assays enabling preliminary identification of potentially hazardous nanomaterials before animal testing [78]. However, cell line-specific variations in sensitivity to nanoparticles can lead to inconsistent or non-reproducible results. For example, epithelial cells may respond differently to the same nanoparticle compared to macrophages [48]. Moreover, nanoparticle interference with assay reagents (such as optical scattering in MTT) can confound results, requiring careful assay optimization and appropriate controls. Therefore, standardization of protocols, including cell type selection, exposure duration, and nanoparticle concentrations, is essential for reliable interpretation. A combination of assays is often recommended to capture a comprehensive cytotoxic profile and reduce false positives or negatives.

2.2 Hemocompatibility and Genotoxicity

Nanomaterials intended for biomedical use come into direct contact with blood, making hemocompatibility testing a critical safety parameter [18]. Hemolysis assays assess the potential of nanoparticles to rupture red blood cells, leading to haemoglobin release and possible anaemia or kidney damage. Additionally, platelet aggregation and activation tests evaluate whether nanoparticles induce clot formation, which can result in thrombotic complications [88]. The hemocompatibility assessment technique involves exposing blood to nanoparticles under controlled conditions. The released hemoglobin from damaged red blood cells is measured as cyanmethemoglobin using spectrophotometry to determine the extent of hemolysis [54]. The example of hemocytotoxicity of AgNPs was assessed by examining hemolysis, lymphocyte proliferation, platelet aggregation, coagulation, and complement activation. AgNPs induced hemolysis and inhibited lymphocyte viability at all investigated concentrations (10, 20, 40 μg/mL), while showing no significant effects on platelet function, coagulation, or complement activation, highlighting selective but concerning blood cell toxicity [38]. These interactions are highly influenced by the surface chemistry, size, charge, and concentration of the nanoparticles. For instance, positively charged nanoparticles often exhibit higher hemolytic activity due to electrostatic interactions with negatively charged cell membranes, emphasizing charge-dependent biological effects [29].

In parallel, genotoxicity assessments are to ensure NPs do not induce mutations or DNA damage, which may lead to carcinogenesis [71]. Techniques such as the comet assay [13], micronucleus test [27], and chromosomal aberration analysis are most commonly used to detect single or double strand DNA breakage, DNA fragmentation, chromosomal abberations and mitotic errors. Genotoxic responses are also modulated by nanoparticle composition and their ability to enter the nucleus or interfere with DNA repair mechanisms. Mechanisms of Genotoxicity includes oxidative stress, direct DNA interaction, and epigenetic modifications, that affects gene expression and stability [71].

In vitro, 3D models enhance genotoxicity testing by mimicking human tissues, improving accuracy over traditional 2D cultures. High-throughput techniques, including organ-on-chip systems, enable rapid screening of nanomaterials for DNA damage assessment [45]. As with hemocompatibility, standardized testing under physiologically relevant conditions is essential for accurate risk assessment, particularly for nanoparticles intended for long-term or systemic administration as in the case of Diabetes Mellitus.

2.3 In Vivo Toxicology and Organ-Specific Effects

While in vitro assays offer preliminary safety data, in vivo toxicology studies provide a more comprehensive understanding of how nanomaterials behave in the complex environment of a living organism. Rodent models, particularly mice and rats, are widely used to study biodistribution, metabolism [36], and clearance pathways of nanoparticles following various routes of administration—oral, intravenous, inhalational, or dermal [42]. These studies reveal organ-specific accumulation, with the liver, spleen, lungs, and kidneys being primary sites of nanoparticle deposition due to their roles in filtration and immune surveillance. The reticuloendothelial system (RES) plays a central role in nanoparticle sequestration, affecting both therapeutic efficacy and potential toxicity.

Histopathological examination, blood biochemistry, and organ weight analysis help identify acute and subacute toxic effects. Importantly, long-term exposure studies are crucial to uncover chronic toxicity, immune modulation, or carcinogenic potential. Persistent inflammation, fibrosis, or oxidative stress in vital organs can manifest over time, even if short-term toxicity is minimal. Factors such as particle shape, surface functionalization, and biodegradability significantly influence these outcomes. Therefore, a thorough in vivo evaluation, including both acute and chronic endpoints, is indispensable before advancing nanoparticle-based therapies to clinical trials. Ethical practices and adherence to regulatory guidelines remain essential throughout these assessments.

3 Biocompatibility Considerations in Metabolic Syndrome Models

3.1 Obese and Diabetic Animal Models

Obese and diabetic animal models are indispensable tools for studying metabolic syndrome, but they also introduce complexities in nanoparticle behaviour. These models typically exhibit altered pharmacokinetics and pharmacodynamics, which can significantly impact nanoparticle biodistribution, metabolism, and clearance [86]. For example, in high-fat diet-induced obese mice, impaired hepatic function and altered lipid metabolism may reduce the liver's ability to process and clear nanoparticles, leading to unexpected organ accumulation or prolonged systemic circulation [2, 22].

Additionally, chronic inflammation and vascular endothelial dysfunction—common features of metabolic syndrome—can exacerbate nanotoxicity [35]. Inflammation increases vascular permeability, allowing more nanoparticles to cross endothelial barriers and enter tissues. While this may improve drug delivery to inflamed areas, it also raises the risk of non-specific accumulation and local tissue damage [35, 55]. Furthermore, elevated levels of reactive oxygen species (ROS) and pro-inflammatory cytokines in diabetic models can enhance oxidative stress induced by some nanoparticles, compounding their cytotoxic effects [16].

Therefore, biocompatibility assessments in these models must consider disease-specific physiological changes by emphasizing on the factors involved in the biocompatibility of the nanoparticles (Fig. 11.2). Using control animals alone is insufficient; comparative studies in both healthy and disease models are essential to fully understand how metabolic syndrome alters nanoparticle interactions and safety profiles.

3.2 Tissue Engineering and Nanocarrier Implantation

In metabolic syndrome models, tissue engineering approaches—such as scaffolds, nanofibre implants, or hydrogel-based delivery systems—must be evaluated for both functionality and long-term biocompatibility [34]. These engineered constructs are often designed to promote regeneration in damaged tissues (e.g., adipose, vascular, or pancreatic), but their integration into the host environment can be complicated by the inflammatory and fibrotic tendencies observed in obesity and diabetes [23].

Macrophage activation refers to the immune response triggered by foreign bodies such as implants or nanoparticles. It plays a central role in implant rejection, as over-activated macrophages can release pro-inflammatory cytokines, leading to chronic inflammation and fibrosis.

Nanopatches are nanoscale or micrometre-thin materials designed to adhere to biological surfaces for local and controlled drug delivery. They are commonly used to treat localized inflammation or deliver insulin transdermally.

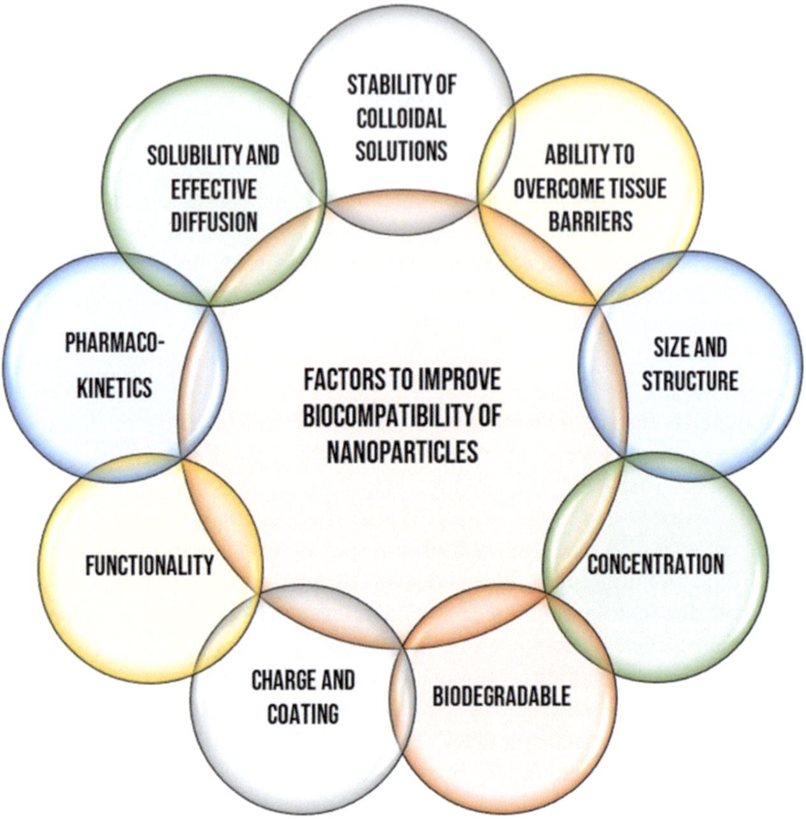

Fig. 11.2 Factors involved in improving the biocompatibility of the nanomaterials

Successful biocompatibility in this context hinges on the scaffold's ability to integrate without eliciting fibrotic encapsulation, chronic inflammation, or foreign body reactions [6]. Nanomaterial-based scaffolds with carefully controlled porosity, degradation rates, and surface chemistry can enhance cell adhesion and vascularization [91]. However, the degradation products of such scaffolds—especially if they are polymeric or metallic—may provoke immune responses, particularly in a host already experiencing metabolic and inflammatory stress [82].

Furthermore, nanocarrier implantation for localized drug delivery (e.g., insulin-releasing nanogels or anti-inflammatory nanopatches) must be assessed for both mechanical compatibility and immunological tolerance. Metabolic syndrome-associated immune dysregulation may lead to enhanced macrophage activation, increased cytokine production, and delayed healing around the implant site [9]. Thus, preclinical evaluation of nanocarrier-based implants must include histological analysis, local immune cell profiling, and fibrosis markers in disease-relevant models to predict long-term performance and safety.

3.3 Sex, Age, and Comorbidities

Individual physiological differences such as sex, age, and coexisting conditions play a significant role in determining the biocompatibility of nanomaterials, particularly in the complex landscape of metabolic syndrome. These factors influence hormonal regulation, immune responses, and metabolic processing, all of which can alter nanoparticle interactions with tissues and cells [47]. Sex-based differences are increasingly recognized as critical variables. For example, oestrogen is known to modulate immune responses and endothelial function, potentially reducing nanoparticle-induced inflammation in premenopausal females[83]. Conversely, males may exhibit different biodistribution profiles due to higher baseline oxidative stress and metabolic rates. Ignoring sex as a biological variable could lead to biased interpretations of nanoparticle safety.

Age is another essential determinant. Aging is associated with a decline in immune function (immunosenescence), changes in organ function, and slower metabolism, all of which affect how nanoparticles are cleared or retained [32]. Elderly individuals may have an increased risk of chronic toxicity due to reduced renal [25] hepatic clearance [11] and greater susceptibility to oxidative damage [50]. Moreover, comorbidities such as hypertension, atherosclerosis, or renal dysfunction—which frequently coexist with metabolic syndrome—can further complicate biocompatibility outcomes. These conditions may impair vascular integrity or organ filtration, facilitating undesirable accumulation of nanoparticles in sensitive tissues like the brain, heart, or kidneys [80].

Therefore, a personalized approach is needed in both preclinical and clinical evaluation. Animal studies should include models of varying age and sex, and, where feasible, incorporate comorbidity simulations. This stratified assessment can enhance the predictive power of safety studies and ensure that nanotherapeutics are safe for use across diverse patient populations.

4 Regulatory Guidelines and Standardization Challenges

4.1 International Safety Standards for Nanomedicine

The clinical translation of nanomaterials in medicine is governed by evolving safety standards issued by leading global regulatory agencies such as the U.S. Food and Drug Administration (FDA), the European Medicines Agency (EMA), and the International Organization for Standardization (ISO). These organizations acknowledge that nanomedicines possess unique physicochemical and biological properties that deviate significantly from conventional pharmaceuticals, requiring adapted regulatory approaches [20]. Difficulties arise due to the vast variety of nanoparticles, exposure routes, characterization issues, and limitations in research methodologies [84]. A study by Barbara D, summarizes over 200 reports that had offered recommendation for standardized testing methods, biological response evaluations, and safe-by-design nanomaterials to improve regulatory approvals [21].

The FDA provides guidance through its Nanotechnology Regulatory Science Program, which encourages the submission of detailed nanoparticle characterization data and toxicity studies tailored to specific nanomaterial types [12]. The EMA similarly provides recommendations for the development and quality assessment of nanoparticle-based drugs, emphasizing the importance of safety, quality control, and risk management [63]. Meanwhile, ISO has developed a set of standards (e.g., ISO/TR 10993 series) related to biological evaluation of medical devices containing nanomaterials, including testing for cytotoxicity, sensitization, and hemocompatibility [64]. An organized method was created by DuPont and the Environmental Defense Fund to assess nanoparticle hazards over the course of their lifecycle [84]. As compared to the discovered, created, and preclinically tested nanomaterials, the number of nanopharmaceuticals approved by the European Medicines Agency (EMA) or the Food and Drug Administration (FDA) is still remarkably low [79].

Despite these advancements, one of the major challenges in the global regulatory landscape is the lack of harmonized testing protocols. Regulatory frameworks across different countries often rely on different endpoints, validation methods, and testing matrices, making it difficult to compare results and align global product development strategies. For instance, variations in cytotoxicity testing protocols (e.g., use of MTT vs. LDH assays), differences in animal model selection, and inconsistencies in exposure durations or dose metrics contribute to conflicting safety profiles across regions.

Moreover, current regulatory documents often lack specificity for the wide range of nanomaterial types and surface chemistries, which hinders consistent risk evaluation. Regulatory ambiguity creates uncertainty for industry stakeholders, slowing innovation and increasing development costs. To overcome these barriers, international collaborations such as the International Pharmaceutical Regulators Programme (IPRP) and joint workshops by the OECD Working Party on Manufactured Nanomaterials (WPMN) aim to facilitate consensus on nanotoxicology testing standards and terminology [51]. Standardized testing and risk assessment are crucial for ensuring nanoparticle safety in medical and industrial applications [71].

In conclusion, while substantial progress has been made, the absence of globally harmonized regulatory standards continues to pose a significant obstacle to the development and commercialization of nanomedicines. A unified regulatory approach, supported by validated and reproducible methodologies, is urgently needed to ensure safe and efficient translation of nanotechnology-based therapeutics.

4.2 Preclinical to Clinical Transition Requirements

Translating nanomedicines from laboratory innovation to clinical application necessitates a comprehensive preclinical evaluation followed by stringent clinical safety trials. Unlike conventional small-molecule drugs, nanomaterials often exhibit complex interactions with biological systems due to their nanoscale size, surface properties, and tendency to form protein coronas in physiological fluids. This complexity complicates the extrapolation of in vitro findings to in vivo scenarios and ultimately to human safety profiles, making preclinical standardization a critical requirement.

A key step in the transition from bench to bedside is Good Laboratory Practice (GLP)-compliant toxicological assessment. Regulatory agencies, including the FDA and EMA, mandate that nanomedicine developers conduct a full spectrum of toxicity studies under GLP conditions. These include acute, subchronic, and chronic toxicity evaluations, genotoxicity, immunotoxicity, reproductive toxicity, and organ-specific histopathology. Importantly, these studies must also account for nanoparticle-specific behaviours such as aggregation, surface reactivity, and bioaccumulation, which can significantly affect toxicity profiles [3].

The predictive value of in vitro studies, such as cytotoxicity or oxidative stress assays, is often limited due to oversimplified culture conditions that do not replicate the complex dynamics of living organisms. In many cases, nanoparticles that appear safe in vitro exhibit unexpected immunogenicity or organ toxicity in animal models. Therefore, bridging this gap requires iterative validation between in vitro and in vivo results, using physiologically relevant models that mimic human pathophysiology—especially in disease contexts such as metabolic syndrome.

Furthermore, the pharmacokinetic (PK) and pharmacodynamic (PD) properties of nanomedicines are highly variable. Absorption, distribution, metabolism, and excretion (ADME) patterns differ significantly from those of conventional drugs. These aspects must be comprehensively studied in multiple species to estimate human dosing regimens and to design first-in-human trials. In addition, the formation of protein coronas, which alters nanoparticle identity and cellular uptake, must be studied under physiological conditions to predict potential immune responses or off-target effects [39].

For clinical translation, developers must submit a robust Investigational New Drug (IND) or Clinical Trial Application (CTA) dossier, including validated data from all preclinical studies, quality assurance documentation, and a detailed risk management plan. Failures at this stage are often due to insufficient or non-reproducible safety data, underscoring the necessity for early regulatory engagement and continuous dialogue with agencies throughout the development lifecycle.

5 Advances in Safer Nanomaterial Design

5.1 Surface Functionalization and Coatings

Surface functionalization plays a pivotal role in enhancing biocompatibility, and circulatory stability, and targeting the efficiency of nanomaterials. One widely adopted approach is PEGylation, which involves the covalent attachment of polyethylene glycol (PEG) chains to the nanoparticle surface[77]. PEGylation confers a hydrophilic corona around the nanoparticle, thereby reducing protein adsorption (opsonization), preventing recognition by the mononuclear phagocyte system, and extending circulation time. This "stealth" property significantly decreases immunogenicity and nonspecific uptake by off-target tissues [74].

Homotypic targeting refers to a nanoparticle's ability to selectively bind to cells of the same origin as the membrane or cell type used for coating. This enhances targeting specificity and minimizes off-target effects. In addition to synthetic coatings, biomimetic strategies such as cell membrane cloaking have emerged as advanced surface engineering tools. Coating nanoparticles with membranes derived from red blood cells, leukocytes, platelets, or cancer cells provides a biological interface that mimics natural cell interactions [92]. These biomimetic coatings can facilitate immune evasion, prolong systemic circulation, and enhance homotypic targeting—allowing for selective accumulation in diseased tissues such as tumours or inflamed vasculature, common in metabolic syndrome [85].

Furthermore, surface ligand modification, such as conjugating peptides, antibodies, or aptamers, allows for receptor-mediated targeting, enhancing nanoparticle uptake in specific cells, including adipocytes, hepatocytes, or inflamed endothelium. These engineered interfaces not only improve drug delivery specificity but also reduce off-target toxicity, enhancing the therapeutic index of nanomedicines [87]. Overall, surface functionalization transforms inert nanocarriers into bio-interactive platforms that can navigate complex biological environments safely and effectively. Optimizing this nanoparticle–biological interface is critical for reducing immune clearance, improving targeting, and minimizing systemic side effects in patients with metabolic disorders.

5.2 Green and Biodegradable Nanomaterials

The increasing concern over the environmental and biological impact of synthetic nanomaterials has accelerated the development of green and biodegradable nanoparticles. These materials are synthesized using environmentally benign methods and are derived from renewable, biocompatible sources such as plant extracts, natural polysaccharides (e.g., chitosan, alginate, cellulose), and proteins (e.g., silk fibroin, albumin, casein) [57]. Such nanoparticles are not only less toxic but also degrade into non-harmful by-products, minimizing the risk of long-term bioaccumulation [62].

Green synthesis techniques replace hazardous chemicals with plant-based reducing agents and stabilizers, avoiding toxic solvents and high-energy processes. For instance, plant phytochemicals such as flavonoids, tannins, and terpenoids have been employed to synthesize metallic and polymeric nanoparticles under mild conditions. These eco-friendly nanoparticles often exhibit antioxidant, anti-inflammatory, or antimicrobial properties, making them particularly useful in the context of metabolic syndrome, where oxidative stress and inflammation are central pathological features [44].

Biodegradable polymer-based nanocarriers, such as those composed of poly (lactic-co-glycolic acid) (PLGA), polycaprolactone (PCL), or natural gum resins, undergo enzymatic or hydrolytic degradation into metabolizable monomers

[49]. Their degradation profiles can be tailored by adjusting polymer composition, molecular weight, and crosslinking density, enabling controlled drug release aligned with disease progression [31].

The use of plant-derived or biodegradable nanomaterials also aligns with principles of sustainability and regulatory preference for safer, cleaner production methods [62]. As a result, they are increasingly favoured in pharmaceutical and biomedical applications, particularly for chronic conditions such as obesity, diabetes, and cardiovascular disease, where long-term exposure demands maximum biocompatibility with minimal systemic burden.

5.3 Stimuli-Responsive and Smart Nanoplatforms

Stimuli-responsive or "smart" nanoplatforms represent a cutting-edge advancement in nanomedicine design, offering on-demand drug release triggered by specific physiological or pathological cues [53]. These environment-sensitive systems respond to alterations in pH, enzymatic activity, redox state, or specific biomolecule concentrations (e.g., glucose), allowing for spatially and temporally controlled therapeutic delivery [26]. This level of precision significantly reduces systemic toxicity and enhances therapeutic efficacy, particularly important in the management of complex metabolic diseases. Safe-by-Design (SbD) nanomaterials are the focus of current research, which emphasizes early hazard screening, tailored in vitro assays, and lifecycle risk assessment to improve safety, lower exposure, and increase regulatory compliance in applications involving nanotechnology [66].

One of the most promising applications is in the development of glucose-responsive nanoparticles for diabetes treatment [81]. These systems often incorporate glucose oxidase or phenylboronic acid moieties that can detect elevated glucose levels and trigger insulin or antidiabetic drug release accordingly. Such platforms mimic pancreatic function and hold the potential for automated glycaemic control, minimizing the risks associated with frequent dosing and hypoglycaemic episodes [52]. Nanoparticles ensure safe oral insulin delivery in diabetes by optimizing encapsulation, particle size, and excipient selection while minimizing toxicity risks [65].

Similarly, pH-sensitive nanoparticles exploit the acidic microenvironments of inflamed adipose tissue or ischemic regions, which are characteristic of metabolic syndrome [30]. These systems remain stable at physiological pH but rapidly release their payload in acidic conditions, concentrating therapeutic action at the disease site. Enzyme-responsive systems, triggered by matrix metalloproteinases or lipases upregulated in obesity or atherosclerosis, enable highly selective delivery to pathological tissues while sparing healthy ones [8].

Advanced designs also integrate multi-responsive behaviour or incorporate imaging agents, enabling theranostic capabilities—simultaneous diagnosis and therapy. These smart platforms are being refined for greater specificity, tunability, and biocompatibility, making them attractive candidates for chronic disease management, where personalized and adaptive therapies are critically needed.

6 Future Directions and Research Gaps

6.1 Long-Term Biocompatibility and Bioaccumulation Studies

As nanomaterials become increasingly integrated into therapies for metabolic syndrome, understanding their long-term biocompatibility remains a critical challenge. Most current toxicological studies focus on short-term exposure, while chronic effects on metabolic, reproductive, and immune systems are inadequately characterized. Particularly in patients with metabolic syndrome—who often require lifelong therapy—the risk of nanoparticle bioaccumulation in organs such as the liver, kidney, and adipose tissue poses potential safety concerns. Additionally, the influence of long-term exposure on reproductive health and transgenerational inheritance of toxicity remains largely unexplored. Addressing these gaps requires longitudinal and multi-generational animal studies, incorporating metabolic disease models to better simulate clinical realities.

6.2 Personalized Nanomedicine in Metabolic Syndrome

Emerging advances in omics technologies (genomics, proteomics, metabolomics) and artificial intelligence (AI) offer opportunities to tailor nanomedicine to individual patient profiles. Personalized nanotherapies could account for variations in metabolic pathways, drug metabolism, and immune responses. However, biocompatibility testing must also evolve to consider genetic and epigenetic variability, including sex, age, and ethnicity, which can influence nanoparticle interactions and therapeutic outcomes. Developing predictive models that integrate patient-specific data with nanoparticle characteristics will be essential to advance precision nanomedicine in complex diseases like metabolic syndrome.

6.3 Ethical and Societal Implications

The clinical deployment of nanomedicine raises important ethical and societal questions, particularly concerning equitable access, informed consent, and risk-benefit communication. Advanced nanotherapies are often expensive, raising concerns about healthcare disparities and availability in low-resource settings. Furthermore, patients must be fully informed about potential long-term risks, even in the absence of complete toxicological data. Transparent regulatory processes and robust post-marketing surveillance will be essential to maintain public trust. Finally, public perception of nanotechnology is heavily influenced by media, education, and regulatory transparency. Efforts to engage communities through science communication and policy dialogue will be necessary to align technological advances with societal expectations and ethical responsibility.

7 Conclusion

Nanomaterials present transformative opportunities for diagnosing and treating metabolic syndrome, offering precision, targeted delivery, and novel therapeutic mechanisms. However, these benefits must be carefully weighed against potential risks, including toxicity, immunogenicity, and long-term bioaccumulation. Stringent safety evaluations—from in vitro assays to in vivo models—are essential to ensure clinical viability. Moving forward, safe-by-design strategies, coupled with interdisciplinary collaboration across nanotechnology, biology, and regulatory science, will be crucial. Bridging innovation with robust safety frameworks and ethical oversight will ultimately determine the successful translation of nanomedicine from the bench to the bedside in metabolic health care. Despite technological advances, the application of nanotechnology in metabolic syndrome faces several hurdles. The chronic nature of this condition, coupled with systemic inflammation and comorbidities, complicates safety assessments. Tailored safety frameworks and disease-specific regulatory pathways are urgently needed to ensure effective translation from bench to bedside.

References

1. Abadeer, N. S., & Murphy, C. J. (2016). Recent progress in cancer thermal therapy using gold nanoparticles. *The Journal of Physical Chemistry C, 120*(9), 4691–4716. https://doi.org/10.1021/acs.jpcc.5b11232
2. Abulikemu, A., Zhao, X., Xu, H., Li, Y., Ma, R., Yao, Q., Wang, J., Sun, Z., Li, Y., & Guo, C. (2023). Silica nanoparticles aggravated the metabolic associated fatty liver disease through disturbed amino acid and lipid metabolisms-mediated oxidative stress. *Redox Biology, 59*(February), 102569. https://doi.org/10.1016/j.redox.2022.102569
3. Adamo, J. E., Bauer, G., Berro, M., Burnett, B. K., Hartman, K. A., Masiello, L. M., Moorman-White, D., Rubinstein, E. P., & Schuff, K. G. (2012). A roadmap for academic health centers to establish good laboratory practice-compliant infrastructure. *Academic Medicine, 87*(3), 279–284. https://doi.org/10.1097/ACM.0b013e318244838a
4. Ahamed, M., Alsalhi, M. S., & Siddiqui, M. K. J. (2010). Silver nanoparticle applications and human health. *Clinica Chimica Acta; International Journal of Clinical Chemistry, 411*(23–24), 1841–1848. https://doi.org/10.1016/j.cca.2010.08.016
5. Ali, M. (2023). What function of nanoparticles is the primary factor for their hyper-toxicity? *Advances in Colloid and Interface Science, 314*(April), 102881. https://doi.org/10.1016/j.cis.2023.102881
6. Amer, L. D., Saleh, L. S., Walker, C., Thomas, S., Janssen, W. J., Alper, S., & Bryant, S. J. (2019). Inflammation via myeloid differentiation primary response gene 88 signaling mediates the fibrotic response to implantable synthetic poly(ethylene glycol) hydrogels. *Acta Biomaterialia, 100*(December), 105–117. https://doi.org/10.1016/j.actbio.2019.09.043
7. Beloqui, A., Solinís, M. Á., Rodríguez-Gascón, A., Almeida, A. J., & Préat, V. (2016). Nanostructured lipid carriers: Promising drug delivery systems for future clinics. *Nanomedicine: Nanotechnology, Biology, and Medicine, 12*(1), 143–161. https://doi.org/10.1016/j.nano.2015.09.004
8. Cao, Z., Li, W., Liu, R., Li, X., Li, H., Liu, L., Chen, Y., Lv, C., & Liu, Y. (2019). pH- and enzyme-triggered drug release as an important process in the design of anti-tumor drug delivery systems. *Biomedicine & Pharmacotherapy, 118*(October), 109340. https://doi.org/10.1016/j.biopha.2019.109340

9. Caturano, A., Nilo, R., Nilo, D., Russo, V., Santonastaso, E., Galiero, R., Rinaldi, L., et al. (2024). Advances in nanomedicine for precision insulin delivery. *Pharmaceuticals, 17*(7), 945. https://doi.org/10.3390/ph17070945
10. Chen, S., Hou, Y., Cheng, G., Zhang, C., Wang, S., & Zhang, J. (2013). Cerium oxide nanoparticles protect endothelial cells from apoptosis induced by oxidative stress. *Biological Trace Element Research, 154*(1), 156–166. https://doi.org/10.1007/s12011-013-9678-8
11. Cohen, E. B., Patwardhan, M., Raheja, R., Alpers, D. H., Andrade, R. J., Avigan, M. I., Lewis, J. H., et al. (2024). Drug-induced liver injury in the elderly: Consensus statements and recommendations from the IQ-DILI initiative. *Drug Safety, 47*(4), 301–319. https://doi.org/10.1007/s40264-023-01390-5
12. Commissioner, Office of the. (2019). *FDA's approach to regulation of nanotechnology products*. FDA. https://www.fda.gov/science-research/nanotechnology-programs-fda/fdas-approach-regulation-nanotechnology-products
13. Cordelli, E., Bignami, M., & Pacchierotti, F. (2021). Comet assay: A versatile but complex tool in genotoxicity testing. *Toxicology Research, 10*(1), 68–78. https://doi.org/10.1093/toxres/tfaa093
14. Couvreur, P., & Vauthier, C. (1991). Polyalkylcyanoacrylate nanoparticles as drug carrier: Present state and perspectives. *Journal of Controlled Release, 17*(2), 187–198. https://doi.org/10.1016/0168-3659(91)90058-L
15. Croissant, J. G., Fatieiev, Y., Almalik, A., & Khashab, N. M. (2018). Mesoporous silica and organosilica nanoparticles: Physical chemistry, biosafety, delivery strategies, and biomedical applications. *Advanced Healthcare Materials, 7*(4), 1700831. https://doi.org/10.1002/adhm.201700831
16. Dai, Y., Guo, Y., Tang, W., Chen, D., Xue, L., Chen, Y., Guo, Y., et al. (2024). Reactive oxygen species-scavenging nanomaterials for the prevention and treatment of age-related diseases. *Journal of Nanobiotechnology, 22*(May), 252. https://doi.org/10.1186/s12951-024-02501-9
17. Danhier, F., Ansorena, E., Silva, J. M., Coco, R., Le Breton, A., & Préat, V. (2012). PLGA-based nanoparticles: An overview of biomedical applications. *Journal of Controlled Release, 161*(2), 505–522. https://doi.org/10.1016/j.jconrel.2012.01.043
18. de la Harpe, K. M., Kondiah, P. P. D., Choonara, Y. E., Marimuthu, T., du Toit, L. C., & Pillay, V. (2019). The hemocompatibility of nanoparticles: A review of cell–nanoparticle interactions and hemostasis. *Cells, 8*(10), 1209. https://doi.org/10.3390/cells8101209
19. Derfus, A. M., Chan, W. C. W., & Bhatia, S. N. (2004). Probing the cytotoxicity of semiconductor quantum dots. *Nano Letters, 4*(1), 11–18. https://doi.org/10.1021/nl0347334
20. Đorđević, S., Gonzalez, M. M., Conejos-Sánchez, I., Carreira, B., Pozzi, S., Acúrcio, R. C., Satchi-Fainaro, R., Florindo, H. F., & Vicent, M. J. (2022). Current hurdles to the translation of nanomedicines from bench to the clinic. *Drug Delivery and Translational Research, 12*(3), 500–525. https://doi.org/10.1007/s13346-021-01024-2
21. Drasler, B., Sayre, P., Steinhäuser, K. G., Petri-Fink, A., & Rothen-Rutishauser, B. (2017). *In vitro* approaches to assess the hazard of nanomaterials. *NanoImpact, 8*(October), 99–116. https://doi.org/10.1016/j.impact.2017.08.002
22. Eisinger, K., Krautbauer, S., Hebel, T., Schmitz, G., Aslanidis, C., Liebisch, G., & Buechler, C. (2014). Lipidomic analysis of the liver from high-fat diet induced obese mice identifies changes in multiple lipid classes. *Experimental and Molecular Pathology, 97*(1), 37–43. https://doi.org/10.1016/j.yexmp.2014.05.002
23. El-Sherbiny, I. M., & Yacoub, M. H. (2013). Hydrogel scaffolds for tissue engineering: Progress and challenges. *Global Cardiology Science & Practice, 2013*(3), 316–342. https://doi.org/10.5339/gcsp.2013.38
24. Elzoghby, A. O., Samy, W. M., & Elgindy, N. A. (2012). Albumin-based nanoparticles as potential controlled release drug delivery systems. *Journal of Controlled Release, 157*(2), 168–182. https://doi.org/10.1016/j.jconrel.2011.07.031
25. Fang, Y., Gong, A. Y., Haller, S. T., Dworkin, L. D., Liu, Z., & Gong, R. (2020). The ageing kidney: Molecular mechanisms and clinical implications. *Ageing Research Reviews, 63*-(November), 101151. https://doi.org/10.1016/j.arr.2020.101151

26. Fatima, M., Almalki, W. H., Khan, T., Sahebkar, A., & Kesharwani, P. (2024). Harnessing the power of stimuli-responsive nanoparticles as an effective therapeutic drug delivery system. *Advanced Materials, 36*(24), 2312939. https://doi.org/10.1002/adma.202312939
27. Fenech, M. (2008). The micronucleus assay determination of chromosomal level DNA damage. *Methods in Molecular Biology (Clifton, N.J.), 410*, 185–216. https://doi.org/10.1007/978-1-59745-548-0_12
28. Firdhouse, M. J., & Lalitha, P. (2015). Biosynthesis of silver nanoparticles and its applications. *Journal of Nanotechnology, 2015*(1), 829526. https://doi.org/10.1155/2015/829526
29. Fröhlich, E. (2012). The role of surface charge in cellular uptake and cytotoxicity of medical nanoparticles. *International Journal of Nanomedicine, 7*, 5577–5591. https://doi.org/10.2147/IJN.S36111
30. Gao, W., Chan, J., & Farokhzad, O. C. (2010). pH-responsive nanoparticles for drug delivery. *Molecular Pharmaceutics, 7*(6), 1913–1920. https://doi.org/10.1021/mp100253e
31. Geszke-Moritz, M., & Moritz, M. (2024). Biodegradable polymeric nanoparticle-based drug delivery systems: Comprehensive overview, perspectives and challenges. *Polymers, 16*(17), 2536. https://doi.org/10.3390/polym16172536
32. Goyani, P., Christodoulou, R., & Vassiliou, E. (2024). Immunosenescence: Aging and immune system decline. *Vaccine, 12*(12), 1314. https://doi.org/10.3390/vaccines12121314
33. Gupta, A. K., & Gupta, M. (2005). Synthesis and surface engineering of iron oxide nanoparticles for biomedical applications. *Biomaterials, 26*(18), 3995–4021. https://doi.org/10.1016/j.biomaterials.2004.10.012
34. Han, F., Wang, J., Ding, L., Hu, Y., Li, W., Yuan, Z., Guo, Q., et al. (2020). Tissue engineering and regenerative medicine: Achievements, future, and sustainability in Asia. *Frontiers in Bioengineering and Biotechnology, 8*, 83. https://doi.org/10.3389/fbioe.2020.00083
35. Havelikar, U., Ghorpade, K. B., Kumar, A., Patel, A., Singh, M., Banjare, N., & Gupta, P. N. (2024). Comprehensive insights into mechanism of nanotoxicity, assessment methods and regulatory challenges of nanomedicines. *Discover Nano, 19*(1), 165. https://doi.org/10.1186/s11671-024-04118-1
36. Hickman, D. L., Johnson, J., Vemulapalli, T. H., Crisler, J. R., & Shepherd, R. (2017). Commonly used animal models. In *Principles of animal research for graduate and undergraduate students* (pp. 117–175). Academic Press. https://doi.org/10.1016/B978-0-12-802151-4.00007-4
37. Hillegass, J. M., Shukla, A., Lathrop, S. A., MacPherson, M. B., Fukagawa, N. K., & Mossman, B. T. (2010). Assessing nanotoxicity in cells in vitro. *Wiley Interdisciplinary Reviews. Nanomedicine and Nanobiotechnology, 2*(3), 219–231. https://doi.org/10.1002/wnan.54
38. Huang, H., Lai, W., Cui, M., Liang, L., Lin, Y., Fang, Q., Liu, Y., & Xie, L. (2016). An evaluation of blood compatibility of silver nanoparticles. *Scientific Reports, 6*(1), 25518. https://doi.org/10.1038/srep25518
39. Jain, A., & Bhise, K. (2025). Nano-pharmacokinetics and pharmacodynamics of green-synthesized ZnO nanoparticles: A pathway to safer therapeutic applications. *Xenobiotica, 55*(4), 265–276. https://doi.org/10.1080/00498254.2025.2505062
40. Kannan, R. M., Nance, E., Kannan, S., & Tomalia, D. A. (2014). Emerging concepts in dendrimer-based nanomedicine: From design principles to clinical applications. *Journal of Internal Medicine, 276*(6), 579–617. https://doi.org/10.1111/joim.12280
41. Karlsson, H. L., Gustafsson, J., Cronholm, P., & Möller, L. (2009). Size-dependent toxicity of metal oxide particles—A comparison between nano- and micrometer size. *Toxicology Letters, 188*(2), 112–118. https://doi.org/10.1016/j.toxlet.2009.03.014
42. Kaur, J., Singh, H., Gupta, N. K., Puri, S., & Khatri, M. (2021). In-vivo models for toxicity screening of nanohazards: Opportunities and challenges. *Materials International, 3*(4), 8. https://doi.org/10.33263/Materials34.008
43. Khan, I., Saeed, K., & Khan, I. (2019). Nanoparticles: Properties, applications and toxicities. *Arabian Journal of Chemistry, 12*(7), 908–931. https://doi.org/10.1016/j.arabjc.2017.05.011

44. Khedulkar, A. P., Bobade, R. G., Doong, R.-a., Pandit, B., Ky, N. M., Ambare, R. C., Hoang, T.-D., & Kumar, K. J. (2025). Bio-based nanomaterials as effective, friendly solutions and their applications for protecting water, soil, and air. *Materials Today Chemistry, 46*(June), 102688. https://doi.org/10.1016/j.mtchem.2025.102688
45. Kohl, Y., Rundén-Pran, E., Mariussen, E., Hesler, M., El Yamani, N., Longhin, E. M., & Dusinska, M. (2020). Genotoxicity of nanomaterials: Advanced in vitro models and high throughput methods for human hazard assessment—A review. *Nanomaterials, 10*(10), 1911. https://doi.org/10.3390/nano10101911
46. Kumari, A., Yadav, S. K., & Yadav, S. C. (2010). Biodegradable polymeric nanoparticles based drug delivery systems. *Colloids and Surfaces. B, Biointerfaces, 75*(1), 1–18. https://doi.org/10.1016/j.colsurfb.2009.09.001
47. Kyriakides, T. R., Raj, A., Tseng, T. H., Xiao, H., Nguyen, R., Mohammed, F. S., Halder, S., et al. (2021). Biocompatibility of nanomaterials and their immunological properties. *Biomedical Materials (Bristol, England), 16*(4), 042005. https://doi.org/10.1088/1748-605X/abe5fa
48. Lee, C. E., & Rezaee, F. (2024). Nanoparticles and airway epithelial cells: Exploring the impacts and methodologies in toxicity assessment. *International Journal of Molecular Sciences, 25*(14), 7885. https://doi.org/10.3390/ijms25147885
49. Makadia, H. K., & Siegel, S. J. (2011). Poly lactic-co-glycolic acid (PLGA) as biodegradable controlled drug delivery carrier. *Polymers, 3*(3), 1377–1397. https://doi.org/10.3390/polym3031377
50. Maldonado, E., Morales-Pison, S., Urbina, F., & Solari, A. (2023). Aging hallmarks and the role of oxidative stress. *Antioxidants, 12*(3), 651. https://doi.org/10.3390/antiox12030651
51. Malloy, T. F. (2011). Nanotechnology regulation: A study in claims making. *ACS Nano, 5*(1), 5–12. https://doi.org/10.1021/nn103480e
52. Mi, P. (2020). Stimuli-responsive Nanocarriers for drug delivery, tumor imaging, therapy and Theranostics. *Theranostics, 10*(10), 4557–4588. https://doi.org/10.7150/thno.38069
53. Nag, S., Mohanto, S., Ahmed, M. G., & Subramaniyan, V. (2024). 'Smart' stimuli-responsive biomaterials revolutionizing the theranostic landscape of inflammatory arthritis. *Materials Today Chemistry, 39*(July), 102178. https://doi.org/10.1016/j.mtchem.2024.102178
54. Neun, B., & Dobrovolskaia, M. (2011). Method for analysis of nanoparticle hemolytic properties in vitro. *Methods in Molecular Biology (Clifton, N.J.), 697*, 215–224. https://doi.org/10.1007/978-1-60327-198-1_23
55. Nong, J., Glassman, P. M., & Muzykantov, V. R. (2022). Targeting vascular inflammation through emerging methods and drug carriers. *Advanced Drug Delivery Reviews, 184*(May), 114180. https://doi.org/10.1016/j.addr.2022.114180
56. Nouri, Z., Hajialyani, M., Izadi, Z., Bahramsoltani, R., Farzaei, M. H., & Abdollahi, M. (2020). Nanophytomedicines for the prevention of metabolic syndrome: A pharmacological and biopharmaceutical review. *Frontiers in Bioengineering and Biotechnology, 8*(May), 425. https://doi.org/10.3389/fbioe.2020.00425
57. Osman, A. I., Zhang, Y., Farghali, M., Rashwan, A. K., Eltaweil, A. S., Abd El-Monaem, E. M., Mohamed, I. M. A., et al. (2024). Synthesis of green nanoparticles for energy, biomedical, environmental, agricultural, and food applications: A review. *Environmental Chemistry Letters, 22*(2), 841–887. https://doi.org/10.1007/s10311-023-01682-3
58. Pardeike, J., Hommoss, A., & Müller, R. H. (2009). Lipid nanoparticles (SLN, NLC) in cosmetic and pharmaceutical dermal products. *International Journal of Pharmaceutics, 366*(1–2), 170–184. https://doi.org/10.1016/j.ijpharm.2008.10.003
59. Peppas, N. A., Bures, P., Leobandung, W., & Ichikawa, H. (2000). Hydrogels in pharmaceutical formulations. *European Journal of Pharmaceutics and Biopharmaceutics, 50*(1), 27–46. https://doi.org/10.1016/s0939-6411(00)00090-4
60. Poland, C. A., Duffin, R., Kinloch, I., Maynard, A., Wallace, W. A. H., Seaton, A., Stone, V., Brown, S., Macnee, W., & Donaldson, K. (2008). Carbon nanotubes introduced into the abdominal cavity of mice show asbestos-like pathogenicity in a pilot study. *Nature Nanotechnology, 3*(7), 423–428. https://doi.org/10.1038/nnano.2008.111

61. Poleboina, S., Sheth, V. G., Sharma, N., Sihota, P., Kumar, N., & Tikoo, K. (2022). Selenium nanoparticles stimulate osteoblast differentiation via BMP-2/MAPKs/β-catenin pathway in diabetic osteoporosis. *Nanomedicine (London, England), 17*(9), 607–625. https://doi.org/10.2217/nnm-2021-0401
62. Rajak, P. (2025). Green nanomaterial-based sustainable analysis of contaminant-remediation in wastewater: A bird's-eye view on recent advances and limitations. *Sustainable Chemistry for the Environment, 10*, 100238. https://doi.org/10.1016/j.scenv.2025.100238
63. Ramos, T. I., Villacis-Aguirre, C. A., López-Aguilar, K. V., Padilla, L. S., Altamirano, C., Toledo, J. R., & Vispo, N. S. (2022). The Hitchhiker's guide to human therapeutic nanoparticle development. *Pharmaceutics, 14*(2), 247. https://doi.org/10.3390/pharmaceutics14020247
64. Rodríguez-Gómez, F. D., Monferrer, D., Penon, O., & Rivera-Gil, P. (2025). Regulatory pathways and guidelines for nanotechnology-enabled health products: A comparative review of EU and US frameworks. *Frontiers in Medicine, 12*(March), 1544393. https://doi.org/10.3389/fmed.2025.1544393
65. Romero-Carmona, C. E., Chávez-Corona, J. I., Lima, E., Cortés, H., Quintanar-Guerrero, D., Bernad-Bernad, M. J., Ramos-Martínez, I., Peña-Corona, S. I., Sharifi-Rad, J., & Leyva-Gómez, G. (2024). Nanoparticle and microparticle-based systems for enhanced oral insulin delivery: A systematic review and meta-analysis. *Journal of Nanobiotechnology, 22*-(December), 802. https://doi.org/10.1186/s12951-024-03045-8
66. Ruijter, N., Soeteman-Hernández, L. G., Carrière, M., Boyles, M., McLean, P., Catalán, J., Katsumiti, A., et al. (2023). The state of the art and challenges of in vitro methods for human hazard assessment of nanomaterials in the context of safe-by-design. *Nanomaterials (Basel), 13*(3), 472. https://doi.org/10.3390/nano13030472
67. Sangeetha, G., Rajeshwari, S., & Venckatesh, R. (2011). Green synthesis of zinc oxide nanoparticles by *Aloe barbadensis Miller* leaf extract: Structure and optical properties. *Materials Research Bulletin, 46*(12), 2560–2566. https://doi.org/10.1016/j.materresbull.2011.07.046
68. Sharma, V., Shukla, R. K., Saxena, N., Parmar, D., Das, M., & Dhawan, A. (2009). DNA damaging potential of zinc oxide nanoparticles in human epidermal cells. *Toxicology Letters, 185*(3), 211–218. https://doi.org/10.1016/j.toxlet.2009.01.008
69. Shi, Z., Neoh, K. G., Kang, E. T., Poh, C., & Wang, W. (2008). Bacterial adhesion and osteoblast function on titanium with surface-grafted chitosan and immobilized RGD peptide. *Journal of Biomedical Materials Research. Part A, 86*(4), 865–872. https://doi.org/10.1002/jbm.a.31648
70. Shi, H., Magaye, R., Castranova, V., & Zhao, J. (2013). Titanium dioxide nanoparticles: A review of current toxicological data. *Particle and Fibre Toxicology, 10*(April), 15. https://doi.org/10.1186/1743-8977-10-15
71. Shukla, R. K., Badiye, A., Vajpayee, K., & Kapoor, N. (2021). Genotoxic potential of nanoparticles: Structural and functional modifications in DNA. *Frontiers in Genetics, 12*-(September), 728250. https://doi.org/10.3389/fgene.2021.728250
72. Singh, N., Jenkins, G. J. S., Asadi, R., & Doak, S. H. (2010). Potential toxicity of superparamagnetic iron oxide nanoparticles (SPION). *Nano Reviews, 1*, 5358. https://doi.org/10.3402/nano.v1i0.5358
73. Singh, P., Santosh Pandit, V. R. S. S., Mokkapati, A. G., Ravikumar, V., & Mijakovic, I. (2018). Gold nanoparticles in diagnostics and therapeutics for human cancer. *International Journal of Molecular Sciences, 19*(7), 1979. https://doi.org/10.3390/ijms19071979
74. Singh, R., Srinivas, S. P., Kumawat, M., & Daima, H. K. (2024). Ligand-based surface engineering of nanomaterials: Trends, challenges, and biomedical perspectives. *Open Nano, 15*(January), 100194. https://doi.org/10.1016/j.onano.2023.100194
75. Slowing, I. I., Vivero-Escoto, J. L., Chia-Wen, W., & Lin, V. S.-Y. (2008). Mesoporous silica nanoparticles as controlled release drug delivery and gene transfection carriers. *Advanced Drug Delivery Reviews, 60*(11), 1278–1288. https://doi.org/10.1016/j.addr.2008.03.012
76. Sogias, I. A., Williams, A. C., & Khutoryanskiy, V. V. (2008). Why is chitosan mucoadhesive? *Biomacromolecules, 9*(7), 1837–1842. https://doi.org/10.1021/bm800276d

77. Suk, J. S., Qingguo, X., Kim, N., Hanes, J., & Ensign, L. M. (2016). PEGylation as a strategy for improving nanoparticle-based drug and gene delivery. *Advanced Drug Delivery Reviews, 99*(Pt A), 28–51. https://doi.org/10.1016/j.addr.2015.09.012
78. Tirumala, M. G., Pratibha, A., Susmitha, R., Mahesh, R., & Chandraiah, G. (2021). Novel methods and approaches for safety evaluation of nanoparticle formulations: A focus towards in vitro models and adverse outcome pathways. *Frontiers in Pharmacology, 12*(September), 612659. https://doi.org/10.3389/fphar.2021.612659
79. Tkachenko, A. (2024). Hemocompatibility studies in nanotoxicology: Hemolysis or eryptosis? (a review). *Toxicology In Vitro, 98*(June), 105814. https://doi.org/10.1016/j.tiv.2024.105814
80. Tune, J. D., Goodwill, A. G., Sassoon, D. J., & Mather, K. J. (2017). Cardiovascular consequences of metabolic syndrome. *Translational Research: The Journal of Laboratory and Clinical Medicine, 183*(May), 57–70. https://doi.org/10.1016/j.trsl.2017.01.001
81. Volpatti, L. R., Matranga, M. A., Cortinas, A. B., Delcassian, D., Daniel, K. B., Langer, R., & Anderson, D. G. (2020). Glucose-responsive nanoparticles for rapid and extended self-regulated insulin delivery. *ACS Nano, 14*(1), 488–497. https://doi.org/10.1021/acsnano.9b06395
82. Wang, F., Cai, X., Shen, Y., & Meng, L. (2023). Cell–scaffold interactions in tissue engineering for oral and craniofacial reconstruction. *Bioactive Materials, 23*(May), 16–44. https://doi.org/10.1016/j.bioactmat.2022.10.029
83. Wardhani, K., Yazzie, S., McVeigh, C., Edeh, O., Grimes, M., Jacquez, Q., Dixson, C., et al. (2024). Systemic immunological responses are dependent on sex and ovarian hormone presence following acute inhaled Woodsmoke exposure. *Particle and Fibre Toxicology, 21*(May), 27. https://doi.org/10.1186/s12989-024-00587-5
84. Warheit, D. B. (2018). Hazard and risk assessment strategies for nanoparticle exposures: How far have we come in the past 10 years? *F1000Research, 7*(March), 376. https://doi.org/10.12688/f1000research.12691.1
85. Xuan, M., Shao, J., & Li, J. (2019). Cell membrane-covered nanoparticles as biomaterials. *National Science Review, 6*(3), 551–561. https://doi.org/10.1093/nsr/nwz037
86. Xuan, L., Zhao, J., Skonieczna, M., Zhou, P.-K., & Huang, R. (2023). Nanoparticles-induced potential toxicity on human health: Applications, toxicity mechanisms, and evaluation models. *MedComm (London), 4*(4), e327. https://doi.org/10.1002/mco2.327
87. Yan, S., Na, J., Liu, X., & Wu, P. (2024). Different targeting ligands-mediated drug delivery systems for tumor therapy. *Pharmaceutics, 16*(2), 248. https://doi.org/10.3390/pharmaceutics16020248
88. Yedgar, S., Barshtein, G., & Gural, A. (2022). Hemolytic activity of nanoparticles as a marker of their hemocompatibility. *Micromachines, 13*(12), 2091. https://doi.org/10.3390/mi13122091
89. Zhang, Y., Ali, S. F., Dervishi, E., Yang, X., Li, Z., Casciano, D., & Biris, A. S. (2010). Cytotoxicity effects of graphene and single-wall carbon nanotubes in neural phaeochromocytoma-derived PC12 cells. *ACS Nano, 4*(6), 3181–3186. https://doi.org/10.1021/nn1007176
90. Zhang, X., Yin, J., Peng, C., Weiqing, H., Zhu, Z., Li, W., Fan, C., & Huang, Q. (2011). Distribution and biocompatibility studies of graphene oxide in mice after intravenous administration. *Carbon, 49*(3), 986–995. https://doi.org/10.1016/j.carbon.2010.11.005
91. Zhao, X., Hu, D. A., Wu, D., He, F., Wang, H., Huang, L., Shi, D., et al. (2021). Applications of biocompatible scaffold materials in stem cell-based cartilage tissue engineering. *Frontiers in Bioengineering and Biotechnology, 9*(March), 603444. https://doi.org/10.3389/fbioe.2021.603444
92. Zhu, L., Yuan, Z., Wu, S., Yan, M., Cao, Y., Mou, N., Wang, G., Sun, D., & Wu, W. (2022). Cell membrane camouflaged biomimetic nanoparticles: Focusing on tumor theranostics. *Materials Today Bio, 14*(March), 100228. https://doi.org/10.1016/j.mtbio.2022.100228
93. Zielińska, A., Costa, B., Ferreira, M. V., Miguéis, D., Louros, J. M. S., Durazzo, A., Lucarini, M., et al. (2020). Nanotoxicology and nanosafety: Safety-by-design and testing at a glance. *International Journal of Environmental Research and Public Health, 17*(13), 4657. https://doi.org/10.3390/ijerph17134657

Navigating the Future Landscape and Overcoming Challenges in Nanomaterials for Metabolic Syndrome

12

Nitish Kumar Singh, Jaikee Kumar Singh, Rishika Munjal, Vivek Chandra Verma, and Aprajita Singh

Abstract

Following COVID-19, people's lives have altered significantly, with people spending less time exercising and more time on electronic devices. Furthermore, urbanization and globalization have changed lifestyles, leading to the emergence of several lifestyle disorders that are generally referred to as metabolic syndrome (MetS). Diabetes, glucose intolerance, insulin resistance, dyslipidemia, obesity, non-alcoholic fatty liver disease (NAFLD), hypertension, and other metabolic abnormalities are hallmarks of metabolic syndrome, which predisposes people to cardiovascular and neurological disorders. The pathophysiological causes of metabolic syndrome are widely understood. However, a few drawbacks, including limited effectiveness, low bioavailability, lack of target selectivity, and dose-dependent adverse effects, make the currently available medicines ineffective. Fresh hope in the fight against metabolic syndrome is provided by targeted

N. K. Singh (✉)
Department of Biosciences Manipal University, Jaipur, Rajasthan, India
e-mail: nitish.2510160010@muj.manipal.edu

J. K. Singh
Department of Bioinformatics, Faculty of Engineering and Technology, Marwadi University, Rajkot, Gujarat, India
e-mail: jaikee.singh@marwadieducation.edu.in

Rishika Munjal
AIIMS, New Delhi, India

V. C. Verma
Department of Biochemistry, Punjab University, Chandigarh, India

A. Singh
Department of Microbiology, Institute of Medical Sciences (Banaras Hindu University), Varanasi, Uttar Pradesh, India

V. K. Chaturvedi et al. (eds.), *Nanotheranostics in Metabolic Syndrome*,
https://doi.org/10.1007/978-981-95-3746-4_12

nanotherapy. Nanomaterials are constantly changing, and efficient drug delivery strategies that manage the intricate relationships between molecules and technology guarantee medication safety, effectiveness, and targeting. A few nano-drug delivery systems have been developed for bioactive substances, including liposomes, hydrogel nanocomposites, nanomicelles, nanoemulsions, core-shell nanoparticles, and metallic nanoparticles. Bioactive substances' nano-encapsulation has transformed the clinical and pharmaceutical sectors and aided in the creation of innovative treatments for metabolic syndrome and related conditions like Alzheimer's disease. Metabolic syndrome may be diagnosed and treated with carbon nanoparticles. In those with metabolic syndrome, it can be utilized as a nanosensor to monitor blood pressure, blood glucose levels, and dyslipidemia. Nanozymes can be tailored in such a way as to enhance their catalytic efficiency, targeting ability, and ability to eliminate ROS and alleviate oxidative stress, thereby promising potential for preventing metabolic syndrome and improving treatment efficacy. The major challenges for nanomaterials used for drug delivery systems are their optimization to check their cellular uptake and/or targeting the desired tissue without any side effects. This chapter briefly assesses metabolic syndrome, and the potential use of nanomaterial coupled with targeted drug delivery of bioactive compounds as emerging therapies and challenges.

Keywords

Metabolic syndrome · Nanomaterials · Bioactive compounds · Enzyme · Nano-encapsulation

Abbreviations

AD	Alzheimer's disease
AMPK	AMP-activated protein kinase
BBB	Blood-brain barrier
CNMs	Carbon nanomaterials
CNTs	Carbon nanotubes
CQDs	Carbon quantum dots
CTGF	Connective tissue growth factor
CVDs	Cardiovascular diseases
DAG	Diacylglycerol
EGCG	Epigallocatechin gallate
eNOS	Endothelial nitric oxide synthase
FDG-PET	Fluorodeoxyglucose Positron Emission Tomography
FFAs	Free fatty acids
GO	Graphene oxide
GPx	Glutathione peroxidase
HDL	High-density lipoprotein

HIF-1α	Hypoxia-inducible factor 1-alpha.
HPA	Hypothalamic-pituitary-adrenal
IDE	Insulin-degrading enzyme
IDF	International Diabetes Federation
LDL	Low-density lipoprotein
LOX	Lipoxygenase
LPS	Lipopolysaccharides
MetS	Metabolic syndrome
MRs	Mineralocorticoid receptors
NAFLD	Non-alcoholic fatty liver disease
NCEP	National Cholesterol Education Program
NPs	Nanoparticles
PKC	Protein kinase C
RAS	Renin-angiotensin system
ROS	Reactive oxygen species
SCFAs	Short-chain fatty acids
SeNPs	Selenium nanoparticles
SOD	Superoxide dismutase
T2DM	Type 2 diabetes mellitus
TLR4	Toll-like receptor 4
UPR	Unfolded protein response
VAT	Visceral adipose tissue
WHO	World Health Organization

1 Introduction

The phrase "Metabolic Syndrome" first appeared in the latter half of the twentieth century to describe a group of metabolic risk factors that frequently co-occurred and markedly increased the risk of both type 2 diabetes and cardiovascular disease. MetS, which Reaven first described in full in 1988 under the name "Syndrome X," attracted notice for its hormonal and systemic inflammatory consequences [1]. Over time, the World Health Organization (WHO), the International Diabetes Federation (IDF), and the National Cholesterol Education Program (NCEP) came up with consensus definitions that incorporated central obesity and insulin resistance as essential elements. By the early 2000s, scientists started speculating about common pathways such as chronic inflammation, endothelial dysfunction, and poor insulin signalling as potential links between MetS and neurodegenerative diseases. Together, obesity, insulin resistance, hypertension, hyperglycemia, and dyslipidemia raise the risk of cardiovascular disease, type 2 diabetes mellitus, and stroke [2]. This complex, multifactorial illness is called metabolic syndrome (MetS) (Fig. 12.1). MetS has become a major public health problem because of its rising global incidence, which is fuelled by sedentary lifestyles, imbalanced diets, and

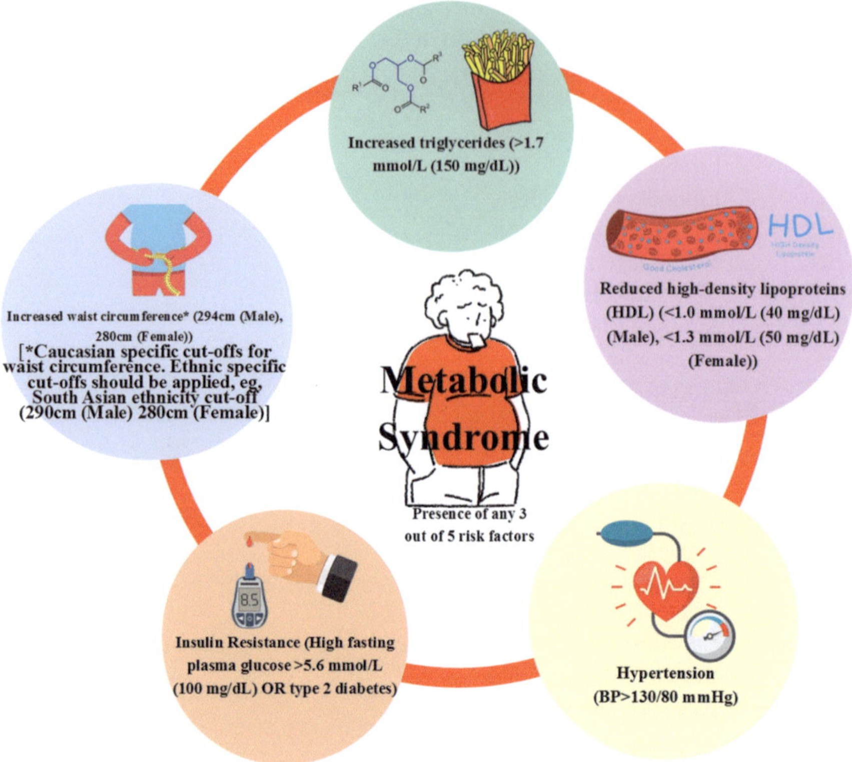

Fig. 12.1 Overview of metabolic syndrome

ageing populations. Existing therapies frequently lack holistic effectiveness and are hampered by systemic adverse effects, poor compliance, and limited bioavailability, even if there are pharmaceutical and lifestyle interventions that target specific syndrome components [3]. By creating and using nanomaterials with at least one dimension in the nanoscale range (1–100 nm)—nanotechnology has become a game-changing method that offers potential pathways for diagnosing, monitoring, and treating MetS. These materials' distinct physical and chemical characteristics, such as increased surface area, reactivity, adjustable surface functions, and the capacity to pass across biological membranes, allow for innovative biomedical applications. By encasing bioactive molecules, nanomaterials can serve as tailored drug delivery systems in therapeutic settings, allowing site-specific release, increased bioavailability, and less systemic toxicity. To improve the administration of medications that lower blood pressure, lower cholesterol, reduce inflammation, and treat diabetes, liposomes, polymeric nanoparticles, dendrimers, metal-organic frameworks, and inorganic nanostructures like silica and gold nanoparticles have all been investigated. Furthermore, in response to pathophysiological signals particular to MetS, stimuli-responsive nanocarriers which react to pH, oxidative stress, or enzymatic triggers are being created to release their payloads, providing precision medicine approaches that precisely match the characteristics of individual diseases [4]. Using quantum dots, magnetic nanoparticles, and biosensor-integrated platforms, nanomaterials transform biomarker detection in the diagnostic field by enabling the ultrasensitive and quick identification of lipid abnormalities, insulin resistance, glucose levels, and inflammatory cytokines. These features assist in the customization of treatment plans and facilitate early identification and illness monitoring. Furthermore, by creating nano-formulations that deliver probiotics or prebiotics with improved stability and efficiency, nanomaterials show potential for non-pharmacological therapies like modifying the gut microbiota, a crucial regulator of metabolic health. This multidisciplinary approach, which connects materials science, molecular biology, chemistry, pharmacology, and clinical medicine, makes it possible to create nanostructures suited to therapeutic needs and biological interactions rationally [5]. However, despite these developments, many obstacles remain to overcome. Since the long-term consequences of nanoparticle accumulation, degradation, and immune system interaction are still poorly known, safety is still paramount. The absence of standardized toxicity tests and regulatory frameworks unique to nanomedicine hampers clinical translation. Furthermore, a significant technological obstacle still exists in producing consistent-quality, scalable, and repeatable nanomaterials, which are essential for commercial viability. Restricted reimbursement policies, high manufacturing costs, and regulatory concerns further complicate the development pipeline. Ethical and social concerns must be addressed for public trust and broad acceptance, including patient privacy, informed consent, and fair access to nanomedicine [6]. Biodegradable and bioinspired nanomaterials that reduce toxicity and environmental impact are still being researched; however, the combination of AI and machine learning is starting to improve safety profiling, predict biological interactions, and optimize nanoparticle design. Innovations in green synthesis, 3D printing, and microfluidics are increasing the sustainability and accuracy of nanoparticle production. Proteomics, metabolomics, and genomics technologies combine with nanomedicine to create precision nanotherapeutics suited to specific metabolic phenotypes. This could change the clinical approach to MetS from reactive symptom management to proactive, individualized prevention and treatment. Nanotechnology may soon play a key role in tackling one of our most widespread health issues with careful policy development, thorough safety evaluations, and strategic research [7]. This could lead to safer, more effective, and more individualized interventions that address the syndrome in all its complexity.

2 Pathogenesis of Metabolic Syndrome

A mix of metabolic, genetic, and environmental variables that upset homeostasis might result in metabolic syndrome (MetS). Chronic inflammation and poor insulin signalling are caused by central obesity and insulin resistance. Adipokines and pro-inflammatory mediators are secreted by visceral adipose tissue, exacerbating insulin resistance [8]. The control of metabolism is further compromised by oxidative stress and mitochondrial malfunction. In addition to lifestyle variables, including poor nutrition and inactivity, these disturbances are exacerbated by hypothalamic-pituitary-adrenal (HPA) axis changes [9]. Even with advancements in our knowledge of MetS, the complete intricacy of its pathophysiology is still unknown, and further study is required to find novel treatment targets.

2.1 Insulin Resistance

Insulin resistance, which interferes with normal glucose and lipid metabolism, is a key pathogenic characteristic of MetS. The PI3K-Akt signalling pathway, necessary for glucose uptake and metabolic control, is activated when insulin binds to its receptor and activates insulin receptor substrates (IRS-1 and IRS-2) [10]. However, this route is disrupted in obesity and inflammation by kinases such as JNK, IKKβ, and mTOR/S6K1 that phosphorylate IRS proteins on serine residues. As a result, insulin signalling is compromised, which hinders the liver, skeletal muscle, and adipose tissue from efficiently absorbing and metabolizing glucose. The accumulation of lipid intermediates, particularly ceramides and diacylglycerol (DAG), exacerbates insulin resistance. DAG activates certain isoforms of protein kinase C (PKC), which phosphorylate IRS proteins at inhibitory areas. Ceramides contribute to further insulin signalling system failure by promoting apoptosis, inhibiting Akt activation, and inducing inflammatory responses. Endoplasmic reticulum (ER) stress is also crucial in adipocytes and hepatocytes. Chronic ER stress increases systemic inflammation and further impairs insulin action by activating the unfolded protein response (UPR) and inflammatory pathways, including IRE1α, JNK, and NF-κB [11]. The development of metabolic syndrome, type 2 diabetes, and cardiovascular disease are accelerated by cumulative insulin signalling disruptions, resulting in a series of metabolic abnormalities such as hyperglycemia, hyperinsulinemia, increased hepatic gluconeogenesis, and enhanced lipogenesis.

2.2 Adipose Tissue Dysfunction

Closely linked to insulin resistance is the dysfunction of adipose tissue, particularly visceral fat depots. Visceral adipose tissue (VAT) is not merely a passive fat store. Still, it functions as an active endocrine organ, secreting various bioactive molecules, including pro-inflammatory cytokines such as TNF-α, IL-6, and MCP-1.VAT expansion during obesity creates a hypoxic microenvironment that stimulates the upregulation of these inflammatory mediators, disrupting insulin signalling and promoting systemic insulin resistance [12]. TNF-α interferes with insulin signalling by inducing serine phosphorylation of IRS-1, while IL-6 encourages the expression of SOCS3, a negative regulator of insulin receptor signalling. Furthermore, VAT growth uses chemokines like MCP-1 to attract immune cells, especially

M1-polarized macrophages. Hypertrophic adipocytes produce free fatty acids (FFAs), which cause macrophages' toll-like receptor 4 (TLR4) to become activated. This exacerbates inflammatory reactions and further reduces insulin sensitivity. By upregulating the genes for collagen and connective tissue growth factor (CTGF), hypoxia in increasing VAT stabilizes hypoxia-inducible factor 1-alpha (HIF-1α), which encourages fibrosis [13]. The ensuing adipose tissue fibrosis exacerbates insulin resistance, dyslipidemia, and systemic metabolic dysfunction by limiting adipose tissue expandability, impairing adipocyte function, and causing ectopic fat deposition in the liver and muscle.

2.3 Neurohormonal Regulation

Neurohormonal pathways significantly changed in MetS and are important in regulating metabolic processes. Adipocytes release the hormone leptin, essential for controlling hunger and energy balance. Leptin resistance and hyperleptinemia paradoxically coexist in obesity, mostly due to elevated SOCS3 and PTP1B expression that disrupts leptin receptor signalling. This disrupts the hypothalamic regulation of energy expenditure and satiety, exacerbating metabolic dysfunction and encouraging weight gain [14]. Through AMPK activation, adiponectin, another adipokine, typically promotes fatty acid oxidation and glucose metabolism. Adiponectin's insulin-sensitizing and anti-inflammatory properties are diminished in obesity, though, which exacerbates the metabolic abnormalities of MetS. Additionally, obesity causes adipose tissue's renin-angiotensin system (RAS) to become active, which raises the synthesis of angiotensin II (Ang II). Reactive oxygen species (ROS) are produced when Ang II binds to AT1 receptors, activating NADPH oxidase and contributing to insulin resistance, hypertension, and vascular dysfunction [15]. Further raising the risk of cardiovascular disease and exacerbating insulin resistance, high aldosterone levels also activate mineralocorticoid receptors (MRs), promoting salt retention, hypertension, and fibrotic remodelling in adipose tissue and cardiovascular organs.

2.4 Oxidative Stress

One of the main pathogenic mechanisms in MetS is oxidative stress, which exacerbates insulin resistance and inflammation. NADPH oxidase activation in immune and endothelial cells, as well as mitochondrial malfunction in metabolic organs, are the causes of elevated reactive oxygen species (ROS) [16]. Vascular dysfunction and oxidative stress are exacerbated when uncoupled endothelial nitric oxide synthase (eNOS) switches from generating nitric oxide (NO) to superoxide. Overproduction of ROS damages lipids, proteins, and DNA by oxidizing LDL, which macrophages then absorb through LOX-1 receptors to create foam cells and start atherosclerosis. ROS also increase inflammation and endothelial dysfunction by activating transcription factors, including AP-1 and NF-κB [17]. This oxidative imbalance contributes to hypertension and cardiovascular problems in MetS by decreasing NO bioavailability, increasing vascular resistance, and impairing arterial compliance.

2.5 Gut Microbiota Dysbiosis

The pathophysiology of MetS is significantly influenced by dysbiosis of the gut microbiota. Metabolic dysfunction and obesity are associated with changes in the microbial composition, particularly an elevated Firmicutes-to-Bacteroidetes ratio. Intestinal permeability is increased by dysbiosis, which interferes with tight junction proteins like occludin and claudins [18]. This "leaky gut" permits the movement of microbial products such as lipopolysaccharides (LPS) into the bloodstream, triggering inflammatory pathways controlled by TLR4 and NF-κB that encourage insulin resistance and endothelial dysfunction [19]. Furthermore, dysbiosis inhibits the synthesis of short-chain fatty acids (SCFAs), such as butyrate, acetate, and propionate, which typically maintain gut integrity, control metabolism, and inhibit inflammation. This exacerbates systemic metabolic disorders.

2.6 Mitochondrial Dysfunction

One important mechanism underlying the metabolic abnormalities of MetS is mitochondrial malfunction. Obesity reduces mitochondrial oxidative phosphorylation, which hinders the oxidation of glucose and fatty acids to produce ATP [20]. Lipid buildup and the production of lipotoxic intermediates such as ceramides and diacylglycerols, which interfere with insulin signalling, cause cellular stress, and encourage inflammation, are caused by reduced fatty acid oxidation. Additionally, obesity decreases mitochondrial mass and metabolic flexibility by downregulating PGC-1α, a crucial regulator of mitochondrial biogenesis [21]. Furthermore, oxidative alteration of insulin signalling components and impaired glucose absorption are caused by the excessive ROS produced by malfunctioning mitochondria. These changes feed a vicious cycle of inflammation, oxidative stress, insulin resistance, and fat buildup. Metabolic, inflammatory, hormonal, and cellular abnormalities interact intricately in the pathophysiology of metabolic syndrome (Fig. 12.2). In a vicious loop, central obesity, insulin resistance, oxidative stress, neurohormonal imbalances, gut dysbiosis, and mitochondrial dysfunction contribute to systemic metabolic dysfunction [22]. Developing successful preventative and treatment interventions requires a deeper knowledge of these systems.

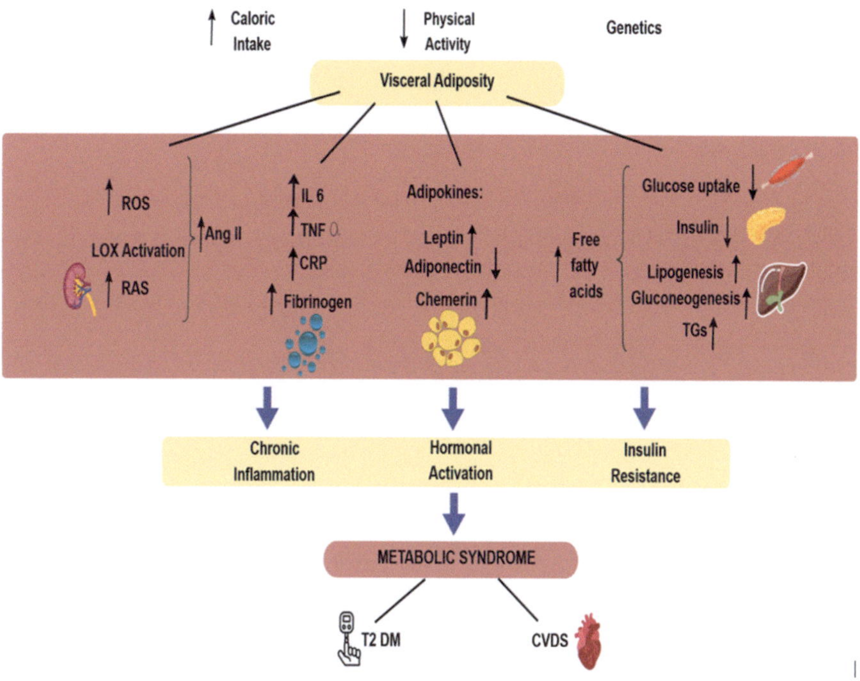

Fig. 12.2 Molecular and physiological mechanisms contributing to the development of metabolic syndrome. Caloric excess, reduced physical activity, and genetic predisposition promote visceral adiposity, which leads to increased production of reactive oxygen species (ROS), activation of the renin-angiotensin system (RAS) and lipoxygenase (LOX) pathways, pro-inflammatory cytokines (IL-6, TNF-α, CRP), dysregulation of adipokines (increased leptin and chemerin, decreased adiponectin), and elevated free fatty acids (FFAs). These alterations result in chronic inflammation, hormonal dysregulation, and insulin resistance, ultimately culminating in metabolic syndrome and its associated complications, including type 2 diabetes mellitus (T2DM) and cardiovascular diseases (CVDs)

3 Pathogenesis of Alzheimer's Disorder Linked with Metabolic Syndrome

Alois Alzheimer initially identified Alzheimer's disease in 1906 after noticing odd histological characteristics in the brain of a 51-year-old lady who was experiencing gradual memory loss. These traits included intracellular neurofibrillary tangles and extracellular amyloid plaques. The cause of AD has been unknown for many years. The "amyloid cascade hypothesis," which postulated that the deposition of Aβ peptides set off a series of events that resulted in tau disease, synaptic failure, and neuronal death, gained popularity in the 1980s and 1990s [23]. Despite little clinical success, this concept dominated the field and directed drug development tactics. However, epidemiological research started to raise doubts about whether AD had a solely neurological cause in the early 2000s. Researchers found a robust association between a higher risk of cognitive deterioration in later life and midlife obesity, type 2 diabetes mellitus (T2DM), and hypertension [24]. These correlations led to a more thorough investigation of systemic effects on brain function, which prepared the way for the connection between AD and metabolic syndrome. As a brain-specific insulin-resistance illness, Alzheimer's disease was dubbed "type 3 diabetes" by de la Monte and Wands in 2005. According to their research, insulin resistance in the brain interferes with survival signalling pathways and negatively affects neuronal glucose uptake. Additionally, insulin-resistant states cause an overload of the insulin-degrading enzyme (IDE), which breaks down insulin and amyloid-beta, lowering Aβ clearance and promoting plaque buildup. Years before cognitive problems manifested, FDG-PET scans conducted later showed that people with T2DM and MetS had reduced glucose metabolism in areas of the brain commonly impacted by AD. These results changed the way that Alzheimer's disease is conceptualized and provided compelling evidence for the metabolic role in the illness's pathophysiology. In the past, "vascular dementia," which was thought to be separate from AD, was blamed for the vascular component of cognitive loss. But as studies went on, it became clear that vascular damage is also common in Alzheimer's brains and could work in concert with tau and amyloid disease. Research conducted in the early 2010s showed that a weakened blood-brain barrier (BBB), a common consequence of MetS, makes it easier for pro-inflammatory lipids and cytokines to enter the brain, aggravating tau hyperphosphorylation and neuroinflammation [25]. Accordingly, AD and vascular dementia are not mutually exclusive; rather, they are both part of a spectrum of neurovascular illness that is greatly impacted by systemic metabolic health. An additional significant historical turning point in the connection between AD and MetS was the realization that chronic inflammation is a cause of neurodegeneration. The development of AD was linked to microglia, the brain's innate immune cells, by the early 2000s. Normally, dormant microglia are triggered by systemic inflammatory signals and Aβ buildup. Chronic diseases like MetS can cause microglia to go into a persistently pro-inflammatory state, generating cytokines and ROS that harm neurons and keep the environment hazardous [26]. It is now widely acknowledged that peripheral metabolic abnormalities can cause and maintain neuroinflammation, which is a defining feature of Alzheimer's disease. Metabolic syndrome is now considered a pathogenic driver of Alzheimer's disease rather than just a risk factor. Common pathophysiological pathways between metabolic and neurodegenerative illnesses demonstrate their interdependence. The bidirectional association has been shown by neuroimaging, genetic research, and longitudinal epidemiological data: MetS raises the risk and speeds AD advancement, whereas early brain alterations may exacerbate metabolic control through stress and behavioural changes [27]. This growing understanding has made new approaches to diagnosis, prevention, and therapy possible.

4 Green Synthesis of Nanoparticles and Its Therapeutic Role

Nanotechnology has revolutionized modern science by enabling the development and application of materials at the nanoscale (1–100 nm), where unique physicochemical properties emerge. Among the most significant advancements are nanoparticles (NPs), engineered materials with a wide range of applications in imaging, drug delivery, treatment, and diagnostics [28]. However, because traditional methods of creating nanoparticles might include hazardous chemicals, a significant energy input, and environmentally dangerous byproducts, there are substantial concerns about sustainability and biocompatibility. These challenges have given rise to the concept of producing green nanoparticles. Green synthesis uses biological agents like plant extracts, bacteria, fungi, and algae to produce nanoparticles in a method that is safe for the environment, profitable, and biocompatible. These biological agents replace harsh chemicals as stabilizing and reducing physical and chemical synthesis agents [29]. Due to their enhanced biocompatibility and adherence to green chemistry principles, green-synthesized nanoparticles are perfect for medicinal uses. Green synthesis is guided by the 12 principles of green chemistry, emphasizing reduced toxicity, energy efficiency, and renewable resources. The green technique uses natural ingredients and ambient circumstances to create nanoparticles without using chemical reductants. Following reduction, the same biomolecules act as capping agents, stabilizing the nanoparticles by preventing agglomeration and limiting their size and form. Furthermore, stabilizing processes enhance biological functionality, especially for drug administration and biocompatibility. Variables, including pH, temperature, incubation time, and biomolecule concentration, greatly influence the nanoparticles' size, shape, and stability. For instance, an alkaline pH frequently promotes smaller particle sizes because phenolic compounds have a stronger propensity to decrease.

4.1 Biological Systems Used in Green Synthesis

Plant extracts are the most often employed biological systems because of their accessibility, quick reaction times, and variety of phytochemicals. For instance, gold and silver nanoparticles with potent antibacterial and anticancer effects have been created using an extract from *Azadirachta indica*, or neem. Polyphenols found in *Camellia sinensis*, or green tea, yield extremely stable gold nanoparticles that may have anticancer properties. Although microbial synthesis requires strict sterile conditions, it has the advantages of facile scaling and genetic modifiability. Metal ions can be reduced by microbes extracellularly or intracellularly [30]. For instance, Metal salts can be reduced into NPs by bacteria such as *Pseudomonas stutzeri* and *Bacillus subtilis*. Fungi: *Fusarium oxysporum* can produce gold, silver, and selenium nanoparticles. Actinomycetes and yeasts are effective sources of silver and selenium nanoparticles. Polysaccharides from algae such as *Sargassum wightii* and *Chlorella vulgaris* aid in the creation of nanoparticles. Biopolymers such as chitosan, dextran, and alginate are also used to make biocompatible and biodegradable nanocarriers.

4.2 Therapeutic Roles of Green-Synthesized Nanoparticles in Metabolic Syndrome

Green-synthesized nanoparticles (NPs), which are produced utilizing environmentally friendly techniques and biological sources like bacteria, fungi, and plant extracts, have shown great promise in treating the complex pathophysiology of metabolic syndrome (MetS). There is a growing need for safe, efficient, and multipurpose treatment agents because of the chronic and systemic character of MetS, which includes obesity, insulin resistance, hypertension, dyslipidemia, and hyperglycemia. Such promise is provided by green-synthesized nanoparticles, which combine improved therapeutic efficacy with biocompatibility. Green-synthesized nanoparticles' antidiabetic properties, which include improving insulin sensitivity and lowering blood glucose levels, are among their most promising uses in MetS [31]. In streptozotocin-induced diabetic rats, for example, silver nanoparticles (AgNPs) made with neem (*Azadirachta indica*) extract have been shown to dramatically lower fasting blood glucose and raise insulin levels. Pancreatic β-cell activity is improved, oxidative stress is suppressed, and glucose transporters are modulated as part of the mechanism. Similarly, *Syzygium cumini* (jamun) seed extract-derived gold nanoparticles (AuNPs) have demonstrated anti-hyperglycemic properties, which are ascribed to their capacity to modify insulin receptor signalling and lower systemic inflammation. Additionally, green-synthesized nanoparticles exhibit promises in lipid control, a crucial component of MetS. For instance, in animal models, zinc oxide nanoparticles (ZnONPs) made from the leaves of *Moringa oleifera* have shown lipid-lowering qualities, successfully lowered levels of triglycerides, total cholesterol, and low-density lipoprotein (LDL) while raised levels of high-density lipoprotein (HDL). Antioxidant activity frequently coexists with this lipid-modulatory action, preventing lipid peroxidation and safeguarding vascular integrity. Certain plant-mediated nanoparticles have anti-adipogenic qualities in managing obesity by preventing preadipocytes from differentiating into adipocytes. For example, in models of obesity caused by a high-fat diet, green tea-mediated AuNPs have been shown to decrease adipose tissue mass and body weight increase. These effects are associated with increased mitochondrial biogenesis, fatty acid oxidation, and downregulation of important adipogenic transcription factors such as PPAR-γ and C/EBP-α. Additionally, green-synthesized nanoparticles aid in the reduction of oxidative stress, a prevalent factor in the development of MetS [32]. These nanoparticles frequently have strong antioxidant qualities because of their high surface reactivity and the presence of phytochemicals as capping agents. In animal models of diabetes and hypertension, selenium nanoparticles (SeNPs) made from *Allium sativum* (garlic) have demonstrated effectiveness in raising levels of antioxidant enzymes such glutathione peroxidase (GPx), superoxide dismutase (SOD), and catalase. These nanoparticles lower the risk of cardiovascular problems linked to MetS and maintain endothelial function by reducing oxidative stress. The anti-inflammatory properties of green-synthesized nanoparticles specifically target the persistent low-grade inflammation seen in MetS, represent another significant field. As an illustration, in obese mice, copper oxide nanoparticles (CuONPs) made from *Ocimum sanctum* (tulsi) significantly reduce pro-inflammatory cytokines such as TNF-α, IL-6, and CRP. Two of the main pathogenic characteristics of metabolic syndrome, insulin resistance and vascular inflammation, can be avoided because of this immunomodulatory effect. Green nanoparticles have also demonstrated potential in treating hypertension, which is frequently associated with endothelial dysfunction and oxidative stress in MetS [33]. Iron oxide nanoparticles (FeONPs) made from *Emblica officinalis* (amla) have shown vasorelaxant effects and blood pressure-lowering action in rats with spontaneous hypertension. These effects are thought to result from decreased oxidative damage to the vascular endothelium and increased nitric oxide bioavailability. Additionally, green-synthesized nanoparticles can be used as MetS therapeutic drug delivery vehicles. Because of their biocompatibility and functionalized surfaces, antidiabetic, antihypertensive, and lipid-lowering medications can be delivered precisely, improving the therapeutic index and lowering adverse effects [34]. For example, in animal models of MetS, curcumin-loaded chitosan nanoparticles, which are made from *Curcuma longa* extract, have shown higher oral bioavailability and increased anti-inflammatory and insulin-sensitizing effects. To summarize, green-synthesized nanoparticles offer a flexible and encouraging treatment approach to the intricate pathophysiology of metabolic syndrome. Their potential as integrative tools in managing metabolic syndrome are highlighted by their multifunctional qualities, which include anti-inflammatory, hypoglycemic, antihyperlipidemic, antihypertensive, and antioxidant qualities. More research on standardized production, toxicity assessment, and clinical translation is essential to utilize the therapeutic potential of these environmentally safe nanomaterials fully.

5 Nanomaterials for Metabolic Syndrome

A revolutionary class of materials, nanomaterials are found at the nanoscale, usually in at least one dimension between 1 and 100 nm. Compared to their bulk counterparts, materials of this scale have distinct and frequently improved physical, chemical, and biological characteristics. These characteristics include enhanced mechanical strength, surface area, reactivity, quantum effects, and the capacity to interact at the molecular level with biological systems accurately. Because of these properties, nanomaterials are now essential tools in a few fields, including electronics, environmental research, energy storage, and, most notably, biomedicine [35]. Nanomaterials have transformed medication delivery, therapy, diagnosis, and regenerative medicine in the biomedical field. Their adjustable surface chemistry makes functionalization with targeted ligands, medications, imaging agents, and stabilizers possible. Nanomaterials can, therefore, be designed to transport therapeutic chemicals to certain cells or tissues precisely, increasing treatment effectiveness while lowering systemic toxicity and off-target effects. This is especially important for long-term, complex illnesses like cancer, neurological conditions, and metabolic syndrome (MetS). Metallic nanoparticles (such as gold, silver, and zinc oxide), polymeric nanoparticles (such as PLGA and chitosan), lipid-based nanocarriers (such as liposomes and solid lipid nanoparticles), carbon-based materials (such as graphene and carbon nanotubes), and biomolecule-conjugated nanostructures are some of the general categories into which nanomaterials can be divided [36]. Depending on their characteristics, each of these classes has distinct biomedical uses. Lipid-based systems imitate biological membranes for effective cellular uptake, polymeric nanoparticles are perfect for prolonged drug release, and metallic nanoparticles are frequently employed for imaging, antibacterial action, and oxidative stress management. Significantly, the safety and sustainability of nanotechnology have been enhanced by developments in the green production of nanomaterials, which use microorganisms, plant extracts, or biopolymers. Because of their natural stabilizers or phytochemical coatings, these environmentally friendly nanomaterials are more likely to be biocompatible and possess intrinsic therapeutic bioactivity [37]. As a result, they are becoming more popular in applications that call for long-term use or contact with delicate biological settings. Nanomaterials' ability to treat the various pathophysiological features of metabolic syndrome, such as insulin resistance, oxidative stress, inflammation, obesity, dyslipidemia, and endothelial dysfunction, is the subject of ongoing research. Therapeutic effects can be improved by using nanocarriers to deliver anti-inflammatory medications, antioxidants, anti-obesity chemicals, and antidiabetic medicines directly to metabolic tissues such as the liver, pancreas, or adipose tissue [38]. Furthermore, nanosensors and nano-imaging technologies make early diagnosis and real-time metabolic biomarker monitoring possible, which support personalized medication strategies. Nanomaterials' multidisciplinary nature unites physics, chemistry, biology, and engineering, opening new medical possibilities. They are particularly well-suited to treat complicated disorders like metabolic syndrome, where traditional medicines frequently fail, because of their precise cellular and subcellular interactions. Nanomaterials are expected to be a key component of integrative and precision medicine in the future due to continuous advancements in synthesis, functionalization, and clinical translation (Table 12.1).

6 Management of Metabolic Syndrome by Nanomaterials Loaded with Phytochemicals

A collection of interconnected metabolic irregularities characterizes metabolic syndrome (MetS), a multifactorial condition. Conventional pharmaceutical treatments for MetS frequently target individual symptoms and are linked to adverse effects, low bioavailability, and problems with patient compliance. A new and promising strategy for integrating and targeted care for metabolic syndrome has arisen in response to these constraints: nanomaterials loaded with phytochemicals [42]. Bioactive substances originating from plants, such as polyphenols, flavonoids, alkaloids, terpenoids, saponins, and lignans, are known as phytochemicals. Because of their many pharmacological actions, including lipid-lowering, anti-inflammatory, antidiabetic, antihypertensive, and antioxidant qualities, these substances are excellent choices for managing MetS. However, their low stability, quick metabolism, restricted bioavailability, and poor water solubility have hampered their therapeutic relevance.

Table 12.1 Nanomaterial and its therapeutic role [39–41]

S. no	Nanomaterial	Source/ composition	Therapeutic role in MetS	Mechanism of action
1.	Gold nanoparticles (AuNPs)	Utilizing plant polyphenols for green synthesis	Lipid regulation, anti-inflammatory, and anti-obesity	Inhibits adipogenesis by downregulating inflammatory cytokines and PPARγ
2.	Silver nanoparticles (AgNPs)	Microbial or plant synthesis	Antioxidants improve insulin sensitivity	Stimulates GLUT4 translocation and ROS-mediated signalling
3.	Zinc oxide nanoparticles (ZnONPs)	Chemical and green synthesis using zinc salts	Antihyperglycemic and insulin-mimetic	Increases β-cell activity and activates insulin receptor pathways
4.	Selenium nanoparticles (SeNPs)	Plant-based (garlic, amla, etc.)	Antioxidant, anti-inflammatory	increases the activity of antioxidant enzymes (GPx, SOD) and scavenges ROS.
5.	Chitosan nanoparticles	Crustacean natural polysaccharide	Enhanced insulin absorption and glucose regulation	Mucoadhesive: tight connection manipulation improves insulin absorption
6.	PLGA nanoparticles	Poly (lactic-co-glycolic acid) is a synthetic polymer	Continuous delivery of insulin sensitizers and anti-obesity medications	It keeps medications safe from deterioration and releases them gradually through hydrolysis
7.	Liposomes	Phospholipid bilayers	Insulin, curcumin, resveratrol delivery	Interferes with cell membranes and releases medications that are hydrophilic or lipophilic
8.	Solid-lipid nanoparticles (SLNs)	Biodegradable lipids (e.g., glyceryl monostearate)	Increased bioavailability of drugs and regulated release	Slow release of lipophilic drugs in the gastrointestinal tract or bloodstream
9.	Graphene oxide (GO)	Functionalized carbon nanosheets	Anti-inflammatory, theranostic agent	Bioactive adsorption, immune response modulation, and imaging utility
10.	Carbon nanotubes (CNTs)	Cylindrical carbon-based structures	Biosensing, gene/ drug delivery	High surface area facilitates electrochemical sensing and translocates across membranes

Nanotechnology provides an efficient answer to these problems. By encasing phytochemicals in nanocarriers, including liposomes, solid lipid nanoparticles, polymeric nanoparticles, and nanomicelles, they become more soluble, are shielded from enzymatic breakdown, can be released under regulated conditions, and are easier to target to specific tissues. Another example is Berberine, an isoquinoline alkaloid with antidiabetic and lipid-lowering properties. However, berberine has a famously low oral bioavailability; berberine-loaded nanoemulsions or nanoparticles have demonstrated markedly improved absorption and metabolic benefits, such as activation of AMP-activated protein kinase (AMPK), which is essential for maintaining energy balance [43]. To combat oxidative stress and obesity, green tea polyphenols, particularly epigallocatechin gallate (EGCG), have been nanoformulated into lipid-based carriers. In preclinical settings, these phytochemical-loaded nanocarriers decrease visceral fat deposition, enhance thermogenesis, and prevent adipogenesis. Crucially, by addressing the underlying pathophysiological mechanisms—oxidative stress, chronic inflammation, mitochondrial dysfunction, and disruption of insulin signalling—these phytochemical-nanomaterial systems not only treat metabolic dysfunctions that already exist but also stop the progression of the disease. In summary, a promising avenue for the all-encompassing treatment of metabolic syndrome is the combination of nanotechnology and plant bioactives. Nanomaterials loaded with phytochemicals offer a targeted, sustained delivery method that overcomes pharmacokinetic restrictions, making them a natural, biocompatible, and efficient way to treat and prevent the many interrelated problems linked to MetS (Table 12.2). More clinical research and delivery system modification are necessary to turn this promise into standard medical practice.

7 Therapeutic Role of Nanozymes in Metabolic Disease and its Management

Artificial nanoparticles known as "nanozymes," which have catalytic qualities like enzymes, have quickly become a ground-breaking development in nanomedicine. They hold promise for detecting and treating a few metabolic disorders, including metabolic syndrome (MetS). Obesity, insulin resistance, hypertension, and dyslipidemia are all part of the metabolic syndrome, a group of linked illnesses that increase the risk of type 2 diabetes and heart disease. Chronic inflammation, oxidative stress, mitochondrial dysfunction, and altered glucose and lipid metabolism are key components of the pathophysiology of MetS [45]. Because they mirror the activities of natural antioxidant enzymes such as superoxide dismutase (SOD), catalase, peroxidase, and oxidase, nanozymes have demonstrated great promise in moderating these pathological processes. Reactive oxygen species buildup is one of the most important causes of metabolic failure. Increased ROS levels worsen hepatic steatosis, harm insulin-sensitive tissues, and encourage adipocyte malfunction. Superoxide and hydrogen peroxide are efficiently scavenged by nanozymes like cerium oxide (CeO_2) nanoparticles, which are recognized for their regenerative cycling between Ce^{3+} and Ce^{4+} oxidation states. This reduces oxidative stress in hepatic tissues and pancreatic β-cells. These nanoparticles enhance insulin signalling pathways and restore mitochondrial function [46]. Similar antioxidant benefits have also been shown by manganese oxide (MnO_4) and iron oxide (FeO_4) nanozymes, which in animal models have increased systemic insulin sensitivity, decreased hepatic fat buildup, and reduced blood glucose levels. Nanozymes have anti-inflammatory properties in addition to their antioxidant function. For instance, nanozymes can change the polarization of macrophages from the pro-inflammatory

Table 12.2 Phytochemical-loaded nanomaterials and their sustained delivery [33, 44]

S. no	Phytochemical	Plant source	Nanoformulation type	Targeted MetS component
1.	Curcumin	*Curcuma longa* (turmeric)	PLGA, chitosan NPs, liposomes	Insulin resistance, inflammation
2.	Resveratrol	Grapes, red wine, berries	Solid-lipid nanoparticles, liposomes	Obesity, endothelial dysfunction
3.	Berberine	*Berberis vulgaris*, goldenseal	Nanoemulsion, SLNs	Hyperglycemia, dyslipidemia
4.	Epigallocatechin Gallate (EGCG)	*Camellia sinensis* (green tea)	Liposomes, PEGylated NPs	Obesity, oxidative stress
5.	Quercetin	Onions, apples, berries	PLGA, chitosan NPs	Hypertension, insulin resistance
6.	Naringenin	Oranges, grapefruit	Chitosan nanoparticles	Dyslipidemia, hepatic steatosis
7.	Genistein	Soybeans	Liposomes, niosomes	Obesity, glucose intolerance
8.	Kaempferol	Kale, tea, broccoli	AuNPs, polymeric NPs	Inflammation, adipogenesis
9.	Thymoquinone	*Nigella sativa* (black seed)	SLNs, Nanoemulsion	Obesity, systemic inflammation
10.	Apigenin	Parsley, celery, chamomile	PLGA nanoparticles	Oxidative stress, insulin resistance
11.	Silymarin	Milk thistle (*Silybum marianum*)	Nanoemulsion, lipid carriers	Liver dysfunction, dyslipidemia
12.	Piperine	Black pepper	Chitosan NPs, lipid NPs	Bioenhancement, insulin resistance
13.	Boswellic acid	*Boswellia serrata*	Solid lipid nanoparticles	Inflammation, insulin resistance
14.	Luteolin	Celery, thyme	Liposomes, nanocrystals	Cardiometabolic risk, oxidative stress
15.	Hesperidin	Citrus fruits	Nanosuspensions, polymeric NPs	Dyslipidemia, hypertension

M1 phenotype to the anti-inflammatory M2 phenotype by lowering ROS levels and modifying immune cell responses. This change promotes tissue repair and metabolic equilibrium while reducing systemic inflammation, a defining feature of MetS. Carbon nanodots and platinum-based enzymes also show peroxidase-like activity, which helps to detoxify ROS and reduce inflammatory cytokines, including TNF-α and IL-6. Additionally, nanozymes are incorporated into intelligent delivery systems that enable stimuli-responsive release and disease-specific targeting. Some nanozymes can be engineered to activate in high glucose or oxidative conditions to ensure focused therapy with low systemic toxicity [40]. Their biocompatibility, biodistribution, and cell-specific uptake are all improved by combining polymers or ligands. Because of their many uses, nanozymes are well suited for theranostic

Table 12.3 Therapeutic applications of nanozymes in managing metabolic syndrome [48, 49]

S. no	Nanozyme type	Enzyme-mimicking activity	Targeted metabolic function	Therapeutic role in MetS
1	Cerium oxide (CeO_2) NPs	Catalase, SOD	Oxidative stress in the pancreas, liver, and adipose tissue	Lessens hepatic fat accumulation, enhances insulin sensitivity, shields β-cells and lowers ROS
2	Manganese oxide (Mn_3O_4) NPs	SOD, peroxidase	Mitochondrial oxidative stress, glucose metabolism	It improves insulin resistance, lowers hyperglycemia, and strengthens mitochondria
3	Iron oxide (Fe_3O_4) NPs	Peroxidase	Systemic inflammation, oxidative stress	Increases the polarization of M2 macrophages and inhibits inflammatory cytokines
4	Platinum NPs	Peroxidase, oxidase	Inflammatory signalling, endothelial dysfunction	Reduces cardiovascular risk factors, improves endothelial function, and detoxifies ROS
5	Carbon nanodots	Peroxidase	Inflammation and oxidative injury	Promotes metabolic homeostasis, suppresses cytokine release, and lessens oxidative damage

applications involving treatment and real-time diagnostics [47]. To sum up, nanozymes provide a powerful and versatile platform for treating the complex pathophysiology associated with metabolic syndrome (Table 12.3). In addition to neutralizing dangerous ROS and reducing chronic inflammation, they also assist focused and customized therapy by imitating the actions of natural enzymes, which might revolutionize how metabolic illnesses are treated.

8 Carbon Nanomaterials in Metabolic Syndrome

In recent years, carbon nanomaterials (CNMs), such as carbon nanotubes (CNTs), graphene, graphene oxide (GO), fullerenes, and carbon quantum dots (CQDs), have drawn a lot of attention due to their distinct physicochemical characteristics and expanding range of biological uses. Because of their large surface area, biocompatibility, electrical conductivity, mechanical strength, and functional ability, CNMs have demonstrated promising therapeutic and diagnostic potential in metabolic syndrome (MetS) [50]. A cluster of conditions such as obesity, insulin resistance, hypertension, and dyslipidemia characterize this complex disorder. Because of these characteristics, CNMs may be used in various ways, including medication delivery, biosensing, anti-inflammatory treatment, and oxidative stress modulation—all critical for managing MetS. Excessive formation of reactive oxygen species (ROS) damages pancreatic β-cells, promotes insulin resistance, and contributes to chronic inflammation, making oxidative stress one of the main causes of MetS [51]. Because of its capacity to receive electrons, CNMs like fullerenes (C60) and carbon quantum dots (CQDs) have intrinsic antioxidant qualities. These nanomaterials can help insulin-sensitive tissues regain redox equilibrium by neutralizing ROS and scavenging free radicals. For example, in diabetic models, C60 fullerenes have shown promise in reducing inflammation and protecting islet cells, which may help people with MetS who experience similar metabolic dysfunctions. Another family of carbon-based nanomaterials that have been the subject of much research is graphene oxide (GO) and its derivatives, which have been studied for their potential to deliver drugs precisely and reduce inflammation. The conjugation of medicinal substances like insulin, curcumin, or metformin is made possible by GO's large surface area and profusion of functional groups. These GO-based delivery systems enhance the bioavailability and targeted distribution of medications to metabolic organs such as the liver or adipose tissue [52]. Furthermore, it has been shown that functionalized GO sheets may change the polarization of macrophages from the pro-inflammatory M1 phenotype to the anti-inflammatory M2 phenotype, which lowers systemic inflammation, a major factor in the development of metabolic syndrome. The main uses of carbon nanotubes (CNTs), both single-walled (SWCNTs) and multi-walled (MWCNTs), are in medication delivery and biosensing applications for the treatment of MetS. CNTs effectively pass through biological membranes to deliver genetic material or small chemicals. For instance, insulin-conjugated CNTs exhibit promise for non-invasive delivery methods, whereas metformin-loaded CNTs have been investigated for improved glycemic management. Furthermore, to help with early detection and individualized treatment of MetS, functionalized CNTs have been included in electrochemical biosensors that can identify glucose, cholesterol, or oxidative stress indicators in real time. Ongoing surface functionalization and green synthesis developments enhance carbon nanomaterial biocompatibility and safety profile despite persistent worries about their toxicity and biodegradability [53]. PEGylation, carboxylation, and coating with biopolymers like chitosan or dextran often lessen toxicity and improve physiological compatibility. To sum up, carbon nanomaterials provide a strong, versatile platform for treating several pathogenic features of metabolic syndrome. Their promise as next-generation tools in the complete therapy and monitoring of metabolic diseases are highlighted by their flexibility in drug administration, biosensing, immunological modulation, and antioxidant protection.

9 Nanomaterials Challenges in Metabolic Syndrome

Although the use of nanomaterials in diagnosing, treating, and managing metabolic syndrome (MetS) offers intriguing and novel opportunities, there are significant obstacles to overcome. A complicated, multifactorial illness, metabolic syndrome includes several disorders, including central obesity, insulin resistance, dyslipidemia, and hypertension. Nanotechnology provides answers through enhanced medication transport, controlled release, biosensing, and focused therapy. However, a few scientific, clinical, regulatory, and ethical obstacles must be overcome before nanomaterials can be safely and successfully incorporated into clinical treatment for metabolic diseases. The issues of toxicity and biocompatibility are among the most urgent ones. The potential toxicity of nanomaterials is further influenced by their special physicochemical characteristics, which include tiny size, large surface area, and surface reactivity. Unpredictable interactions between nanoparticles and biological membranes, proteins, and organelles can result in oxidative stress, inflammatory reactions, or cell death. For example, although cerium oxide nanoparticles (CeO_2) have antioxidant properties, surface changes or incorrect doses might cause pro-oxidant activity, particularly in organs with impaired metabolism like the kidneys or liver [54]. Furthermore, chronic toxicity might result from the long-term buildup of non-biodegradable nanomaterials in tissues, which is especially problematic for MetS patients who need ongoing or lifetime therapy. Targeting accuracy and biodistribution is another significant obstacle. Therapeutic effectiveness and limiting off-target effects depend on the precise distribution of nanomaterials to certain organs or cell types, such as liver hepatocytes, adipose tissue, or pancreatic β-cells. However, biological constraints such as non-specific protein binding, renal filtration, and the mononuclear phagocyte system (MPS) frequently shorten the period that nanoparticles circulate and encourage sequestration in the liver or spleen. This can increase the risk of side effects by compromising medication absorption and requiring greater dosages. Although techniques like ligand-based targeting and PEGylation have been established, not all patients or nanoparticle types respond well to these approaches. Another major challenge in the synthesis of nanomaterials is scalability and repeatability. Under Good Manufacturing Practices (GMP), many nanoparticle formulations that exhibit promise in animal or laboratory investigations cannot be scaled up for clinical-grade manufacturing. Despite being economical and environmentally benign, green synthesis techniques frequently have issues with loading efficiency, surface chemistry, and size distribution variability, all of which can affect biological behaviour [55]. Uniformity and consistency are essential for clinical applications, especially in chronic conditions like MetS, where treatment is continuous. The categorization, authorization, and oversight of nanomedicine for metabolic disorders are still in their infancy from a regulatory and ethical standpoint. There are concerns about risk-benefit analyses, labelling, and post-market surveillance since current pharmaceutical laws are frequently inadequate to examine the complicated behaviour of nanomaterials. Furthermore, few long-term investigations are currently conducted on the behaviour of nanomaterials, including metabolism, excretion, and possible endocrine-disrupting effects. In conclusion, even though nanomaterials have great promise for treating metabolic syndrome in a customized and precise manner, it is crucial to address these complex issues. Future studies should prioritize biocompatibility testing, focused delivery methods, scalable production, and thorough regulatory frameworks to guarantee nanotechnology's secure and efficient incorporation into the clinical setting.

10 Conclusion and Future Perspectives

A paradigm change in contemporary medicine is represented using nanomaterials in treating metabolic syndrome (MetS). Because Metabolic Syndrome is a complicated illness that involves obesity, insulin resistance, dyslipidemia, hypertension, and systemic inflammation, it is difficult to treat with typical monotherapies. Targeted distribution, controlled release, improved solubility, and multifunctional therapeutic effects are made possible by nanotechnology, which provides a special set of tools to get beyond these restrictions. It has been shown that several classes of nanomaterials, such as metallic nanoparticles, carbon-based materials, lipid-based carriers, polymeric nanostructures, and nanozymes, are essential for reducing chronic inflammation, enhancing insulin sensitivity, controlling lipid metabolism, and modulating oxidative stress—all of which are key components of the pathophysiology of MetS. Furthermore, the use of green synthesis methods in manufacturing nanoparticles has brought forth economical, biocompatible, and environmentally friendly methods that align with sustainable therapeutic development. Phytochemical-loaded nanoformulations, including EGCG, resveratrol, berberine, and curcumin, have demonstrated improved pharmacokinetic profiles and synergistic effectiveness against various metabolic targets. Additionally, novel pathways for redox control and metabolic balance have been made possible by the emergence of nanozymes as enzyme mimetics. Early diagnosis, illness monitoring, and individualized treatment plans are made possible by the growing use of carbon nanomaterials and functionalized nanocarriers in biosensing and diagnostics. Clinical translation is still scarce despite these encouraging developments. Regarding toxicity, immunogenicity, biodistribution, and long-term safety, the intricate interactions between biological systems and nanoparticle properties (size, shape, surface charge, and functionalization) continue to provide difficulties. Moving from lab research to practical medical applications will require addressing these issues through high-throughput screening, bioinformatics modelling, and systematic in vivo experiments. Furthermore, regulatory frameworks for nanomedicine must change to keep up with scientific advancements. Precise standards for nanoparticle characterization, safety evaluation, clinical trial design, and post-market surveillance are required to guarantee safe and efficient patient usage. Looking ahead, the confluence of diverse technologies holds the key to the future of nanomedicine in metabolic syndrome. Machine learning, artificial intelligence, and systems biology developments will make more accurate nanoparticle design and patient-specific therapeutic models possible. Real-time, adaptive therapy may be possible using stimuli-responsive nanocarriers, which release therapeutic chemicals in response to environmental signals like pH, temperature, or glucose levels. How we treat chronic metabolic illnesses will be revolutionized by hybrid nanostructures combining therapeutic and diagnostic capabilities (theranostics), enabling simultaneous monitoring and intervention. Because of their exceptional functional capabilities and adaptability, nanoparticles have already started to change the landscape of managing metabolic syndrome. Even while there is still much to be done to close the gap between experimental promise and clinical application, the ongoing advancement of nanoscience might lead to safer, more intelligent, and more efficient ways to address one of the most important global health issues of our day.

References

1. Al-Daghri, N. M., Al-Attas, O. S., Alokail, M. S., Alkharfy, K. M., & Sabico, S. (2022). Metabolic syndrome: Updates on pathophysiology and management in 2022. *International Journal of Molecular Sciences, 23*(2), 786.
2. Bhalwar, R. (2020). Metabolic syndrome: The Indian public health perspective. *Medical Journal Armed Forces India, 76*(1), 8–16.
3. Krishnamoorthy, Y., Rajaa, S., Murali, S., Sahoo, J., & Kar, S. S. (2022). Association between anthropometric risk factors and metabolic syndrome among adults in India: A systematic review and meta-analysis of observational studies. *Preventing Chronic Disease, 19*, E24.
4. Patial, R., Batta, I., Thakur, M., Sobti, R. C., & Agrawal, D. K. (2024). Aetiology, pathophysiology, and treatment strategies in the prevention and management of metabolic syndrome. *Archives of Internal Medicine Research, 7*(4), 273–283.
5. Devi, U., & Kapila, S. (2024). Metabolic syndrome: A comprehensive review on understanding pathophysiology, clinical implications, and management strategies. *Journal of Metabolic Syndrome, 13*(1), 1–5.
6. Shrivastava, S., Singh, K., Chakma, T., & Kavishwar, A. (2023). Metabolic syndrome in Indian tribes: Challenges to reveal its true status. *Frontiers in Clinical Diabetes and Healthcare, 4*, 1194471.
7. Krishnan, A., & Joshi, S. R. (2020). The epidemiological profile of metabolic syndrome in Indian population: A systematic review. *Clinical Epidemiology and Global Health, 8*(1), 1–7.
8. Gupta, R., Deedwania, P. C., Gupta, A., Rastogi, S., Panwar, R. B., & Kothari, K. (2004). Prevalence of metabolic syndrome in an Indian urban population. *International Journal of Cardiology, 97*(2), 257–261.
9. Accili, D., Deng, Z., & Liu, Q. (2025). Insulin resistance in type 2 diabetes mellitus. *Nature Reviews Endocrinology, 21*, 413–426.
10. Reyes-Farias, M., Fos-Domenech, J., Serra, D., Herrero, L., & Sanchez-Infantes, D. (2021). White adipose tissue dysfunction in obesity and aging. *Biochemical Pharmacology, 192*, 114723.
11. Cho, Y. K., Lee, S., Lee, J., Doh, J., Park, J. H., Jung, Y. S., & Lee, Y. H. (2023). Lipid remodeling of adipose tissue in metabolic health and disease. *Experimental & Molecular Medicine, 55*(9), 1955–1973.
12. Manolis, A. A., Manolis, T. A., & Manolis, A. S. (2023). Neurohumoral activation in heart failure. *International Journal of Molecular Sciences, 24*(20), 15472.
13. Jomova, K., et al. (2023). Reactive oxygen species, toxicity, oxidative stress, and antioxidants: chronic diseases and aging. *Archives of Toxicology, 97*, 2499–2574.
14. Joshi, S. R., Saboo, B., Vadivale, M., Dani, S. I., Mithal, A., Kaul, U., Badgandi, M., Iyengar, S. S., Viswanathan, V., Sivakadaksham, N., & Chattopadhyaya, P. S. (2012). Prevalence of diagnosed and undiagnosed diabetes and hypertension in India—Results from the Screening India's Twin Epidemic (SITE) study. *Diabetes Technology & Therapeutics, 14*(1), 8–15.
15. Mohan, V., Deepa, R., Rani, S. S., & Premalatha, G. (2005). Prevalence of coronary artery disease and its relationship to lipids in a selected population in South India. *Journal of the American College of Cardiology, 38*(3), 682–687.
16. Matsuda, S. (2024). Understanding the pathophysiology of metabolic syndrome: A comprehensive overview. *Journal of Metabolic Syndrome, 13*(1), 368.

17. Kaur, J. (2014). A comprehensive review on metabolic syndrome. *Cardiology Research and Practice, 2014*, 943162.
18. Mottillo, S., Filion, K. B., Genest, J., Joseph, L., Pilote, L., Poirier, P., et al. (2010). The metabolic syndrome and cardiovascular risk: A systematic review and meta-analysis. *Journal of the American College of Cardiology, 56*(14), 1113–1132.
19. Grundy, S. M. (2008). Metabolic syndrome pandemic. *Arteriosclerosis, Thrombosis, and Vascular Biology, 28*(4), 629–636.
20. Reaven, G. M. (2005). The metabolic syndrome: Requiescat in pace. *Clinical Chemistry, 51*(6), 931–938.
21. Grundy, S. M., Brewer, H. B., Cleeman, J. I., Smith, S. C., & Lenfant, C. (2004). Definition of metabolic syndrome: Report of the National Heart, Lung, and Blood Institute/American Heart Association conference on scientific issues related to definition. *Circulation, 109*(3), 433–438.
22. Alberti, K. G. M. M., Zimmet, P., & Shaw, J. (2005). The metabolic syndrome—A new worldwide definition. *The Lancet, 366*(9491), 1059–1062.
23. Eckel, R. H., Grundy, S. M., & Zimmet, P. Z. (2005). The metabolic syndrome. *The Lancet, 365*(9468), 1415–1428.
24. Wang, Z., Zhang, L., & Qin, C. (2025). Alzheimer's disease pathogenesis: Standing at the crossroads of lipid metabolism and immune response. *Molecular Neurodegeneration, 20*, 67.
25. Ibrahim, H., Kamil, M., & Kamal, M. (2022). Alzheimer's disease: An update and insights into pathophysiology. *Frontiers in Aging Neuroscience, 14*, 742408.
26. Shubina, L. (2025). Mechanisms of the pathogenesis of Alzheimer's disease: The role of neuroinflammation and oxidative stress. *Neuroscience and Behavioral Physiology, 55*(3), 345–356.
27. Zipfel, P., Rochais, C., Baranger, K., Rivera, S., & Dallemagne, P. (2021). Matrix metalloproteinases as new targets in Alzheimer's disease: Opportunities and challenges. *Journal of Neurochemistry, 157*(4), 1234–1245.
28. Kareva, I. (2025). The many roads to dementia: A systems view of Alzheimer's disease. *arXiv preprint*, arXiv:2504.05441.
29. Kirubakaran, D., Abdul Wahid, J. B., Karmegam, N., Jeevika, R., Sellapillai, L., Rajkumar, M., & SenthilKumar, K. J. (2025). A comprehensive review on the green synthesis of nanoparticles: Advancements in biomedical and environmental applications. *Biomedical Materials & Devices., 4*, 388.
30. Mayegowda, S. B., Sarma, G., Gadilingappa, M. N., Alghamdi, S., Aslam, A., Refaat, B., Almehmadi, M., Allahyani, M., Alsaiari, A. A., Aljuaid, A., & Al-Moraya, I. S. (2023). Green-synthesized nanoparticles and their therapeutic applications: A review. *Green Processing and Synthesis, 12*(1), 1–25.
31. Izadi, R., Bahramikia, S., & Akbari, V. (2024). Green synthesis of nanoparticles using medicinal plants as an eco-friendly and therapeutic potential approach for neurodegenerative diseases: A comprehensive review. *Frontiers in Neuroscience, 18*, 1453499.
32. Haridas, E. S. H., Bhattacharya, S., Varma, M. K. R., & Chandra, G. K. (2022). Green synthesis of eco-friendly silver nanoparticles using coffee arabica leaf extract and development of a cost-effective biosensor for cysteine. *arXiv preprint*, arXiv:2209.03823.
33. Singh, P., Dan, A., Kannan, P. P., Dhanka, M., Bhatia, D., & Saha, J. (2024). Non-toxic fabrication of fluorescent carbon nanoparticles from medicinal plants/sources with their antioxidant assay. *arXiv preprint*, arXiv:2401.10673.
34. Mukherjee, P., Senapati, S., Mandal, D., Ahmad, A., Khan, M. I., Kumar, R., & Sastry, M. (2002). Extracellular synthesis of gold nanoparticles by the fungus *Fusarium oxysporum*. *Chembiochem, 3*(5), 461–463.
35. Rajak, K. K., Pahilani, P., Patel, H., Kikani, B., Desai, R., & Kumar, H. (2023). Green synthesis of silver nanoparticles using *Curcuma longa* flower extract and antibacterial activity. *arXiv preprint*, arXiv:2304.04777.
36. Noah, N., & Ndangili, P. (2021). Green synthesis of nanomaterials from sustainable materials for biosensors and drug delivery. *arXiv preprint*, arXiv:2112.04740.

37. El-Fallal, A. A., Elfayoumy, R. A., & El-Zahed, M. M. (2023). Antibacterial activity of biosynthesized zinc oxide nanoparticles using Kombucha extract. *Polymers, 15*(1), 100.
38. Manoj, K., Rakesh, R., Dandapat, S., Srivastava, R., Bharti, S. R., & Manoranjan, P. S. (2023). Green nanotechnology: Synthesis of silver nanoparticles using aqueous leaf extracts of *Swertia chirayita* and *Punica granatum*. *The Bioscan, 18*(3), 167–176.
39. Kaur, J., & Singh, R. (2020). Role of nanoparticles in the management of metabolic disorders. In S. K. Sharma & R. K. Gaur (Eds.), *Nanomaterials in health care* (pp. 589–605). Springer.
40. Yousaf, I. (2024). The current and future perspectives of zinc oxide nanoparticles in the treatment of diabetes mellitus. *arXiv preprint*, arXiv:2409.04486.
41. El-Naggar, M. E., Al-Joufi, F., Anwar, M., & El-Wakil, N. A. (2023). Curcumin-loaded PLA-PEG copolymer nanoparticles for treatment of liver inflammation in streptozotocin-induced diabetic rats. *Colloids and Surfaces B: Biointerfaces, 177*, 389–398.
42. Wu, K., Su, D., Liu, J., Saha, R., & Wang, J.-P. (2018). Magnetic nanoparticles in nanomedicine. *arXiv preprint*, arXiv:1811.01418.
43. Sharma, R., & Singh, R. (2023). Nanomedicine in the treatment of metabolic diseases. In *Nanotechnology in the life sciences* (pp. 123–145). Springer.
44. Adetunji, C. O., Michael, O. S., Rathee, S., Singh, K. R. B., Ajayi, O. O., Adetunji, J. B., Ojha, A., Singh, J., & Singh, R. P. (2021). Potentialities of nanomaterials for the management and treatment of metabolic syndrome: A new insight. *Materials Today Advances, 12*, 100198.
45. Taghipour, B., Ghaedi, M., & Asghari, M. (2021). Nanomedicine of plant origin for the treatment of metabolic syndrome: A review. *Frontiers in Bioengineering and Biotechnology, 9*, 811917.
46. Ghaffari, M., Dehghan, G., & Baradaran, B. (2019). Nanoformulations of natural products for management of metabolic syndrome. *International Journal of Nanomedicine, 14*, 5309–5321.
47. Wang, Q., Wei, H., Zhang, Z., Wang, E., & Dong, S. (2018). Nanozyme: An emerging alternative to natural enzyme for biosensing and immunoassay. *TrAC Trends in Analytical Chemistry, 105*, 218–224.
48. Jiang, D., Ni, D., Rosenkrans, Z. T., Huang, P., Yan, X., & Cai, W. (2019). Nanozyme: New horizons for responsive biomedical applications. *Chemical Society Reviews, 48*(14), 3683–3704.
49. Zandieh, M., & Liu, Y. (2022). Nanozymes: Definition, activity, and mechanisms. *Advanced Materials, 34*(10), 2211041.
50. Fan, K., Xi, J., Fan, L., Wang, P., Zhu, C., Tang, Y., et al. (2018). In vivo guiding nitrogen-doped carbon nanozyme for tumor catalytic therapy. *Nature Communications, 9*, 1440.
51. Zhang, R., Fan, K., & Yan, X. (2020). Nanozymes: Created by learning from nature. *Science China Life Sciences, 63*(9), 1183–1200.
52. Zhang, C., & Du, X. (2020). Electrochemical sensors based on carbon nanomaterial used in diagnosing metabolic disease. *Frontiers in Chemistry, 8*, 651.
53. Kim, S., Kim, C., Lee, S., & Park, S. (2024). Carbon materials and their metal composites for biomedical applications: A short review. *Nanoscale, 16*, 16313–16328.
54. Nouri, Z., Hajialyani, M., Izadi, Z., Bahramsoltani, R., Farzaei, M. H., & Abdollahi, M. (2020). Nanophytomedicines for the prevention of metabolic syndrome: A pharmacological and biopharmaceutical review. *Frontiers in Bioengineering and Biotechnology, 8*, 425.
55. Ullah, M., Awan, U. A., Ali, H., Wahab, A., Khan, S. U., Naeem, M., Ruslin, M., Mustopa, A. Z., & Hasan, N. (2025). Carbon dots: New rising stars in the carbon family for diagnosis and biomedical applications. *Journal Of Nanotheranostics, 6*(1), 1.

3D-Printed Nanotechnology-Based Drug Delivery in Diagnosis and Therapeutics of Metabolic Syndrome

13

Pankaj Kalia, Pooja Thakur, Devinder Kumar, Sunil Dutt, Ashutosh Kumar, Brijesh Kumar, and Rajesh Kumar

Abstract

The metabolic syndrome, a group of modifiers such as obesity, diabetes, hypertension, and dyslipidemia, is a major worldwide health issue. Metabolic syndrome is the diagnosis of multiple physiological and biochemical abnormalities, including abdominal obesity, dyslipidemia, raised blood pressure and insulin resistance, atherosclerotic vascular disease, and the presence of other risk factors for cardiovascular disease, such as the metabolic syndrome described above, in a patient with multiple conditions requiring separate treatments, such as obesity, hypertension, type 2 diabetes, atherosclerosis, cardiovascular disease, smoking, and sedentary Incorporation of these systems with nanotechnology impressively improves the delivery of therapeutic agents to target sites with minimal side effects. Liposomes, polymeric nanoparticles, and metal-organic frameworks are incorporated into 3D-printed scaffold to achieve drug targeting, better bioavailability, and stability of the drug. For instance, a hydrogel or scaffold loaded with

P. Kalia · P. Thakur · S. Dutt · R. Kumar (✉)
Maharaja Agrasen School of Pharmacy, Maharaja Agrasen University, Solan, Himachal Pradesh, India

D. Kumar
Maharaja Agrasen School of Pharmacy, Maharaja Agrasen University, Solan, Himachal Pradesh, India

Amar Shaheed Baba Ajit Singh Jujhar Singh Memorial College of Pharmacy (An Autonomous College), Bela, Ropar, Punjab, India

A. Kumar
Department of Ophthalmology, University of California Los Angeles, Los Angeles, CA, USA

B. Kumar
Department of Pharmacology, Institute of Medical Sciences, Banaras Hindu University, Varanasi, India

V. K. Chaturvedi et al. (eds.), *Nanotheranostics in Metabolic Syndrome*,
https://doi.org/10.1007/978-981-95-3746-4_13

nanoparticles can be 3D-printed for a targeted delivery system for inflammation, insulin resistance, or vascular complications of metabolic syndrome. Moreover, diagnostic purposes are based on 3D-bioprinted biosensors with different nanomaterial such as graphene or gold nanoparticles to identify biomarkers sensitive and selective. These technologies facilitate proper screening of metabolic syndrome progression, with an occasion to intervene at appropriate time. These fields complement each other to promote a personalized medicine concept that involves tailored clinical actions based on patient's genetic and metabolic characteristics. Nonetheless, issues such as regulatory issues, capacity issues, and long-term safety of products and their applications are among the outstanding problems of the enormous potential academic discipline. These are challenges that will need intervention in order to actualize this revolutionary technology into practice in the management of metabolic syndrome and positively affect patient outcomes all over the world.

Keywords

Nanotechnology · Metabolic syndrome · Liposomes · Polymeric nanoparticles · Nanomaterial

1 Introduction

The field of nanotechnology has profoundly influenced modern medicine, offering innovative solutions to complex medical challenges. Among its most transformative applications, three-dimensional (3D) printing of nanotechnology-based materials has opened new frontiers in the development of advanced drug delivery systems. This multidisciplinary approach, combining the precision of 3D printing with the unique properties of nanomaterials, enables enhanced drug delivery, precise targeting, improved bioavailability, and increased stability [1]. These advancements are particularly relevant in addressing metabolic syndrome, a multifaceted condition comprising obesity, dyslipidemia, raised blood pressure, insulin resistance, and associated complications such as atherosclerotic vascular disease and the exacerbating impact of smoking. Metabolic syndrome represents a significant global health concern due to its association with severe chronic conditions like type 2 diabetes, cardiovascular disease, and stroke. The complex pathophysiology of this syndrome requires therapeutic interventions capable of overcoming numerous biological and pharmacological barriers [1, 2]. Traditional treatment modalities are often limited by issues such as poor drug solubility, lack of specificity, systemic side effects, and limited bioavailability. Addressing these challenges demands innovative approaches, and the convergence of 3D printing and nanotechnology offers unprecedented possibilities. By integrating 3D-printed scaffolds, hydrogels, and nanoparticles into therapeutic strategies, it is now possible to develop systems that not only deliver drugs with high precision but also improve therapeutic efficacy and reduce adverse effects [2]. 3D-printed scaffolds, which are biomimetic structures designed to replicate the extracellular matrix, are instrumental in localized and sustained drug delivery. These scaffolds are engineered using advanced techniques that provide unparalleled control over their architecture, porosity, and drug-loading capabilities. Their design enables the targeted release of therapeutic agents, such as lipid-lowering drugs for obesity or anti-inflammatory agents for atherosclerosis, directly to the site of pathology [3]. The biocompatibility and degradability of scaffold materials, such as polylactic acid and polycaprolactone, enhance their integration into biological systems, reducing the risk of adverse reactions. The incorporation of functional nanomaterials such as graphene oxide nanoparticles, gold nanoparticles, silver nanoparticles, iron oxide nanoparticles, silica nanoparticles, zinc oxide nanoparticles, titanium dioxide nanoparticles, cerium oxide nanoparticles, quantum dots, carbon nanotubes, magnetic nanoparticles, nanodiamonds, lipid nanoparticles, polymeric nanoparticles, and mesoporous silica nanoparticles into these scaffolds further amplifies their drug delivery potential [4, 5]. These nanomaterials make the scaffolds responsive to specific physiological triggers such as pH changes or enzymatic activity, which are often characteristic of inflamed or diseased tissues [6]. Recent studies have demonstrated that graphene oxide enhances scaffold mechanical strength and allows for sustained release of drugs like metformin, an antidiabetic agent critical in managing insulin resistance. Hydrogels, as hydrophilic polymeric networks, serve as another versatile platform for drug delivery in the management of metabolic syndrome. Their ability to encapsulate large quantities of therapeutic agents, combined with their tunable mechanical and chemical properties, makes them particularly suited for personalized medicine [4, 7]. Hydrogels can be designed to release their payload in response to physiological cues, such as glucose levels, which is especially beneficial in addressing insulin resistance and diabetes. For example, glucose-responsive hydrogels can deliver insulin in a controlled manner, minimizing the risk of hypoglycemia. The inclusion of metallic nanoparticles such as silver nanoparticles within hydrogel matrices has demonstrated improved antimicrobial properties, a beneficial feature in addressing vascular inflammation and infection associated with atherosclerosis. Functionalized hydrogels incorporating graphene oxide have shown promise in regulating oxidative stress, a critical contributor to metabolic syndrome. Studies conducted in 2023 highlighted the effectiveness of these hydrogels in reducing oxidative stress markers in obese animal models [3–5, 7, 8].

Nanoparticles themselves are at the forefront of targeted drug delivery systems due to their small size, high surface area, and tunability [9]. Their role in managing metabolic syndrome is extensive, encompassing the delivery of drugs, genes, and diagnostic agents. For example, polymeric nanoparticles can be loaded with appetite-suppressing agents to manage obesity or with statins to reduce vascular inflammation [5, 10, 11]. Metallic nanoparticles, such as gold or iron oxide, offer additional functionalities, including imaging and theranostic applications. Gold nanoparticles, owing to their unique surface Plasmon resonance properties, have been utilized for the co-delivery of anti-inflammatory and lipid-lowering agents, enhancing their bioavailability and reducing systemic toxicity [12]. Iron oxide nanoparticles, functionalized with ligands targeting vascular endothelium, have been employed for magnetic resonance imaging (MRI)-based detection of early atherosclerotic plaques. Studies published in 2024 reported that gold nanoparticles conjugated with peptide ligands selectively inhibited pro-inflammatory cytokine activity in endothelial cells, a breakthrough for mitigating vascular inflammation. Similarly, mesoporous silica nanoparticles have been explored for co-delivery systems involving curcumin and quercetin, targeting inflammation and oxidative stress in metabolic syndrome [10, 11]. One of the most significant challenges in managing metabolic syndrome is the poor bioavailability and stability of many therapeutic agents. Drugs such as peptides, proteins, and hydrophobic small molecules are particularly vulnerable to enzymatic degradation and rapid clearance, which limits their therapeutic efficacy [13]. The encapsulation of these agents within 3D-printed hydrogels or nanoparticles provides a protective environment that enhances their stability and prolongs their circulation time. This is particularly relevant for drugs targeting chronic conditions, where sustained therapeutic levels are essential for effective management. For instance, lipid-based nanoparticles, including nanostructured lipid carriers (NLCs), have been used to stabilize hydrophobic drugs like pioglitazone, enhancing their therapeutic efficacy in reducing insulin resistance [2, 3, 11]. Moreover, 3D-printed systems integrated with carbon nanodots have shown enhanced photo thermal properties, enabling localized therapy for adipose tissue reduction in obesity [14]. The integration of 3D-printed nanotechnology extends beyond drug delivery to diagnostic and theranostic applications. The ability to encapsulate imaging agents within nanoparticles or hydrogels allows for precise visualization of metabolic disturbances. For instance, quantum dots functionalized with specific antibodies have been utilized for fluorescence imaging of inflammatory markers in obesity. In 2024, researchers demonstrated the use of bimetallic nanoparticles, combining gold and palladium, in dual imaging and therapy for atherosclerosis, offering a high degree of specificity and reduced off-target effects. Theranostic systems, which combine therapeutic and diagnostic functionalities, represent a promising frontier in the management of metabolic syndrome [5, 7, 10]. These platforms enable real-time monitoring of therapeutic efficacy, allowing for adaptive treatment strategies tailored to the patient's evolving condition [15]. Despite the immense potential of 3D-printed nanotechnology-based systems, several challenges remain. The scalability and reproducibility of these technologies need to be addressed to enable widespread clinical application. Regulatory hurdles, particularly concerning the safety and long-term effects of nanomaterials, also pose significant barriers [16]. However, continuous advancements in bio printing techniques, material science, and nanotechnology are expected to overcome these limitations, paving the way for the broader adoption of these systems in clinical practice. The application of 3D-printed scaffolds, hydrogels, and nanoparticles in the management of metabolic syndrome demonstrates the power of combining precision engineering with advanced materials science (Fig. 13.1). These technologies address the limitations of traditional therapies, offering improved drug delivery, enhanced targeting, and greater bioavailability. By tackling the multifactorial challenges of metabolic syndrome, these innovations have the potential to transform its management, offering patients more

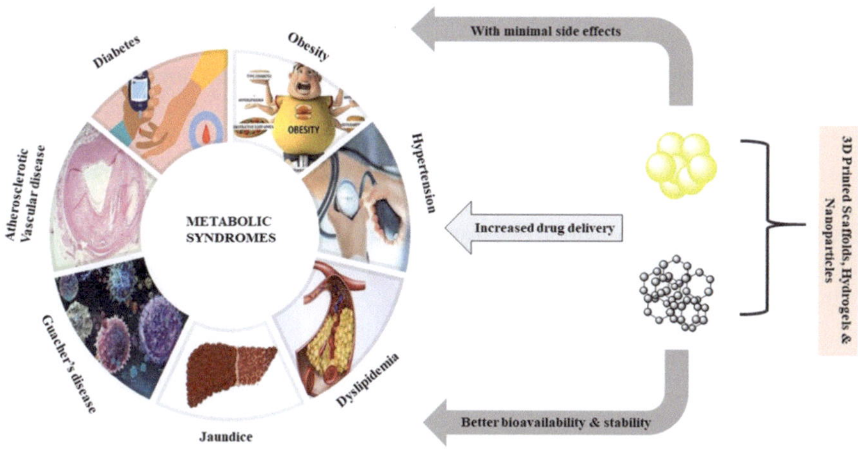

Fig. 13.1 3D nanoparticles in metabolic syndromes

effective and personalized treatment options while reducing the risk of complications. As research in this field progresses, the integration of 3D-printed nanotechnology into medical practice is likely to become a cornerstone of metabolic syndrome diagnosis and therapy, representing a significant step toward the realization of precision medicine [3, 4, 8, 10].

The future of 3D-printed nanotechnology in managing metabolic syndrome is poised to redefine precision medicine through innovative applications [17]. Recent studies have emphasized the potential of 4D-printed nanocomposites, which dynamically respond to physiological changes, offering on-demand drug release tailored to patient-specific needs. For example, 2024 research highlighted self-adaptive hydrogels embedded with gold nanoparticles (AuNPs) for enhanced glucose sensing and insulin delivery, demonstrating precise regulation of blood glucose levels in animal models. Similarly, mesoporous silica nanoparticles (MSNs) functionalized with lipid bilayers have shown success in delivering anti-inflammatory drugs for vascular therapy, as evidenced by Phase I trials targeting atherosclerotic plaques [18]. The integration of artificial intelligence (AI) is further accelerating advancements by enabling the design of multi-modal drug delivery platforms. AI-optimized polymeric nanoparticles loaded with GLP-1 receptor agonists for obesity management have exhibited superior bioavailability and patient compliance in recent trials [11]. Moreover, bio printed vascular scaffolds infused with iron oxide nanoparticles (IONPs) are under development for regenerative therapy, promoting angiogenesis and controlled drug delivery simultaneously. While challenges remain in scalability and regulatory approvals, the convergence of bio printing, nanotechnology, and machine learning heralds a transformative era. These systems offer unprecedented control over drug delivery and therapeutic outcomes, holding immense promise for revolutionizing metabolic syndrome management in clinical settings [19].

2 Metabolic Syndromes and Types

Metabolic syndromes encompass a spectrum of interconnected disorders with significant implications for global health, driven by multifactorial etiologies and complex pathophysiology (Table 13.1) [20–22]. Obesity, a cornerstone of metabolic dysfunction, presents in two primary types: central obesity, marked by visceral adiposity and linked to increased pro-inflammatory cytokine secretion such as TNF-α and IL-6, and peripheral obesity, characterized by subcutaneous fat accumulation, often less metabolically active. Diabetes mellitus, a critical metabolic disease, includes Type 1, involving autoimmune β-cell destruction via mechanisms such as HLA gene variations and molecular mimicry, and Type 2, driven by insulin resistance due to impaired insulin receptor signaling and lipid overload [23–25]. According to the International Diabetes Federation (2023), over 537 million adults globally suffer from diabetes, with Type 2 accounting for 90% of cases, highlighting its epidemic proportions [26]. Hypertension, a significant cardiovascular risk factor, manifests as primary (essential) hypertension, influenced by genetic polymorphisms in the renin-angiotensin-aldosterone system, and secondary hypertension, arising from renal or endocrine disorders [27]. Dyslipidemia includes hyperlipidemia, characterized by elevated LDL-C and hypertriglyceridemia linked to apolipoprotein B dysfunction, and hypolipidemia, marked by low HDL-C, atheroprotective in nature, exacerbating atherosclerotic plaque development [28].

Table 13.1 Metabolic syndromes and types

S. no.	Metabolic syndrome	Type 1	Type 2
1.	Obesity	Central (visceral fat accumulation)	Peripheral (subcutaneous fat accumulation)
2.	Diabetes mellitus	Type 1 (autoimmune destruction of β-cells)	Type 2 (insulin resistance)
3.	Hypertension	Primary (essential, no identifiable cause)	Secondary (due to underlying conditions)
4.	Dyslipidemia	Hyperlipidemia (high LDL, triglycerides)	Hypolipidemia (low HDL levels)
5.	Nonalcoholic fatty liver disease	Simple steatosis (fat accumulation without inflammation)	NASH (nonalcoholic steatohepatitis)
6.	Polycystic ovary syndrome (PCOS)	Insulin-resistant PCOS	Non-insulin-resistant PCOS
7.	Metabolic bone disorders	Osteoporosis (low bone density)	Osteomalacia (defective bone mineralization)
8.	Hyperuricemia and gout	Primary (genetic predisposition)	Secondary (associated with CKD or other conditions)
9.	Lipodystrophy	Congenital (genetic origin)	Acquired (drug-induced or related to diseases like HIV)
10.	Atherosclerosis	Coronary artery disease	Peripheral arterial disease

Nonalcoholic fatty liver disease (NAFLD), an emerging hepatic condition, spans simple steatosis, involving ectopic fat deposition, and nonalcoholic steatohepatitis (NASH), featuring hepatocyte ballooning and inflammatory infiltration. Recent studies (Nature Reviews Gastroenterology, 2024) reveal that NAFLD prevalence has surged to 25% globally, with significant implications for liver cirrhosis and hepatocellular carcinoma [26–28]. Polycystic ovary syndrome (PCOS), a multifaceted endocrine disorder, is classified into insulin-resistant PCOS, associated with hyperinsulinemia and upregulated androgen biosynthesis, and non-insulin-resistant PCOS, driven by dysregulated hypothalamic-pituitary-ovarian axis signaling. Metabolic bone disorders include osteoporosis, characterized by reduced bone mineral density due to altered RANK/RANKL signaling, and osteomalacia, caused by vitamin D deficiency and hypophosphatemia, affecting hydroxyapatite deposition. Hyperuricemia and gout involve excessive uric acid levels, with primary gout linked to SLC2A9 gene variants and secondary forms associated with chronic kidney disease and tumor lysis syndrome [23, 26, 29].

Lipodystrophy, a rare disorder of adipose tissue distribution, manifests as congenital lipodystrophy, driven by mutations in genes such as LMNA or AGPAT2, and acquired forms, often resulting from protease inhibitors in HIV therapy. Atherosclerosis, a chronic inflammatory condition, involves lipid-laden foam cell accumulation, with subtypes including coronary artery disease (CAD), leading to myocardial infarction via plaque rupture, and peripheral arterial disease (PAD), causing critical limb ischemia [30–32]. These syndromes share underlying mechanisms such as chronic low-grade inflammation, oxidative stress, mitochondrial dysfunction, and gut dysbiosis, amplifying their severity and prevalence. Recent advancements in transcriptomics and metabolomics have unraveled biomarkers like adipokines, microRNAs, and metabolite signatures for early detection and stratified therapy. The interconnected nature of these syndromes necessitates comprehensive management approaches integrating lifestyle interventions, pharmacological modulation (e.g., GLP-1 receptor agonists for obesity and SGLT2 inhibitors for diabetes), and emerging technologies such as AI-driven diagnostic tools and CRISPR-based gene therapies to mitigate their growing global burden [33, 34].

3 Role of Nanotechnology in Diagnosis and Therapeutics

Recent advancements in 3D-printed nanotechnology have significantly enhanced the diagnosis of metabolic disorders, offering innovative, noninvasive, and personalized approaches (Fig. 13.2). In 2024, researchers at the Mayo Clinic utilized artificial intelligence (AI) in conjunction with an advanced 3D body-volume scanner to predict the risk and severity of metabolic syndrome. This method provides a more precise alternative to traditional measures like body mass index (BMI) and waist-to-hip ratio [35–37]. The study demonstrated that combining 3D body scans with AI algorithms could accurately identify individuals with metabolic syndrome, which affects over a third of the U.S. population and a quarter of people globally. This approach offers a reliable, repeatable measure of metabolic syndrome risk and

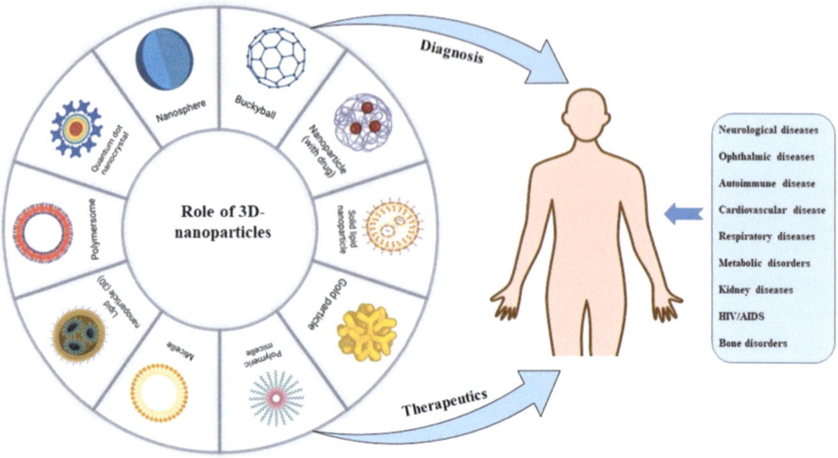

Fig. 13.2 Role of 3D nanoparticles in diagnosis and therapeutics

severity, potentially guiding clinicians and patients toward more effective interventions [38]. In another 2024 study, researchers developed a 3D pancreatic cancer model integrated with optical sensors for noninvasive metabolism monitoring and drug screening. This alginate-based 3D model embedded ratiometric optical pH sensors and cocultures of tumor and stromal cells to monitor extracellular pH variations during drug testing. The model allowed for real-time, noninvasive monitoring of metabolic changes, providing insights into tumor response to therapy and highlighting the importance of studying cellular metabolic mechanisms in 3D tumor models [39]. Additionally, in 2023, researchers from the Birla Institute of Technology & Science (BITS) in Hyderabad, India, unveiled a novel approach to detect and monitor diabetes using 3D printing technology. They developed a portable device capable of analyzing glucose and lactate concentrations in sweat samples. This noninvasive system eliminates the need for blood samples to gauge glucose levels. The device operates based on electro chemi-luminescence (ECL), emitting an electrical signal upon the introduction of sweat, leading to a chemical reaction that results in light emission. By measuring light intensity, the device can ascertain lactate concentration [40, 41]. The researchers also employed machine learning to determine precise lactate concentrations, enhancing the device's accuracy. They are developing a wearable version for ongoing monitoring, with plans to launch the device in collaboration with an industry partner. These developments underscore the transformative potential of 3D-printed nanotechnology in diagnosing metabolic disorders [42]. By leveraging advanced materials and fabrication techniques, researchers are creating innovative diagnostic tools that offer greater precision, sensitivity, and patient comfort. As these technologies continue to evolve, they hold promise for improving early detection and management of metabolic disorders, ultimately enhancing patient outcomes [43].

In another observation nanotechnology has emerged as a transformative force in the field of therapeutics, particularly in the management of metabolic syndromes, which are a cluster of conditions including obesity, insulin resistance, dyslipidemia, and hypertension, all of which increase the risk of cardiovascular diseases, diabetes, and stroke. The therapeutic applications of nanotechnology hold immense potential in addressing the multifaceted nature of metabolic syndromes, offering novel approaches for diagnosis, treatment, and prevention. Through the precise manipulation of materials at the nanoscale, nanotechnology enables the development of innovative drug delivery systems, diagnostic tools, and biomaterials, enhancing the efficacy and specificity of therapeutic interventions in metabolic diseases [38–40, 43].

In recent years, several studies have explored the role of nanotechnology in the management of metabolic syndromes. A comprehensive review published in 2020 examined the use of nanoformulations of herbal extracts in the prevention of metabolic syndrome. The study highlighted that nanoscale delivery systems, such as Nanoemulsions, solid lipid nanoparticles, and green-synthesized silver, gold, and zinc oxide nanoparticles, have been developed using herbal extracts. These formulations demonstrated the ability to decrease lipid profiles, inflammation, oxidative damage, and insulin resistance in in vitro and in vivo models of metabolic syndrome-related complications [41, 43, 44]. The review emphasized the importance of preparation methods and related parameters, such as temperature and type of polymer, in affecting the physicochemical stability and therapeutic activity of the final product. The authors concluded that further studies are needed to confirm the safety and efficacy of these novel herbal formulations for clinical application.

Another study published in 2022 focused on plant extract-based nanosystems for the treatment of metabolic disorders. The review discussed various nano-drug delivery systems, including liposomes, hydrogel nanocomposites, nanoemulsions, micelles, solid lipid, and core-shell nanoparticles, designed using plant extracts. The findings indicated that these nanoformulations successfully reduced oxidative stress, insulin resistance, chronic inflammation, and lipid profiles in in vitro and in vivo studies [45]. The study concluded that these novel plant-based nanosystems could act as promising candidates for clinical applications in the management of metabolic disorders [39, 46]. Furthermore, in 2022 the potentialities of nanomaterials for the management and treatment of metabolic syndrome. The study highlighted that nanotechnology utilizes nanoscale drugs and therapeutics that can reach specific target sites, improving drug delivery systems, early disease detection, and diagnosis using nano-imaging techniques. The review emphasized that nanotechnology increases bioavailability, permeability, compatibility, and the effectiveness of treatment while reducing adverse effects. The authors discussed the role of nanomaterial-based therapeutics in managing metabolic disorders such as nonalcoholic fatty liver disease, obesity, insulin resistance, and glucose intolerance, all of which increase the risk of developing cardiovascular diseases and diabetes [47]. These studies underscore the growing interest and potential of nanotechnology in the therapeutic management of metabolic syndromes [48]. The development of nanoformulations and nanomaterials offers promising avenues for enhancing the efficacy and specificity of treatments, addressing the complex nature of metabolic disorders. However, further research is necessary to confirm the safety and efficacy of these novel approaches for clinical application [44, 46, 49].

4 3D-Printed Nanotechnology-Based Drug Delivery

Advancements in 3D printing and nanotechnology have revolutionized drug delivery and diagnostic applications, providing highly precise, customizable systems capable of addressing complex medical challenges [50]. One of the significant innovations is the development of 3D-printed nanoporous scaffolds, which enable the controlled, sustained release of therapeutic agents. These scaffolds are particularly effective in the targeted delivery of nanoparticles for applications such as biosensors for disease biomarker detection (Table 13.2). Furthermore, nanoengineered hydrogels, designed for diffusion-controlled drug release, have proven useful as imaging contrast agents in MRI and fluorescence imaging, enabling enhanced diagnostic imaging with precise localization [51, 52]. Liposome-loaded 3D-printed implants are another breakthrough, offering both burst and controlled drug release modes that are particularly beneficial in tumor imaging via nanocarriers. Nanosphere-embedded devices have emerged as a promising solution for pulsatile drug delivery, mimicking the natural secretion cycles of endogenous systems, and also function as optical and electrochemical sensors for glucose monitoring, contributing to advancements in diabetes management [53, 54]. Additionally, bioprinted nanocomposite patches, which provide localized drug delivery via diffusion or erosion, show promise in biosensing applications such as the detection of infections or inflammation markers. Quantum dot-infused 3D materials are leveraged for real-time, high-sensitivity imaging of tumor progression and disease monitoring, enhancing the accuracy of diagnostic processes [55]. Moreover, nanozyme-based 3D-printed devices, designed to activate drugs through catalytic processes, provide targeted therapeutic efficacy at the cellular level by detecting oxidative stress markers, facilitating both diagnostic and therapeutic applications [56].

In the realm of magnetic drug delivery, magnetic nanoparticle-infused systems are employed for magnetic field-responsive drug release, offering noninvasive, precise control over therapeutic applications, particularly as MRI contrast agents for cardiovascular and neuroimaging. 3D-printed exosome-loaded devices enable slow-release delivery of exosomes, which act as biomarkers for neurodegenerative diseases, mimicking natural biological processes for therapeutic monitoring [57–59]. Gold nanoparticle-integrated structures utilize light-responsive drug release, offering enhanced plasmonic imaging for tumor tracking with minimal side effects. Silver nanoparticle-coated devices are designed for antimicrobial drug release, making them effective biosensors for infection detection in clinical settings. The integration of biodegradable metallic nanostructures for ion-mediated drug release presents a safer alternative in medical imaging, ensuring minimal long-term toxicity. Nanofiber mats engineered for gradual drug release allow for sustained therapeutic

Table 13.2 3D-printed nanotechnology-based drug delivery in therapeutics and diagnosis

S. no.	Name of 3D-printed nanotechnology	Mode of drug delivery	Diagnosis applications	Therapeutic applications	Challenges
1.	3D-printed nanoporous scaffolds	Sustained release of nanoparticles	Biosensors for disease marker detection	Bone regeneration (drug-releasing scaffolds), wound healing	Ensuring nanoparticle stability in scaffolds
2.	Nanoengineered hydrogels	Diffusion-controlled drug release	Imaging contrast agents for MRI, fluorescence imaging	Chemotherapy (targeted drug release), controlled insulin delivery	Precise manufacturing of responsive materials
3.	Liposome-loaded 3D-printed implants	Burst and controlled drug release modes	Tumor imaging via nanocarriers	Targeted chemotherapeutics, site-specific antibiotic delivery	Maintaining liposome integrity during printing
4.	Nanosphere-embedded devices	Pulsatile drug delivery	Optical and electrochemical sensors for glucose monitoring	Hormonal therapy, pain management (opioid delivery systems)	Integration of nanospheres in printed matrices
5.	Bio printed nanocomposite patches	Localized drug delivery via diffusion or erosion	Biosensing for infection or inflammation markers	Anti-inflammatory drug delivery, postoperative pain relief	Cost and complexity of bioprinting
6.	Quantum dot-infused 3D materials	Intravenous or topical drug administration	Real-time imaging of tumor or disease progression	Gene therapy, photodynamic therapy	Quantum dot toxicity
7.	Nanozyme-based 3D-printed devices	Catalytic drug activation at the target site	Detection of oxidative stress markers	Antioxidant therapies, enzyme replacement therapy	Ensuring biocompatibility and safety
8.	Porous polymer nanoparticle systems	Controlled release through degradation	Diagnostic microfluidic devices	Localized cancer therapies, immunomodulatory treatments	Polymer degradation in physiological conditions
9.	Magnetic nanoparticle-infused models	Magnetic field-responsive release	MRI contrast agents for cardiovascular diseases	Hyperthermia cancer therapy, site-specific drug targeting	Stability under magnetic fields
10.		Slow-release delivery of biomimetic exosomes		Regenerative medicine, gene therapy delivery	Limited scalability and stability of exosomes

(continued)

Table 13.2 (continued)

S. no.	Name of 3D-printed nanotechnology	Mode of drug delivery	Diagnosis applications	Therapeutic applications	Challenges
	3D-printed exosome-loaded devices		Exosome-based biomarkers for neurodegenerative diseases		
11.	Gold nanoparticle-integrated structures	Light-responsive drug delivery	Plasmonic imaging for real-time tumor tracking	Photothermal therapy, precision surgery	Cost of gold nanoparticles
12.	Silver nanoparticle-coated devices	Antimicrobial drug release	Infection detection (biosensors)	Wound healing, antibacterial coatings for implants	Cytotoxicity at higher concentrations
13.	Biodegradable metallic nanostructures	Ion-mediated drug release	Imaging enhancement (metal-based contrast agents)	Cardiovascular stent coatings, bone repair	Limited manufacturing techniques
14.	3D-printed nanofiber mats	Gradual release of encapsulated drugs	Point-of-care diagnostic devices	Skin regeneration, localized cancer treatment	Limited scalability for large-scale applications
15.	Nanocluster-enhanced hydrogels	Stimuli-responsive drug delivery	Fluorescence imaging for microbial infections	Immunotherapy, vaccines (e.g., mRNA delivery)	Complex synthesis protocols
16.	Carbon nanotube-embedded devices	Thermal-activated drug delivery	High-resolution biosensing	Targeted cancer therapies, neural repair applications	Biocompatibility concerns, regulatory hurdles
17.	Zinc oxide nanoparticle infusions	pH-responsive drug release	UV-light diagnostic devices	Antimicrobial treatments, cancer therapies	Potential oxidative stress in cells
18.	Titanium dioxide nanostructures	UV-light-induced drug activation	Photocatalytic sensors for cancer markers	Bone implant coatings, antimicrobial surfaces	Poor degradability in biological systems
19.	Graphene oxide-based 3D prints	Responsive to electrical stimuli	Electrochemical biosensors for diabetes management	Neurostimulation devices, targeted drug delivery	Risk of cytotoxicity
20.	Polymer-nanoparticle hybrid systems	Swelling-controlled drug delivery	Multi-analyte biosensors	Anti-cancer and anti-inflammatory therapies	Complex fabrication process

21.	Nanodiamond-incorporated devices	Gradual diffusion-based drug release	High-resolution imaging in oncology	Localized drug delivery for resistant tumors	Expensive synthesis methods
22.	Metal-organic framework-based systems	Gas-triggered drug release	Microfluidic devices for disease detection	Oxygen delivery in hypoxic tumors, drug detoxification	Scalability and stability challenges
23.	Silica nanoparticle-embedded devices	Sustained release of hydrophilic drugs	Fluorescence biosensors for infectious disease detection	Enzyme delivery, controlled drug release	Integration into 3D printing systems
24.	Iron oxide nanoparticle-loaded systems	Magnetic field-responsive drug release	MRI contrast agents for neuroimaging	Magnetic hyperthermia therapy, drug delivery in vascular diseases	Managing toxicity at higher doses
25.	Nanostructured lipid carriers	Diffusion-controlled drug release	Lipid-based imaging for metabolic diseases	Enhanced drug bioavailability, vaccine delivery	Controlling lipid oxidation
26.	Polymeric microneedle arrays	Transdermal drug delivery	Biosensors for skin-related diseases	Painless vaccine administration, localized treatment of psoriasis	Mechanical fragility of microneedles

effects in point-of-care diagnostics [60]. Carbon nanotube-embedded devices, utilizing thermal-activated drug delivery, provide high-resolution biosensing for various diseases, while zinc oxide nanoparticle systems enable pH-responsive drug release, beneficial in tumor microenvironments [53, 55, 61]. Titanium dioxide nanostructures, activated by UV light, offer photocatalytic sensors for cancer marker detection, ensuring controlled and localized drug delivery. Graphene oxide-based 3D prints responsive to electrical stimuli are applied in electrochemical biosensors for precise diabetes management [62]. Finally, polymer-nanoparticle hybrid systems designed for swelling-controlled drug release and nanodiamond-incorporated devices for high-resolution imaging further advance the efficacy and precision of drug delivery and diagnostic applications [62]. These technologies, along with gas-triggered metal-organic frameworks, silica nanoparticle-embedded devices, and iron oxide nanoparticle-loaded systems, represent a transformative shift in precision medicine, allowing for targeted, controlled, and sustained drug delivery, as well as highly sensitive diagnostics with reduced side effects [63].

5 Screening of Metabolic Syndrome Progression

The application of 3D nanoparticles for screening metabolic syndrome progression represents a groundbreaking intersection of nanotechnology and medicine. Their versatility in detecting biomarkers, enabling high-resolution imaging, and providing therapeutic insights makes them invaluable tools for advancing precision medicine and metabolic health management. By addressing existing challenges and continuing to refine their design and functionality, 3D nanoparticles like gold nanoparticles (AuNPs), liposomes, polymeric micelles (PEG-b-PCL), and dendrimers (PAMAM) have the potential to revolutionize the diagnosis and treatment of metabolic syndrome, offering hope for more effective and personalized approaches to managing this complex condition [60, 64, 65].

Screening metabolic syndrome progression using 3D nanoparticles is an innovative and interdisciplinary approach that integrates nanotechnology, biomedical imaging, and molecular biology. Metabolic syndrome encompasses a cluster of conditions, including obesity, insulin resistance, dyslipidemia, hypertension, and systemic inflammation, that significantly increase the risk of cardiovascular diseases and type 2 diabetes [66]. Early and accurate detection of its progression is critical for effective management and timely intervention. Three-dimensional nanoparticles, such as gold nanoparticles (AuNPs), silver nanoparticles (AgNPs), iron oxide nanoparticles (Fe_3O_4 NPs), and silica nanoparticles (SiO_2 NPs), provide unique opportunities to study metabolic syndrome due to their nanoscale size, high surface area, and ability to be functionalized for targeted interactions. Their applications can be categorized into three main areas: biomarker detection, imaging and monitoring, and delivery systems for probing [67].

Biomarker detection using 3D nanoparticles involves their functionalization with ligands, antibodies, or aptamers that target specific biomarkers associated with metabolic syndrome, such as C-reactive protein (CRP), adipokines (leptin and adiponectin), or inflammatory cytokines like IL-6 and TNF-α. Functionalized nanoparticles, including gold nanoparticles (AuNPs) and silver nanoparticles (AgNPs), enable highly sensitive detection of these biomarkers, significantly enhancing diagnostic accuracy [64, 65, 68]. For instance, gold nanoparticles (AuNPs) and silver nanoparticles (AgNPs) amplify signals in biosensors, facilitating the precise quantification of biomarkers even at extremely low concentrations. Furthermore, 3D nanoparticles enable multiplexed detection, where multiple biomarkers can be simultaneously identified, offering a comprehensive screening of the metabolic abnormalities underlying the syndrome. This capability is particularly valuable, as metabolic syndrome involves interconnected pathological processes that require a holistic diagnostic approach [69, 70].

In the area of imaging and monitoring, 3D nanoparticles functionalized with fluorescent dyes, quantum dots, or magnetic resonance imaging (MRI) contrast agents enable noninvasive imaging of tissues affected by metabolic syndrome. This molecular imaging capability allows for the visualization of dynamic changes at the cellular and tissue levels, providing valuable insights into disease progression. For example, quantum dots (CdSe/ZnS QDs) and iron oxide nanoparticles (Fe_3O_4 NPs) can be used for imaging the liver, pancreas, or adipose tissues, highlighting the extent of lipid accumulation or inflammation, both hallmarks of metabolic syndrome. Additionally, these nanoparticles enable dynamic monitoring, where real-time tracking of metabolic changes provides clinicians with actionable data on disease trajectories. Targeted imaging is another significant advancement made possible by 3D nanoparticles [71, 72]. By designing lipid-coated or peptide-functionalized nanoparticles such as liposomes or polymeric nanoparticles (e.g., PLGA NPs) that selectively accumulate in specific tissues, researchers can enhance the specificity of diagnostic imaging, reducing false positives and improving the overall reliability of metabolic syndrome screening. These advancements underscore the pivotal role of 3D nanoparticles in bridging the gap between basic research and clinical applications [66, 71].

Another transformative application of 3D nanoparticles is their role as delivery systems for probing metabolic syndrome. Smart probes designed to interact with specific metabolic pathways—such as glucose metabolism or lipid accumulation—can provide diagnostic feedback, elucidating the molecular underpinnings of the syndrome. These nanoparticles can also function as theranostic tools, combining therapeutic and diagnostic capabilities in a single platform. For instance, dendrimers (e.g., PAMAM dendrimers) or polymeric micelles (PEG-b-PCL) loaded with therapeutic agents can simultaneously treat metabolic imbalances and monitor therapeutic efficacy, offering a dual-purpose solution for managing the condition. Such integrated systems not only streamline the diagnostic and therapeutic processes but also pave the way for personalized medicine approaches tailored to individual patient needs [73, 74].

The technological advancements driving these applications are remarkable. Advances in nanoparticle engineering have enabled precise control over their size, shape, and surface chemistry, significantly enhancing their targeting capabilities and specificity. For example, nanoparticles like magnesium hydroxide nanoparticles (Mg

$(OH)_2$ NPs) and chitosan nanoparticles (Chitosan NPs) with tailored surface modifications can evade immune recognition, improving their circulation time in the bloodstream and increasing their efficacy in reaching target tissues. Additionally, integrating these nanoparticles with artificial intelligence (AI) and machine learning algorithms has further expanded their diagnostic potential. By analyzing the complex data generated by nanoparticle-based screening, AI can identify subtle patterns and predict disease trajectories, offering unprecedented insights into metabolic syndrome progression [75]. Despite these advancements, there are challenges that must be addressed to fully realize the potential of 3D nanoparticles in clinical settings. Ensuring the biocompatibility of nanoparticles, such as titanium dioxide nanoparticles (TiO_2 NPs), is paramount, as their use in vivo must not elicit adverse immune responses or toxicity. Developing biodegradable nanoparticles made from materials such as silica nanoparticles (SiO_2 NPs) or polymers can mitigate long-term toxicity concerns, offering a safer alternative for clinical applications. Standardization is another critical challenge, as consistent protocols for nanoparticle synthesis and use are necessary to ensure reproducibility and reliability across studies. Moreover, regulatory hurdles associated with gaining approval for clinical applications of nanoparticles must be navigated carefully to facilitate their translation from research to practice [76].

6 Patient Genetics and Clinical Applications

The emergence of 3D-printed nanoparticles has dramatically transformed the landscape of personalized medicine, offering promising solutions across diverse fields such as drug delivery, gene therapy, tissue engineering, immunotherapy, and diagnostics, with an unprecedented ability to tailor treatments to individual genetic profiles. The use of 3D printing technology enables the precise design and fabrication of nanoparticles, such as poly(lactic-co-glycolic acid) (PLGA), polycaprolactone (PCL), chitosan-based nanoparticles, gold nanoparticles (AuNPs), and lipid nanoparticles, providing a customizable approach that addresses patient-specific needs. This customization is essential for improving the therapeutic efficacy of treatments, as genetic differences between patients can significantly influence how drugs are absorbed, metabolized, and eliminated from the body [77]. For example, variations in drug-metabolizing enzymes, such as cytochrome P450, or genetic mutations in receptors and transporters can impact the efficiency of drug delivery, making it crucial to design nanoparticles that are optimized for specific genetic profiles. In the context of cancer treatment, nanoparticles designed with genetic considerations in mind can be used to deliver chemotherapeutic agents with greater precision, minimizing off-target effects while enhancing the drug's potency at the tumor site. Similarly, in the case of genetic disorders, nanoparticles can be tailored to carry and release therapeutic agents such as nucleic acids, proteins, or small molecules in a manner that overcomes genetic defects, as seen in the use of gene therapies for conditions like cystic fibrosis [78, 79]. Furthermore, by utilizing genetic data, nanoparticles can be engineered to address the unique genetic markers of different patients, ensuring a more targeted approach to therapy that maximizes efficacy and minimizes adverse effects. This approach is particularly advantageous in personalized medicine, where pharmacogenomic testing can be used to identify genetic markers that determine how an individual will respond to specific treatments. For example, variants in the P-glycoprotein gene can influence drug resistance in cancer therapy, and customizing nanoparticles to bypass such resistance can lead to better treatment outcomes. The role of 3D-printed nanoparticles in tissue engineering is another area where genetic considerations are vital. Nanoparticles incorporated into scaffolds can promote cell growth, differentiation, and tissue regeneration, with careful attention paid to the genetic makeup of the patient [77, 80, 81]. For instance, in regenerative medicine, scaffolds made from materials like hydroxyapatite (HA)-based nanoparticles, collagen-coated PLGA nanoparticles, and silk fibroin nanoparticles can be customized to enhance tissue repair and regeneration in genetically predisposed individuals. Genetic factors such as mutations in collagen synthesis genes or immune response genes can affect how tissues regenerate, and by incorporating these genetic insights into the design of nanoparticles, the effectiveness of tissue engineering approaches can be significantly enhanced. Additionally, the incorporation of 3D-printed nanoparticles in gene therapy, such as in CRISPR-based therapies, RNA interference (RNAi), or mRNA vaccine delivery, presents a major advancement in the ability to deliver genes and genetic material with high precision [82]. Lipid nanoparticles (LNPs) have emerged as a popular choice for mRNA delivery, and polyethylenimine (PEI)-based nanoparticles are widely used for gene delivery due to their ability to condense and protect nucleic acids. By integrating genetic information into the design of these nanoparticles, such as using patient-specific data on gene mutations and immune response variations, the efficiency of gene delivery can be significantly improved. For example, genetic factors such as the presence of immune system mutations, like variations in the MHC or HLA genes, can influence how a patient responds to gene therapies, making it essential to design nanoparticles that are compatible with these genetic variations. By considering these factors, gene therapies can be made more effective and tailored to the genetic needs of individual patients. In diagnostic applications, 3D-printed nanoparticles have been shown to significantly enhance the sensitivity and specificity of molecular diagnostics, particularly when genetic screening is involved [79, 80, 83]. Nanoparticles such as magnetic nanoparticles (Fe_3O_4), quantum dots (CdSe), and silica nanoparticles functionalized with antibodies can be used to detect genetic mutations associated with various diseases, including cancer, cardiovascular disease, and genetic disorders. By utilizing nanoparticles that are specifically designed to bind to genetic markers, early detection of diseases such as cancer can be achieved with greater precision, enabling earlier intervention and more personalized treatment options. For example, the detection of BRCA1/2 mutations, which are associated with a higher risk of breast and ovarian cancers, can be improved with the use of nanoparticles that are engineered to selectively bind to these mutations, allowing for faster and more accurate genetic screening. This genetic specificity also enhances the ability of nanoparticles to be used in point-of-care diagnostics, where rapid and accurate results are essential for effective disease management [81–83]. In immunotherapy, which has gained significant attention in cancer treatment, 3D-printed nanoparticles play a critical role in modulating the immune system. By engineering nanoparticles to carry immunomodulatory agents, such as checkpoint inhibitors or cytokines, these particles can enhance the immune system's ability to recognize and target cancer cells. The design of these nanoparticles takes into account genetic factors influencing immune function, such as variations in immune checkpoint genes or mutations in T-cell receptors, which can affect how the immune system responds to cancer. By personalizing the design of nanoparticles to these genetic differences, it is possible to create more effective immunotherapies that enhance the body's ability to fight cancer [84]. Additionally, immunotherapy nanoparticles can be engineered to suppress autoimmune reactions in individuals with autoimmune diseases by modulating the immune response to restore immune system balance. Finally, biodegradable nanoparticles made from materials like PLGA and poly (lactic acid) (PLA) offer a safe and effective alternative to nonbiodegradable particles, as they naturally degrade in the body after performing their intended therapeutic function. The degradation rate of these nanoparticles can be influenced by genetic factors, such as variations in enzymes that metabolize biodegradable polymers, including cytochrome P450 enzymes. By considering these genetic differences, the rate of nanoparticle degradation can be optimized for individual patients, ensuring that the therapeutic benefits are delivered in a controlled and safe manner, while minimizing the risk of toxic accumulation [81–83]. Personalized approaches to biodegradable nanoparticle design can significantly enhance the safety and efficacy of nanoparticle-based drug delivery systems. In conclusion, the integration of 3D-printed nanoparticles in personalized medicine represents a major step forward in the development of tailored therapeutic strategies that take into account individual genetic profiles. By leveraging genetic data, nanoparticles can be precisely designed to address specific patient needs, whether for drug delivery, gene therapy, tissue engineering, immunotherapy, or diagnostics. As research continues to uncover the complexities of genetics and nanotechnology, the potential for personalized nanoparticle-based treatments to improve patient outcomes and revolutionize the future of medicine is immense [84, 85].

As shown in Fig. 13.3, 3D nanoparticles play a transformative role in patient genetics and clinical applications, leveraging their unique properties for advanced diagnostics, therapeutics, and genetic studies. Quantum dots (QDs) enable high-resolution imaging of genetic mutations, proving invaluable in cancer diagnostics and real-time bioimaging. Gold nanoparticles (AuNPs), with their plasmonic properties, facilitate precise gene sequencing and marker identification while serving as carriers in targeted therapies and biosensors. Magnetic nanoparticles assist in the separation and purification of nucleic acids, enhancing genetic analyses, and are pivotal in MRI imaging and hyperthermia-based cancer treatments [82, 83, 85].

Lipid nanoparticles (LNPs) are essential for delivering genetic material, such as mRNA, enabling breakthroughs in mRNA vaccines and gene therapies. Similarly, silica nanoparticles stabilize genetic payloads for efficient delivery in treating genetic disorders. Carbon nanotubes (CNTs) improve transfection efficiency, supporting effective gene editing and delivery in cancer therapy. Polymer

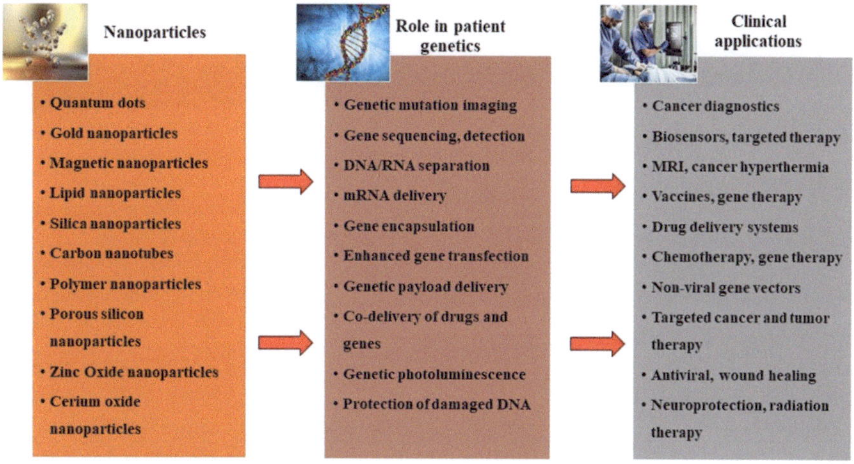

Fig. 13.3 Patient genetics and clinical application of nanoparticles

nanoparticles, with their biocompatibility, act as nonviral vectors for safer, controlled gene therapy. Porous silicon nanoparticles (pSiNPs) co-deliver drugs and genetic material, enhancing efficacy in targeted treatments such as siRNA and CRISPR-based therapies.

Zinc oxide nanoparticles aid in genetic research through photoluminescence and oxidative stress modulation, while cerium oxide nanoparticles protect DNA from oxidative damage, benefiting neurodegenerative disease management and radiation therapy. These nanoparticles enable precise genetic studies and advanced clinical interventions, driving progress in personalized medicine, nanomedicine, and therapeutic innovations for complex diseases. Their continued development holds immense promise for future healthcare solutions [86].

7 Future Applications

The future applications of 3D-printed nanotechnology in the diagnosis and therapeutics of metabolic syndrome represent a transformative frontier in biomedical innovation, leveraging advanced materials science, nanotechnology, and precision engineering (Table 13.3). In diagnostics, 3D-printed lab-on-a-chip platforms, integrated with nanosensors like zinc oxide (ZnO) nanoparticles and gold nanoshells (AuNS), are enabling ultra-sensitive detection of critical biomarkers such as glucose, triglycerides, low-density lipoprotein (LDL) cholesterol, and inflammatory cytokines like interleukin-6 (IL-6) and tumor necrosis factor-alpha (TNF-α) [87]. These devices utilize advanced techniques such as surface plasmon resonance (SPR) and electrochemical sensing, providing rapid, accurate, and real-time monitoring capabilities for metabolic syndrome markers. Similarly, nanostructured contrast agents composed of gadolinium-doped carbon quantum dots (Gd-CQDs) and

Table 13.3 Future applications of 3D-nanoparticles explaining the solicitation of various therapeutic and diagnostic techniques

S. no.	Category	Applications	Description
1.	Diagnosis	Personalized diagnostic platforms	3D-printed nanotechnology enables the fabrication of lab-on-a-chip devices with integrated nanosensors for detecting biomarkers related to metabolic syndrome (e.g., glucose, cholesterol, inflammatory markers)
2.	Imaging probes	Nanostructured contrast agents	Nanoparticles produced via 3D printing offer enhanced imaging for metabolic tissues using techniques like MRI, CT, and PET scans by targeting specific organs (e.g., liver, pancreas, adipose tissue)
3.	Therapeutics	Controlled drug delivery systems	Customizable 3D-printed nanocarriers (liposomes, hydrogels, or microneedles) for targeted and controlled drug release, particularly for antidiabetic, lipid-lowering, and anti-inflammatory drugs
4.	Regenerative medicine	Bio-printed pancreatic islets	Fabrication of 3D-printed bioengineered tissues for replacing damaged pancreatic islets, promoting insulin secretion and glucose homeostasis
5.	Wearable devices	3D-printed smart sensors for continuous monitoring	Nanostructured wearable devices printed to monitor glucose, lipid levels, and blood pressure in real time, linked to patient health databases for personalized therapeutic adjustments
6.	Gene therapy	3D-printed nanocarriers for delivering CRISPR/Cas9 or RNA-based therapies	Custom nanocarriers enable precise delivery of genetic editing tools to target tissues involved in metabolic syndrome, e.g., editing genes linked to obesity, diabetes, or dyslipidemia
7.	Vaccination	Nano-vaccine platforms for metabolic syndrome-associated diseases	3D-printed nanoparticle systems tailored to deliver antigens for vaccines targeting inflammation or metabolic disorders such as obesity or insulin resistance
8.	Tissue engineering	Bio-printed adipose or liver tissue models for drug testing	Creation of 3D-printed nanostructured tissue models that replicate metabolic tissues for preclinical drug evaluation and metabolic syndrome research
9.	Microbiome modulation	Engineered probiotics in nanocarriers	Custom nanocarriers for delivering live probiotics targeted to specific

(continued)

Table 13.3 (continued)

S. no.	Category	Applications	Description
			gut regions to modulate the microbiome for improving metabolic health
10.	Artificial organs	3D-printed bioengineered liver or kidney constructs	Bio-printed artificial organs with nanotechnology for metabolic function restoration, particularly in patients with fatty liver disease or renal complications of metabolic syndrome
11.	In silico modeling	3D-printed nanostructures for computational modeling	Fabrication of physical nanostructures to validate computational models related to the progression of metabolic syndrome, aiding in predictive studies
12.	Anti-obesity treatments	Injectable nanoscaffolds for fat tissue remodeling	Nanoscaffolds created via 3D printing that stimulate lipolysis or adipose tissue remodeling, helping to reduce fat accumulation
13.	Bioactive implants	3D-printed implants releasing bioactive molecules	Implants with embedded nanostructures that release antidiabetic or anti-inflammatory agents locally to manage metabolic syndrome symptoms
14.	Neuroprotection	Nanostructured delivery systems for addressing brain-metabolism links	3D-printed nanocarriers to deliver neuroprotective agents targeting metabolic syndrome-related cognitive decline or neuroinflammation
15.	Antioxidant therapies	Nanoencapsulation of antioxidants	3D-printed systems for encapsulating antioxidants like resveratrol or polyphenols to combat oxidative stress associated with metabolic syndrome
16.	Bioinformatics tools	Platforms integrating 3D-printed sensors and AI for data analysis	Systems combining 3D-printed nanosensors and AI algorithms for analyzing patient-specific data and optimizing metabolic syndrome therapies
17.	Nanobots	3D-printed nanorobots for metabolic repair	Fabrication of autonomous nanobots for targeted delivery of drugs, repair of damaged tissues, or removal of toxic metabolites in metabolic syndrome patients
18.	Hybrid systems	3D-printed multi-functional nanostructures	Systems combining diagnosis, treatment, and monitoring into a single 3D-printed nanosystem for comprehensive metabolic syndrome management

iron oxide nanoparticles (Fe_3O_4 NPs) are enhancing imaging modalities like magnetic resonance imaging (MRI), computed tomography (CT), and positron emission tomography (PET) by improving specificity and resolution. These nanoparticles can be conjugated with targeting ligands such as folic acid or monoclonal antibodies to home in on metabolic tissues, including the liver, pancreas, and adipose tissue, thus enabling earlier and more precise disease diagnosis [88, 89].

In therapeutics, the development of 3D-printed nanocarriers, including poly(lactic-co-glycolic acid) (PLGA) nanoparticles and mesoporous silica nanoparticles (MSNs), is revolutionizing drug delivery by enabling controlled and targeted release of therapeutics like metformin, pioglitazone, and atorvastatin. These systems are particularly advantageous in addressing the multifaceted pathophysiology of metabolic syndrome, as they allow the co-delivery of drugs to combat hyperglycemia, dyslipidemia, and chronic inflammation simultaneously [90]. Furthermore, regenerative medicine applications such as bio-printed pancreatic islets are being explored to restore insulin secretion and glucose homeostasis. These constructs often incorporate biomaterials such as gelatin methacryloyl (GelMA) hydrogels and nanoscale cerium oxide (CeO_2) particles to mimic the native extracellular matrix and provide antioxidant protection. Wearable devices integrated with 3D-printed graphene-based sensors and titanium dioxide (TiO_2) nanostructures are facilitating continuous monitoring of glucose levels, lipid profiles, and blood pressure, linked to real-time patient data through artificial intelligence (AI) platforms for personalized therapeutic adjustments [89, 91].

Gene therapy is being enhanced through the use of 3D-printed lipid nanoparticles (LNPs) and dendrimer-based nanocarriers for the delivery of CRISPR/Cas9 systems or RNA-based therapies. For instance, poly(amidoamine) (PAMAM) dendrimers functionalized with polyethylene glycol (PEG) enable precise editing of genes associated with obesity, insulin resistance, and dyslipidemia. Nano-vaccine platforms utilizing 3D-printed chitosan nanoparticles and calcium phosphate nanocarriers are also being developed to deliver antigens targeting inflammation and immune dysregulation in metabolic syndrome, offering potential solutions to mitigate disease progression. Tissue engineering is advancing with bio-printed models of adipose and liver tissues, constructed using nanocomposites like hydroxyapatite (HA) and collagen-mimetic peptide scaffolds, which provide physiologically relevant platforms for drug screening and disease modeling [91, 92].

Microbiome modulation represents another frontier, with engineered probiotics encapsulated in 3D-printed alginate nanoparticles being targeted to specific gut regions. This approach aims to restore gut microbiota balance, reduce systemic inflammation, and improve metabolic outcomes. Artificial organs, such as bioengineered liver and kidney constructs using nanoscale silicon dioxide (SiO_2) and graphene oxide (GO), are being explored to address metabolic complications, including nonalcoholic fatty liver disease (NAFLD) and renal dysfunction. The integration of 3D-printed nanostructures in in silico modeling is also enabling the validation of computational models for metabolic syndrome progression, particularly by replicating the biomechanical and biochemical environments of affected tissues [93].

Anti-obesity treatments are leveraging injectable nanoscaffolds made from materials like polycaprolactone (PCL) and nanoscale silver (Ag) particles to stimulate lipolysis or promote adipose tissue remodeling. Similarly, bioactive implants incorporating magnesium oxide (MgO) nanoparticles and layered double hydroxides (LDHs) are being designed to release antidiabetic and anti-inflammatory agents locally. Neuroprotective strategies are being developed using 3D-printed nanocarriers, such as cerium oxide nanospheres and lipid-based nanovesicles, to deliver neuroprotective agents across the blood-brain barrier (BBB), addressing metabolic syndrome-related cognitive decline and neuroinflammation. Antioxidant therapies are being optimized through the nanoencapsulation of bioactive compounds like resveratrol, curcumin, and quercetin using 3D-printed polydopamine nanoparticles and silk fibroin matrices, which enhance stability, bioavailability, and therapeutic efficacy [91, 92, 94].

Bioinformatics tools integrating 3D-printed nanosensors and AI algorithms are enabling the real-time analysis of patient-specific data, facilitating personalized treatment strategies for metabolic syndrome. These tools often incorporate nanoscale semiconductors like molybdenum disulfide (MoS_2) and boron nitride nanotubes for high-throughput data collection and processing. Nanobots fabricated using 3D printing with materials such as zinc ferrite ($ZnFe_2O_4$) and gold nanorods (AuNRs) are being developed for metabolic repair, offering applications in the targeted removal of toxic metabolites, delivery of bioactive molecules, and tissue repair. Hybrid systems that integrate diagnostic, therapeutic, and monitoring functionalities into a single 3D-printed nanosystem are also being explored to provide holistic management solutions for metabolic syndrome [95].

These advancements underscore the transformative potential of 3D-printed nanotechnology in addressing the complexities of metabolic syndrome through innovative diagnostic and therapeutic approaches. By combining the precision of nanotechnology with the flexibility of 3D printing, researchers are paving the way for personalized, efficient, and multi-modal healthcare solutions. While significant challenges remain, including regulatory compliance, scalability, and ensuring biocompatibility, the convergence of advanced materials, nanotechnology, and bioengineering is poised to redefine the landscape of metabolic syndrome management [96, 97].

8 Conclusion

The integration of 3D-printed nanotechnology into the diagnosis and treatment of metabolic syndrome represents a groundbreaking advancement in precision medicine, offering innovative solutions to the multifaceted challenges posed by this complex disorder. Metabolic syndrome, characterized by a range of interconnected pathologies such as obesity, insulin resistance, hypertension, dyslipidemia, and nonalcoholic fatty liver disease, requires highly personalized and adaptive therapeutic strategies to address its diverse etiology and pathophysiology. The convergence of nanotechnology and 3D printing facilitates the development of multifunctional diagnostic and therapeutic systems with unprecedented precision. In diagnostics, the combination of nanosensors, lab-on-a-chip platforms, and advanced nanostructured contrast agents significantly enhances the sensitivity and specificity of biomarker detection, enabling earlier and more accurate identification of metabolic syndrome and its associated complications. For example, the use of gold and zinc oxide nanoparticles as biosensors, along with contrast agents like gadolinium-functionalized carbon quantum dots and iron oxide nanoparticles, enables real-time, high-resolution imaging and monitoring of disease progression. These diagnostic systems offer not only enhanced detection but also the ability to monitor the therapeutic outcomes of interventions. On the therapeutic front, 3D-printed nanocarriers and bioengineered tissues provide targeted, controlled release of therapeutic agents such as anti-inflammatory, hypoglycemic, and lipid-lowering drugs. The use of polymeric nanoparticles like PLGA, mesoporous silica nanoparticles (MSNs), and lipid-based nanostructures, along with bio-printed tissues, has shown immense promise in providing localized and sustained drug delivery, minimizing systemic side effects while enhancing drug bioavailability and therapeutic efficacy. Furthermore, the application of 3D-printed systems in gene therapy and regenerative medicine, such as the bio-printing of pancreatic islets and nanoscale tissue constructs, offers the potential to restore glucose homeostasis and address tissue dysfunction. Innovations like these, coupled with nanotechnology's ability to modulate genetic, immune, and microbiome factors, highlight the transformative potential of this multidisciplinary approach. The development of hybrid systems that integrate diagnostic, therapeutic, and monitoring functionalities within a single 3D-printed platform further expands the therapeutic possibilities, providing a comprehensive solution for managing metabolic syndrome. Additionally, AI-driven analytics and machine learning platforms enhance the adaptability and responsiveness of these systems by enabling patient-specific therapeutic strategies, informed by real-time data from wearable sensors and other diagnostic tools. Despite these promising advancements, several challenges remain, including the scalability, reproducibility, and regulatory approval of these novel technologies. Issues related to the long-term biocompatibility and stability of 3D-printed nanomaterials also need to be addressed to ensure their safe application in clinical settings. The need for standardization and the development of comprehensive regulatory frameworks to guide the clinical translation of these innovations is critical. However, ongoing interdisciplinary research and collaboration across fields such as materials science, bioengineering, and AI are likely to overcome these challenges, paving the way for the widespread clinical adoption of 3D-printed nanotechnology. As these technologies continue to evolve, they hold the potential to significantly improve the management of metabolic syndrome, offering personalized, efficient, and sustainable healthcare solutions. By addressing the underlying mechanisms of the syndrome through targeted drug delivery, real-time diagnostics, and regenerative medicine, these innovations promise to reduce the global burden of metabolic syndrome, ultimately enhancing patient outcomes and quality of life. Thus, the convergence of 3D printing and nanotechnology in metabolic syndrome management marks a critical step forward in the development of more effective, patient-centered treatments, heralding a new era of precision healthcare.

References

1. Kumah, E. A., Fopa, R. D., Harati, S., Boadu, P., Zohoori, F. V., & Pak, T. (2023). Human and environmental impacts of nanoparticles: A scoping review of the current literature. *BMC Public Health, 23*(1). https://doi.org/10.1186/s12889-023-15958-4
2. Mahmoud, K., Swidan, S., El-Nabarawi, M., & Teaima, M. (2022). Lipid based nanoparticles as a novel treatment modality for hepatocellular carcinoma: A comprehensive review on targeting and recent advances. *Journal of Nanobiotechnology, 20*(1), 1–42. https://doi.org/10.1186/s12951-022-01309-9
3. Kankala, R. K., Han, Y. H., Xia, H. Y., Bin Wang, S., & Chen, A. Z. (2022). Nanoarchitectured prototypes of mesoporous silica nanoparticles for innovative biomedical applications. *Journal of Nanobiotechnology, 20*(1), 1–67. https://doi.org/10.1186/s12951-022-01315-x
4. Tapia-Arellano, A., Cabrera, P., Cortés-Adasme, E., Riveros, A., Hassan, N., & Kogan, M. J. (2024). Tau- and α-synuclein-targeted gold nanoparticles: applications, opportunities, and future outlooks in the diagnosis and therapy of neurodegenerative diseases. *Journal of Nanobiotechnology, 22*(1), 1–29. https://doi.org/10.1186/s12951-024-02526-0
5. Majeed, H. F., Hamad, T. I., & Bairam, L. R. (2024). Enhancing 3D-printed denture base resins: A review of material innovations. *Science Progress, 107*(3), 1–20. https://doi.org/10.1177/00368504241263484
6. Singh, A., Prasad, L. B., Shiv, K., Kumar, R., & Garai, S. (2023). Synthesis, characterization, and in vitro antibacterial and cytotoxic study of Co(II), Ni(II), Cu(II), and Zn(II) complexes of N-(4-methoxybenzyl) N-(phenylethyl) dithiocarbamate ligand. *Journal Of Molecular Structure, 1288*(May), 135835. https://doi.org/10.1016/j.molstruc.2023.135835
7. Saini, R. S., et al. (2024). Mapping the research landscape of nanoparticles and their use in denture base resins: a bibliometric analysis. *Discover Nano, 19*(1). https://doi.org/10.1186/s11671-024-04037-1
8. Madni, A., et al. (2021). Mechanistic approaches of internalization, subcellular trafficking, and cytotoxicity of nanoparticles for targeting the small intestine. *AAPS PharmSciTech, 22*(1). https://doi.org/10.1208/s12249-020-01873-z
9. Kumar, A., et al. (2022). Acute and subacute toxicity study of ethanolic extract of Calotropis procera (Aiton) Dryand flower in Swiss albino mice. *Phytomedicine Plus, 2*(2), 100224. https://doi.org/10.1016/j.phyplu.2022.100224
10. Heo, J., et al. (2023). Method for 3D atomic structure determination of multi-element nanoparticles with graphene liquid-cell TEM. *Scientific Reports, 13*(1), 1–10. https://doi.org/10.1038/s41598-023-28492-5
11. Liu, Z., et al. (2023). Janus nanoparticles targeting extracellular polymeric substance achieve flexible elimination of drug-resistant biofilms. *Nature Communications, 14*(1), 1–18. https://doi.org/10.1038/s41467-023-40830-9
12. Harsh Pandey, Y. B. T., Srivastava, S., & Kumar, R. (2018). Preclinical acute and repeated dose toxicity of Pueraria tuberosa (PTWE) on charles foster rats. *International Journal of Pharmaceutical Sciences and Research, 9*(11), 4572–4581. https://doi.org/10.13040/IJPSR.0975-8232.9(11).4572-81
13. Singh, A., et al. (2024). Synthesis, crystal structure and screening for anticonvulsant and antianxiety activities of three new Ni(II), Cu(II), and Zn(II) dithiocarbamate complexes. *Journal of Molecular Structure, 1298*(P2), 137052. https://doi.org/10.1016/j.molstruc.2023.137052
14. Kumar, A., et al. (2022). Toxicity (acute and subacute) assessment and in-vivo antiurolithiatic activity of ethanolic extract of Caesalpinia bonducella seed in albino Wistar rat. *Journal Of Applied Pharmaceutical Science, 12*(2), 187–197. https://doi.org/10.7324/JAPS.2022.120220
15. Pandey, H., Srivastava, S., Kumar, A., Kumar, R., & Tripathi, Y. B. (2019). Pharmacokinetics study of Puerarin absorption in blood after consumption of Pueraria tuberosa water extract (PTWE) by rats. *Exploration of Animal Medicine Research, 9*(2), 188–196.
16. Dhiman, R., Kumar, D., Kumar, B., & Pandey, B. L. (2016). Quantitative determination of atorvastatin, ortho-hydroxy atorvastatin, para-hydroxy atorvastatin in human plasma using

rosuvastatin as internal standard by LC-MS/MS. *International Journal of Pharmaceutical Chemistry Science, 4*(4), 487–500.
17. Vijaya Durga, R. D. (2015). Determination of S-amlodipine in human plasma using tizanidine as internal standard by Lc/Ms/Ms method. *Journal of Bioanalysis and Biomedicine, 07*(03), 91–96. https://doi.org/10.4172/1948-593x.1000130
18. Kumar, R., Kumar, B., Kumar, A., Kumar, A., & Singh, M. (2022). GC-MS analysis of phytocomponents in the methanol extract of Premna latifolia Roxb. *Pharmacognosy Research, 14*(1), 19–23. https://doi.org/10.5530/pres.14.1.4
19. El-Naggar, N. E. A., Shiha, A. M., Mahrous, H., & Mohammed, A. B. A. (2024). A sustainable green-approach for biofabrication of chitosan nanoparticles, optimization, characterization, its antifungal activity against phytopathogenic Fusarium culmorum and antitumor activity. *Scientific Reports, 14*(1), 1–23. https://doi.org/10.1038/s41598-024-59702-3
20. Passos, J. S., Lopes, L. B., & Panitch, A. (2023). Collagen-binding nanoparticles for paclitaxel encapsulation and breast cancer treatment. *ACS Biomaterials Science & Engineering, 9*(12), 6805–6820. https://doi.org/10.1021/acsbiomaterials.3c01332
21. Kumar, A., et al. (2021). Antiurolithiasis, antioxidant, anti-inflammatory, analgesic, and diuretic activity of ethanolic extract of seeds of Caesalpinia bonducella. *International Journal of Pharmaceutical Investigation, 11*(3), 306–311. https://doi.org/10.5530/ijpi.2021.3.54
22. G. R., et al. (2022). COVID-19: A challenge towards the sustainability of health in platform era. *NeuroQuantology, 20*(11), 2503–2519. https://doi.org/10.14704/NQ.2022.20.11.NQ66250
23. Lu, Q., et al. (2024). Nanoparticles in tumor microenvironment remodeling and cancer immunotherapy. *Journal of Hematology & Oncology, 17*(1). https://doi.org/10.1186/s13045-024-01535-8
24. Priwitaningrum, D. L., et al. (2023). Evaluation of paclitaxel-loaded polymeric nanoparticles in 3D tumor model: Impact of tumor stroma on penetration and efficacy. *Drug Delivery and Translational Research, 13*(5), 1470–1483. https://doi.org/10.1007/s13346-023-01310-1
25. Xu, H., Li, S., & Liu, Y. S. (2022). Nanoparticles in the diagnosis and treatment of vascular aging and related diseases. *Signal Transduction and Targeted Therapy, 7*(1). https://doi.org/10.1038/s41392-022-01082-z
26. Iskuzhina, L., et al. (2024). Advances in the toxicity assessment of silver nanoparticles derived from a Sphagnum fallax extract for monolayers and spheroids. *Biomolecules, 14*(6). https://doi.org/10.3390/biom14060611
27. Kucharz, K., et al. (2021). Post-capillary venules are the key locus for transcytosis-mediated brain delivery of therapeutic nanoparticles. *Nature Communications, 12*(1). https://doi.org/10.1038/s41467-021-24323-1
28. Nkosi, N. C., Basson, A. K., Ntombela, Z. G., Dlamini, N. G., & Pullabhotla, R. V. S. R. (2024). A review on bioflocculant-synthesized copper nanoparticles: Characterization and application in wastewater treatment. *Bioengineering, 11*(10), 1–35. https://doi.org/10.3390/bioengineering11101007
29. Song, Q., Zheng, Y., Zhong, G., Wang, S., He, C., & Li, M. (2024). Application of nanoparticles in the diagnosis and treatment of colorectal cancer. *Anti-Cancer Agents in Medicinal Chemistry, 24*(18), 1305–1326. https://doi.org/10.2174/0118715206323900240807110122
30. Crawley, J. W. M., et al. (2022). Heterogeneous trimetallic nanoparticles as catalysts. *Chemical Reviews, 122*(6), 6795–6849. https://doi.org/10.1021/acs.chemrev.1c00493
31. Singh, A., et al. (2024). Dithiocarbamate-based novel anti-histaminic agents: synthesis, characterization, crystal structure and thermal study. *Dalton Transactions*. https://doi.org/10.1039/d4dt01777c
32. Singh, M., et al. Pharmacokinetic screening to estimate the drug likeliness characteristics of selected herbal anticancer drugs. *Research Journal of Pharmacy and Technology, 16*(7), 3422–3426., 2023. https://doi.org/10.52711/0974-360X.2023.00566

33. Liu, L., & Corma, A. (2018). Metal catalysts for heterogeneous catalysis: From single atoms to nanoclusters and nanoparticles. *Chemical Reviews, 118*(10), 4981–5079. https://doi.org/10.1021/acs.chemrev.7b00776
34. Dhiman, R., Durga, V., & Rio, A. J. (2015). Research Article Bioanalytical method for lornoxicam determination in human plasma by using piroxicam as internal standard by LC-MS/MS. *Journal of Chemical and Pharmaceutical Research, 7*(8), 54–60.
35. Zhang, W., Mehta, A., Tong, Z., Esser, L., & Voelcker, N. H. (2021). Development of polymeric nanoparticles for blood–brain barrier transfer – Strategies and challenges. *Advancement of Science, 8*(10), 1–32. https://doi.org/10.1002/advs.202003937
36. Colak, B., & Ertas, Y. N. (2024). Implantable, 3D-printed alginate scaffolds with bismuth sulfide nanoparticles for the treatment of local breast cancer via enhanced radiotherapy. *ACS Applied Materials & Interfaces, 16*(13), 15718–15729. https://doi.org/10.1021/acsami.3c17024
37. Dhiman, R., Kumar, B., Kumar, D., & Kumar, A. (2016). Research Article Liquid-liquid extraction method for Ziprasidone (ZRS) bioanalysis by using ZRS-D 8 (stable isotope) as internal standard. *Journal of Chemical and Pharmaceutical Research, 8*(3), 722–727.
38. Astaneh, M. E., & Fereydouni, N. (2024). Silver nanoparticles in 3D printing: A new frontier in wound healing. *ACS Omega*. https://doi.org/10.1021/acsomega.4c04961
39. Bhattacharya, S., Anjum, M. M., & Patel, K. K. (2022). Gemcitabine cationic polymeric nanoparticles against ovarian cancer: Formulation, characterization, and targeted drug delivery. *Drug Delivery, 29*(1), 1060–1074. https://doi.org/10.1080/10717544.2022.2058645
40. Girma, A., Abera, B., Mekuye, B., & Mebratie, G. (2024). Antibacterial activity and mechanisms of action of inorganic nanoparticles against foodborne bacterial pathogens: A systematic review. *IET Nanobiotechnology, 1*, 2024. https://doi.org/10.1049/2024/5417924
41. Morgan, R. N., & Aboshanab, K. M. (2024). Green biologically synthesized metal nanoparticles: Biological applications, optimizations and future prospects. *Future Science OA, 10*(1). https://doi.org/10.2144/fsoa-2023-0196
42. Kumar, R., Dutt, S., Tripathi, A. D., Singh, A. K., Chaturvedi, V. K., & Singh, S. K. (2024). Navigating safety and toxicity challenges in nanomedicine: Strategies, assessment, and mitigation BT – Nanomedicine: Innovations, applications, and breakthroughs in the quest for health and medicine's future. In V. Gautam, R. Kumar, K. Das Manandhar, & S. C. Kamble (Eds.), (pp. 15–37). Springer. https://doi.org/10.1007/978-3-031-72467-1_2
43. Huynh, K. H., et al. (2020). Synthesis, properties, and biological applications of metallic alloy nanoparticles. *International Journal of Molecular Sciences, 21*(14), 1–29. https://doi.org/10.3390/ijms21145174
44. Kulkarni, D., et al. (2023). Biofabrication of nanoparticles: Sources, synthesis, and biomedical applications. *Frontiers in Bioengineering and Biotechnology, 11*(May), 1–26. https://doi.org/10.3389/fbioe.2023.1159193
45. Upreti, P., Devhare, L. D., Abdulmageed, L. H., Kumar, Y. G., Kumar, R., & Dharmamoorthy, G. (2024). Combatting antibiotic resistance: Leveraging fecal microbial transplantation for gut health BT. In M. Gangwar & G. Nath (Eds.), *Emerging paradigms for antibiotic-resistant infections: Beyond the pill* (pp. 211–232). Springer. https://doi.org/10.1007/978-981-97-5272-0_10
46. Dhir, R., et al. (2023). Plant-mediated synthesis of silver nanoparticles: unlocking their pharmacological potential – A comprehensive review. *Frontiers in Bioengineering and Biotechnology, 11*(January), 1–24. https://doi.org/10.3389/fbioe.2023.1324805
47. A. Hamid et al., "Artificial intelligence (AI) based colonoscopy," 2024. https://doi.org/10.1088/978-0-7503-6134-7CH12.
48. R. Kumar, S. Dutt, A. Goyal, A. Kumar, and B. Kumar, "Nanotechnology in gastrointestinal endoscopy," 2024, doi: https://doi.org/10.1088/978-0-7503-6134-7CH5.
49. Jangid, H., Singh, S., Kashyap, P., Singh, A., & Kumar, G. (2024). Advancing biomedical applications: An in-depth analysis of silver nanoparticles in antimicrobial, anticancer, and wound healing roles. *Frontiers in Pharmacology, 15*(August), 1–25. https://doi.org/10.3389/fphar.2024.1438227

50. Xie, W., & Xu, Z. (2024). (Nano)biotechnological approaches in the treatment of cervical cancer: Integration of engineering and biology. *Frontiers in Immunology, 15*(September), 1461894. https://doi.org/10.3389/fimmu.2024.1461894
51. Fifere, N., et al. (2023). Phytomediated-assisted preparation of cerium oxide nanoparticles using plant extracts and assessment of their structural and optical properties. *International Journal of Molecular Sciences, 24*(10). https://doi.org/10.3390/ijms24108917
52. Gatto, M. S., & Najahi-Missaoui, W. (2023). Lyophilization of nanoparticles, does it really work? Overview of the current status and challenges. *International Journal of Molecular Sciences, 24*(18). https://doi.org/10.3390/ijms241814041
53. Radulescu, D. M., Surdu, V. A., Ficai, A., Ficai, D., Grumezescu, A. M., & Andronescu, E. (2023). Green synthesis of metal and metal oxide nanoparticles: A review of the principles and biomedical applications. *International Journal of Molecular Sciences, 24*(20). https://doi.org/10.3390/ijms242015397
54. Kumar, R., Singh, Y., Piplani, M., Bhateja, P., Garg, R., & Kumar, B. (2024). The chemistry and pharmacological study of Copaifera spp. oils. *The Chemistry inside Spices & Herbs: Research and Development, 4*, 79–116. https://doi.org/10.2174/9789815196832124040006
55. Nicolae-Maranciuc, L. M., Chicea, A., & Chicea, D. (2022). Approche de la composition biochimique de la menthe verte (Mentha spicata Z) dans la région de Ouargla. *University of Ouargla*, 7–8.
56. Inam, H., Sprio, S., Tavoni, M., Abbas, Z., Pupilli, F., & Tampieri, A. (2024). Magnetic hydroxyapatite nanoparticles in regenerative medicine and nanomedicine. *International Journal of Molecular Sciences, 25*(5). https://doi.org/10.3390/ijms25052809
57. Mundekkad, D., & Cho, W. C. (2022). Nanoparticles in clinical translation for cancer therapy. *International Journal of Molecular Sciences, 23*(3), 1–29. https://doi.org/10.3390/ijms23031685
58. Caizer, I. S., & Caizer, C. (2022). Superparamagnetic hyperthermia study with cobalt ferrite nanoparticles covered with γ-Cyclodextrins by computer simulation for application in alternative cancer therapy. *International Journal of Molecular Sciences, 23*(8). https://doi.org/10.3390/ijms23084350
59. A. Kumar, P. Singh, R. Kumar, and S. Dutt, "Novel drug delivery systems for inflammatory bowel disease," 2024, doi: https://doi.org/10.1088/978-0-7503-6134-7CH4.
60. Kumar, A., et al. (2022). Assessment of pain-ameliorative potential of ethanolic extract of Calotropis procera and its active components characterization by HPTLC and GC-MS Analysis, *13*(09), 581–586. https://doi.org/10.31858/0975-8453.13.6.659-664
61. Kučuk, N., Primožič, M., Knez, Ž., & Leitgeb, M. (2023). Sustainable biodegradable biopolymer-based nanoparticles for healthcare applications. *International Journal of Molecular Sciences, 24*(4). https://doi.org/10.3390/ijms24043188
62. Kumar, R., et al. (2021). HPTLC and GC-MS analyses of biologically active extracts and fractions from Premna latifolia Roxb leaves reveal Anti-inflammatory and analgesic effects, *12*(10), 811–818.
63. Hasan, A., Morshed, M., Memic, A., Hassan, S., Webster, T. J., & Marei, H. E. S. (2018). Nanoparticles in tissue engineering: Applications, challenges and prospects. *International Journal of Nanomedicine, 13*, 5637–5655. https://doi.org/10.2147/IJN.S153758
64. Qiu, T., et al. (2021). In vivo multi-scale photoacoustic imaging guided photothermal therapy of cervical cancer based on customized laser system and targeted nanoparticles. *International Journal of Nanomedicine, 16*, 2879–2896. https://doi.org/10.2147/IJN.S301664
65. Yao, H., et al. (2020). A novel therapeutic sirna nanoparticle designed for dual-targeting cd44 and gli1 of gastric cancer stem cells. *International Journal of Nanomedicine, 15*, 7013–7034. https://doi.org/10.2147/IJN.S260163
66. Miyazawa, T., Itaya, M., Burdeos, G. C., Nakagawa, K., & Miyazawa, T. (2021). A critical review of the use of surfactant-coated nanoparticles in nanomedicine and food nanotechnology. *International Journal of Nanomedicine, 16*, 3937–3999. https://doi.org/10.2147/IJN.S298606

67. Bîrcă, A. C., et al. (2024). Infection-free and enhanced wound healing potential of alginate gels incorporating silver and tannylated calcium peroxide nanoparticles. *International Journal of Molecular Sciences, 25*(10). https://doi.org/10.3390/ijms25105196
68. Niżnik, Ł., et al. (2024). Gold nanoparticles (AuNPs) – Toxicity, safety and green synthesis: A critical review. *International Journal of Molecular Sciences, 25*(7). https://doi.org/10.3390/ijms25074057
69. Petrovic, S., Bita, B., & Barbinta-Patrascu, M. E. (2024). Nanoformulations in pharmaceutical and biomedical applications: Green perspectives. *International Journal of Molecular Sciences, 25*(11). https://doi.org/10.3390/ijms25115842
70. Kumar, R., Kumar, B., & Kumar, A. (2017). Pharmacokinetic profile of antitubercular drugs and their clinical applications in therapeutic drug monitoring. *Asian Journal of Pharmaceutical Sciences, 3*(5), 145–154. [Online]. Available: http://ajpp.in/uploaded/p71.pdf
71. Rónavári, A., et al. (2021). Green silver and gold nanoparticles: Biological synthesis approaches and potentials for biomedical applications. *Molecules, 26*(4), 1–39. https://doi.org/10.3390/molecules26040844
72. Dhiman, R., Durga, V., Jayasankar, K., & Raj, S. (2015). Solid phase extraction: Potential carbocisteine bioanalysis using carbocisteine 13c3 stable isotope: LC-MS/MS technique with pharmacokinetics on healthy human subjects. *Der Pharmacia Lettre, 7*(8), 126–134.
73. Stamer, K. S., et al. (2022). Water saturated with pressurized CO2 as a tool to create various 3D morphologies of composites based on chitosan and copper nanoparticles. *Molecules, 27*(21). https://doi.org/10.3390/molecules27217261
74. Dhiman, R., et al. (2015). Novel LC-MS/MS bioanalytical method validation for gabapentin in human plasma in accordance with current regulatory guidances. *International Journal of Pharmaceutical Chemistry Science, 4*(4), 462–468. [Online]. Available: http://www.ijpcsonline.com/files/08-10-15/07-1035.pdf
75. Nowak-Jary, J., Płóciennik, A., & Machnicka, B. (2024). Functionalized magnetic Fe3O4 nanoparticles for targeted methotrexate delivery in ovarian cancer therapy. *International Journal of Molecular Sciences, 25*(16). https://doi.org/10.3390/ijms25169098
76. Marshall, S. K., et al. (2024). Cytotoxicity enhancement in osteosarcoma with multifunctional I-131 radiotherapeutic nanoparticles: In vitro three-dimensional spheroid model and release kinetics modeling. *Molecules, 29*(3). https://doi.org/10.3390/molecules29030630
77. Eker, F., et al. (2024). A comprehensive review of nanoparticles: From classification to application and toxicity. *Molecules, 29*(15), 1 83. https://doi.org/10.3390/molecules29153482
78. Wang, G., et al. (2024). Advances in engineered nanoparticles for the treatment of ischemic stroke by enhancing angiogenesis. *International Journal of Nanomedicine, 19*(May), 4377–4409. https://doi.org/10.2147/IJN.S463333
79. Wang, D., Li, Q., Xiao, C., Wang, H., & Dong, S. (2024). Nanoparticles in periodontitis therapy: A review of the current situation. *International Journal of Nanomedicine, 19*(June), 6857–6893. https://doi.org/10.2147/IJN.S465089
80. Devaraji, M., Thanikachalam, P. V., & Elumalai, K. (2024). The potential of copper oxide nanoparticles in nanomedicine: A comprehensive review. *Biotechnology Notes, 5*(March), 80–99. https://doi.org/10.1016/j.biotno.2024.06.001
81. Wong, K. H., Lu, A., Chen, X., & Yang, Z. (2020). Natural ingredient-based polymeric nanoparticles for cancer treatment. *Molecules, 25*(16), 1–24. https://doi.org/10.3390/molecules25163620
82. Li, X., Bhullar, A. S., Binzel, D. W., & Guo, P. (2022). The dynamic, motile and deformative properties of RNA nanoparticles facilitate the third milestone of drug development. *Advanced Drug Delivery Reviews, 186*, 1–39. https://doi.org/10.1016/j.addr.2022.114316
83. Gloria, S. S., Kang, G. J., Ewing-Nelson, S. R., Mackey, L., Schlitt, J. T., Marathe, A., & Abbas, K. M. (2018). 乳鼠心肌提取 HHS public access. *Physiology & Behavior, 176*(1), 139–148. https://doi.org/10.23736/S1824-4785.17.02969-7.Radiolabeled
84. Brokowski, A. M. C. (2019). 乳鼠心肌提取 HHS public access. *Physiology & Behavior, 176*(5), 139–148. https://doi.org/10.1016/j.biomaterials.2021.120871.3D

85. Zhang, Q., Toprakcioglu, Z., Jayaram, A. K., Guo, G., Wang, X., & Knowles, T. P. J. (2023). Formation of protein nanoparticles in microdroplet flow reactors. *ACS Nano, 17*(12), 11335–11344. https://doi.org/10.1021/acsnano.3c00107
86. Zhong, C., et al. (2025). Inhibition of endothelial cell tube formation by anti-vascular endothelial growth factor/anti-angiopoietin-2 RNA nanoparticles. *Pharmaceutics, 17*(1), 1–17. https://doi.org/10.3390/pharmaceutics17010055
87. Elmowafy, M., et al. (2023). Polymeric nanoparticles for delivery of natural bioactive agents: Recent advances and challenges. *Polymers (Basel), 15*(5), 1–34. https://doi.org/10.3390/polym15051123
88. Van Zundert, I., Fortuni, B., & Rocha, S. (2020). From 2d to 3d cancer cell models – The enigmas of drug delivery research. *Nanomaterials, 10*(11), 1–30. https://doi.org/10.3390/nano10112236
89. Micha, R. (2017). 乳鼠心肌提取 HHS public access. *Physiology & Behavior, 176*(1), 100–106. https://doi.org/10.1177/0022146515594631.Marriage
90. Curcio, A., et al. (2024). Elucidating the dynamics of biodegradation and biosynthesis of magnetic nanoparticles in human stem cells. *Small, 2407034*, 1–14. https://doi.org/10.1002/smll.202407034
91. Bejarano, J., et al. (2018). Nanoparticles for diagnosis and therapy of atherosclerosis and myocardial infarction: Evolution toward prospective theranostic approaches. *Theranostics, 8*(17), 4710–4732. https://doi.org/10.7150/thno.26284
92. Huang, H. Y., et al. (2020). Collagenase IV and clusterin-modified polycaprolactone-polyethylene glycol nanoparticles for penetrating dense tumor tissues. *Theranostics, 11*(2), 906–924. https://doi.org/10.7150/thno.47446
93. Kouhpanji, M. R. Z., & Stadler, B. J. H. (2020). A guideline for effectively synthesizing and characterizing magnetic nanoparticles for advancing nanobiotechnology: A review. *Sensors (Switzerland), 20*(9), 1–37. https://doi.org/10.3390/s20092554
94. Li, N., Khan, S. B., Chen, S., Aiyiti, W., Zhou, J., & Lu, B. (2023). Promising new horizons in medicine: Medical advancements with nanocomposite manufacturing via 3D printing. *Polymers (Basel), 15*(20). https://doi.org/10.3390/polym15204122
95. Krishnan, S. K., Singh, E., Singh, P., Meyyappan, M., & Nalwa, H. S. (2019). A review on graphene-based nanocomposites for electrochemical and fluorescent biosensors. *RSC Advances, 9*(16), 8778–8781. https://doi.org/10.1039/c8ra09577a
96. Ifijen, I. H., et al. (2024). Advancements in tantalum based nanoparticles for integrated imaging and photothermal therapy in cancer management. *RSC Advances, 14*(46), 33681–33740. https://doi.org/10.1039/d4ra05732e
97. Cao, Z., et al. (2023). Cross-scale tracing of nanoparticles and tumors at the single-cell level using the whole-lung atlas. *Science Advances, 9*(31), 1–14. https://doi.org/10.1126/sciadv.adh7779

Ethical and Social Implications of Nanotechnology in the Management of Metabolic Syndrome

14

Bhaskar Sharma, Anjali Yadav, Srishti Jain, and Ravi Kumar Yadav

Abstract

This chapter delves into the ethical and social implications of employing nanotechnology in the management of various strategies such as prevention, diagnosis, and therapeutics of metabolic syndrome (MetS). MetS is obesity, insulin resistance, dyslipidaemia, and hypertension. Though nanotechnology provides innovative approaches relevant to drug delivery, tissue regeneration and diagnostics, it's paramount to address the concerns about its safety, challenges and social view. The present chapter explains insights about how nanoscale innovations are transforming clinical outcomes along with their associated risks. In addition, readers gain insight into the unique ethical challenges that arise in the context of metabolic health, especially related to data privacy, accessibility, informed consent, and public perception. Real-world studies, clinical trial metrics, and statistical evidence are presented to underscore both the promise and potential pitfalls of nanotechnology for its applications in MetS. The chapter concludes with actionable recommendations for responsible innovation in this emerging field.

Keywords

Bioethics · Clinical Nanomedicine · CRISPR-Cas9 · Data Privacy · Nanotechnology · Magnetic Nanoparticles (MNPs) · Metabolic Syndrome (MetS) · Nanosensors · Precision Medicine · Public Health

B. Sharma (✉) · A. Yadav · S. Jain
Department of Anatomy, All India Institute of Medical Sciences, New Delhi, India

R. K. Yadav
Department of Botany, Kashi Naresh Government Post Graduate College, Gyanpur, Bhadohi, Uttar Pradesh, India

V. K. Chaturvedi et al. (eds.), *Nanotheranostics in Metabolic Syndrome*,
https://doi.org/10.1007/978-981-95-3746-4_14

Purpose and Scope of the Chapter

Examining the ethical and social ramifications of treating MetS, a collection of diseases that includes insulin resistance, obesity, hypertension, and dyslipidaemia, with nanotechnology is the aim of this chapter. Nano-technology raises concerns around informed consent, transparency, and potential long-term health impacts, even if it provides novel approaches to treating these associated diseases, such as tailored drug delivery and nanoparticle-based diagnostics. The objective of this chapter is to assess how the scientific potential of nanomedicine interacts with the need for ethical considerations to prevent misuse and unintended harm. The scope covers public perception, regulatory frameworks, and policy-making in addition to technical applications. As nanomedical interventions grow more prevalent, it is critical to comprehend how these technologies may impact different socioeconomic groups, especially in cases of chronic metabolic diseases that need lifelong management. This chapter evaluates the risk-benefit ratio with a focus on how to fairly integrate nanotechnology in healthcare settings without exacerbating pre-existing health inequities. This chapter focuses on MetS, a disorder that is growing indefinitely and is more prevalent globally.

1 Introduction

The term metabolic syndrome (MetS) refers to a group of related metabolic risk factors [24] or metabolic dysregulations [27] that together raise the risk of type 2 diabetes mellitus (T2DM) and atherosclerotic cardiovascular disease (ASCVD). Also, this term refers to a collection of risk factors rather than a disease in and of itself [70]. The combination of multiple factors, including metabolic, environmental, and/or genetic causes, is a common characteristic of MetS. Glucose intolerance or insulin resistance involves the liver in producing more glucose and hinders cells' ability to absorb it [32, 47]. Abdominal obesity, also referred to as visceral obesity or central adiposity [85], is commonly determined by waist circumference [1]. Excessive visceral fat accumulation leads to malfunctioning of adipose tissue [62]. Low levels of high-density lipoproteins (HDL) cholesterol, low-density lipoproteins (LDL) particles, and high triglycerides (TGs) are the hallmarks of dyslipidaemia. Insulin resistance affects the inhibition of hepatic very low-density lipoprotein (VLDL) synthesis, increasing triglycerides [47]. Insulin resistance and hyperinsulinaemia can both contribute to hypertension [108]. Prothrombotic and pro-inflammatory states are other frequently observed features. Excess visceral fat secretes pro-inflammatory cytokines [96]. A person must normally possess a specific number (at least three) of these risk factors from a predetermined list to be diagnosed with MetS [95]. The National Cholesterol Education Program (NCEP), the International Diabetes Federation (IDF), and the World Health Organization (WHO) have all created somewhat distinct diagnosis criteria. For example, the NCEP ATP III criteria can diagnose MetS if at least 3 out of 5 existing criteria, which include high BP, increased fasting glucose, elevated TGs, decreased HDL cholesterol, and higher waist circumference, as shown in Table 14.1 [32]. Abdominal obesity is given

Table 14.1 Diagnostic criteria for metabolic syndrome and environmental concerns associated with nanotechnology

Diagnostic Criteria for Metabolic Syndrome (Based on NCEP ATP III Guidelines)	
Component	**Threshold**
Waist circumference	>102 cm (men), >88 cm (women)
Triglycerides	≥150 mg/dl
HDL cholesterol	<40 mg/dl (men), <50 mg/dl (women)
Blood pressure	≥130/85 mm hg
Fasting glucose	≥110 mg/dl
Environmental concerns associated with nanotech	
Concern	**Description**
Nano-waste	Toxic waste from manufacturing and disposal
Ecotoxicity	Impact on aquatic and soil ecosystems
Bioaccumulation	Potential uptake in the food chain
Public health impact	Long-term exposure effects through air/water

Abbreviations: *HDL* high density lipoproteins, *cm* centimetre, *mg* microgram, *dl* decilitre, *mm Hg* millimetres of mercury

priority in the IDF criteria as a necessary condition for diagnosis [33]. Insulin resistance is a major concern for the WHO [84]. Other names for MetS that have been used in the past are syndrome X, insulin resistance syndrome (IRS), and the "deadly quartet" [70].

MetS is regarded as one of the major health risks of the modern era and has spread to epidemic proportions worldwide [73, 88]. The prevalence of obesity and type-2 diabetes frequently co-exists [42]. In addition, the identification of MetS compounds at an early age in the lifetime of an individual can aid in the prediction of its development in the future. The estimation of the prevalence of MetS in young people is important in decreasing the disease burden later. About one-third of American individuals suffer from MetS [74]. Children in Western nations are increasingly exhibiting MetS, particularly due to a shift in lifestyle [69]. Prevalence estimates exhibit variability contingent on the diagnostic criteria used. For instance, the nationwide survey in Iran revealed varying incidence percentages based on ATP III, IDF, and ATP III/AHA/NHLBI categories [62]. About 6.4–7.5% of MetS prevalence was reported in Iranian young people [43]. In Middle Eastern countries, prevalence is often found to be higher in women than men [65]. In India, the age-adjusted prevalence of MetS is 25%, including an 18.5% prevalence in men and 31% in women. A study reported that 13–15% of adults in India are affected by MetS, where women are more prevalent in comparison to other international studies [7]. In the United States, the incidence of MetS was 35% to 37% higher in the 60–69 age group than in the 20–29 age group [2, 101]. Some ethnic groups, such as African Americans and Mexican Americans, are also reported to have higher rates of MetS [71, 86].

MetS is a significant risk factor for cardiovascular disease (CVD), which encompasses peripheral vascular disease, coronary artery disease, and stroke [37]. The individuals with MetS were found to have double the risk of CVD events

[12, 97]. The components of the syndrome, such as insulin resistance and dyslipidaemia, have a direct impact on the pathophysiology of atherosclerosis and persistent hyperglycaemia, which are crucial processes in CVD and type 2 diabetes. If individuals with MetS do not take the therapy, it leads to an elevated risk of type-2 diabetes and atherogenic cardiovascular diseases (CVDs) [11, 95]. Also, individuals with MetS have a higher risk of type 2 diabetes than those without it. Epidemiological studies have shown that individuals with MetS have a five-fold increased risk of developing type-2 diabetes (T2DM) [40]. The MetS is associated with a 1.5-fold increase in all-cause mortality and a two-fold increase in the risk of CVD [24]. Furthermore, the incidence of MetS is increasing worldwide along with obesity and sedentary lifestyles, increasing the risk of CVD and type-2 diabetes [61]. Numerous studies reported that a sedentary lifestyle and intake of a high-fat diet elevate the risk of MetS [73]. MetS is also reported to be linked with sleep apnoea, polycystic ovary syndrome, fatty liver, and possibly cancer [56]. Public health must address MetS through efficient management and prevention strategies due to its strong correlation with major causes of morbidity and mortality. The key to reducing the high burden of this common illness and reducing cardiovascular and metabolic risks is early detection and therapy, mostly through lifestyle modifications. Nanotechnology is the process of creating functional systems at the nanoscale, typically between 1 and 100 nm. As a result, the medical industry has seen revolutionary breakthroughs like tailored drug delivery systems, tissue engineering for regenerative medicine, and nanosensors for disease diagnosis. Nanoparticles can directly carry drugs to diseased cells, minimising negative effects and increasing therapeutic efficacy. For instance, liposomal and polymeric nanoparticles have shown promising results in the treatment of cancer and chronic diseases due to their ability to get past biological barriers and release therapeutic chemicals in a regulated manner [80]. Furthermore, nanotechnology enables the detection of early illness with highly sensitive biosensors that can detect biomarkers at extremely low concentrations. This is particularly helpful for illnesses whose early symptoms are moderate. For metabolic disorders like diabetes, nanodevices such as glucose nanosensors allow for real-time monitoring and customised therapy regimens. These advancements improve patient outcomes by enabling faster and more accurate diagnosis. As the technology develops, nanomedicine is expected to fundamentally alter the healthcare paradigm from reactive to preventive and personalised [67]. Nanotechnology presents unique ethical problems in healthcare systems due to its complexity, novelty, and ability to completely alter paradigms for diagnosis and treatment. Safeguarding patient privacy, ensuring informed consent in clinical trials utilising nanomaterials, and addressing possible toxicity or undiscovered adverse effects are ethical considerations in the context of nano-enabled biosensors and diagnostic instruments [102]. Furthermore, because more nanomedicines may remain in the body for longer periods, research into their long-term interactions with biological systems is crucial for risk assessment and the safety of patients [46]. From a societal perspective, limiting nanomedicine to affluent groups could unintentionally widen the gap in access to medical treatment. There are concerns about distributive justice and equitable access, especially in

low-resource settings where MetS is on the rise [14]. Given their promise to revolutionise customised treatment, transparent communication, and public involvement are essential to fostering faith in nanotechnology solutions. Policymakers and healthcare professionals must work together to create ethical standards to guarantee that these advances do not compromise social justice or human dignity [100].

2 Nanotechnology Applications in Metabolic Syndrome

Nanotechnology has significantly improved biomarkers for the identification of MetS, enabling personalised monitoring and early diagnosis (Table 14.2). The applications of nanotechnology in diagnostics, therapeutics, monitoring and regenerative medicines are given in Table 14.3. Nanosensors such as carbon nanotubes, gold nanoparticles, and quantum dots allow for the detection of inflammatory cytokines like TNF-α and IL-6, as well as crucial biomarkers like glucose, insulin, and leptin, at extremely low concentrations. By quickly and precisely assessing blood, saliva, or interstitial fluids, these sensors offer a real-time, non-invasive alternative to traditional diagnostic technologies [78]. To increase the accuracy of diagnosing insulin resistance and prediabetes, for instance, fluorescence and electrochemical biosensors based on nanoparticles have been utilised to track cellular

Table 14.2 Benefits and ethical concerns in nanotechnology in metabolic syndrome

Benefits of nanotechnology in metabolic syndrome			
Domain	**Nanotechnology application**	**Benefits**	**Concern**
Diagnostics	Nanosensors for biomarkers	Early detection, non-invasive, high sensitivity	Data interpretation, false positives
Therapeutics	Nano-drug delivery	Targeted therapy, few side effects	Toxicity, off-target effects
Monitoring	Wearable nanodevices	Continue glucose and lipid monitoring	Privacy, reliability
Regenerative medicine	Nano-scaffolds for islet cell regeneration	Potential diabetes reversal	Ethical use of stem cells
Ethical concerns in nanotechnology in metabolic syndrome			
Ethical issues	**description**	**Examples/challenges**	
Safety and risks	Unknown long-term effects of nanomaterials	Bioaccumulation or toxicity from repeated nano-exposure	
Informed consent	Complexity in explaining nanomedical procedures	Patients' misunderstanding of the risks of nano-therapies	
Equity and accessibility	High cost and limited access for vulnerable populations	Urban v/s rural health tech availability	
Regulatory oversight	Lack of clear guidelines specific to nanomedicine	Approval delays or ethical grey zones in clinical trials	
Data privacy	Wearables collecting real-time metabolic data	Potential misuse or hacking of personal health data	

Table 14.3 Examples of nanotechnology applications in metabolic syndrome

S. no.	Nanotech applications	Outcome/matric	References
1	GNP in insulin delivery	2.8-fold increase in insulin absorption	[21]
2	Silica nanoparticles delivering metformin	increased cellular uptake in liver cells	[79]
3	Cerium oxide nanoparticles for oxidative stress in metabolic syndrome (Obese rat model)	reduction in oxidative stress markers (52% MDA, 48% ROS)	[9]
4	PEGylated liposomes for hypertension therapy	Reduction in SBP (38%) and DBP (34%) with nano-formulation	[104]
5	Oral chitosan-Eudragit S100 nanoparticles of liraglutide	Approx 41% encapsulation efficiency, 61% sustained release, and 71% bioavailability with liraglutide compared to subcutaneous injection	[109]
6	Dual-targeted Rosiglitazone nanoparticles (prohibitin + CPP)	Approx. 10% reduction in body weight gain	[38]
7	Prohibitin targeted KLA peptide nanoparticles	Body weight gain reduction of about 10% after 25 days; two-fold increase in angiogenesis in adipose tissue	[39]
8	CeO_2 nanoparticles in adipocytes	Improved insulin sensitivity; adiponectin (Adipoq) and IRS-1 increased by 35–45%, while anti-inflammatory IL-10 increased by 35%.	[57]
9	CeO_2 nanoparticles in a rat model of NAFLD	Hepatoprotective and anti-oxidative effects by reducing key lipid peroxidation by 21–43% and 50% in hepatic fat accumulation	[20]
10	Nifedipine-loaded PLGA nanoparticles in vitro characterisation	Moderate encapsulation efficacy (13–19%) and excellent sustained release capability (100% release over 22–38 days)	[75]
11	Hyaluronate -Hallow gold nanosphere-adipocyte- targeting peptides in C57BL/6 obese mice within just 2 hours of treatment	60% reduction in adipose imaging signal and 20–25% reduction in localised fat	[55]
12	Zinc oxide nanoparticles on obesity-induced CVDs, metabolic syndrome markers and iron overload in high-fat diet rats	Normalised metabolic syndrome markers (reduced BW gain & BMI, TGs, cholesterol, leptin, etc.) and cardioprotective action (reduced CK-MB, LDH, BP, etc.)	[10]

Abbreviations: *GNP* gold nano-particles, *CeO_2* cerium oxide, *MDA* malonaldehyde, *ROS* reactive oxygen species, *SBP* systolic blood pressure, *DBP* diastolic blood pressure, *CPP* cell penetrating peptides, *KLA* pro-apoptotic peptide sequence, *IRS-1* insulin receptor substrate 1, *IL-10* interleukin 10, *NAFLD* non-alcoholic fatty liver disease, *PLGA* poly (lactic-co-glycolic acid), *CVD* cardiovascular disease, *BW* body weight, *BMI* body mass index, *TG* triglycerides, *CK-MB* creatine kinase-myocardial band, *LDH* lactate dehydrogenase, *BP* blood pressure changes in metabolic indications [22]. Additionally, point-of-care (POC) diagnostics based on lab-on-a-chip (LOC) have emerged as useful tools for the treatment of metabolic disorders. These microfluidic systems integrate nanosensors onto tiny platforms that may provide comprehensive metabolic profiles from a single drop of blood or other body fluids. They are particularly useful in remote and underprivileged locations with little access to centralised laboratory facilities. LOC systems can detect biomarkers such as triglycerides and LDL cholesterol, providing fast results and enabling timely action [99]. Due to their ease of use and quick turnaround, nano-enabled POC tests are ideal for routine screening and monitoring of individuals who are at risk of developing MetS. Nanotechnology has revolutionised therapeutic techniques for MetS by enabling targeted and accurate drug delivery. Nano-particles such as liposomes, dendrimers, and polymeric micelles can encapsulate drugs that prevent diabetes or decrease cholesterol and carry them directly to the liver or fat cells. This tailored administration reduces the required dosage and minimises systemic side effects, improving the therapeutic efficacy of medications such as metformin, statins, or thiazolidinediones [106]. Certain nano-carriers are also made to respond to physiological signals like pH or glucose content, so the drug would only be released when needed [23]. Additionally, nanotechnology facilitates the delivery of gene-editing technologies like CRISPR-Cas9 to treat the genetic foundation of MetS. Lipid-based nanoparticles or gold nano-clusters are being used as carriers to introduce CRISPR designs into specific cells to modify genes linked to insulin signalling, lipid metabolism, or inflammation [54]. This approach has the potential to treat or even prevent MetS at the molecular level by correcting underlying genetic defects. Although currently, this approach is in the experimental stage, gene therapy aided by nanotechnology may provide long-term solutions for those who are genetically susceptible [68]. By offering continuous, non-invasive monitoring of critical health markers, wearables incorporating nanotechnology are transforming the treatment of MetS. Nanosensor-equipped smartwatches and skin patches allow for real-time monitoring of blood pressure, heart rate variability, glucose, and hydration levels. To detect changes in perspiration, interstitial fluids, or skin conductivity, these biosensors use nanomaterials like graphene or nano-wires to wirelessly transmit data to mobile apps or healthcare practitioners [49]. This enables timely interventions and empowers patients to take control of their health [48]. In addition to wearables, implantable nanodevices are being developed for dynamic illness management. Smart implants can release insulin or other hormones in response to metabolic stimuli detected by integrated nanosensors. For instance, hormone-regulating devices or insulin pumps with nano-engineered construction that are placed subcutaneously can autonomously adjust their output to maintain metabolic homeostasis without the user's assistance. These closed-loop systems lessen the strain on patients and healthcare systems, marking a significant advancement towards completely automated chronic disease management (Table 14.3).

3 Ethical Implications

One of the most urgent ethical issues regarding the long-term safety of nanomaterials for their use in clinics is still a concern. Table 14.2 represents the ethical concerns associated with the use of nanotechnology in MetS. Unlike traditional drugs, nanomedicines interact in ways not entirely known at the molecular and cellular levels. Concern is growing over the potential for long-term exposure to designed nano-particles to have unforeseen consequences, such as immunological dysregulation, inflammation, or disruption of regular cellular signalling [8, 93]. The small size of nano-particles raises questions about their long-term biocompatibility because it enables them to evade biological barriers and accumulate in organs like the brain, liver, or lungs. Because toxicological assessments have not kept pace with nanotechnology advancements, safety data are insufficient. Numerous nanoparticles exhibit peculiar surface properties that could result in oxidative stress or unintended interactions with biomolecules, potentially affecting medication formulations. Given the potential for bioaccumulation and prolonged retention in tissues, extensive preclinical and long-term post-market surveillance investigations are required [51]. Ethical responsibility requires that these risks be thoroughly evaluated and documented before the use of nanotherapies in vulnerable groups, such as individuals with MetS who may have impaired organ function [31]. The complexity of nanomedicine often makes it difficult for patients to fully comprehend the nature of the interventions they are consenting to. Informed consent must go beyond straightforward disclosures to ensure that patients understand the dangers, how nanotechnology operates, and the remaining uncertainties [89]. Conventional consent forms may not adequately address the nuances of nanotechnological risk, such as unpredictable long-term effects or unpredictable immunological responses [100]. This makes it ethically challenging to get proper informed permission, particularly from populations with limited health literacy. Maintaining patient autonomy while developing cutting-edge technologies presents another difficulty [53, 58]. Although patients have the right to make informed decisions regarding their treatments, they may feel pressured to adopt new and experimental technologies that promise better outcomes. Patients and providers may be discreetly coerced by power imbalances when nanotechnology is promoted as the only innovative answer. Healthcare systems must create transparent communication plans and ethical review procedures that safeguard patient rights and ethically foster innovation [76]. A 2022 pilot trial using lipid-based nanoparticles for insulin delivery in type 2 diabetes patients (a major component of MetS) showed a 32% improvement in glycaemic control but had an opt-in rate of only 56%, including apprehensions over unfamiliar technology. A transparent explanation of long-term risks, nanoscale interactions, and irreversible cellular modifications must be conveyed to patients [63].

The possibility of making already-existing disparities worse is a significant worry with the use of nanotechnology in healthcare [41]. Access to modern nano-diagnostics and treatments may be limited for economically disadvantaged people due to their high cost. For conditions like MetS, this might lead to a two-tiered

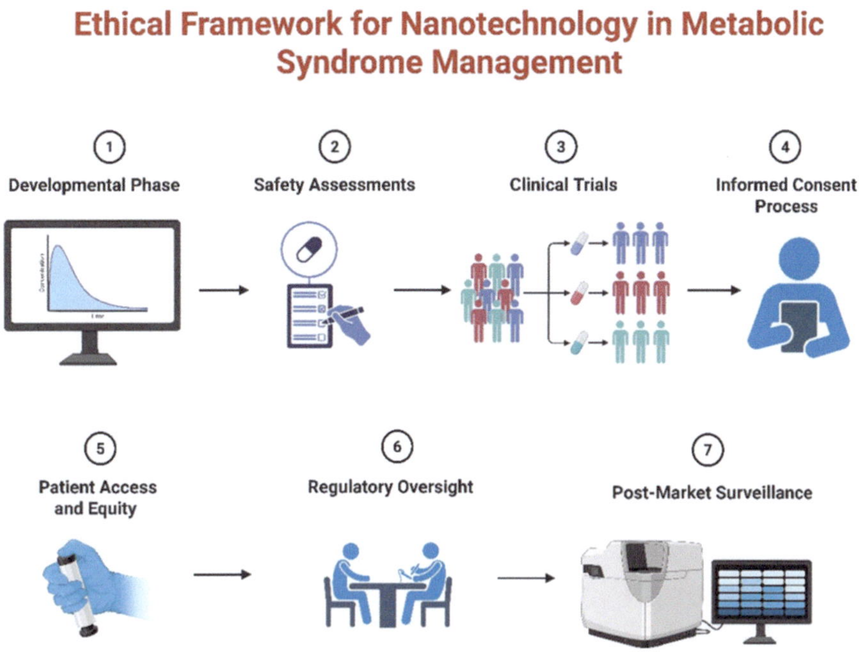

Fig. 14.1 This figure presents a structured ethical framework for nanotechnology-based clinical interventions. It highlights key phases including safety assessments, clinical trials, informed consent, equitable access, and regulatory oversight. Post-market surveillance ensures long-term safety and accountability

system in which only the wealthy have access to early identification, customised medications, or real-time monitoring technologies. Failure to address these differences could lead to a wider health gap between developed and developing regions [13]. For nanomedicine to be used ethically, accessibility and affordability must be integrated into the development process (Fig. 14.1). Public-private partnerships, government subsidies, and open-access platforms can all help democratise access to these innovative technologies. Developing successful and equitable treatments will also be aided by considering the incidence of diseases in a particular location and including a variety of demographics in clinical trials. Without these actions, the promise of nanotechnology may be undermined if it is unable to reach those who need it the most [100]. Despite the rapid advancement of nanomedicine, regulatory frameworks are still in their infancy in many countries. Current biomedical rules often ignore the unique properties and risks of nanoscale materials. This regular ambiguity leaves manufacturers and physicians feeling uneasy, which could result in either overly cautious delays in bringing helpful drugs to market or early approvals without adequate safety data [92]. To evaluate nanotechnology on its terms, regulatory agencies must establish interdisciplinary review panels and precise rules [4]. Clinical trials incorporating nanomedicine present additional ethical challenges, particularly in first-in-human studies. The novelty of these technologies makes risk assessment more challenging for Institutional Review Boards (IRBs), which struggle to weigh the prospective benefits against the unknown risks. Furthermore, oversight systems must ensure that participants are not overwhelmed by risk, particularly when vulnerable groups are involved. Stronger international cooperation and unified policies can help create universal ethical standards for the creation and use of nanomedicine [59]. Nanosensors embedded in wearables or implants offer real-time monitoring of metabolic parameters like blood glucose, cholesterol, and triglycerides. While these enhance predictive care, they also create persistent data flows that may be misused [26]. For example, continuous glucose monitoring nanosystems used in 8000+ prediabetic individuals in the United States raised ethical concerns over third-party data access by insurance companies and employees. Ethical frameworks should enforce stringent data protection measures specific to chronic metabolic disorders [105] (Fig. 14.2).

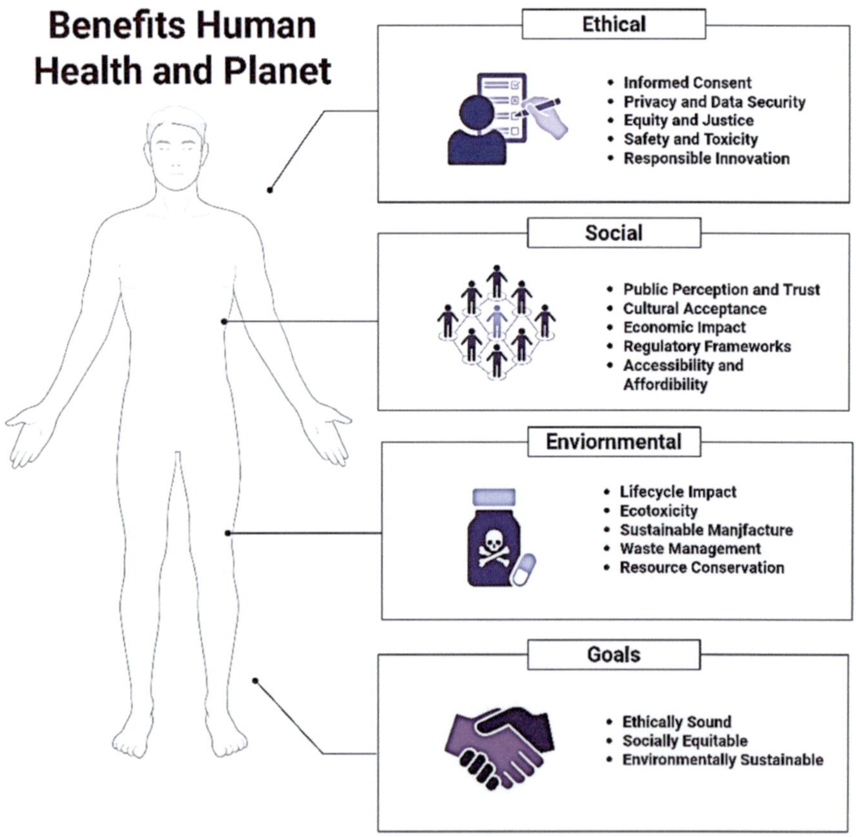

Fig. 14.2 Ethical-social-environmental interplay in nanomedicine for metabolic syndrome. Diagram showing the overlap of ethical, social, and environmental factors in sustainable nanotechnology development for the benefit of human health and the planet as an outcome

4 Social Implications

Determining the pharmacokinetics of nanomedicines is one of the most difficult regulatory tasks. Unlike conventional small-molecule drugs, nanomedicines often deviate from predictable absorption, distribution, metabolism, and excretion (ADME) patterns. This unpredictability can lead to prolonged systemic bioavailability even after a single dose. If such formulations were made available over the counter without professional supervision, long-term exposure or accidental accumulation could present a significant risk to public health. Both regulatory review and public acceptance of nano-based treatments for MetS, which includes diabetes, obesity, and other chronic conditions, are impacted by these safety concerns [29]. A survey of 458 people with Type 1 Diabetes (PWD) and 54 caregivers revealed major impacts, including fatigue (50%), daily hindrance (43%), and anxiety (40%). Caregivers reported higher anxiety about complications than PWD, and distress was linked to hypo/hyperglycaemia. Most PWD (68%) desired more psychological support, and over half lacked sufficient T1D knowledge. Key coping strategies included acceptance, positivity, and self-organisation. To sum up, T1D continues to place a heavy burden on PWDs and caregivers, necessitating the provision of psychological and educational support for T1D management [81]. Public education campaigns focused on metabolic health applications are essential to build trust.

The safety and toxicity of the nanoparticles utilised are major concerns that affect public trust in nanomedicine. Contrary to popular assumption, not all biodegradable nanomaterials are intrinsically safe for use in medical contexts. For instance, although being biodegradable, poly-alkyl-cyanoacrylate (PAC) breaks down into potentially hazardous metabolites. This demonstrates how physicochemical properties like particle size, surface charge, and chemical composition can cause unexpected negative effects even in materials intended for safe disintegration [3]. Compared to conventional nanoparticles, biodegradable ones may be less dangerous. Their long-term impacts and any potentially dangerous payloads delivered during biodegradation are still unknown, though. If we do not have enough safety rules and robust toxicity tests in place, public trust may erode. Consequently, producers and regulators have increased moral obligations to guarantee the creation and use of reliable, safe, and efficient nanomedicines for long-term conditions such as MetS [3]. The emergence of social media is noted to have drastically changed how the public absorbs information. Following trends in newspaper coverage, the talk about nanotechnology on social media tends to be positive. Attention to social media is suited alongside newspapers to understand how they might affect an individual's nanotechnology knowledge and opinions. Most research on media discourse and social media influencing nanotechnology knowledge is seen from the perspective of Western countries [83]. Nanotechnology is the preferred method for improving metabolic syndrome, which is a significant global health challenge, increasing the risk of type 2 diabetes and cardiovascular diseases. This technology aims to improve the treatment options. In the case of Semaglutide, a GLP-1 receptor agonist (Ozempic) TikTok trend, social media posts highlighting personal experiences led to a surge in demand for the drug for off-label weight loss use. This trend is driven by widespread visibility (hundreds of millions of views), compromising the drug's availability for people with type 2 diabetes, for whom it is approved, and leading to global shortages [17]. Twitter can amplify negative sentiment and attention around specific events or concerns related to nano-silver during the Ebola scare, and outrage/negative sentiment around the use of titanium dioxide (nanomaterial) in food (like Dunkin Doughnuts) or its classification as a carcinogen. These instances demonstrate how social media can quickly create a "Twitterstorm" around perceived risks, overwhelming expert voices, and potentially leading to real-world consequences like regulatory action or impacting public confidence [66].

Managing and preventing metabolic syndrome increasingly relies on the "Big Data" approaches. These approaches utilise vast amounts of information from diverse sources. This data includes, but is limited to, omics data (such as genomics, proteomics, epigenomics, and metabolomic information). Nanotechnology-based breath analysis using nanosensors contributes to gathering biological markers, including those related to metabolic products [52]. Clinical practice data, including electronic health records, medical imaging (like CT, MRI, and Ultrasound), and laboratory results [45]. Data from wearable devices and IoT, including flexible gaseous sensors, are used for real-time monitoring of physiological parameters and biomarkers, which feed into this "Big Data" landscape. Nanotechnology is key in developing these sensors [107]. The collection, storage, processing, and analysis of this "Big Data", which is inherently personal and sensitive, poses new ethical and legal challenges. Major technical and infrastructural challenges are data protection and information risk. The information leak is described as an essential challenge in "Big Data" research, whether in omics studies or clinical data mining, and it can cause a significant violation of privacy [77]. Various security and privacy technologies are adopted to protect healthcare data. These measures include authentication, encryption, data masking, access control, and monitoring and auditing. However, despite the adoption of these technologies, there is information leakage in omics studies, such as genomics information, despite efforts at data sanitisation. Challenges exist with clinical data from clinical records, including potential confounding factors and problematic missing data, which can impact security and privacy efforts. A current significant problem is the lack of a unified protocol for the integration of data from different dimensions within "Big Data" approaches for metabolic syndrome. The diversity and absence of standardized protocols can hinder the success of overall security initiatives. Public attitudes towards the reuse of health data show support when it benefits the common good, but concerns arise regarding the commercialisation of data, data security, and the potential for data to be used against the interests of the individuals providing it. The government need to establish affirmative policies to protect health data confidentiality, privacy, and security, while also enabling scientific advancements. Achieving effective governance of health-related data is considered essential for future healthcare systems [64].

Wearable health monitoring devices (HMWDs) collect continuous data on physiological parameters like heart rate, respiration, and physical activity. Nanotechnology-enabled sensors are used in breath analysis to detect biomarkers such as volatile organic compounds (VOCs) for diseases like diabetes (e.g., acetone) [52]. This generates a significant amount of personal health data [15]. Although more convenient, technologies like Bluetooth, WiFi, and NFC also raise the possibility of signal interception and hacking. Furthermore, if data protection measures are insufficient, wearable sensors with flexible circuits and digital signal processing units may process raw health data locally or remotely, increasing the hazards. Maintaining public confidence in nanotechnology-based health monitoring systems and protecting individual privacy requires robust encryption, secure wireless protocols, and open data governance frameworks. A major concern is the risk of information leaks and personal data breaches, including unintentional disclosures or unauthorised access by third parties. Data security is considered a critical technical and infrastructural challenge. There is a risk of data commercialisation, exploitation, and misuse. This includes data being sold or used for data mining to create customer profiles for targeted advertising without user consent. It can also lead to potential discriminatory policies regarding employment, credit, or health insurance. Concerns about surveillance are also noted, particularly for marginalised groups. This ambiguity between commercial and medical domains can challenge data security and trust [19]. Users may have inadequate awareness and control over the use of their data. Wearables manufacturers should provide clear privacy policies so users are informed and can withdraw their data. Information on how data is collected, classified, and interpreted is often not transparently shared, even with researchers [18]. Existing regulations may not specifically cover all types of wearable data unless a healthcare partner is involved.

Integration of AI (machine learning and deep learning) and big data (genomics, transcriptomics, epigenomics, proteomics, and metabolomics) in nanomedicine raises specific ethical issues related to patient safety, accuracy, and personalisation [52]. It is crucial to guarantee data accuracy, traceability, and informed consent due to the substantial datasets and nanoscale manipulations involved. Transparency is necessary when using algorithmic decision-making ethically, particularly when using AI to develop or administer treatments based on nanoparticles. The extensive collection of genetic and physiological data raises privacy concerns. To avoid inequalities and maintain patient confidence, it is also crucial to prioritise proper data management and fair access to nanomedicine advancements (Harishbhai [36]). In the context of omics data analysis using deep learning (DL), the lack of interpretability ("black box") of many machine learning (ML) techniques is a different issue, hindering their growth in healthcare. Final users, such as clinicians, need to understand why a prediction is made to build trust and make decisions. Furthermore, many studies using DL in omics studies analysis lack supportive validation from a medical perspective, relying only on theoretical pieces of evidence [64].

Religiosity and spirituality (R/S) have been investigated for their association with MetS. Some R/S traditions offer faith-based guidance on health behaviours like diet and exercise that aligns with lifestyle recommendations relevant to MetS management. R/S may also provide coping mechanisms for stress, which can physiologically impact health [60]. Studies showing the link between R/S and MetS have been conducted, particularly in Hispanic/Latino populations where R/S is an integral aspect of the culture. Findings have been mixed; one study in Hispanic women found that high religious faith was associated with lower incidence rates of MetS compared to low faith, while another study in a broader Hispanic/Latino sample found largely non-significant associations with MetS, although some weak inverse associations with BP were observed [16]. Culturally sensitive approaches are essential for effective diabetes self-management education and support (DSMES), particularly as Type 2 Diabetes (T2D) disproportionately affects racial and ethnic minorities. Cultural and social determinants of health influence lifestyle modifications, access to technology, and patient-provider communications in diabetes management. Providers need cultural competency and humanity to effectively engage with diverse patients [87]. Religious perspectives have a significant influence on how the general public perceives nanotechnology; extremely religious people frequently see it as "playing God" or disrupting the natural order, which can lead to moral dilemmas and the designation of the technology as immoral. Compared to more secular nations like Sweden or Germany, studies indicate that more religious societies-such as the US, Italy, and Ireland, are more likely to consider nanotechnology to be immortal. When assessing the risks linked with nanotechnology, people often utilise their religious beliefs as heuristic cues, particularly among ordinary populations. Nonetheless, research from Germany and Malaysia revealed that religiosity had little to no impact on risk perception, indicating that regional variations exist in the impact of religiosity [83].

5 Environmental Considerations

The rapid advancement of nanotechnology and the production and disposal of nanomedicines result in significant amounts of nano-waste. Table 14.1 represents the environmental concerns associated with the use of nanotechnology. It is estimated that between 63% and 91% of engineered nanomaterials produced globally end up in landfills, where they present unknown environmental and health risks. Research labs and industries are having issues as a result. Challenges in working with different, often untested nanomaterials that have complex physicochemical properties. Certain guidelines for the classification, labelling, storage, and disposal of nano-waste may be necessary because conventional waste management practices may not be adequate. With an emphasis on regulation based on the potential for hazards and the behaviour of nanomaterials in the environment, a cautious approach is essential [90]. According to recent studies, nanowaste may be ecotoxic and bioaccumulative to humans. Nanowaste, which includes engineered nanomaterials released during the production, application, and disposal phases, can enter the human body through ingestion, inhalation, and skin contact. Once inside, nanoparticles may accumulate in organs like the liver, lungs, and kidneys, which could be harmful to health. Among these effects are oxidative stress, inflammation, DNA damage, and metabolic pathway disruption. Bundschuh et al. reviewed the toxicological effects of nanoparticles, including their bioaccumulation and associated health risks. However, according to some research, nanoplastics can disrupt metabolic functions and eventually cause organ dysfunction. This emphasises the need for comprehensive risk assessments and regulatory measures to limit exposure to nanowaste. The disposal of micro- and nanoplastics (MNPs) prioritises preserving public health and reducing adverse environmental impacts. Scientists are searching for effective ways, like employing substances and living things that can decompose plastics. For example, marine fungi like *Zalerion maritimum* and mixed bacterial cultures have shown promise in the breakdown of plastic. Enzymatic treatments, such as those involving cutinase variants, have been demonstrated to accelerate the degradation of polyester plastics like PET, suggesting that they could be a useful plastic treatment technique. Although biodegradable plastics help reduce long-term pollution, they still contribute to MNPs during composting, so more research is required. MNPs like nano-Fe_3O_4 particles and micro polyacrylate styrene can reduce indoor air pollution through vacuum-gasification-condensation techniques. However, in rapidly developing, densely populated countries like China and India, implementing comprehensive sewage treatment processes is essential to lowering MNP emissions. These disposal methods contribute to sustainable development by lessening the adverse effects of MNPs on the environment and public health [44]. The effects of these technologies extend beyond human health, even though the primary ethical and social considerations about nanomedicines in healthcare are their safety and acceptability. As the use of nanomaterials increases, it is necessary to conduct a thorough assessment of their environmental impact, particularly concerning waste management and ecosystem impacts. Understanding these environmental risks is essential to ensuring that the benefits of nanotechnology do not come at a higher ecological cost.

6 Ethical Frameworks and Guidelines

The global regulatory landscape for nanomedicine (NTc) is diverse and ever-evolving, with different regions implementing distinct approaches to the regulation of nanomaterials (NMs) and nanomedicines (NMc). In the USA, the FDA has published guidelines for NMc that prioritise safety and effectiveness on a case-by-case basis. The European Medicines Agency (EMA) has proposed special nanomedicine guidelines in Europe that emphasise physicochemical characterisation, toxicology, and exposure conditions. In the UK, NMs are governed by general chemical laws, which emphasise the potential unique toxicity of nanomaterials. Australia regulates NMc under the current therapeutic goods laws, placing a strong emphasis on safety and risk management. Countries in Asia, including Japan and India, are creating NTc regulatory frameworks. Japan is focusing on limiting exposure, while India has provided guidelines for the efficacy and safety of nanopharmaceuticals. Canada's regulatory approach involves risk assessment and collaboration with international organisations through Health Canada and other ministries. International organisations like the ICH, ISO, and WHO are attempting to harmonise regulatory standards to address common issues in nanomedicine and promote coordinated efforts for global safety. Biomedical ethics is based on four core principles: beneficence (promoting the patient's welfare), autonomy (respecting a patient's right to make informed decisions), nonmaleficence (avoiding harm), and justice (ensuring an equitable distribution of healthcare resources). These principles guide ethical clinical judgement and promote patient-centred care. They have their roots in both ancient traditions and modern philosophical concepts, and they serve as the moral cornerstone for physicians navigating complex medical and social situations [98]. A neat and tidy, ethical approach to creating nanomedicine is offered by the Frameworks for Responsible Research and Innovation (RRI): Early involvement of a variety of stakeholders, including scientists, ethicists, patients, and legislators, is highly valued by RRI. This ensures that technologies align with societal values and requirements. Part of ethical design is anticipating potential privacy and health risks, like those posed by AI-enabled nanodevices that collect private health data. Recognising that exclusivity and patents may lead to unjust monopolies in the field of nanomedicine, RRI advocates for a fair distribution of benefits. Novel approaches, like decentralised data storage and portable nanodevices, are proposed to safeguard accessibility and privacy [103].

7 Applications of Nanotech Based on Quantitative Evidence Through Case Studies and Real-World Examples

Metabolic diseases are clinical outcomes of several diseases such as obesity, insulin resistance, high blood pressure, and dyslipidaemia (Table 14.4). Usually, drugs are used for the treatment of blood pressure, hypoglycaemia, and other factors; however, these drugs cause side effects such as bloating in patients, resulting in reduced ability for drug tolerance [5]. However, nanotechnology is a choice of method due to its high stability, biological distribution, and increased solubility of compounds. For instance, thymoquinone, which was found in *Nigella sativa* (Liliaceae family), was

Table 14.4 Success and failure in nanotechnology applied to metabolic syndrome

Disease	Mechanism	Improved success rate	References
T2DM	Polymeric nanocapsules of thymoquinone and metformin	Yes	[82]
Dyslipidaemia	Nano-curcumin, nano-TiO, and GDNP	Yes	[6]
Diabetes and diabetic wounds	QCT-AgNPs	Yes	[7]
Obesity	SC-AuNPs	Yes	[30]
Cardiovascular	PLGA, IOs, AuNPs, CNTF, and GO	Yes	[72]
Myocardial infarction	IONPs	Yes	[35]

Abbreviations: *T2DM* Type 2 diabetes mellitus, *TiO* titanium oxide, *GDNP* ginger-extracted nanoparticles, *QCT-AgNPs* silver nanoparticles created quercetin, *SC-AuNPs* Salacia chinensis-containing gold nanoparticles, *PLGA* poly lactic-co-glycolic acid, *IOs* iron oxide nanoparticles, *AuNPs* nanoscale gold particles, *CNTF* carbon nanotube fibres, *GO* graphene oxide, *IONPs* iron oxide nanoparticles found effective for respiratory and diabetes diseases, specifically for coronary artery disease, using in vivo and in vitro experiments [28]. However, this drug was not qualified for clinical trials; by employing the nanoprecipitation method, the nanocapsules were found to have a better hypoglycaemic effect than thymoquinone alone [82]. In addition, one of the severe consequences of diabetes is amputation due to wounds and foot ulcers, affecting 15% of diabetic cases. In these cases, nanotechnology has been applied for enhanced performance with the application of control of the uniform distribution of stable and specific compounds at a defined site to cure wounds by employing hydrogels with silver nanoparticles created by Quercetin (QCT-AgNPs). An in vivo study by Badhwar et al. demonstrated that approximately. 92.09% of QCT was entrapped in QCT-AgNPs hydrogel matrices that significantly ($p < 0.001$) diminished the wound gap and enhanced the percentage of re-epithelialisation in an excisional diabetic wound model [7]. Furthermore, obesity is another vital cause of many metabolic disorders. A study by Zu et al. also demonstrated the targeted delivery of resveratrol using ligand-coated R-encapsulated nanoparticles (L-Rnano) in obese mice, which significantly stimulates ASCs differentiation into beige adipocytes, resulting in reduced fat mass by up to 40% along with improved glucose [110]. Dyslipidaemia is caused by atypical changes in the composition of lipids, such as low-density lipoproteins (LDL), high-density lipoproteins (HDL), triglycerides and cholesterol [94]. The drugs used for dyslipidaemia are simvastatin, pravastatin, lovastatin, and atorvastatin; however, the use of these drugs is limited by their side effects, including myalgia and rhabdomyolysis. Also, ginger-extracted nanoparticles (GDNP) and nano-curcumin have been employed by other studies for the treatment of dyslipidaemia [6, 50]. Another study was developed in connection with a participatory action research (PAR) program to enhance the health behaviours of adult women with MetS risk factors. The use of PAR led to significant improvements in health-related quality of life, social support, and empowerment. Health behaviours like reading nutrition labels and exercising also improved after the intervention. Numerous indices related to metabolism also showed notable improvements. According to these results, PAR can help people who are at risk for Mets improve their health outcomes and encourage changes in behaviours related to their health [25].

8 Conclusion and Future Directions

The diagnosis, management, and treatment of MetS could be revolutionised by nanotechnology. To guarantee that it benefits humanity fairly and sustainably, its implementation must be directed by strong ethical frameworks, inclusive laws, and active societal involvement. Future nanomedicine research should concentrate on targeted nanoparticle (NP) design challenges, such as enhancing the efficacy, predictability, and reproducibility of NP conjugation. The development of flexible, multifunctional platforms with improved targeting moieties and effective delivery systems is crucial. Artificial intelligence (AI) and machine learning (ML) together have the potential to revolutionise the design process by predicting NP behaviour, off-target effects, and therapeutic efficacy. Standardising experimental data and promoting cross-centre collaboration will accelerate the development of targeted nanomedicine.

9 Summary

This study explores the ethical and social dimensions of applying nanotechnology to the management of metabolic syndrome (MetS), a combination of interrelated metabolic disorders including obesity, insulin resistance, dyslipidaemia, and hypertension. With its potential to revolutionise early diagnosis, targeted drug delivery, and real-time biomonitoring, nanotechnology presents a promising avenue for personalised interventions. However, the rapid development and deployment of nano-based medical solutions raise significant ethical challenges, such as the need for informed consent, patient autonomy, safety evaluation, and data protection. Moreover, social implications such as unequal access to advanced therapies, cost barriers, and lack of awareness in low-resource settings further complicate the adoption of these technologies. The study underscores the need for interdisciplinary collaboration, involving clinicians, bioethicists, policymakers, and patient communities to ensure that the integration of nanotechnology in metabolic syndrome management is ethically sound, socially equitable, and environmentally responsible.

Glossary

Accessibility Making sure everyone, including people with disabilities or limited resources, can use healthcare products or services.

Artificial Intelligence (AI) Technology that enables machines to think, learn, and make decisions like humans.

Atherosclerotic Cardiovascular Disease (ASCVD) A condition where cholesterol and other substances build up in blood vessels, increasing the risk of heart attacks and strokes.

Bioaccumulation When harmful substances (like chemicals or toxins) build up in living organisms over time.

Biocompatibility How well a material or device works in the human body without causing harm.

Biosensors A device that detects and measures substances (like glucose or toxins) using biological parts like enzymes or antibodies.

CRISPR-Cas9 A powerful tool used to edit genes (DNA) accurately and efficiently.

Cardiovascular Disease (CVD) Health conditions that affect the heart and blood vessels.

Data Privacy Protecting people's personal and medical information from being shared or misused.

Ecotoxicity Harmful effects of chemicals or materials on the environment, especially on animals and plants.

Electrochemical biosensors Sensors that use electrical signals to detect biological or chemical substances.

Encapsulation A method of covering drugs or particles to protect them and control how they are released in the body.

Equity Fair access to healthcare and resources for all, regardless of background or status.

Ethics Principles of right and wrong used to guide actions and decisions.

Fluorescence biosensors Sensors that use fluorescence light to detect specific substances in the body.

Informed Consent Making sure patients fully understand a treatment or research before agreeing to take part.

International Conference on Harmonisation (ICH) A global group that makes rules for testing medicines to ensure safety and quality.

International Standardisation Organisation (ISO) An organisation that creates international standards for products and services, including health and safety.

Insulin Resistance Syndrome (IRS) A condition where the body does not respond well to insulin, leading to high blood sugar levels.

Lab-on-a-Chip (LOC) Tiny devices that can perform many laboratory tests quickly on a small chip.

Magnetic nanoparticles (MNPs) These are tiny magnetic nanoparticles made from either iron, cobalt, or nickel and having a size of less than 100 nm.

Metabolic Syndrome (MetS) MetS is a network of interconnected factors linked with an increased risk of cardiovascular diseases (CVD), coronary heart disease (CHD), and type 2 diabetes mellitus (T2DM).

Microfluidic system A small system that moves tiny amounts of fluid through channels, often used for quick lab tests or diagnostics.

Nanosensors surveillance Using tiny sensors to monitor health, environment, or safety in real-time.

Nanosensors Nanosensors are small sensors size of 1–100 nm utilised for the detection and measurement of different chemical, physical and biological parameters.

Nanotechnology Science of working with materials at a very small scale (nanometre) to create new medical or industrial products.

Patient Autonomy The rights of patients to make their own decisions about their health and treatments.

Point-of-Care (POC) diagnostics Medical tests done at or near where the patient is (e.g., in clinics, homes), giving fast results.

Policy Rules and guidelines created by governments or organisations to manage healthcare or research.

Post-market surveillance Monitoring the safety and performance of the product after it is released to the public.

Preclinical The stage of research done in a laboratory or on animals before testing a drug or device on humans.

Prevalence How common a disease or condition is in a certain population at a certain time.

Regenerative Medicines Treatments that help repair or replace damaged tissues or organs using cells, genes, or engineered materials.

Type 1 Diabetes Mellitus (T1DM) A condition where the immune system of the body destroys insulin-producing cells in the pancreas.

Type 2 Diabetes Mellitus (T2DM) A condition where the body does not use insulin properly, causing high blood sugar levels.

World Health Organisation (WHO) A global health organisation that works to improve public health and set international health standards.

References

1. Abou Ziki, M. D., & Mani, A. (2016). Metabolic syndrome: Genetic insights into disease pathogenesis. *Current Opinion in Lipidology, 27*(2), 162–171.
2. Ahluwalia, N., Raghavan, R., Zhang, G., Talegawkar, S. A., & Jacques, P. F. (2023). Vitamin D status and prevalence of metabolic syndrome by race and Hispanic origin in US adults: Findings from the 2007–2014 NHANES. *The American Journal of Clinical Nutrition, 116*(5), 1400.
3. Albalawi, F., Hussein, M. Z., Fakurazi, S., & Masarudin, M. J. (2021). Engineered nanomaterials: The challenges and opportunities for nanomedicines. *International Journal of Nanomedicine, Volume 16*, 161–184.
4. Amutha, C., Gopan, A., Pushbalatatha, I., Ragavi, M., & Reneese, J. A. (2024). Nanotechnology and governance: Regulatory framework for responsible innovation. In *Nanotechnology in societal development* (pp. 481–503). Springer Nature Singapore.
5. Anderson, M., Powell, J., Campbell, K. M., & Taylor, J. R. (2014). Optimal management of type 2 diabetes in patients with increased risk of hypoglycemia. *Diabetes, Metabolic Syndrome and Obesity: Targets and Therapy, 7*, 85–94.
6. Ashtary-Larky, D., Rezaei Kelishadi, M., Bagheri, R., Moosavian, S. P., Wong, A., Davoodi, S. H., et al. (2021). The effects of nano-curcumin supplementation on risk factors for cardiovascular disease: A GRADE-assessed systematic review and meta-analysis of clinical trials. *Antioxidants, 10*(7), 1015.
7. Badhwar, R., Mangla, B., Neupane, Y. R., Khanna, K., & Popli, H. (2021). Quercetin loaded silver nanoparticles in hydrogel matrices for diabetic wound healing. *Nanotechnology, 32*(50), 505102.
8. Barik, P., & Mondal, S. (2025). Immunomodulatory effects of metal nanoparticles: Current trends and future prospects. *Nanoscale, 17*, 10433.
9. Bashandy, S. A., Elbaset, M. A., Ibrahim, F. A., Abdelrahman, S. S., Moussa, S. A. A., & El-Seidy, A. M. (2025). Management of cardiovascular disease by cerium oxide nanoparticles via alleviating oxidative stress and adipokine abnormalities. *Scientific Reports, 15*(1), 5709.
10. Bashandy, S. A., El-Seidy, A. M., Ibrahim, F. A., Abdelrahman, S. S., Abdelmottaleb Moussa, S. A., & ElBaset, M. A. (2023). Zinc nanoparticles ameliorated obesity-induced cardiovascular disease: Role of metabolic syndrome and iron overload. *Scientific Reports, 13*(1), 16010.
11. Bauer, J., Hegewald, J., Rossnagel, K., Jankowiak, S., Prigge, M., Chalabi, J., et al. (2025). Incidence of type 2 diabetes and metabolic syndrome by occupation–10-year follow-up of the Gutenberg Health Study. *BMC Public Health, 25*(1), 502.

12. Bazmandegan, G., Abbasifard, M., Nadimi, A. E., Alinejad, H., & Kamiab, Z. (2023). Cardiovascular risk factors in diabetic patients with and without metabolic syndrome: A study based on the Rafsanjan cohort study. *Scientific Reports, 13*(1), 559.
13. Benfer, E. A., Mohapatra, S., Wiley, L. F., & Yearby, R. (2019). Health justice strategies to combat the pandemic: Eliminating discrimination, poverty, and health disparities during and after COVID-19. *Yale Journal of Health Policy, Law & Ethics, 19*, 122.
14. Bhattacharya, B., Roy, P., Bhattacharya, S., Prasad, B., & Mandal, A. K. (2023). Nanotechnology and sustainable development: Overcoming the obstacles by adopting ethical practices for future farming. In *Engineered nanomaterials for sustainable agricultural production, soil improvement and stress management* (pp. 431–445). Academic Press.
15. Bouderhem, R. (2023). Privacy and regulatory issues in wearable health technology. *Engineering Proceedings, 58*(1), 87.
16. Brintz, C. E., Birnbaum-Weitzman, O., Llabre, M. M., Castañeda, S. F., Daviglus, M. L., Gallo, L. C., et al. (2017). Spiritual well-being, religious activity, and the metabolic syndrome: results from the Hispanic Community Health Study/Study of Latinos Sociocultural Ancillary Study. *Journal of Behavioral Medicine, 40*, 902–912.
17. Burki, T. (2022). Social media and misinformation in diabetes and obesity. *The Lancet Diabetes & Endocrinology, 10*(12), 845.
18. Canali, S., Schiaffonati, V., & Aliverti, A. (2022). Challenges and recommendations for wearable devices in digital health: Data quality, interoperability, health equity, fairness. *PLOS Digital Health, 1*(10), e0000104.
19. Capulli, E., Druda, Y., Palmese, F., Butt, A. H., Domenicali, M., Macchiarelli, A. G., et al. (2025). Ethical and legal implications of health monitoring wearable devices: A scoping review. *Social Science & Medicine, 370*, 117685.
20. Carvajal, S., Perramón, M., Oró, D., Casals, E., Fernández-Varo, G., Casals, G., et al. (2019). Cerium oxide nanoparticles display antilipogenic effect in rats with non-alcoholic fatty liver disease. *Scientific Reports, 9*(1), 12848.
21. Chen, G., Kang, W., Li, W., Chen, S., & Gao, Y. (2022). Oral delivery of protein and peptide drugs: From non-specific formulation approaches to intestinal cell targeting strategies. *Theranostics, 12*(3), 1419.
22. Das, R., Nag, S., & Banerjee, P. (2023). Electrochemical nanosensors for sensitization of sweat metabolites: from concept mapping to personalized health monitoring. *Molecules, 28*(3), 1259.
23. Dastidar, D. G., Ghosh, D., Ghosh, S., & Chakrabarti, G. (2023). Current therapeutic strategies and possible effective drug delivery strategies against COVID-19. *Current Drug Delivery, 20*(10), 1441–1464.
24. Dhondge, R. H., Agrawal, S., Patil, R., Kadu, A., & Kothari, M. (2024). A comprehensive review of metabolic syndrome and its role in cardiovascular disease and type 2 diabetes mellitus: Mechanisms, risk factors, and management. *Cureus, 16*(8).
25. Eo, Y. S. (2021). Effects of the participatory action research on reducing the risk of metabolic syndrome in adult women. *International Journal of Environmental Research and Public Health, 18*(21), 11103.
26. Erdem, A., Eksin, E., Senturk, H., Yildiz, E., & Maral, M. (2024). Recent developments in wearable biosensors for healthcare and biomedical applications. *TrAC Trends in Analytical Chemistry, 171*, 117510.
27. Fahed, G., Aoun, L., Bou Zerdan, M., Allam, S., Bou Zerdan, M., Bouferraa, Y., & Assi, H. I. (2022). Metabolic syndrome: updates on pathophysiology and management in 2021. *International Journal of Molecular Sciences, 23*(2), 786.
28. Farkhondeh, T., Samarghandian, S., & Borji, A. (2017). An overview on cardioprotective and anti-diabetic effects of thymoquinone. *Asian Pacific Journal of Tropical Medicine, 10*(9), 849–854.

29. Foulkes, R., Man, E., Thind, J., Yeung, S., Joy, A., & Hoskins, C. (2020). The regulation of nanomaterials and nanomedicines for clinical application: current and future perspectives. *Biomaterials Science, 8*(17), 4653–4664.
30. Gao, L., Hu, Y., Hu, D., Li, Y., Yang, S., Dong, X., et al. (2020). Anti-obesity activity of gold nanoparticles synthesized from Salacia chinensis modulates the biochemical alterations in high-fat diet-induced obese rat model via AMPK signaling pathway. *Arabian Journal of Chemistry, 13*(8), 6589–6597.
31. Garg, A., Pandey, P., Singh, V., & Ashique, S. (2025). Ethical considerations of cancer immunotherapy and patient perspectives on the benefits and risks of treatment. In *Nanotechnology based strategies for cancer immunotherapy: Concepts, design, and clinical applications* (pp. 419–438). Springer Nature Singapore.
32. Ginsberg, H. N., & MacCallum, P. R. (2009). The obesity, metabolic syndrome, and type 2 diabetes mellitus pandemic: Part I. Increased cardiovascular disease risk and the importance of atherogenic dyslipidemia in persons with the metabolic syndrome and type 2 diabetes mellitus. *Journal of the Cardiometabolic Syndrome, 4*(2), 113–119.
33. Gundogan, K., Bayram, F., Gedik, V., Kaya, A., Karaman, A., Demir, Ö., et al. (2013). Metabolic syndrome prevalence according to ATP III and IDF criteria and related factors in Turkish adults. *Archives of Medical Science, 9*(2), 243–253.
34. Anwer, A. H., Saadaoui, M., Mohamed, A. T., Ahmad, N., & Benamor, A. (2024). State-of-the-art advances and challenges in wearable gas sensors for emerging applications: Innovations and future prospects. *Chemical Engineering Journal, 502*, 157899.
35. Han, J., Kim, B., Shin, J. Y., Ryu, S., Noh, M., Woo, J., et al. (2015). Iron oxide nanoparticle-mediated development of cellular gap junction crosstalk to improve mesenchymal stem cells' therapeutic efficacy for myocardial infarction. *ACS Nano, 9*(3), 2805–2819.
36. Tilala, M. H., Chenchala, P. K., Choppadandi, A., Kaur, J., Naguri, S., Saoji, R., et al. (2024). Ethical considerations in the use of artificial intelligence and machine learning in health care: a comprehensive review. *Cureus, 16*(6).
37. Hayden, M. R. (2023). Overview and new insights into the metabolic syndrome: risk factors and emerging variables in the development of type 2 diabetes and cerebrocardiovascular disease. *Medicina, 59*(3), 561.
38. Hiradate, R., Khalil, I. A., Matsuda, A., Sasaki, M., Hida, K., & Harashima, H. (2021). A novel dual-targeted rosiglitazone-loaded nanoparticle for the prevention of diet-induced obesity via the browning of white adipose tissue. *Journal of Controlled Release, 329*, 665–675.
39. Hossen, M. N., Kajimoto, K., Akita, H., Hyodo, M., & Harashima, H. (2012). Vascular-targeted nanotherapy for obesity: unexpected passive targeting mechanism to obese fat for the enhancement of active drug delivery. *Journal of Controlled Release, 163*(2), 101–110.
40. Huang, D. Q., Wong, V. W., Rinella, M. E., Boursier, J., Lazarus, J. V., Yki-Järvinen, H., & Loomba, R. (2025). Metabolic dysfunction-associated steatotic liver disease in adults. *Nature Reviews Disease Primers, 11*(1), 14.
41. Invernizzi, N., & Foladori, G. (2005). Nanotechnology and the developing world: Will nanotechnology overcome poverty or widen disparities. *Nanotechnology Law & Business, 2*, 294.
42. Islam, M. S., Wei, P., Suzauddula, M., Nime, I., Feroz, F., Acharjee, M., & Pan, F. (2024). The interplay of factors in metabolic syndrome: Understanding its roots and complexity. *Molecular Medicine, 30*(1), 279.
43. Jamali, Z., Ayoobi, F., Jalali, Z., Bidaki, R., Lotfi, M. A., Esmaeili-Nadimi, A., & Khalili, P. (2024). Metabolic syndrome: A population-based study of prevalence and risk factors. *Scientific Reports, 14*(1), 3987.
44. Jiang, B., Kauffman, A. E., Li, L., McFee, W., Cai, B., Weinstein, J., et al. (2020). Health impacts of environmental contamination of micro-and nanoplastics: A review. *Environmental Health and Preventive Medicine, 25*, 1–15.
45. Jiang, X., Yang, Z., Wang, S., & Deng, S. (2022). "Big data" approaches for prevention of the metabolic syndrome. *Frontiers in Genetics, 13*, 810152.

46. Kardani, S. L. (2024). Nanocarrier-based formulations: Regulatory challenges, ethical and safety considerations in pharmaceuticals. *Asian Journal of Pharmaceutics, 18*(02).
47. Kawai, T., Autieri, M. V., & Scalia, R. (2021). Adipose tissue inflammation and metabolic dysfunction in obesity. *American Journal of Physiology-Cell Physiology, 320*(3), C375–C391.
48. Khonina, S. N., & Kazanskiy, N. L. (2025). Trends and advances in wearable plasmonic sensors utilizing surface-enhanced Raman spectroscopy (SERS): A comprehensive review. *Sensors, 25*(5), 1367.
49. Kim, J., Yoo, S., Liu, C., Kwak, S. S., Walter, J. R., Xu, S., & Rogers, J. A. (2023). Skin-interfaced wireless biosensors for perinatal and paediatric health. *Nature Reviews Bioengineering, 1*(9), 631–647.
50. Kumar, A., Ren, Y., Sundaram, K., Mu, J., Sriwastva, M. K., Dryden, G. W., et al. (2021). miR-375 prevents high-fat diet-induced insulin resistance and obesity by targeting the aryl hydrocarbon receptor and bacterial tryptophanase (tnaA) gene. *Theranostics, 11*(9), 4061.
51. Kumar, R. D. (2024). The phases of clinical trials: From preclinical studies to post-market surveillance. *Global Journal of Medical and Clinical Case Reports*, 055–065.
52. Lagopati, N., Valamvanos, T. F., Proutsou, V., Karachalios, K., Pippa, N., Gatou, M. A., et al. (2023). The role of nano-sensors in breath analysis for early and non-invasive disease diagnosis. *Chem, 11*(6), 317.
53. Lankshear, G., & Mason, D. (2001). Technology and ethical dilemmas in a medical setting: Privacy, professional autonomy, life and death. *Ethics and Information Technology, 3*, 223–233.
54. Lee, H., Rho, W. Y., Kim, Y. H., Chang, H., & Jun, B. H. (2025). CRISPR-Cas9 gene therapy: Non-viral delivery and stimuli-responsive Nanoformulations. *Molecules, 30*(3), 542.
55. Lee, J. H., Jeong, H. S., Lee, D. H., Beack, S., Kim, T., Lee, G. H., et al. (2017). Targeted hyaluronate–hollow gold nanosphere conjugate for anti-obesity photothermal lipolysis. *ACS Biomaterials Science & Engineering, 3*(12), 3646–3653.
56. Li, M., Li, X., & Lu, Y. (2018). Obstructive sleep apnea syndrome and metabolic diseases. *Endocrinology, 159*(7), 2670–2675.
57. Lopez-Pascual, A., Urrutia-Sarratea, A., Lorente-Cebrián, S., Martinez, J. A., & González-Muniesa, P. (2019). Cerium oxide nanoparticles regulate insulin sensitivity and oxidative markers in 3T3-L1 adipocytes and C2C12 Myotubes. *Oxidative Medicine and Cellular Longevity, 2019*(1), 2695289.
58. Lu, H., Alhaskawi, A., Dong, Y., Zou, X., Zhou, H., Ezzi, S. H. A., et al. (2024). Patient autonomy in medical education: navigating ethical challenges in the age of artificial intelligence. *INQUIRY: The Journal of Health Care Organization, Provision, and Financing, 61*, 00469580241266364.
59. Marchant, G. E., Sylvester, D. J., Abbott, K. W., & Danforth, T. L. (2010). International harmonization of regulation of nanomedicine. *Studies in Ethics, Law, and Technology, 3*(3).
60. Masters, K. S., Wilson, C. L., & Morozink Boylan, J. (2025). Associations between religiosity/spirituality with insulin resistance and metabolic syndrome in the midlife in the United States (MIDUS) study. *PLoS One, 20*(2), e0319002.
61. Mensah, G. A., Mokdad, A. H., Ford, E., Narayan, K. V., Giles, W. H., Vinicor, F., & Deedwania, P. C. (2004). Obesity, metabolic syndrome, and type 2 diabetes: Emerging epidemics and their cardiovascular implications. *Cardiology Clinics, 22*(4), 485–504.
62. Mohamed, S. M., Shalaby, M. A., El-Shiekh, R. A., El-Banna, H. A., Emam, S. R., & Bakr, A. F. (2023). Metabolic syndrome: Risk factors, diagnosis, pathogenesis, and management with natural approaches. *Food Chemistry Advances, 3*, 100335.
63. Mohammadpour, F., Kamali, H., Gholami, L., McCloskey, A. P., Kesharwani, P., & Sahebkar, A. (2022). Solid lipid nanoparticles: A promising tool for insulin delivery. *Expert Opinion on Drug Delivery, 19*(12), 1577–1595.
64. Mohammed, M. A., Abdulkareem, K. H., Dinar, A. M., & Zapirain, B. G. (2023). Rise of deep learning clinical applications and challenges in omics data: A systematic review. *Diagnostics, 13*(4), 664.

65. Mousa, M., Al-Jefout, M., Alsafar, H., Kirtley, S., Lindgren, C. M., Missmer, S. A., et al. (2021). Prevalence of common gynecological conditions in the middle east: Systematic review and meta-analysis. *Frontiers in Reproductive Health, 3*, 661360.
66. Murphy, F., Alavi, A., Mullins, M., Furxhi, I., Kia, A., & Kingston, M. (2022). The risk perception of nanotechnology: Evidence from twitter. *RSC Advances, 12*(18), 11021–11031.
67. Muthu, M. S., Leong, D. T., Mei, L., & Feng, S. S. (2014). Nanotheranostics· application and further development of nanomedicine strategies for advanced theranostics. *Theranostics, 4*(6), 660.
68. Nazir, A., Hussain, F. H. N., & Raza, A. (2024). Advancing microbiota therapeutics: The role of synthetic biology in engineering microbial communities for precision medicine. *Frontiers in Bioengineering and Biotechnology, 12*, 1511149.
69. Noubiap, J. J., Nansseu, J. R., Lontchi-Yimagou, E., Nkeck, J. R., Nyaga, U. F., Ngouo, A. T., et al. (2022). Global, regional, and country estimates of metabolic syndrome burden in children and adolescents in 2020: A systematic review and modelling analysis. *The Lancet Child & Adolescent Health, 6*(3), 158–170.
70. Oda, E. (2018). Historical perspectives of the metabolic syndrome. *Clinics in Dermatology, 36*(1), 3–8.
71. Okosun, I. S., Okosun, B., Lyn, R., & Airhihenbuwa, C. (2020). Surrogate indexes of insulin resistance and risk of metabolic syndrome in non-Hispanic White, non-Hispanic Black and Mexican American. *Diabetes & Metabolic Syndrome: Clinical Research & Reviews, 14*(1), 3–9.
72. Omidian, H., Babanejad, N., & Cubeddu, L. X. (2023). Nanosystems in cardiovascular medicine: Advancements, applications, and future perspectives. *Pharmaceutics, 15*(7), 1935.
73. O'Neill, S., & O'Driscoll, L. J. O. R. (2015). Metabolic syndrome: A closer look at the growing epidemic and its associated pathologies. *Obesity Reviews, 16*(1), 1–12.
74. Ostrominski, J. W., Arnold, S. V., Butler, J., Fonarow, G. C., Hirsch, J. S., Palli, S. R., et al. (2023). Prevalence and overlap of cardiac, renal, and metabolic conditions in US adults, 1999–2020. *JAMA Cardiology, 8*(11), 1050–1060.
75. Ozakar, E., Cetin, M., Ates, O., & Hacimuftuoglu, A. (2019). Nifedipine-loaded polymeric nanoparticles: Preparation and in vitro characterization. *Pakistan Journal of Pharmaceutical Sciences, 32*(2), 547.
76. Pacifico Silva, H., Lehoux, P., Miller, F. A., & Denis, J. L. (2018). Introducing responsible innovation in health: A policy-oriented framework. *Health Research Policy and Systems, 16*, 1–13.
77. Pastorino, R., De Vito, C., Migliara, G., Glocker, K., Binenbaum, I., Ricciardi, W., & Boccia, S. (2019). Benefits and challenges of Big Data in healthcare: An overview of the European initiatives. *European Journal of Public Health, 29*(Supplement_3), 23–27.
78. Pathak, K., Saikia, R., Sarma, H., Pathak, M. P., Das, R. J., Gogoi, U., et al. (2023). Nanotheranostics: application of nanosensors in diabetes management. *Journal of Diabetes & Metabolic Disorders, 22*(1), 119–133.
79. Patiño-Herrera, R., Louvier-Hernández, J. F., Escamilla-Silva, E. M., Chaumel, J., Escobedo, A. G. P., & Pérez, E. (2019). Prolonged release of metformin by SiO2 nanoparticles pellets for type II diabetes control. *European Journal of Pharmaceutical Sciences, 131*, 1–8.
80. Patra, J. K., Das, G., Fraceto, L. F., Campos, E. V. R., Rodriguez-Torres, M. D. P., Acosta-Torres, L. S., et al. (2018). Nano based drug delivery systems: Recent developments and future prospects. *Journal of Nanobiotechnology, 16*, 1–33.
81. Penfornis, A., Down, S., Seignez, A., Vives, A., Bonnemaire, M., & Kulzer, B. (2025). European survey on adults with type 1 diabetes and their caregivers: Insights into personal experience and needs for improving diabetes care. *Diabetes Therapy, 16*, 471–484.
82. Rani, R., Dahiya, S., Dhingra, D., Dilbaghi, N., Kim, K. H., & Kumar, S. (2018). Improvement of antihyperglycemic activity of nano-thymoquinone in rat model of type-2 diabetes. *Chemico-Biological Interactions, 295*, 119–132.

83. Rathore, A., & Mahesh, G. (2021). Public perception of nanotechnology: A contrast between developed and developing countries. *Technology in Society, 67*, 101751.
84. Roberts, C. K., Hevener, A. L., & Barnard, R. J. (2013). Metabolic syndrome and insulin resistance: Underlying causes and modification by exercise training. *Comprehensive Physiology, 3*(1), 1.
85. Rochlani, Y., Pothineni, N. V., Kovelamudi, S., & Mehta, J. L. (2017). Metabolic syndrome: Pathophysiology, management, and modulation by natural compounds. *Therapeutic Advances in Cardiovascular Disease, 11*(8), 215–225.
86. Rosenthal, T., Touyz, R. M., & Oparil, S. (2022). Migrating populations and health: Risk factors for cardiovascular disease and metabolic syndrome. *Current Hypertension Reports, 24*(9), 325–340.
87. Saenz, C., Salinas, M., Rothman, R. L., & White, R. O. (2025). Personalized lifestyle modifications for improved metabolic health: The role of cultural sensitivity and health communication in type 2 diabetes management. *Journal of the American Nutrition Association, 44*(3), 198–211.
88. Saklayen, M. G. (2018). The global epidemic of the metabolic syndrome. *Current Hypertension Reports, 20*(2), 1–8.
89. Satalkar, P., Elger, B. S., & Shaw, D. (2016). Naming it 'nano': Expert views on 'nano'terminology in informed consent forms of first-in-human nanomedicine trials. *Nanomedicine, 11*(8), 933–940.
90. Schwab, F., Rothen-Rutishauser, B., Scherz, A., Meyer, T., Karakoçak, B. B., & Petri-Fink, A. (2023). The need for awareness and action in managing nanowaste. *Nature Nanotechnology, 18*(4), 317–321.
91. Sharma, N., Kurmi, B. D., Singh, D., Mehan, S., Khanna, K., Karwasra, R., et al. (2024). Nanoparticles toxicity: an overview of its mechanism and plausible mitigation strategies. *Journal of Drug Targeting, 32*(5), 457–469.
92. Sharma, R., Gulati, A., & Chopra, K. (2025). Era of surrogate endpoints and accelerated approvals: A comprehensive review on applicability, uncertainties, and challenges from regulatory, payer, and patient perspectives. *European Journal of Clinical Pharmacology, 81*, 605–623.
93. Summer, M., Ashraf, R., Ali, S., Bach, H., Noor, S., Noor, Q., et al. (2024). Inflammatory response of nanoparticles: Mechanisms, consequences, and strategies for mitigation. *Chemosphere, 363*, 142826.
94. Sun, Y. E., Wang, W., & Qin, J. (2018). Anti-hyperlipidemia of garlic by reducing the level of total cholesterol and low-density lipoprotein: A meta-analysis. *Medicine, 97*(18), e0255.
95. Swarup, S., Ahmed, I., Grigorova, Y., & Zeltser, R. (2024). Metabolic syndrome. InStatPearls [internet] 2024 Mar 7.
96. Sypniewska, G. (2007). Pro-inflammatory and prothrombotic factors and metabolic syndrome. *EJIFCC, 18*(1), 39.
97. Targher, G., Byrne, C. D., & Tilg, H. (2024). MASLD: A systemic metabolic disorder with cardiovascular and malignant complications. *Gut, 73*(4), 691–702.
98. Varkey, B. (2021). Principles of clinical ethics and their application to practice. *Medical Principles and Practice, 30*(1), 17–28.
99. Vashist, S. K., & Luong, J. H. (Eds.). (2025). Handbook of immunoassay technologies: Approaches, performances, and applications. Elsevier.
100. Verma, V., Gupta, P., Singh, P., & Pandey, N. K. (2024). Considerations in the development and deployment of nanotechnology. In *Nanotechnology in societal development* (pp. 505–540). Springer Nature Singapore.
101. Wang, H. H., Lee, D. K., Liu, M., Portincasa, P., & Wang, D. Q. (2020). Novel insights into the pathogenesis and management of the metabolic syndrome. *Pediatric Gastroenterology, Hepatology & Nutrition, 23*(3), 189.

102. Wang, J., Zhao, W., Zhang, Z., Liu, X., Xie, T., Wang, L., et al. (2024). A journey of challenges and victories: A bibliometric worldview of nanomedicine since the 21st century. *Advanced Materials, 36*(15), 2308915.
103. Wasti, S., Lee, I. H., Kim, S., Lee, J. H., & Kim, H. (2023). Ethical and legal challenges in nanomedical innovations: A scoping review. *Frontiers in Genetics, 14*, 1163392.
104. Yao, M., Wu, M., Yuan, M., Wu, M., Shen, A., Chen, Y., et al. (2024). Enhancing the therapeutic potential of isoliensinine for hypertension through PEG-PLGA nanoparticle delivery: A comprehensive in vivo and in vitro study. *Biomedicine & Pharmacotherapy, 174*, 116541.
105. Zahalka, S. J., Galindo, R. J., Shah, V. N., & Low Wang, C. C. (2024). Continuous glucose monitoring for prediabetes: What are the best metrics? *Journal of Diabetes Science and Technology, 18*(4), 835–846.
106. Zaky, M. Y., Mohamed, E. E., & Ahmed, O. M. (2025). Neurodegenerative disorders: Available therapies and their limitations. In *Nanocarriers in neurodegenerative disorders* (pp. 29–46). CRC Press.
107. Zheng, F., Jiang, H. Y., Yang, X. T., Guo, J. H., Sun, L., Guo, Y. Y., et al. (2024). Reviews of wearable healthcare systems based on flexible gas sensors. *Chemical Engineering Journal, 490*, 151874.
108. Zhou, M. S., Wang, A., & Yu, H. (2014). Link between insulin resistance and hypertension: What is the evidence from evolutionary biology? *Diabetology & Metabolic Syndrome, 6*, 1–8.
109. Ziebarth, J., da Silva, L. M., Lorenzett, A. K. P., Figueiredo, I. D., Carlstrom, P. F., Cardoso, F. N., et al. (2024). Oral delivery of liraglutide-loaded zein/eudragit-chitosan nanoparticles provides pharmacokinetic and glycemic outcomes comparable to its subcutaneous injection in rats. *Pharmaceutics, 16*(5), 634.
110. Zu, Y., Zhao, L., Hao, L., Mechref, Y., Zabet-Moghaddam, M., Keyel, P. A., et al. (2021). Browning white adipose tissue using adipose stromal cell-targeted resveratrol-loaded nanoparticles for combating obesity. *Journal of Controlled Release, 333*, 339–351.

Current Prospective in Nanotheranostics-Based Therapy in Neurometabolic Disorder

15

Devinder Kumar, Sunil Dutt, Pankaj Kalia, and Rajesh Kumar

Abstract

Neurometabolic diseases caused by genetic defects related to metabolism of the nervous system can be effectively treated with nanotheranostics, a combination of diagnosis and therapy through nanotechnology. Nanotheranostic platforms, such as nanoparticles, liposomes, dendrimers, and quantum dots, offer dual functionality: drug release and imaging techniques for looking at the effect of drugs in the body system. For instance, nanoparticles may be designed surface ligands that help them enter the brain by passing through the BBB while at the same time delivering the drugs to neuronal tissue without affecting other tissues. At the same time, these nanoparticles can also incorporate imaging, for instance, by carrying fluorophores or magnetic resonance imaging agents. Some of the advancements under development are stimuli-sensitive nanocarriers that release cargo in response to pH, temperature, or specific enzymes in the injured or pathologic nervous system to provide greater targeting. The current approaches to treating the disease are symptomatic; however, the potential for treating genetic neurometabolic disorders has expanded with the use of traditional gene editing tools like CRISPR/Cas9 combined with nanocarrier structures, which allow for the direct elimination of DNA mutations. In addition to this, Nanotheranostics also strengthens the value of for personalized medicine since the dynamic monitoring of disease pathways of distinct patients is possible. However, current literature shows that important issues such as nanoparticle biocompatibility,

D. Kumar
Amar Shaheed Baba Ajit Singh Jujhar Singh Memorial College of Pharmacy (An Autonomous College), Bela, Ropar, Punjab, India

S. Dutt · P. Kalia · R. Kumar (✉)
Maharaja Agrasen School of Pharmacy, Maharaja Agrasen University, Solan, Himachal Pradesh, India

V. K. Chaturvedi et al. (eds.), *Nanotheranostics in Metabolic Syndrome*,
https://doi.org/10.1007/978-981-95-3746-4_15

immune responses, and long-term safety profiles remain unsolved to allow broad clinical application of the referred platforms.

Keywords

Nanotheranostics · Metabolism · As nanoparticles · Liposomes · Dendrimers

1 Introduction

Neurometabolic disorders (NMD) include a diverse and complex range of genetic and acquired disorders that result from metabolic disorders of the nervous system. Such disorders result from a lack of enzymes, transporters, or cofactors required for the metabolism of amino acids, carbohydrates, lipids, or neurotransmitters [1]. In addition, these disorders include mitochondrial disorders, lysosomal storage diseases, peroxisomal disorders, and amino acid metabolism (NHAs) disorders such as phenylketonuria (PKU) and maple syrup disorder (MSUD). The clinical presentation of NMDs is heterogeneous and has varying degrees of developmental delays and intellectual disability, neurodegeneration, epilepsy, movement disorders, and psychiatric characteristics [2]. They have a complex etiology and a multifactorial pathophysiology, which is why it is difficult to treat them effectively. Traditional therapies such as dietary changes, enzyme replacement therapies, and small molecule agents have successfully alleviated symptoms in some patients but do not address the underlying pathophysiology and often do not prevent the disease from progressing [3]. In recent years, new strategies for diagnosing and treating neurometabolic disorders have focused on the field of nanomedicine, which has led to the development of the modern discipline of nanotheranotics at the nano level. Nanotheranostics is the combination of diagnosis and therapy using multifunctional nanoparticles (NPs), which enables the diagnosis, monitoring, and treatment of diseases using nanotechnology. This method has drawn much attention as it can increase drug delivery, bioavailability, selective targeting of cells and tissues, and reduce systemic toxicity [4].

Nanotheranostics technology has great potential to circumvent the key therapeutic limitations encountered by specially developed nanomicelles to treat neurometabolic disorders that otherwise render them therapeutically useless, such as blood-brain barrier (BBB) permeation problems, extreme drug clearance rates, and targeted effects. The blood-brain barrier (BBB), a highly selective endothelial barrier, limits the entry of most drugs into the central nervous system (CNS), making drug development and treatment of neurological disorders extremely challenging. Nanoparticle (NP)-based drug delivery systems, such as liposomes, polymeric nanoparticles, dendrimers, solid lipid NPs, and metal-based NPs, have also been extensively studied to improve delivery [5]. These nanocarriers are also suitable for ligand functionalization to promote receptor-mediated transeucotic brain health and deliver therapeutic agents directly to target neurons or glial cells. One of the most important benefits of nanotherapy in the context of neurometabolic disorders is its ability to monitor the course of the disease and the effectiveness of treatment in real time [6].

Traditional diagnostic techniques including magnetic resonance imaging (MRI), positron emission tomography (PET), and enzyme testing are often not sensitive or specific enough, leading to delays in diagnosis and treatment. For example, nanoparticles equipped with imaging agents such as quantum dots, SPIONs, or gold nanoparticles can be used as contrast agents for MRI, fluorescence, or CT scans to detect metabolic disorders in the brain early and precisely [7]. In addition, using functionalized nanoparticles, the administration of drugs can even be adapted so that they preferentially bind to specific disease markers, which enables both noninvasive and targeted diagnosis. This could be particularly important for diseases with emerging disorders such as lysosomal storage diseases and mitochondrial disorders, in which neuronal damage is largely irreversible as soon as symptoms occur [8]. In addition to drug administration and imaging, nanotheranostics also enables the development of new gene and enzyme replacement therapies. Gene therapy strategies, such as gene transfer mediated by adeno-associated viruses (AAV), have the potential to directly correct pathogenic mutations that underlie incompletely penetrant neurometabolic disorders [9]. However, immune responses, side effects, and inefficient gene expression in the CNS limit the effectiveness of these therapies. Gene delivery systems based on nanoparticles are safer and more effective. They protect nucleic acids from degradation, promote cell internalization and enable targeted administration to a region of the central nervous system (CNS) [10, 11]. Therefore, traditional enzyme replacement therapies (ERTs) administered for lysosomal storage diseases were hampered by inefficient biodistribution and the breakdown of exogenous enzymes in circulation. Nano-carrier-based enzyme delivery systems can increase the stability of enzymes, extend their circulation time, and facilitate their uptake into the CNS, ultimately increasing the therapeutic effects of ERTs in treating diseases such as Gaucher disease, Fabry disease, and Pompe disease. Nanotherapy-based therapy depends on the biocompatibility and safety of nanomaterials [8, 12]. While nanoparticles have many benefits in treating neurometabolic disorders, they also raise concerns about long-term toxicity, immunogenicity, and potential accumulation in the body. Intensive efforts are therefore focused on developing biodegradable and biocompatible nanomaterials that can be safely metabolized and eliminated from the body. In addition, precision medicine and personalized therapy have paved the way for patient-specific nanotheranostic strategies that adapt treatment (as far as possible) to the individual's genetic and metabolic context. Such a strategy not only increases the effectiveness of treatment but also reduces the side effects it can cause, while leading to better treatment outcomes [13].

Nanotheranotics show promise for neurometabolic disorders and enable the diagnosis, monitoring and treatment of diseases through personalized medicine (Fig. 15.1). Regulatory approval, mass production, and profitability remain challenges when it comes to moving nanomedicine from laboratory to clinic. In addition, further optimization of nanoparticle formulations is required to improve targeting effectiveness and reduce potential side effects [14].

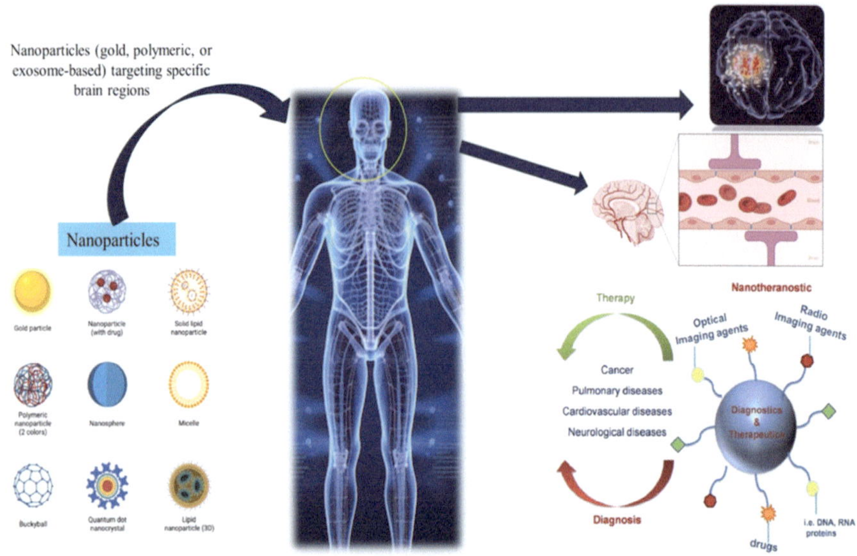

Fig. 15.1 Treatment for neurometabolic disorders-based nanotheranostics

In the future, close collaboration between researchers, clinicians, and pharmaceutical companies will play an important role in turning the greatest potential of nanotheranostic therapies into clinically effective and targeted treatments for people with these devastating diseases. Nanotheranostics is a paradigm shift in the diagnosis and treatment of neurometabolic disorders. It combines targeted drug delivery, real-time imaging, and precision medicine in one system [15, 16]. Nanoparticles have unique properties that enable them to overcome biological barriers, improve therapeutic effectiveness, and enable timely and accurate detection of diseases, making them a promising tool to address the challenges associated with NMD. Research in this area is not stagnating, it is constantly evolving, and if we manage to develop a safe, efficient and clinically translatable nanotheranostic platform, it can revolutionize the management of neurometabolic disorders and significantly improve patients' overall quality of life [7].

Neurometabolic disorders (NDs) are a broad and very heterogeneous group of disorders that often occur early in life and typically manifest early in development. In particular, these diseases have mutations in genes that code for key proteins involved in basic metabolic pathways in multiple parts of the body. The mutations that lead to such genetic changes lead to specific neurological phenotypes and, in extreme cases, early death (for the patient). NDS are considered to be heterogeneous and rare diseases that are generally characterized by low incidence and prevalence in various population groups [17]. Although these findings are encouraging and pave the way for future research, ongoing research on novel gene therapy approaches, new testing treatments, and cutting-edge genetic assays has the potential to further transform the space [18]. Overall, these joint efforts should significantly improve the understanding of these diseases, enable ever faster detection, and enable the planning of dynamic and adaptable personalized treatments that pay close attention to patients' demanding needs and demands. Clinicians around the world are currently focused on early diagnosis and more effective and targeted treatments. And also a growing focus on lipid-based nanotheranostic systems developed specifically for the effective treatment of NDS [4]. With the help of these specialized nanotheranostic systems, doctors will likely be able to provide their patients with more effective and personalized treatment that could lead to a successful cure. These disorders generally result in progressive and irreversible loss of neurological function and early death in patients [2]. In addition, noncommunicable diseases have significant effects outside the central nervous system, as these disorders can also have significant serious effects on other vital organs (e.g. eyes, kidneys, livers), leading to an increased complexity of treatment and treatment strategies. The treatment of these neurodevelopmental disorders therefore requires a high level of international expertise that goes beyond the clinical extremity and an understanding of the system behind every disorder [19].

Early mutations of these genes, which primarily code for proteins involved in the metabolic pathways and homeostasis of neurons and glial cells in the brain, eventually lead to the onset of the disease, which is rich in genetics. Strict regulation of brain metabolism is therefore critical, but the drugs and biomolecules needed to improve these metabolic states face a huge obstacle as they cannot cross the blood-brain barrier (BBB). The biggest challenge faced by strategies for delivering drugs to the brain is due to this physiological and structural barrier in the brain. Only water-soluble therapeutic agents with a molecular weight of less than 400 Da are produced symmetrically from glucose, through facilitated diffusion or by carriers whose transport takes place within the narrow endothelial cells of the blood-brain barrier (BBB) and has free access to the entire brain [20].

2 Genetic Defects Related to the Metabolism of the Nervous System

The sustained metabolism is not only essential for cell function, neurotransmission, and homeostasis in the nervous system. It is well known that genetic defects that target these metabolic pathways result in a range of neurometabolic disorders, many of which result in severe neurological dysfunction, neurological developmental delays, and progressive neurodegeneration [21, 22]. These diseases are usually genetic and arise due to mutations in genes that code for enzymes, transporters, or cofactors needed for metabolism in the brain and peripheral nervous system. These disorders are complex because genetic, biochemical, and environmental interactions that affect nervous system metabolism are anything but simple. Lysosomal storage diseases (LSDs) represent one of the best characterized classes of neurometabolic disorders [23]. These enzymes normally work in lysosomes to break down complex molecules such as glycosaminoglycans, lipids, and glycoproteins, and their mutations result in lysosomal storage disorders. The defective enzyme causes the accumulation of these substrates in the lysosomes, leading to cell dysfunction and progressive neurodegeneration. For example, genetic mutations in the HEXA gene, which codes for beta-hexosaminidase A, can lead to Tay-Sachs disease. This is a hereditary disorder in which GM2 gangliosides build up in neurons, leading to devastating neurodegeneration and death in early childhood [24, 25]. Likewise, mutations in the GBA gene (which codes for glucocerebrosidase) result in the accumulation of glucocerebrosides both in the nervous system and in systemic organs, leading to Gaucher disease. Peroxisomal disorders are another notable group of genetic metabolic disorders that affect the nervous system. The peroxisome function is essential for fatty acid oxidation, lipid metabolism, and the detoxification of reactive oxygen species. Mutations in one of the PEX genes responsible for peroxisome biogenesis cause Zellweger spectrum disorders characterized by severe neurological dysfunction, hypotension, seizures, and developmental delays [26, 27]. Mutations in the ABCD1 gene also cause X-linked adrenoleukodystrophy (X-ALD), which results in the accumulation of very long-chain fatty acids (VLCFAs) in the nervous system, leading to demyelination with progressive motor and cognitive decline [28]. Given the brain's energy requirements, mitochondrial metabolism is also crucial for neuronal function. Mutations of nuclear DNA (nDNA) or mitochondrial DNA (mtDNA) lead to oxidative phosphorylation (OXPHOS) defects, ATP deficiency and dysregulation of reactive oxygen species (ROS) in mitochondrial diseases. Leigh syndrome (LS) is one of the most commonly characterized mitochondrial diseases. It is caused by mutations in KL genes such as the MT-ATP6, SURF1 and ND genes, which code for the subunits of the OXPHOS complex. This leads to progressive neurodegeneration, hypotension, and respiratory failure [29]. Mutations in MT-TL1 cause MELAS syndrome (mitochondrial encephalopathy, lactic acidosis, and strokes) and an impairment of mitochondrial tRNA metabolism, leading to metabolic seizures and cognitive decline. Another important group of genetic defects that affect the nervous system are amino acid metabolism disorders. In phenylketonuria (PKU), mutations in the PAH gene, which codes for phenylalanine hydroxylase, result in phenylalanine accumulation, which is harmful to the developing brain. If left untreated, PKU causes severe mental disabilities, microcephaly, and behavioral disorders. Another, Maple Syrup Urine Disease (MSUD), is caused by mutations in the BCKDHA, BCKDHB, or DBT genes that code for components of the branched-chain alpha-keto acid dehydrogenase complex [30, 31]. This results in the formation of leucine, isoleucine, and valine, which, without treatment, leads to metabolic encephalopathy, seizures, and coma. Genetic mutations can also lead to disorders of neurotransmitter metabolism and thus to disorders of brain function and behavior. Introduction: Dopamine and serotonin metabolism disorders (e.g. aromatic L-amino acid decarboxylase (AADC) deficiency) are caused by mutations in the DDC gene that restrict the synthesis of dopamine and serotonin, which are characterized by hypotension, movement disorders, and developmental delays [32–34]. A tyrosine hydroxylase (TH) deficiency, which impairs dopamine biosynthesis, has Parkinson's characteristics, but in infancy. Purine and pyrimidine metabolism disorders also affect the nervous system. Mutations in HPRT1, which codes for hypoxanthine-guanine phosphoribosyl transferase (HGPRT), result in the accumulation of uric acid and cause severe neurobehavioral disorders, self-harm, and movement disorders, a disease known as Lesch-Nyhan syndrome. Similarly, adenine deaminase (ADA) deficiency is another PNP disorder that results in generalized SCID and neurological disorders as a result of the accumulation of toxic metabolites [35, 36].

Disorders of metal metabolism also affect neurological function. Wilson's disease is an autosomal recessive disorder that results from mutations in the ATP7B gene, which results in the accumulation of copper in the liver and brain, leading to liver dysfunction, typically with neurological symptoms (tremors, dystonia, psychiatric symptoms). Menkes disease, a hereditary copper transport disorder resulting from mutations in ATP7A, is characterized by profound neurodegeneration, hypotension, and connective tissue disorders. GLUT1-DS is an example of a glucose metabolism disorder that affects not only the brain but the entire nervous system and is due to SLC2A1 mutations [24]. This condition impairs the transport of glucose across the blood-brain barrier, leading to epilepsy, developmental delays, and movement disorders. Mutations in an acid alpha-glucosidase gene encoded by GAA also cause glycogen storage disease type II (Pompe disease), which is characterized by glycogen accumulation in neurons and muscles, leading to myopathy and progressive neuromuscular dysfunction. The inability to produce cholesterol due to cholesterol metabolism disorders such as Smith-Lemli-Opitz syndrome (SLOS) due to DHCR7 mutations can lead to intellectual disability, autism spectrum disorders, and congenital malformations. Since the metabolic disorders of the nervous system are diverse, early diagnosis and intervention are essential. Genome sequencing, newborn screening, and biochemical diagnostic techniques are mature and have optimized the ability to identify and thus treat diseases through dietary adjustment, enzyme replacement, gene therapy, and/or pharmacological interventions. All of this has improved over the last century. Through further research on gene editing and targeted molecular therapies, precision medicine approaches are preparing patients suffering from these devastating neurometabolic diseases to be of clinical use in the future [37].

3 Blood-Brain Barrier and Targeted Delivery Constraints

The broad and complex group of rare hereditary metabolic disorders known as neurometabolic disorders comprises around 40 different classified disorders. In fact, these various, often very complex disorders occur in around 1 in 2000 live births and are primarily due to defects in the nervous system systems that are critical to maintaining proper function and homeostasis throughout the body. Therapies that have been (new) developed and (re) designed since time immemorial to treat or at least alleviate these debilitating diseases that trigger the nervous system are unfortunately limited in their effectiveness (due to a whole series of difficult, acute measures) [38]. Some of these confounding variables include the complexity of treatment pathways, general bioavailability, the half-life of the drugs used, and the dose-dependent side effects resulting from the dose-dependent toxicity of classic therapies. In response to these persistent and worrying issues, several vibrant research communities are actively and urgently seeking appropriate alternatives, with the ultimate goal being to achieve better treatment outcomes for patients with these rare diseases [39]. At the same time, scientists face a variety of often terrifying problems associated with finding these alternatives, making the search for workable solutions even more difficult and complicated [40]. In this unique and alarming situation, theranostics is proving to be a promising and ideal approach to overcome and combat these diverse challenges [41]. This novel and innovative paradigmatic approach combines therapeutic and diagnostic functions in a single, uniform and organized system and thus enables the early detection, diagnosis and treatment of a wide range of different diseases. Theranostics enables these and other complex diseases to be identified even earlier, thus enabling better patient care and treatment outcomes for people with such rare and difficult diseases. The use of blood-brain barrier technology is expected to dramatically advance the development of novel drugs (nanotheranotics) that directly address the complex needs of neurometabolic diseases [42, 43]. This incredible shift in science and clinical medicine represents yet another new and rigorous method for clinical care one that is tragically difficult but ultimately rewarding and could well lead to a radical reconceptualization of our treatment paradigms, treatment landscape, and treatment offerings [44]. It is important to note that the blood-brain barrier itself is a heterogeneous selective obstacle that protects the brain from foreign, xenobiotic substances that can enter its sensitive environment from the bloodstream, as well as from the various tissues that can enter plasma as a result of their metabolic processes [45, 46]. In addition, it is the main barrier that limits the use of pharmacotherapy strategies in the complex and vulnerable brain microenvironment. As a result of this protective mechanism, many drugs are inadequate and cannot be brought to market because they penetrate the blood-brain barrier only poorly (Fig. 15.2). This has long been a problem in developing effective therapies [47, 48].

However, both methods are not just one of the many challenging aspects of developing a drug. In view of the gradual revolution in nanotechnology, in particular the ability to produce customized, effective, and optimized drug delivery systems, we now see that exploring new pathways and possibilities within therapeutics enables efficient targeting of the nervous system [49]. These advances are particularly relevant given the ongoing challenges associated with numerous neurometabolic disorders that currently affect millions of people around the world and are therefore having serious effects on public health and wellbeing [50]. Here, we report, discuss and highlight with a view to future efforts the amazing prospect of successfully combining nanotheranotics, which are based on a sophisticated range of advanced and tailored nanoformulations and offer important prospects for overcoming the blood-brain barrier, in the treatment of a wide range of neurometabolic disorders [24]. This in itself is an extension of this area, but we must consider these new technologies and their clinical implications in order to move forward in this ever-stimulating area that can bring therapeutic promise and avoid it being purely academic, and take a step back and validate them and access those who need such treatment the most. Such advances will result in advances that

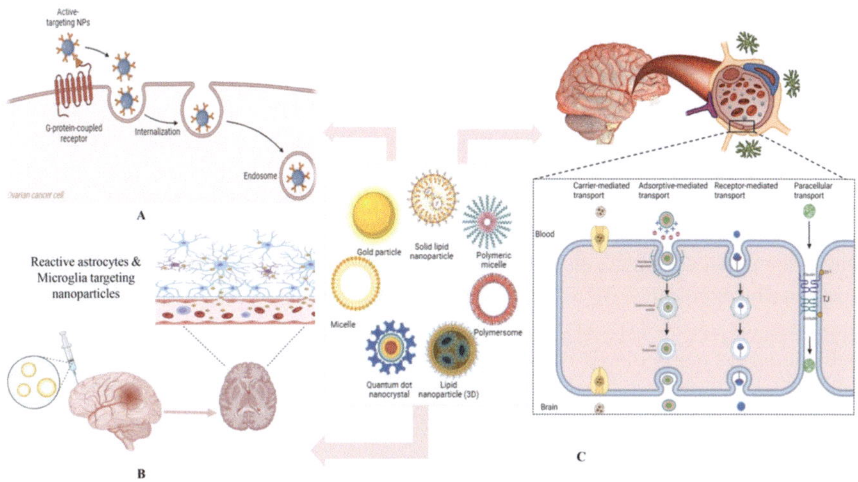

Fig. 15.2 BBB and targeted nanoparticle delivery. (**a**) Active targeting of nanoparticle to cancer cells. (**b**) Aging & Alzheimer's diseases affect nanoparticles delivery across blood brain barrier. (**c**) Schematic representation of the different mechanisms of nanoparticles for crossing the blood–brain barrier

could change the course of treatment for neurometabolic disorders and have the potential to help numerous patients around the world [4].

4 Nanotheranostics

Novel nanotheranostic systems have been successfully developed and characterized for the combined diagnosis and treatment of this important global health challenge. The complex models of advanced theranostic systems, with all their various types, functions, and properties, are attracting widespread interest both from rigorous researchers, who enjoy studying this rapidly developing area, and from doctors who strive to translate these significant achievements into measurable effects in the clinic. This whole concept was really a priority and focused on the various complex multi-metal 3D biomimetic model systems [51]. These complex, advanced systems are able to fully mimic the body's complex and difficult microenvironments, including but not limited to customized cancer stem cell systems specifically tailored to the diverse challenges of brain cancer and its sometimes-complex treatment [52]. In recent decades, significant progress has been made in the wide area of nanomaterials research [37]. The increasing number of novel nanomedicines, which are being investigated in clinical trials and are, being actively integrated into routine clinical practice, illustrates a profound change in this area. While these breakthroughs, insights, and results appear brilliant and glorious in publications and lectures, the path to actually implementing functional, validated, and personalized nanotherapy is still characterized by many and huge obstacles that we encounter on a daily basis [53]. These significant challenges are particularly correlated with the ultimate goal of personalized biomedical applications that are precisely tailored to the individual cancer phenotypes and profiles of individual patients [54]. Arranging and integrating these innovative diagnostics with a relevant next, targeted therapy that is required for this patient's exact cancer problem is important and yet a challenging undertaking in terms of the research and development landscape required for such therapy. And in this article, splinter trajectory synthesis guides the construction of complex integration methods by involving, through careful conjugation, multiple drug compounds, antibodies, or nanoparticles specifically developed for effective cancer treatment [9]. This leads to a specific approach for targeted treatment of the ErbB-2 receptor for novel cancerous effector cell therapies, particularly in situations where breast cancer drugs are frequently administered. Here's the backstory: This data-driven and systematic approach shows that this new delivery strategy doesn't just hit the mark sometimes even better than the traditional Nanoplex delivery group systems that have been used and relied on in the past. Simulations and in vivo studies have proven to be accurate and reproducible time and are already largely consistent with the kinetic model, which is still an active and ongoing area of research pursued by scientists who are passionate about this extremely important and urgent area of study [55].

The medical effects of such nanoparticles, which include therapeutic agents and potentially cytotoxic species, are based on complex kinetic signaling pathways. The associated complexity of reaction-diffusion schemes only adds another level of complexity to the entire landscape, which we must now address in our advanced cancer treatment strategies. Two mushroom-fed animal models worldwide show that significant anatomical differences, including the imprinting of semantic physicality, together with these features, which manifest as unique properties of the entities, each with unique and correctly titrated drug concentrations, which have equivalent and effective physiological effects on the entire body [56]. This indicates the deeply personalized nature of successful treatment in the difficult situation of oncology. In addition, the increased transplantation of nanoparticles resulted in various intact tissues being targeted using advanced reperfusion techniques [57]. This is recognized as an important component of scab for the effective establishment of targeted photothermal therapies for the detection and treatment of cancer. To improve the necessary infrastructure for this new area, a new, simple, cost-effective and user-friendly mathematical model with discrete vascular sources was designed [58, 59]. This summary model is most relevant for current work on the topic of tumor vasculature, which is being carried out with reservations in various new combinations of photothermal therapy. The discussed approach provides a comprehensive and flexible structure to close the knowledge gap about complex dynamics and to refine the effectiveness of targeted therapies in the ongoing fight against cancer. At the same time, attempts are being made to optimize treatment outcomes and reduce the huge burden that this terrible disease places on millions of lives around the world [60].

5 Diagnostic Techniques

The biggest challenges relate primarily to diagnosing and identifying the disorders at their earliest stage, which can be overcome by using various diagnostic modalities used in medicine (Table 15.1). In this context, theranostics, a portmanteau word that means a combination of therapeutic and diagnostic formulations for targeted treatment, is an important factor [37]. Interest in targeting the pathology of neurological diseases and metabolic disorders using the coalescence platform to deliver nanomaterials via the theranotic route is growing rapidly. Achieving nanoparticle therapy suitable for neurometabolic theranosis is a cumulative endeavor that requires combining areas of proven capabilities and resulting insights, as well as the transfer of technologies, tests, and extremely high-tech methods [53]. In addition, these approaches aim to coordinate specific mechanisms, and the bioavailability and elimination of the nanosupporting materials can predict which decisive event is decisive in terms of treatment effectiveness [61]. Properly designed nanomaterials can remain in biological substrates longer than the recommended clinical period, particularly in combination with suitable media, synthesized porosity and physiologically altered sensor physiology [62]. The obvious implementation of this type of Moody and its multi-manifest portals makes it possible to avoid the physical adjustment of the morphological properties of nanomaterials [63]. In addition, this also gives them the ability to adapt to a large extent to any biochemical, theranostic or attractive type of treatment. In fact, through this much more immersive and vague ensemble, which shows the intersectionality of insanity in a collection of intelligent iterations of our neurometabolic modus operandi, researchers and clinicians are simply exploring the intersections of a continuum of subtleties and inequities [64]. It aims to initiate a recent paradigm shift in combining various elements in the search for integral cognition and the treatment of neurological disorders through the use of breathtaking applications of nanotechnology [65]. In this way, an overall strategy is emphasized that integrates both therapeutic benefits and diagnostic safety and thus lays the basis for potentially paradigm-shifting developments towards novel interventions and insights into the biology of neurometabolic disorders [66]. Neurometabolic disorders (NMD) represent important diagnostic and therapeutic challenges due to their heterogeneous etiology and progressive character. Standard diagnostic methods are usually neither specific nor sensitive, resulting in delayed and ineffective interventions that cannot prevent the disease from progressing. Nanotechnology is drastically changing the dynamics of molecular imaging and therapy strategies and enables cutting-edge diagnostics with high-resolution imaging and targeted therapy. The field of nanotheranotics, which combines both therapeutic and imaging aspects using nanomaterials, has developed as a potential method for early detection and treatment of these diseases. Various nanoparticles have been investigated for use in advanced imaging and diagnostic techniques that overcome biological barriers in the central nervous system (CNS) to improve sensitivity, resolution, and therapeutic effectiveness [67].

Magnetic resonance imaging (MRI) is a common, noninvasive imaging method that enables detailed visualization of brain abnormalities. SPIONs are commonly

Table 15.1 Various diagnostic techniques with diverse applications towards therapeutics

S. no.	Technique	Nanomaterial used	Mechanism	Advantages	Applications	Challenges
1	Magnetic Resonance Imaging (MRI)	Superparamagnetic iron oxide nanoparticles (SPIONs)	Enhanced contrast via magnetic properties	High spatial resolution, noninvasive	Imaging brain abnormalities in metabolic disorders (e.g., detecting iron deposition in neurodegeneration)	High cost, limited sensitivity in detecting early-stage disorders
2	Positron Emission Tomography (PET)	Quantum dots (functionalized with targeting moieties)	Detection of positron-emitting radiotracers	High sensitivity, functional imaging	Monitoring glucose metabolism dysregulation in brain disorders	Short half-life of radiotracers, potential toxicity of some nanomaterials
3	Near-Infrared Fluorescence (NIRF)	Gold nanoclusters, upconversion nanoparticles (UCNPs)	Fluorescent signal in NIR region penetrates tissues	High penetration depth, low photobleaching	Detection of amyloid plaques in Alzheimer's disease	Limited targeting in vivo, difficulty in functionalizing nanoparticles
4	Photoacoustic Imaging (PAI)	Gold nanorods, carbon nanotubes	Converts absorbed light into ultrasound signals	High spatial resolution, real-time imaging	Mapping oxidative stress markers in neurometabolic diseases	Limited penetration depth, need for optimization of nanomaterial stability
5	Raman Spectroscopy	Surface-enhanced Raman scattering (SERS) nanoparticles	Enhanced Raman signal for molecular detection	Ultra-high sensitivity, multiplexing capability	Identification of biomolecular changes (lipid, protein alterations in neurodegenerative disorders)	High technical expertise needed, potential toxicity of SERS-active nanoparticles
6	Electrochemical Biosensors	Carbon nanomaterials, gold nanoparticles	Catalytic or conductivity-based signal enhancement	High specificity, rapid detection	Monitoring neurotransmitter imbalances (e.g., dopamine in Parkinson's)	Interference from biological fluids, stability issues with biosensor components
7	Computed Tomography (CT)	Gold nanoparticles (high atomic number contrast agents)	Enhanced X-ray absorption	High resolution for anatomical structures	Detecting structural abnormalities in neurometabolic disorders	Radiation exposure, need for biocompatible contrast agents

8	Optical Coherence Tomography (OCT)	Silica nanoparticles, carbon-based nanomaterials	Light backscattering for high-resolution imaging	Real-time imaging, high depth penetration	Monitoring myelin degeneration and axonal damage in neurometabolic disorders	Requires advanced optical systems, limited tissue penetration
9	Theranostic Nanoprobes	Multifunctional nanoparticles (e.g., dendrimers, liposomes)	Combined imaging and therapeutic delivery	Simultaneous diagnosis and treatment	Monitoring drug distribution and therapeutic response in neurometabolic disorders	Complex synthesis, potential for off-target effects
10	Single-Photon Emission Computed Tomography (SPECT)	Nanoscintillators, radiolabeled nanoparticles	Detection of gamma-emitting isotopes	High sensitivity, functional imaging	Assessing blood-brain barrier (BBB) integrity and metabolic pathways in neurodegenerative diseases	Limited spatial resolution, radiotoxicity concerns
11	Molecular Imaging (MI)	Functionalized nanodots, dendrimer-based nanoparticles	Detects specific biomolecular interactions	High specificity, ability to track molecular changes	Early diagnosis of enzyme deficiencies or oxidative stress markers	Complex preparation of molecular probes, challenges in clinical translation
12	Thermal Imaging	Gold nanoshells, carbon nanotubes	Heat generation for real-time thermal mapping	Noninvasive, rapid results	Detecting inflammation or tissue damage associated with neurometabolic disorders	Limited sensitivity, difficulty in quantifying thermal responses
13	Mass Spectrometry Imaging (MSI)	Nanostructured surfaces (e.g., MALDI-MS nanoparticles)	Ionization of molecules for spatial molecular mapping	High molecular specificity, label-free detection	Mapping lipid metabolism disruptions in brain tissues	Expensive equipment, complex sample preparation
14	Bioluminescence Imaging	Luciferase-functionalized nanoparticles	Detection of bioluminescent signals	High sensitivity, minimal background noise	Tracking mitochondrial dysfunction and oxidative stress in neurometabolic disorders	Limited penetration depth, dependence on specific substrates

used as contrast agents to increase the magnetic properties of tissues and to facilitate imaging of neurometabolic abnormalities [68]. In neurodegenerative diseases, MRI has proven to be particularly useful for detecting iron deposits. However, imaging is expensive and its sensitivity is insufficient for early metabolic disorders, which is a major disadvantage in routine clinical applications [69, 70]. Another highly sensitive imaging method is positron emission tomography (PET), which detects metabolic changes in the brain using radiotracers that emit positrons. Functionalized quantum dots, in which quantum dots are coupled to target molecules, were investigated as attractive PET imaging agents [71]. This method is particularly suitable for monitoring changes in glucose metabolism, an important characteristic of many neurometabolic disorders. Although it is highly sensitive, its clinical use is hampered by the short half-life of radiotracers and the potential toxicity of some nanoparticle formulations [7]. NIRF near infrared fluorescence imaging uses gold nanoclusters or UCNPs (upconversion nanoparticles), which emit fluorescence in the near infrared band. This enabled deeper tissue penetration with less photobleaching and made it an important agent for the detection of amyloid plaques in Alzheimer's disease and other neurodegenerative diseases [72]. Although this method has proven promising in clinical practice, the low efficiency of targeting in vivo and the challenges associated with functionalizing nanoparticles specifically for biomarker detection have only found limited entry into clinical practice [73]. Photoacoustic imaging (PAI) is an innovative imaging method that converts absorbed light into ultrasonic signals, enabling high spatial resolution and real-time imaging of oxidative stress markers in neurometabolic disorders. Gold nanorods and carbon nanotubes with their intensive optical absorption are often used as contrast agents for PAI. While PAI has many advantages, it also has limitations, such as low penetration depth, and optimization of the stability of nanomaterials should be considered for long-term applications. SERS (Surface-Enhanced Raman Scattering) benefits Raman spectroscopy because it amplifies the analyte's molecular signals using nanoparticles [2]. This method enables the highly sensitive detection of biomolecular changes, including changes in lipid and protein composition, that occur in neurodegenerative diseases. Nonetheless, the need for specialized expertise and concerns about the potential toxicity of SERS-active nanoparticles have limited widespread use. Electrochemical biosensors based on either carbon nanomaterials or gold nanoparticles proved to be fast and highly specific to detect neurotransmitter imbalances [74]. Such biosensors are of considerable importance for monitoring dopamine in Parkinson's disease and other neurometabolic disorders [75]. Although promising, clinical performance is affected by issues related to biological fluid disturbances and sensor component stability issues. The use of gold nanoparticles is also beneficial for CT imaging, as gold has a high atomic number, which results in better X-ray absorption. This allows tissue structures to be displayed in high resolution and enables the identification of structural abnormalities associated with neurometabolic disorders. However, safety problems associated with radiation exposure and the need for biocompatible contrast agents prevent the widespread use of imaging methods to monitor minimal residual diseases [76]. We use optical coherence tomography (OCT) with silicon dioxide nanoparticles and carbon nanomaterials to obtain high-resolution imaging based on light backscattering. In addition, it is particularly useful for real-time monitoring of myelin degeneration and axonal damage in neurometabolic disorders at the single axon level. OCT allows deeper tissue penetration, but its clinical use is limited by the need for complex optical systems and its poor ability to visualize larger brain structures [69]. Theranostic nanoprobes (multifunctional nanoparticles (MNPs) such as dendrimers and liposomes) enable simultaneous imaging and therapeutic delivery. This enables precise monitoring of drug distribution and therapeutic response in neurometabolic disorders. Although promising, complex synthesis protocols and the risk of adverse side effects pose significant barriers to their clinical development [77].

Single-photon emission computed tomography (SPECT) uses nanometer-scale scintillation materials and radiolabeled nanoparticles to detect gamma-emitting isotopes, which enables functional imaging with high sensitivity. Especially for evaluating blood-brain barrier (BBB) permeability and metabolic pathways in neurodegenerative diseases. Nonetheless, it has low spatial resolution and concerns about radiotoxicity limit its wider use [78]. Molecular imaging (MI) uses functionalized nanopoints and dendrimer-based nanoparticles to sensitively detect biomolecular interactions. This approach is crucial for the early detection of disorders associated with enzyme defects or increased markers of oxidative stress in neurometabolic diseases. However, there are obstacles to wider application, including the complicated preparation of molecular probes, and barriers to clinical translation. Thermal imaging based on gold nanoshells and carbon nanotubes (CNT) enables real-time mapping of tissue heat generation (inflammation, tissue damage), which is useful for neurometabolic disorders [74]. Although it is a noninvasive and rapid diagnostic procedure, its lack of sensitivity and inability to quantify thermal responses limits its wider clinical utility. Mass spectrometry imaging (MSI) uses nanostructured surfaces such as MALDI-MS nanoparticles to ionize samples for spatial molecular analysis of tissues. This method is particularly suitable for monitoring changes in lipid metabolism in brain tissue. However, the expensive instruments and complex sample preparation requirements represent a major hurdle for clinical implementation. Nanotheranostics has revolutionized the diagnosis of neurometabolic disorders by combining sophisticated imaging techniques with precise therapy. Nanoparticles, with their large surface area and ability to transport different payloads, such as fluorescent dyes, can be used for imaging, tracking disease progression, etc. [61]. However, successful clinical implementation requires consideration of many factors, including toxicity, costs, regulatory aspects, and a balance for optimal nanoparticle delivery. Nanotheranostics show promise for improving the diagnosis, monitoring, and treatment of neurometabolic disorders, and future studies should be focused on developing novel, safer, more efficient, and economical nanotheranostic platforms suitable for treating neurometabolic disorders [78].

6 Recent Development in Nanotheranostics

There are various types of treatment techniques to address neuropsychiatric problems. These are important modalities such as behavioral training, psychotherapy, medical treatment, and rehabilitation techniques that are tailored to the individual patient. Together, these three treatment approaches aim at multimodal long-term recovery of affected patients and target various components of neuropsychiatric disorders [2]. When we carry out a comparative analysis of neuropsychiatric lists of disorders with regard to neuronal metabolism, little comes out: Neurometabolic disorders are primarily due to abnormal chemical reactions in the body. This abnormality can have a devastating effect on the vital biochemical reactions that are crucial for nerve cell functioning. Full knowledge of this fact is crucial, as the clinical symptoms found in most neuronal metabolic disorders correspond very closely to those found in neuropsychiatric disorders [54]. It goes without saying that this alarming similarity in clinical presentation can pose a major dilemma for doctors and alternative practitioners from a diagnostic point of view. What a sad scenario that the causes of most neuronal metabolic disorders are not identified in the expertise of doctors until they are finally atherogenic. Research has continued to focus on the topic of these two signaling pathways, with genetic elements and various changes in normal neuronal function contributing to the pathogenesis of these disorders. This heretic causes various hormonal imbalances that alter the general symptoms of its patients and make them significantly worse [79]. In contrast to the rigorously established and validated treatment protocols for neuropsychiatric disorders, the treatment of neuronal metabolic disorders is still somewhat nebulous because they are not clearly defined and the treatment strategies are not yet accepted by the medical community. Considering the aforementioned challenges, it is imperative to diagnose these cases early and on time to avoid the onset of potentially debilitating neurological symptoms in the patient population. If these neurological symptoms are poorly managed, there is real danger that they will lead to wider and even greater medical problems in the patients [80]. These comorbidities impose a significant socioeconomic toll on patients, as well as their families and on healthcare systems as a whole. The predicament emphasizes the growing and dire requirement for immediate intervention in the identification and management of such alarming health issues. In order to ease the burden and improve the quality of life for affected individuals, it is crucial to offer effective solutions [56]. There is a crucial necessity to thoroughly assess metabolic activity and how metabolism changes in cells. Such insight is a key requirement for enabling timely diagnosis of medical conditions. Cellular metabolism is the overall sequence of chemical reactions that are continually taking place within a living cell. Such a complex intramolecular process consists of multiple few enzymes functioning substrates that synthesizes ultimate real products required for cell growth, functioning and survival in the long run. Glycolysis is one of the metabolic pathways that stands out among the many, as it is the first and most basic pathway of cellular biochemistry that enables the efficient utilization and management of all carbon resources available [64].

MRS, which we rely on to generate accurate metabolic expression patterns and explore the many facets of cell metabolism, we can rely on a certain level of accuracy. In general, MRS has proven to be an extremely powerful and efficient imaging method that is often used to study brain metabolism in healthy volunteers and in various disease-related conditions [75]. However, many of the current methods used in evaluating the MR spectrum are often unsuitable for individual diseases in clinical practice. As a result, there was a demand for a deeper chemical characterization of brain metabolites using a more comprehensive and novel approach [81]. In this context, the development of a novel fluorescent probe made of graphene oxide, which is specifically tailored to the precise detection of metabolic production associated with Alzheimer's pathology, was an important step forward in this process [56]. Interestingly, however, some important players in cellular glucose metabolism also appear to be selectively absorbed by malignant cells. Such a feature can be valuable for insights and diagnostically useful MRI contrasts, which could ultimately be used to better understand and ultimately diagnose these many tumors. All in all, these advances form the basis for better diagnostic approaches and therapeutic strategies, which can serve as important nodes in clinical practice [73].

7 Role of Nanotheranostic Genetic Neurometabolic Disorders

In recent years, nanotheranostic modalities based on the rapidly evolving and revolutionary field of nanotechnology have evolved into some of the most popular and important topics that are attracting the interest of researchers in the scientific community [54]. The combination of these novel, highly efficient nanotheranostic strategies could make a decisive contribution to fundamentally changing the general paradigm of diagnosing and monitoring diseases, particularly with regard to diseases that are typically life-threatening and are associated with high morbidity, which can have a negative impact on patients' quality of life [66]. gNMDs is a broad and heterogeneous group of diseases caused by dysfunction or mutation of the genes that are essential for proteins within critical intermediate metabolic pathways. The strictly protective blood-brain barrier (BBB) not only poses several obstacles to the development of effective drug delivery systems, but also represents a crucial and primary challenge when it comes to transporting essential genetic material through complicated and complex brain structures [47]. Multifunctional nanoparticles that cross the BBB have been developed for nanoimaging and DNA transport as a result of recent advances in nanotechnology. As a result, they also serve as an effective imaging modality for precise diagnosis and long-term monitoring of these GNMDs [41]. In the last year, the incidence of three significant neurodegenerative disorders (NDD), namely Alzheimer's disease, amyotrophic lateral sclerosis (ALS) and Parkinson's disease, has caused major concern and concern in the medical community as their prevalence has increased dramatically. Many feasibility studies in vitro and in vivo have shown and implied that nanoparticles can be very effective and promising in their functions for the diagnosis and treatment of genetic neurometabolic disorders [63]. They provide the basis for such innovative therapeutic solutions and significantly improve treatment outcomes, which literally revolutionizes knowledge about medical treatments. This is especially true when you consider that this complex organ only accounts for around 2% of body weight but uses around 20–30% of the energy that the body processes. These impressive statistics show how efficient our brain is and how much energy it requires in relation to its size. The so-called blood-brain barrier (BBB) is one of the critical functional limiting factors that exist in the interwoven plexus of the body's capillary system. This impenetrable wall provides an essential defense structure and consists of tight connections that are located in the cerebral endothelial cells and have extremely low pinocytotic activity [82]. This makes it highly selective with regard to what it can penetrate. This protective barrier makes the innovative design and application of the Nano-Drug Delivery System (Nano-DDS) difficult and also poses major challenges when it comes to transporting genetic material to this sensitive and sensitive environment in the brain. Genetic neurometabolic disorders (GNMD) is a term that includes complex disorders related to the abnormal expression of genes responsible for synthesizing certain proteins involved in a wide variety of metabolic processes in the central nervous system [20].

Therefore, there is currently no comprehensive or definitive cure for such complex diseases and the therapeutics are unable to cure the disease but only alleviate and alleviate patients' debilitating symptoms. According to reports from anecdotal reports and new studies, several factors, such as lifestyle, genetic mutations and specific brain trauma, are associated with the increased incidence of GNMDs [3]. In fact, at the genetic level, most mutations associated with these disorders are inextricably linked to proteins, which serve as key wheels for the basic needs of neurological and metabolic processes in the brain and centralize the complexity of these disorders [80]. As a result, the range of therapeutic approaches available to patients is chronically limited, as already known drugs and macromolecules represent the penetration of the highly selective and defensive blood-brain barrier. The limited understanding of this issue remains a major obstacle to therapies, in particular for GNMDs, which are designed to actually attack the processes underlying the disease, which underscores the need for new and innovative approaches for targeted delivery of drugs to the brain [3].

8 Biocompatibility and Immune Responses of Nanotheranostic

In recent years, intensive and thorough research efforts have been made to advance the development of a pioneering nanotheranostic platform with great effectiveness for the diagnosis and treatment of various prominent neurometabolic disorders [80]. While these breakthrough advances were taking place, there was a clear focus on the molecular characterization of functional nanomaterials an important component that is essential for continuing to transition to clinical applications and increasing therapeutic capacity [83]. This novel manufacturing strategy was developed specifically with a number of extremely useful properties for biomedical applications, such as biocompatibility, self-organization, blood-brain barrier penetration, suitable surface charge for targeted delivery, and adjustable stimulus-response functions [84].

Recent studies have also described and documented these important components and their properties in detail. The MnO_2-functionalized nanoplatform has proven effective in O_2/A and GSH synergic-enhanced photoluminescence imaging and thus offers the option of further EEG-controlled multiple SDS therapy—an important strategy that should be considered in the sophisticated treatment of neurometabolic disorders of the refectories [84]. In addition, approaches or methods based on the body distribution, biocompatibility and immune effects of the novel nanomaterials were investigated and researched systematically and after careful consideration, which would help to further advance the transition to the indispensable clinical application processes and protocols (Fig. 15.3). With the much more exciting implementation of nanotherapy to combat critical and mass-resistant diseases such as ischemic strokes, gliomas, and multiple neurodegenerative diseases, great and creative nanoplatforms with precise multimode imaging are beginning to flourish [32]. Some of them include the use of synergistically activated partial thromboplastin time (APTT) assays as well as highly efficient and cutting-edge carriers for the CNS (central nervous system) gene and siRNA therapy, which are becoming increasingly important and are sparking research interest in this area. However, most of these promising nanosystems are still in the exploration phase of development [64]. In addition, there are many complex biofunctions associated with these functionalized nanoplatforms, and there are still concerns about potential

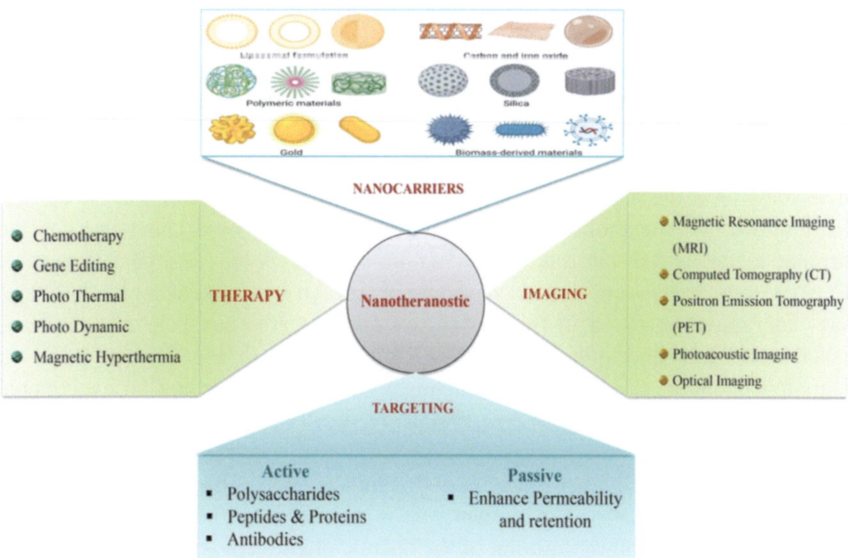

Fig. 15.3 Nanotheranostic biocompatibility and immunological reactions

toxicological issues with their use, which limit their future transfer to real clinical procedures and applications. However, in view of such a dynamic and young field of research, it is only a continuation of ongoing work to find a way to complete clinical development of nanotheranotics to combat neurometabolic diseases [85].

9 Clinical Applications of Nanotheranostics

Therapy based on nanotheranostics is one of the most stimulating concepts in modern medicine. It is an integrated approach that provides effective therapeutic and accurate diagnostic or imaging effects for various diseases within a single comprehensive treatment and analysis platform [54]. This novel method not only significantly simplifies historically complex work processes, but also increases the need to closely monitor the individual aspects. These things are interrelated and include the detailed description of the interplay of drug delivery mechanisms, the controlled release of therapeutic agents, the difference in treatment outcome, and the subtle evolution of the disease over time [55]. In recent months, the pace of progress has really accelerated, and now there is truly an impressive number of approaches and tools, both diagnostic and therapeutic, that are becoming available and can be used. This explosion has led to a completely new understanding and expansion of the number of clinical findings that can be used for a particular patient [60]. Specialized neuroradiology, which addresses specific neurodegenerative diseases, is uniquely positioned to utilize the prudent and strategic application of advanced molecular imaging techniques and functional imaging techniques. These invaluable techniques have proven essential for real-time assessment and evaluation of the complex neurodegenerative processes that occur in the background in connection with these difficult disease states. As a result, this pioneering theranostic strategy, which is used in diagnostic clinical studies, would also be applicable to diseases of the neurodegenerative system and would facilitate early diagnosis. Implementing such techniques can significantly improve management and, as a result, lead to more favorable treatment outcomes [86]. However, it is important to recognize that in vivo brain imaging is the most critical element, particularly when it comes to the uptake, distribution, and metabolic properties of nanoparticles. These characteristics are intrinsic elements that are fundamental to the clinical success and performance of these complex therapeutic approaches that are currently under development. Therefore, continuous research in this novel and exciting area can help to gain more valuable insights and improve the care of patients with the various forms of neurodegeneration [66]. Theranostic nanoparticles (NPs) represent an exceptional opportunity to both demonstrate and advance molecular imaging. It has the potential to revolutionize noninvasive approaches to elucidate a wide range of biological and biochemical events in living organisms in their intact cells (Table 15.2). Noninvasive molecular imaging plays an important role in particular in the early diagnosis of diseases and the careful monitoring of therapeutic effectiveness in vivo [72].

Because of this vital function, it is therefore not surprising that some drugs do not work well in a clinical setting. The theranostic strategy of in vivo drugs is promising

Table 15.2 Clinical applications of nanotheranostic

S. no.	Clinical applications	Nanotheranostics	Scientific mechanisms	Target disorder
1	Imaging and diagnosis of amyloid deposits	Gold nanoparticles (AuNPs)	Plasmon resonance of gold nanoparticles enhances contrast in imaging techniques like MRI and CT for detecting amyloid-beta plaques.	Alzheimer's disease
		Polyethylene glycol-functionalized silica nanoparticles (PEG-SiNPs)	Improve bioavailability and target amyloid-beta deposits for fluorescence imaging.	
2	Targeted drug delivery to neuronal cells	Liposomes composed of phosphatidylcholine (PC)	Liposomes improve blood-brain barrier (BBB) permeability, allowing sustained release of L-DOPA for dopaminergic neuron replenishment.	Parkinson's disease
		Poly(butylcyanoacrylate) nanoparticles (PBCA NPs)	Cross the BBB with drug payloads for sustained release into affected brain regions.	
3	Gene therapy for enzyme deficiencies	Poly(lactic-co-glycolic acid) nanoparticles (PLGA NPs)	Polymeric nanoparticles protect genetic material from degradation, enabling targeted transfection into neurons via endocytosis.	Tay-Sachs disease
		Chitosan-based nanoparticles (deacetylated poly(D-glucosamine))	Enhance transfection efficiency with reduced cytotoxicity for delivering genetic material to lysosomal cells.	
4	Monitoring mitochondrial dysfunction	Cadmium selenide quantum dots (CdSe QDs)	Quantum dots emit fluorescent signals when exposed to dysfunctional mitochondria, enabling real-time imaging of mitochondrial health.	Leigh syndrome
		Triphenylphosphonium-conjugated lipophilic nanoparticles (TPP-LNPs)	Localize to mitochondria and emit fluorescent signals or deliver therapeutics directly to the organelle.	
5	Enzyme replacement therapy delivery	Lipid-based nanoparticles (lecithin and cholesterol)	Nanoparticles stabilize enzymes and target lysosomes in affected cells, restoring metabolic function in glucocerebrosidase-deficient conditions.	Gaucher's disease
		Poly(lactic-co-glycolic acid) nanoparticles (PLGA NPs) with mannose	Enhance lysosomal uptake by targeting mannose receptors on macrophages.	

(continued)

Table 15.2 (continued)

S. no.	Clinical applications	Nanotheranostics	Scientific mechanisms	Target disorder
6	Reduction of oxidative stress	Cerium oxide nanoparticles (CeO_2 NPs)	Cerium oxide nanoparticles mimic superoxide dismutase and catalase activities, neutralizing reactive oxygen species (ROS) in neurodegenerative conditions.	Amyotrophic lateral sclerosis (ALS)
		Selenium nanoparticles (SeNPs)	Scavenge reactive oxygen species and protect neuronal cells from oxidative damage.	
7	Real-time glucose monitoring in brain tissues	Fluorescent boronic acid derivatives	Fluorescent nanomaterials change emission spectra in response to glucose concentrations, enabling dynamic monitoring of glucose in brain microenvironments.	Diabetes-related neuropathy
		Graphene quantum dots (GQDs)	Highly sensitive detection of glucose through fluorescence intensity modulation.	
8	Theranostic for lysosomal storage disorders	Mesoporous silica nanoparticles (MSNs, SiO_2-based)	Mesoporous silica facilitates simultaneous enzyme delivery and siRNA therapy, targeting lysosomal storage defects and modulating sphingomyelin accumulation.	Niemann-Pick disease
		Poly(amidoamine) dendrimers (PAMAMs)	Dendrimers improve lysosomal drug accumulation, enhancing therapeutic outcomes.	
9	Neuroinflammation imaging and suppression	Iron oxide nanoparticles (Fe_3O_4 NPs)	Superparamagnetic properties enhance MRI visualization, and surface-functionalized nanoparticles deliver anti-inflammatory agents to suppress immune activation in CNS.	Multiple sclerosis (MS)
		Liposomes encapsulating dexamethasone (PC and cholesterol)	Deliver corticosteroids to inflamed neural tissues, reducing inflammation and oxidative stress.	
10	Tracking neurotransmitter release	Graphene oxide nanosheets (GO)	Graphene oxide nanosheets are engineered to fluoresce in response to neurotransmitter binding, enabling real-time synaptic activity tracking and dysfunction analysis.	Huntington's disease
		Silicon nanoparticles (SiNPs)	Engineered for neurotransmitter-responsive fluorescence or Raman signals to monitor synaptic transmission.	

for many applications that are revolutionizing the face of personalized medicine, such as: efficient screening of therapeutics, approximation of the drug dose required for therapeutic effectiveness in patients, and monitoring the side effects of these drugs over the course of these therapies [87]. The various theranostic applications of nanomaterials have been roughly divided into two main classes based on their intrinsic physicochemical nature and corresponding imaging properties, which use different imaging methods: morphological and functional imaging. Because of their adjustable biological properties and behavior, nanoparticles have been extensively investigated and have found a variety of applications [80]. These advanced nanomaterials are used as therapeutic and diagnostic agents in a wide range of biomedical applications. Over the last 10 years, these materials have been continuously developed and improved. The theranostic nanoparticles, which have been carefully developed for different types of existing imaging modalities, are also constantly being developed and optimized due to the dynamic requirements of modern medical science and patient care.

10 Conclusion

Numerous studies have been carried out to investigate potential treatments for neurometabolic disorders (NDs) based on innovative bio-nano applications that have just lately surfaced in the medical industry. The term "NDs" refers to a group of uncommon diseases, particularly inborn errors of metabolism (IEMs), which are distinguished by their wide range of phenotypic and genomic variations. Effective treatment strategies are severely hampered by the chronic and progressive neurodegeneration caused by these disorders. Genetic variability among patients, obstacles to appropriate treatment, and the effective delivery of therapeutic agents to the targeted cellular compartments are the main causes of the challenges in providing such treatments for NDs. Some important missing pieces are brought up in this conversation, along with potential future directions for interventions that can support clinical translation of treatment modalities. Pharmacological chaperones, gene therapy, substrate reduction techniques, enzyme replacement strategies, and other treatments that can be applied to inborn errors of metabolism that are aetiologically linked to these neurometabolic disorders are some of the most noteworthy options available. Comprehending these treatment modalities is crucial for future diagnosis and treatment protocols specifically designed for NDs, as well as for current therapeutic approaches.

NDs can be considered one of the most attractive diseases that need to be cured or significantly alleviated by next-generation nanomedicine. When investigating previous treatment methods, disrupting the blood-brain barrier using nanocarriers and artificial nanocapsules represents promising opportunities to replace or supplement the existing small bioactive chemicals and recombinant proteins. The remarkable development of Bionano materials can be driven forward rapidly and with great effectiveness if the diverse scientific and technological support provided by a variety of knowledge-integrated disciplines is continuously advanced and improved. In the face of a variety of unknown potential mechanisms, only expert knowledge, rules and expertise in nanotheranostic approaches, the nano interface and complex systems biology could help to ensure that the ND field is no longer neglected in the face of a variety of unknown potential mechanisms, which would enable revolutionary progress in diagnosis and treatment development. However, it is essential to think ahead of the present in this rapidly evolving area.

References

1. Chu, D., Vu, H., Tien, T., & Vu, T. (2023). Nanotechnology and nucleic acid nanoparticles for treatment of metabolic disorders. *OpenNano, 13*, 100181. https://doi.org/10.1016/j.onano.2023.100181
2. Zorkina, Y., et al. (2020). Nano carrier drug delivery systems for the treatment of neuropsychiatric disorders: Advantages and limitations. *Molecules, 25*(22), 5294. https://doi.org/10.3390/MOLECULES25225294
3. Wilar, G., Suhandi, C., Wathoni, N., Fukunaga, K., Wilar, G., & Suhandi, C. (2024). Nanoparticle-based drug delivery systems enhance treatment of cognitive defects. *International Journal of Nanomedicine, 19*, 11357–11378. https://doi.org/10.2147/IJN.S484838
4. Duan, L., et al. (2023). Nanoparticle-based drug delivery systems: An inspiring therapeutic strategy for neurodegenerative diseases. *Polymers, 15*, 1–19.
5. Zha, S., Liu, H., Li, H., Li, H., Wong, K. L., & All, A. H. (2024). Functionalized nanomaterials capable of crossing the blood-brain barrier. *ACS Nano, 18*(3), 1820–1845. https://doi.org/10.1021/acsnano.3c10674
6. Biswas, A., et al. (2021). Clinico-radiological phenotyping and diagnostic pathways in childhood neurometabolic disorders—A practical introductory guide. *Translational Pediatrics, 10*(4), 1201–1230. https://doi.org/10.21037/tp-20-335
7. Singh, A. P., & Biswas, A. (2019). Targeted therapy in chronic diseases using nanomaterial-based drug delivery vehicles. *Signal Transduction and Targeted Therapy, 4*, 1–21. https://doi.org/10.1038/s41392-019-0068-3
8. Kim, J., Um, H., Kim, N. H., & Kim, D. (2023). Potential Alzheimer's disease therapeutic nano-platform: Discovery of amyloid-beta plaque disaggregating agent and brain-targeted delivery system using porous silicon nanoparticles. *Bioactive Materials, 24*, 497–506. https://pmc.ncbi.nlm.nih.gov/articles/PMC9841037/
9. Yang, Y., & Wang, X. (2024). Nano-drug delivery systems (NDDS) in metabolic dysfunction-associated steatotic liver disease (MASLD): Current status, prospects and challenges. *Frontiers in Pharmacology, 15*, 1–18. https://doi.org/10.3389/fphar.2024.1419384
10. Paprocka, J., & Emich-Widera, E. (2020). Neuroimaging of basal ganglia in neurometabolic diseases in children. *Brain Sciences, 10*, 1–16.
11. Singh, A., Prasad, L. B., Shiv, K., Kumar, R., & Garai, S. (2023). Synthesis, characterization, and in vitro antibacterial and cytotoxic study of Co(II), Ni(II), Cu(II), and Zn(II) complexes of N-(4-methoxybenzyl) N-(phenylethyl) dithiocarbamate ligand. *Journal of Molecular Structure, 1288*, 135835. https://doi.org/10.1016/j.molstruc.2023.135835
12. Kumar, A., et al. (2022). Acute and subacute toxicity study of ethanolic extract of Calotropis procera (Aiton) Dryand flower in Swiss albino mice. *Phytomedicine Plus, 2*(2), 100224. https://doi.org/10.1016/j.phyplu.2022.100224
13. Karimzadeh, P., et al. (2014). Propionic acidemia: Diagnosis and neuroimaging findings of this neurometabolic disorder. *Iranian Journal of Child Neurology, 8*(1), 2–5.
14. Karimzadeh, P., et al. (2013). Methylmalonic acidemia: Diagnosis and neuroimaging findings of this neurometabolic disorder (an Iranian pediatric case series). *Iranian Journal of Child Neurology, 7*(3), 63–66.
15. Ekhator, C., et al. (2023). Advances and opportunities in nanoparticle drug delivery for central nervous system disorders: A review of current advances. *Cureus, 15*(8), e44302. https://doi.org/10.7759/cureus.44302

16. Harsh Pandey, Y. B. T., Srivastava, S., & Kumar, R. (2018). Preclinical acute and repeated dose toxicity of Pueraria tuberosa (PTWE) on charles foster rats. *International Journal of Pharmaceutical Sciences and Research, 9*(11), 4572–4581. https://doi.org/10.13040/IJPSR.0975-8232.9(11).4572-81
17. Kasina, V., Mownn, R. J., Bahal, R., & Sartor, G. C. (2022). Nanoparticle delivery systems for substance use disorder. *Neuropsychopharmacology, 47*, 1431–1439. https://doi.org/10.1038/s41386-022-01311-7
18. Singh, A., et al. (2024). Synthesis, crystal structure and screening for anticonvulsant and antianxiety activities of three new Ni(II), Cu(II), and Zn(II) dithiocarbamate complexes. *Journal of Molecular Structure, 1298*(P2), 137052. https://doi.org/10.1016/j.molstruc.2023.137052
19. Ozceylan, O., & Sezgin-bayindir, Z. (2024). Current overview on the use of nanosized drug delivery systems in the treatment of neurodegenerative diseases. *ACS Omega, 9*(33), 35223–35242. https://doi.org/10.1021/acsomega.4c01774
20. Kumar, R., Chhikara, B. S., Gulia, K., & Chhillar, M. (2021). Review of nanotheranostics for molecular mechanisms underlying psychiatric disorders and commensurate nanotherapeutics for neuropsychiatry: The mind knockout. *Nanotheranostics, 5*(3), 288–308. https://doi.org/10.7150/ntno.49619
21. Willemsen, M. A., Harting, I., & Wevers, R. A. (2016). Neurometabolic disorders. *Neurology Clinical Practice, 6*(4), 348–357. https://doi.org/10.1212/CPJ.0000000000000266
22. Kumar, A., et al. (2022). Toxicity (acute and subacute) assessment and in-vivo antiurolithiatic activity of ethanolic extract of Caesalpinia bonducella seed in albino Wistar rat. *Journal of Applied Pharmaceutical Science, 12*(2), 187–197. https://doi.org/10.7324/JAPS.2022.120220
23. Fernández-Eulate, G., et al. (2022). Diagnostic approach in adult-onset neurometabolic diseases. *Journal of Neurology, Neurosurgery, and Psychiatry, 93*(4), 413–421. https://doi.org/10.1136/jnnp-2021-328045
24. BoAli, A. Y., Alfadhel, M., & Tabarki, B. (2018). Neurometabolic disorders and congenital malformations of the central nervous system. *Neurosciences, 23*(2), 97–103. https://doi.org/10.17712/nsj.2018.2.20170481
25. Pandey, H., Srivastava, S., Kumar, A., Kumar, R., & Bhusan Tripathi, Y. (2019). Pharmacokinetics study of Puerarin absorption in blood after consumption of Pueraria tuberosa water extract (PTWE) by rats. *Exploration of Animal and Medical Research, 9*(2), 188–196.
26. Dhiman, R., Kumar, D., Kumar, B., & Pandey, B. L. (2016). Quantitative determination of atorvastatin, ortho hydroxy atorvastatin, para hydroxy atorvastatin in human plasma using Rosuvastatin as internal standard by LC-MS/MS. *International Journal of Pharmaceutical and Chemical Sciences, 4*(4), 487–500.
27. Vijaya Durga, R. D. (2015). Determination of S-amlodipine in human plasma using Tizanidine as internal standard by Lc/Ms/Ms method. *Journal of Bioanalysis & Biomedicine, 7*(3), 91–96. https://doi.org/10.4172/1948-593x.1000130
28. Tarailo-Graovac, M., et al. (2016). Exome sequencing and the management of neurometabolic disorders. *The New England Journal of Medicine, 374*(23), 2246–2255. https://doi.org/10.1056/nejmoa1515792
29. Kurungottu, P., et al. (2024). Biodegradable nanocomposite of ZnS(Mn) quantum dots immobilized graphene oxide for bioimaging applications. *Nanotheranostics, 8*(2), 150–162. https://doi.org/10.7150/ntno.87536
30. Swanson, P. D. (1995). Diagnosis of inherited metabolic disorders affecting the nervous system. *Journal of Neurology, Neurosurgery, and Psychiatry, 59*(5), 460–470. https://doi.org/10.1136/jnnp.59.5.460
31. Kumar, R., Kumar, B., Kumar, A., Kumar, A., & Singh, M. (2022). GC-MS analysis of phytocomponents in the methanol extract of Premna latifolia Roxb. *Pharmacognosy Research, 14*(1), 19–23. https://doi.org/10.5530/pres.14.1.4
32. Croitoru, G. A., et al. (2024). Nanomaterials in immunology: Bridging innovative approaches in immune modulation, diagnostics, and therapy. *Journal of Functional Biomaterials, 15*(8), 225. https://doi.org/10.3390/jfb15080225

33. Kumar, A., et al. (2021). Antiurolithiasis, antioxidant, anti-inflammatory, analgesic, and diuretic activity of ethanolic extract of seeds of Caesalpinia bonducella. *International Journal of Pharmaceutical Investigation, 11*(3), 306–311. https://doi.org/10.5530/ijpi.2021.3.54
34. Garg, R., et al. (2022). COVID-19: A challenge towards the sustainability of health in platform era. *Neuroquantology, 20*(11), 2503–2519. https://doi.org/10.14704/NQ.2022.20.11.NQ66250
35. Zheng, D., et al. (2020). Her2-targeted multifunctional nano-theranostic platform mediates tumor microenvironment remodeling and immune activation for breast cancer treatment. *International Journal of Nanomedicine, 15*, 10007–10028. https://doi.org/10.2147/IJN.S271213
36. Singh, A., et al. (2024). Dithiocarbamate-based novel anti-histaminic agents: Synthesis, characterization, crystal structure and thermal study. *Dalton Transactions, 53*, 14077–14088. https://doi.org/10.1039/d4dt01777c
37. Sonali, et al. (2018). Nanotheranostics: Emerging strategies for early diagnosis and therapy of brain cancer. *Nanotheranostics, 2*(1), 70–86. https://doi.org/10.7150/ntno.21638
38. Achar, A., Myers, R., & Ghosh, C. (2021). Drug delivery challenges in brain disorders across the blood-brain barrier: Novel methods and future considerations for improved therapy. *Biomedicines, 9*, 1834.
39. Singh, M., et al. (2023). Pharmacokinetic screening to estimate the drug likeliness characteristics of selected herbal anticancer drugs. *Research Journal of Pharmacy and Technology, 16*(7), 3422–3426. https://doi.org/10.52711/0974-360X.2023.00566
40. Dhiman, R., Durga, V., & Rio, A. J. (2015). Bioanalytical method for lornoxicam determination in human plasma by using piroxicam as internal standard by LC-MS/MS. *Journal of Chemical and Pharmaceutical Research, 7*(8), 54–60.
41. Wu, D., & Chen, Q. (2023). The blood-brain barrier: Structure, regulation and drug delivery. *Signal Transduction and Targeted Therapy, 8*, 217. https://doi.org/10.1038/s41392-023-01481-w
42. Lee, I. C. (2021). Approach to neurological channelopathies and neurometabolic disorders in newborns. *Life, 11*(11), 1244. https://doi.org/10.3390/life11111244
43. Dhiman, R., Kumar, B., Kumar, D., & Kumar, A. (2016). Liquid-liquid extraction method for Ziprasidone (ZRS) bioanalysis by using ZRS-D 8 (stable isotope) as internal standard. *Journal of Chemical and Pharmaceutical Research, 8*(3), 722–727.
44. Kumar, R., Dutt, S., Tripathi, A. D., Singh, A. K., Chaturvedi, V. K., & Singh, S. K. (2024). Navigating safety and toxicity challenges in nanomedicine: Strategies, assessment, and mitigation. In V. Gautam, R. Kumar, K. Das Manandhar, & S. C. Kamble (Eds.), *Nanomedicine: Innovations, applications, and breakthroughs in the quest for health and medicine's future* (pp. 15–37). Springer Nature Switzerland. https://doi.org/10.1007/978-3-031-72467-1_2
45. Upreti, P., Devhare, L. D., Abdulmageed, L. H., Kumar, Y. G., Kumar, R., & Dharmamoorthy, G. (2024). Combatting antibiotic resistance: Leveraging fecal microbial transplantation for gut health. In M. Gangwar & G. Nath (Eds.), *Emerging paradigms for antibiotic-resistant infections: Beyond the pill* (pp. 211–232). Springer Nature Singapore. https://doi.org/10.1007/978-981-97-5272-0_10
46. Ciuti, G., et al. (2020). Frontiers of robotic colonoscopy: A comprehensive review of robotic colonoscopes and technologies. *Journal of Clinical Medicine, 9*(6), 1648. https://doi.org/10.3390/jcm9061648
47. Hersh, A. M., Alomari, S., & Tyler, B. M. (2022). Crossing the blood-brain barrier: Advances in nanoparticle technology for drug delivery in neuro-oncology. *International Journal of Molecular Sciences, 23*(8), 4153. https://doi.org/10.3390/ijms23084153
48. Hamid, A., et al. (2024). Artificial intelligence (AI) based colonoscopy. In *Nanobiotechnology and artificial intelligence in gastrointestinal diseases*. IOP Publishing Ltd. https://doi.org/10.1088/978-0-7503-6134-7CH12
49. Kumar, R., Dutt, S., Goyal, A., Kumar, A., & Kumar, B. (2024). Nanotechnology in gastrointestinal endoscopy. In *Nanobiotechnology and artificial intelligence in gastrointestinal diseases*. IOP Publishing Ltd. https://doi.org/10.1088/978-0-7503-6134-7CH5

50. Kumar, R., Singh, Y., Piplani, M., Bhateja, P., Garg, R., & Kumar, B. (2024). The chemistry and pharmacological study of Copaifera spp. oils. In *The chemistry inside spices & herbs: Research and development* (Vol. 4, pp. 79–116). Bentham Science Publishers. https://doi.org/10.2174/9789815196832124040006
51. Rao, S. S., Lannutti, J. J., Viapiano, M. S., Sarkar, A., & Winter, J. O. (2014). Toward 3D biomimetic models to understand the behavior of glioblastoma multiforme cells. *Tissue Engineering Part B: Reviews, 20*(4), 314–327. https://doi.org/10.1089/ten.teb.2013.0227
52. Kumar, A., Singh, P., Kumar, R., & Dutt, S. (2024). Novel drug delivery systems for inflammatory bowel disease. In *Nanobiotechnology and artificial intelligence in gastrointestinal diseases*. IOP Publishing Ltd. https://doi.org/10.1088/978-0-7503-6134-7CH4
53. Kumar, A., et al. (2022). Assessment of pain-ameliorative potential of ethanolic extract of Calotropis procera and its active components characterization by HPTLC and GC-MS analysis. *Systematic Reviews in Pharmacy, 13*(6), 659–664. https://doi.org/10.31858/0975-8453.13.6.659-664
54. Kumar, A., et al. (2020). Nanotheranostic applications for detection and targeting neurodegenerative diseases. *Frontiers in Neuroscience, 14*, 1–11. https://doi.org/10.3389/fnins.2020.00305
55. Padmanabhan, P., et al. (2021). Nanotheranostic agents for neurodegenerative diseases. *Emerging Topics In Life Sciences, 4*(6), 645–675. https://doi.org/10.1042/ETLS20190141
56. Sharma, M., Dube, T., Chibh, S., Kour, A., Mishra, J., & Panda, J. J. (2019). Nanotheranostics, a future remedy of neurological disorders. *Expert Opinion on Drug Delivery, 16*(2), 113–128. https://doi.org/10.1080/17425247.2019.1562443
57. Kumar, R., et al. (2022). HPTLC and GC-MS analyses of biologically active extracts and fractions from Premna latifolia Roxb leaves reveal anti-inflammatory and analgesic effects. *Systematic Reviews in Pharmacy, 13*(58), 544–551.
58. Buchholz, O., et al. (2024). In situ theranostic platform combining highly localized magnetic fluid hyperthermia, magnetic particle imaging, and thermometry in 3D. *Theranostics, 14*(1), 324–340. https://doi.org/10.7150/thno.86759
59. Kumar, R., Kumar, B., Kumar, A., Kumar, A., & Singh, M. (2021). HPTLC and GC-MS analyses of biologically active extracts and fractions from Premna latifolia Roxb leaves. *Journal of Neurological Disorders, 9*, 1.
60. Muthu, M. S., Leong, D. T., Mei, L., & Feng, S. S. (2014). Nanotheranostics – Application and further development of nanomedicine strategies for advanced theranostics. *Theranostics, 4*(6), 660–677. https://doi.org/10.7150/thno.8698
61. Wang, X., et al. (2024). Fluorescent probes for disease diagnosis. *Chemical Reviews, 124*(11), 7106–7164. https://doi.org/10.1021/acs.chemrev.3c00776
62. Kumar, R., Kumar, B., & Kumar, A. (2017). Pharmacokinetic profile of antitubercular drugs and their clinical applications in therapeutic drug monitoring. *Asian Journal of Pharmacy and Pharmacology, 3*(5), 145–154. Available: http://ajpp.in/uploaded/p71.pdf
63. Choi, H. K., Chen, M., Goldston, L. L., & Lee, K. B. (2024). Extracellular vesicles as nanotheranostic platforms for targeted neurological disorder interventions. *Nano Convergence, 11*(1), 19. https://doi.org/10.1186/s40580-024-00426-5
64. Li, W., et al. (2023). Cell membrane-based nanomaterials for theranostics of central nervous system diseases. *Journal of Nanobiotechnology, 21*(1), 1–14. https://doi.org/10.1186/s12951-023-02004-z
65. Dhiman, R., Durga, V., Jayasankar, K., & Raj, S. (2015). Solid phase extraction: Potential carbocisteine bioanalysis using carbocisteine 13c3 stable isotope: LC-MS/MS technique with pharmacokinetics on healthy human subjects. *Der Pharmacia Lettre, 7*(8), 126–134.
66. Ramanathan, S., et al. (2018). Theranostic applications of nanoparticles in neurodegenerative disorders. *International Journal of Nanomedicine, 13*, 5561–5576. https://doi.org/10.2147/IJN.S149022
67. Gautam, A., & Chaudhary, V. (2023). *Theranostic applications of nanotechnology in neurological disorders*. Springer. https://doi.org/10.1007/978-981-99-9510-3

68. Xiao, Y., Paudel, R., Liu, J. U. N., Ma, C., Zhang, Z., & Zhou, S. (2016). MRI contrast agents: Classification and application. *International Journal of Molecular Medicine, 38*, 1319–1326. https://doi.org/10.3892/ijmm.2016.2744
69. Hu, J., et al. (2024). Neurometabolic topography and associations with cognition in Alzheimer's disease: A whole-brain high-resolution 3D MRSI study. *Alzheimer's & Dementia, 20*, 6407–6422. https://doi.org/10.1002/alz.14137
70. Dhiman, R., et al. (2015). Novel LC-MS/MS bioanalytical method validation for gabapentin in human plasma in accordance with current regulatory guidances. *International Journal of Pharmaceutical Chemistry Science, 4*(4), 462–468. Available: http://www.ijpcsonline.com/files/08-10-15/07-1035.pdf
71. Padmanabhan, P., Kumar, A., Kumar, S., Kumar, R., & Gulyás, B. (2016). Nanoparticles in practice for molecular-imaging applications: An overview. *Acta Biomaterialia, 41*, 1–16.
72. Abueva, C. D. (2021). Photo-triggered theranostic nanoparticles in cancer therapy. *Medical Lasers, 10*(1), 7–14. https://doi.org/10.25289/ml.2021.10.1.7
73. Jun, Y. W., et al. (2019). Frontiers in probing Alzheimer's disease biomarkers with fluorescent small molecules. *ACS Central Science, 5*(2), 209–217. https://doi.org/10.1021/acscentsci.8b00951
74. Shabani, L., Abbasi, M., Azarnew, Z., Amani, A. M., & Vaez, A. (2023). Neuro-nanotechnology: Diagnostic and therapeutic nano-based strategies in applied neuroscience. *Biomedical Engineering Online, 22*, 1–41. https://doi.org/10.1186/s12938-022-01062-y
75. Krsek, A., & Baticic, L. (2024). Nanotechnology-driven therapeutic innovations in neurodegenerative disorders: A focus on Alzheimer's and Parkinson's disease. *Future Pharmacology, 4*, 352–379.
76. Faiz, K., Lam, F. C., Chen, J., Kasper, E. M., & Salehi, F. (2022). The emerging applications of nanotechnology in neuroimaging: A comprehensive review. *Frontiers in Bioengineering and Biotechnology, 10*, 1–7. https://doi.org/10.3389/fbioe.2022.855195
77. Huang, Y., Guo, X., Wu, Y., Chen, X., Feng, L., & Xie, N. (2024). Nanotechnology's frontier in combatting infectious and inflammatory diseases: Prevention and treatment. *Signal Transduction and Targeted Therapy, 9*, 34. https://doi.org/10.1038/s41392-024-01745-z
78. Du, W., et al. (2021). Inorganic nanomaterial for biomedical imaging of brain diseases. *Molecules, 26*(23), 7340.
79. Liu, X., Liu, Y., & Liu, Q. (2023). Fluorescent sensing platforms for detecting and imaging the biomarkers of Alzheimer's disease. *Biosensors, 13*(5), 1–15. https://doi.org/10.3390/bios13050515
80. Vashist, A., et al. (2023). Recent advances in nanotherapeutics for neurological disorders. *ACS Applied Bio Materials, 6*(7), 2614–2621. https://doi.org/10.1021/acsabm.3c00254
81. Le, P. G., Le, H. T. N., Kim, H. E., & Cho, S. (2023). SAM-support-based electrochemical sensor for aβ biomarker detection of Alzheimer's disease. *Biosensors, 13*(8), 1–15. https://doi.org/10.3390/bios13080809
82. Teleanu, D. M., Chircov, C., Grumezescu, A. M., Volceanov, A., & Teleanu, R. I. (2018). Blood-brain delivery methods using nanotechnology. *Pharmaceutics, 10*(4), 1–16. https://doi.org/10.3390/pharmaceutics10040269
83. Sharma, I., & Bhattacharjee, S. (2021). Nanotheranostics and biocompatibility. In *Nanobiotechnology: Microbes and plant assisted synthesis of nanoparticles, mechanisms and applications* (pp. 51–60). Elsevier. https://doi.org/10.1016/B978-0-12-822878-4.00003-1
84. Kyriakides, T. R., et al. (2021). Biocompatibility of nanomaterials and their immunological properties. *Biomedical Materials, 16*(4), 1–49. https://doi.org/10.1088/1748-605X/abe5fa
85. Vinod, C., & Jena, S. (2021). Nano-neurotheranostics: Impact of nanoparticles on neural dysfunctions and strategies to reduce toxicity for improved efficacy. *Frontiers in Pharmacology, 12*, 1–10. https://doi.org/10.3389/fphar.2021.612692
86. Chen, F., Ehlerding, E. B., & Cai, W. (2014). Theranostic nanoparticles. *Journal of Nuclear Medicine, 55*(12), 1919–1922. https://doi.org/10.2967/jnumed.114.146019
87. Mistretta, M., Farini, A., Torrente, Y., & Villa, C. (2023). Multifaceted nanoparticles: Emerging mechanisms and therapies in neurodegenerative diseases. *Brain, 146*(6), 2227–2240. https://doi.org/10.1093/brain/awad014

Recent Advances in Nanotheranostics-Based Personalized Medicine

16

Parameswar Patra, Debadatta Mohapatra, Niranjan Panda, Gautam Kumar Jena, and Rakesh Kumar Sahoo

Abstract

Personalized medicine is a promising approach that uses each patient's distinct genetic profile to customize the diagnosis and treatment of diseases. Patient's specific therapy with understanding the inter-individual variability using genomics biomarkers leads to improved therapeutic outcomes and diminishes the side effects. Considering the limitations of conventional diagnosis and treatment, theranostics is a single system that integrates both treatment and diagnostics and could be a potential way for personalized medicine. Nanomaterials possess unique properties. It has been widely reported for targeted drug delivery, improved imaging, and therapeutic action with negligible toxicity. Nanoformulations as theranostics demonstrated their potential use in various diseases, including cancer and metabolic disorders, and could be advantageous in prevention, identification, and treatment by incorporating specific molecular characteristics of the individual patient. The exclusive properties of nanotechnology are utilized to identify the biomarker and drug delivery. Nanotheranostics can be administered noninvasively to diagnose targeted photographs of biomarkers

P. Patra (✉) · N. Panda
School of Pharmacy, The Neotia University, Sarisa, West Bengal, India

D. Mohapatra
Phytomedicine Research Laboratory, Department of Pharmaceutical Engineering & Technology, IIT (BHU), Varanasi, India

Leslie Dan Faculty of Pharmacy, University of Toronto, Toronto, ON, Canada

G. K. Jena
Department of Pharmaceutics, Roland Institute of Pharmaceutical Sciences, Berhampur, Odisha, India

R. K. Sahoo
Department of Ophthalmology and Visual Sciences, University of Illinois, Chicago, IL, USA

V. K. Chaturvedi et al. (eds.), *Nanotheranostics in Metabolic Syndrome*,
https://doi.org/10.1007/978-981-95-3746-4_16

and therapeutics delivery as per the biomarker distribution. The clinical relevance of nanotheranostics assists in earlier detection and assessment of the patients, which potentially responds to the treatment and has greater possibilities for a therapeutic outcome. The concept of nanotheranostics is highly appealing in understanding and deciding the individual therapeutic records to achieve the utmost advantages with a fewer side effects. This chapter discusses the recent advancement of emerging nanotheranostics strategies for personalized medicine.

Keywords

Nanotheranostics · Personalized medicine · Nanomedicine · Cardiovascular · Molecular imaging

1 Introduction

Personalized, also known as precision medicine (PM) is an emerging practice with promising outcomes, primarily in cardiovascular treatment, that promises to improve healthcare at all stages of treatment. Personalized medicine arises due to the recognition that no solo therapeutics have uniform effects on several patients with the same diagnosis. In place of the same prescription for every patient, the PM would customize a remedy that is most appropriate for the individual. It is established on an understanding of the diseases on a genetic level and the customization of the therapy according to the availability of the patient's genes and proteins. Furthermore, by using genome sequencing, it can also aid in perceiving a patient's vulnerability to a disease. It helps in designing the course of medication, which could promptly monitor and prevent the disease. Hence, diagnosis is the pivotal method involved in the treatment and is fundamental to PM, leading to declines in marketable research and development efficiency [1]. The new approach in biomedical research focuses on PM, targeting molecularly identified patients who respond effectively to specific compounds. National Institutes of Health (NIH) and the United States Food and Drug Administration (USFDA) have supported several governments and private centers dealing with projects having potential from bench to bedside [2]. Substantial advances in biomarker identification and theranostics motivate personalized medicine. The combined diagnosis and treatment in a single system is the most suitable approach for PM [3, 4]. Combined molecular imaging and therapy were used in various PM characteristics, including disease staging, therapy selection, treatment planning, and earlier detection of disease and adverse effects.

The ultimate use of the theranostics system in PM for cancer was the type and stage of cancer, and the photography of the heterogeneity of the tumor. The design of customized therapy relies on the diagnosis and imaging findings and the study of therapy effectiveness. Nanotechnology has a significant role in theranostics. The nanosized particles deliver various advances in diagnostics and treatment, headed for the development of nanomedicine and nanosensors. The nanosensors can identify a wide range of biomarkers in a smaller sample size [5–7]. Nanomedicine has the potential to deliver optimal doses with lesser adverse effects with the help of receptor-mediated active targeting directly to the tumor site, not to healthy cells [8]. Nanotechnology is an essential tool in theranostics [9, 10]. This chapter focuses on utilizing nanotheranostics and applying various nanoparticle theranostics in personalized medicine.

2 Nanomaterials

A nanomaterial is well-defined as a material with at least one controllable feature measuring 100–200 nm or less [11]. Simply having a nanoscale structure does not qualify a substance as a nanomaterial, as nearly all materials possess features at this scale. The key distinction lies in the capacity to maintain the structure at the nanoscale deliberately. One could argue that traditional materials with intentionally designed nanoscale components, such as oxide-dispersion-strengthened alloys, can be considered nanomaterials. Though, the current definition typically excludes these classical structural materials. Instead, contemporary nanomaterials are newly engineered substances in which the controlled nanoscale structure significantly influences the material's desired properties or the functionality of a device.

Nanomaterials are classified into three main categories: discrete nanomaterials, nanoscale device materials, and bulk nanomaterials [12, 13]. Discrete nanomaterials (dn materials) are independent materials measuring in the range between 1 and 10 nm having zero or one dimension, such as nanoparticles and nanofibers (carbon nanotubes) [10]. Bulk nanomaterials are produced in large numbers (typically at least mm^3 in volume) with a controlled nanoscale structure [14]. They can be composed of discrete nanomaterials or nanoscale device materials. For instance, a bulk nanomaterial can be made by assembling numerous nanofibers. This classification can also be understood from a dimensional perspective:

- Discrete nanomaterials are generally zero and one dimensional, such as nanoparticles and nanofibers.
- Nanoscale device materials are two dimensional, as seen in nano-thin films.
- Bulk nanomaterials are three-dimensional, encompassing structures that extend in all spatial dimensions.

3 Nanotheranostics

The combination of diagnostic and therapeutic functions in one platform employing nanotechnology is known as nanotheranostics [15]. It is vital in personalized medicine, often known as précised medicine. Even though such systems are currently evolving trends in the clinical field. The current chapter discusses recently developed nanotheranostics systems such as block copolymer, polymeric nanoparticles, polymer gold nanoparticles, polymeric micelles, magnetic-based nanoparticles, silica-based nanoparticles, liposomes, and quantum dots, which are discussed systematically. The recent studies of various nanotheranostics systems are demonstrated in Table 16.1.

3.1 Paramagnetic Manganese-Based Nanoparticles

Paramagnetic nanoagents are highly appealing for their vital role in imaging and therapeutic applications [41]. Their use has expanded beyond traditional magnetic resonance imaging (MRI) contrast agents to multifunctional platforms for diagnostics and theranostics. Nanoparticles (NPs) offer a higher surface area and ease of modification, facilitating the attachment of various ligands. Integrating different components enhances their diagnostic capabilities and imparts therapeutic functionalities based on the specific role of each unit. These advancements have led to the development of novel imaging modalities, advanced therapeutic techniques, and improved precision in imaging-guided therapies.

The integration of diagnostic and therapeutic strategies into a unified "theranostics" platform is a fascinating area of biomedical research. Transition metal manganese (Mn)-based nanoparticle systems have emerged as versatile nanoplatforms by virtue of their distinctive physicochemical properties and biological effects, enabling them to meet stringent biomedical requirements and adopt diverse morphologies [42]. Morphology, a crucial physicochemical attribute, significantly influences nanomaterials' biodistribution, diagnostic accuracy, and therapeutic performance. Despite the considerable progress in Mn-based theranostics, a comprehensive review focusing on the shapes of commonly reported Mn-based nanostructures is lacking.

3.2 Triblock Copolymer-Based Multifunctional Nanotheranostics

Triblock copolymers present great promise for innovative cancer therapy due to their adaptable and multifunctional properties [43]. The ability to respond to various stimuli allows for a targeted drug delivery system with sustained release at the tumor area, minimizing side effects. The innovative design of dual and triple-responsive systems enables precise regulation of drug release kinetics, effectively tackling issues related to multidrug resistance.

These copolymers offer a versatile platform capable of carrying therapeutics, small interfering RNA (siRNA), antibodies, and imaging agents, thus supporting integrated diagnostic and therapeutic approaches [44]. By leveraging triblock self-assembly and receptor-specific ligands, sophisticated site-specific approaches can be developed to improve, specifically, drug delivery to cancer cells.

Table 16.1 Recent investigations on various types of nanotheranostics

Imagining method	Disease	Nanocarrier	Contrast agent	Drug	Function	References
MRI	Wide variety of cancer	Polymersomes	Maghemite nanoparticles	Doxorubicin	Real-time tracking of NPs location and sustained release of drug by the influence of radio frequency and magnetic hyperthermia	[16]
MRI	Nephrogenic systemic fibrosis	Nanobialys	Manganese (III)	Doxorubicin	Targeted detection of fibrin, which is a crucial biochemical characteristic of thrombus	[17]
US	Acute ischemic stroke	Echogenic liposomes	Liposomes	Plasminogen activator	To quantify the drug release from tissue-type plasminogen activator (rt-PA) loaded echogenic liposomes triggered by ultrasound	[18]
MRI	Prostate cancer	Nanoparticles	Iron oxide	Doxorubicin	Nanoparticles are able to detect prostate malignant cells and deliver targeted cytotoxic agents specifically to the prostate cancer cells.	[19]
MRI	Leukemia	Nanoparticles	Iron oxide	Paclitaxel	Evaluation of the effectiveness of paclitaxel entrapped magnetic nanoparticles to determine its cytotoxic effect on Bcr-Abl positive K562 cells	[20]
MRI	Cancer	Nanoparticles	Iron oxide	Docetaxel	Multifunctional nanoparticles visualizing, therapeutic delivery and real-time monitoring of therapeutic effects.	[21]
MRI	Wide range of cancer	Nanoparticles	Iron oxide	Curcumin	Exhibited hemocompatibility, suggesting a nanochemotherapeutic agent for cancer therapy.	[22]
MRI	Restenosis	Nanoparticles	Perfluorooctyl bromide	Paclitaxel/ doxorubicin	The outcome of the study is paramagnetic nanoparticles provide a unique MRI visualization and measurable drug delivery system that leads to restenosis prevention after angioplasty.	[23]
MRI	Cancer	Nanoparticles	Iron oxide	siRNA	The capacity of the developed nanoparticles fabricated to produce precise contrast improvement of glioblastoma cells was confirmed through MR imaging.	[24]

(continued)

Table 16.1 (continued)

Imagining method	Disease	Nanocarrier	Contrast agent	Drug	Function	References
MRI	Wide range of cancer	Liposomes	Gd3+	Doxorubicin	The fast release of MRI contrast agent and doxorubicin from the liposomes happened close to the melting phase transition temperature, making it feasible to image the release of doxorubicin using MRI	[25]
MRI	Lung cancer	Micelles	Iron oxide	Doxorubicin	The combined diagnostic and therapeutic system of micellar nanomedicine potentially supports image-guided, target-specific lung cancer treatment.	[26]
Optical imaging	Wide range of cancer	Nanoparticles	Indocyanine green	Doxorubicin	The double agent encapsulated nano drug delivery system allotted for concurrent delivery of hyperthermia and cytotoxic, agent and this combinational therapy significantly increased cytotoxicity in MES-SA/Dx5 cells	[27]
Optical imaging	Cancer	Upconversion nanoparticles	nanocrystals	Doxorubicin	The potential use of upconversion nanoparticles as interesting nanocarriers for multi-functional cancer therapy and diagnosis.	[28]
Optical imaging	Various types of cancer cells	Dendrimers	Fluorescein	Methotrexate	The outcome exhibits the capacity to design and fabricate polymer-based nanocarriers for the intracellular targeting of drugs and imaging agents,	[29]
Optical imaging	Cancer	Gold nanoparticles	Gold nanoparticles	Paclitaxel	Gold nanoparticles show a substantial role in the diagnosis and treatment of cancer cells and are used as analytic agents.	[30]
MRI	Prostate cancer	Nanoparticles	Iron oxide	Noscapine	The optically and MRI traceable iron oxide-coated nanoparticles have a great ability for image-guided delivery of noscapine for the treatment of prostate cancer.	[31]
MRI	Cancer therapy	Nanoparticles	USPIO	Methotrexate	Nanoprobes can precisely detect and kill overexpressed folate receptors in cancer cells.	[32]

					Although able to perform real-time monitoring of drug treatment via twin-modal fluorescence and magnetic resonance imaging.	
MRI	Various cancer therapies	Nanoparticles	Iron oxide	Paclitaxel	The nanosystem assists both optical/magnetic resonance imaging and targeted killing of folate-expressing cancer cells	[33]
MRI	Squamous cell carcinoma.	Nanoparticles	Iron oxide	Mitoxantrone	The magnetic nanoparticles show a distinctive way to treat malignant tumors locally with reduced toxicity	[34]
MRI	Cancer therapy	Micelles	Gd^{3+}	Paclitaxel	Theranostics uses of nanocarrier systems. The synergistic effects of chemotherapeutics and magnetic resonance imaging systems serve well for their great applications in cancer therapy.	[35]
MRI	Cancer therapy	Micelles	Gd^{3+}	Doxorubicin	The combined systems of diagnostics and treatment of multifunctional pH-disintegrable nano micelles perform well for their practical applications in the field of image-directed cancer therapy.	[36]
MRI	Cancer therapy	Liposomes	Gd^{3+}	Doxorubicin	The in vivo release of doxorubicin from the liposomes can be probed in situ with the simultaneous release of an MRI visualizing agent for the best outcome for cancer therapy.	[37]
US	Cancer therapy	microbubbles	Perfluoro pentane	Doxorubicin	Effective use of microbubbles loaded with chemotherapeutics for the treatment of the MDA MB231 breast cancer has been achieved.	[38]
MRI	Cancer therapy	Hybrid nanoparticles	SPIO	Doxorubicin	The twin targeting mode diagnosis and treatment with the help of hybrid nanoparticles shows early identification and treatment of several types of cancer.	[39]
MRI	Cancer therapy	Micelles	Iron oxide	Doxorubicin	The study result confirmed the high sensitive MRI visualization and alpha(v) beta (3)-specific cytotoxic response of these multifunctional polymeric micelles.	[40]

3.3 Metal-Based Nanotheranostic Agents

Numerous metals and metallic elements were used for the preparation of nanoparticles that are valuable for the diagnosis and treatment of cancer. Metal nanostructures are particularly promising in medical applications because of their distinct physicochemical properties, like magnetic, optical, thermal and electrical characteristics, which set them apart from other types of nanoparticles [45]. The metal nanoparticles were prepared with various metals, such as gold, bismuth, tungsten, gadolinium, silver, iron, platinum, and lead. These metal nanoparticles have been explored in different forms such as solid, poriferous, antibody and folic acid functionalized and core-shell structures. These nanoparticles have been efficiently utilized as imaging modalities, comprising MRI, computed tomography (CT), nuclear medicine and radiation therapy for disease diagnosis and therapy [46].

However, realizing the full potential of triblock copolymers requires overcoming challenges related to exact synthesis, detailed evaluations and effective clinical translation. Continued research focused on optimizing the design and properties of triblock systems is essential to harness their transformative impact on cancer nanomedicine. Gaining a comprehensive understanding of the copolymers and their interactions will lead to the developing of personalized treatments and improved patient outcomes.

3.3.1 Gold Nanotheranostics

Gold nanoparticles (AuNPs) have become prominent in the area of nanotheranostics, ascribed to their unique physicochemical features [47]. Usually, the size range of the gold nanoparticles is between 2 and 100 nm and were prepared with various shapes, including spheres, diamonds, rods prisms and cages. These configurations can exist either as solid particles or in a core-shell arrangement. The unique blend of size and shape influences their properties, which can be harnessed for theranostic applications. These include vibrant optical properties, effectiveness as contrast agents, photothermal capabilities in the infrared (IR) and near-infrared (NIR) regions, capability for biofunctionalization, and toxicity considerations. While most nano theranostic systems utilizing AuNPs have traditionally focused on spherical nanoparticles, altering their shape can adjust their absorption and scattering properties towards the NIR range [47]. Nanostructured particles such as hollow nanoparticles, nanorods and nanostars exhibit strong peaks around the NIR region that support substantially suitable for phototherapy use. For instance, hollow gold nanospheres with a diameter of 44 nm were among the first nanotheranostics systems tested in vivo and demonstrated comparable photothermal efficiency at concentrations 100 times lower than those required for superparamagnetic iron oxide silica-coated nanoparticles. Additionally, hollow nanoparticles and nanocages offer the advantage of serving as encapsulating vectors, enabling a higher loading capacity of therapeutic agents per particle. Gold nanoshells comprise a dielectric core covered by a tinny nanoscale gold shell. It was mainly prepared with various metals such as silica, magnetic, and silver, which enhance the functionalities of the nanoshells. The properties like plasmon resonance, optical absorption, gold shell's thickness and the core's diameter can be fine-tuned to the NIR range, making them effective photo absorbers for NIR photothermal ablation therapy.

Recently, Huang and colleagues developed bellflower-shaped AuNPs (GBFs) using a liquid-liquid-gas triphase interface system created by ultrasound-induced vacuum bubbles in a two-phase liquid-liquid system [48]. These GBFs exhibited a strong plasmon band around 800 nm, indicating their potential as effective photothermal antennas for nanotheranostic applications. Intravenous administration of GBFs was shown to delay tumor growth or induce complete tumor regression within 2 weeks using low laser power (0.5 W/cm^2) while also enabling photoacoustic imaging.

3.4 Magnetic Nanoparticle Theranostics

The multifunctional magnetic nanoparticles were prepared by a mixture of chemotherapeutic agents, i.e., doxorubicin and magnetic nanoparticles (Fe_3O_4) of size less than 100 nm [49]. The Fe_3O_4 is used as an MRI imaging agent, and biodegradable block copolymers are used to formulate multifunctional nanoparticles. The multifunctional nanoparticles were surface-modified with antibodies (HER, Herceptin) to specifically target the human epidermal growth receptor 2 over-expressed in breast cancer tumors. HER2/neu cancer marker NIH3T6.7 cells were injected into the mice to develop the breast cancer animal model. The effectiveness of the developed surface-modified multifunctional magnetic nanoparticles was evaluated for theranostics purposes. The MRI image was taken at predetermined time points to understand the biodistribution of surface-modified magnetic nanoparticles after being injected into the in vivo system. MRI observed a clear and black color image in the tumor area after treatment with surface-modified magnetic nanoparticles compared to pretreatment. The accumulation of Herceptin-modified magnetic nanoparticles was higher compared to the free doxorubicin and doxorubicin+Herceptin solution. The significant inhibition of tumor growth confirmed that the active targeting was due to receptor-mediated cellular internalization, which leads to highly contrasting images and therapeutic effectiveness.

3.5 Polymeric Nanotheranostics

Polymeric nanocarrier systems are able to carry various therapeutic and fluorescent agents for therapeutic and diagnostic applications. Polymer-based theranostic materials generally consist of three essential factors: (i) a polymer matrix which provides stability and biodegradability, (ii) a pharmacological active such as a small or macro molecule (iii) an imaging agent, which could be an MRI contrast [50]. Polymer-based nanomedicine, such as polymer conjugates, nanoparticles, micelles, and dendrimers, has already been in the commercial market for treating various cancers. Polymers in polymeric nanocarrier systems enable either matrix or capsule systems in which the therapeutic agents accurately load and control the release of active compounds over a specific interval of time [51]. The polymeric nanoparticles are generally synthesized using amphiphilic polymers, encapsulated with hydrophobic agents such as antineoplastic and diagnostic agents, and the outer surface is available for ligand conjugation to target the over-expressed biomarker of the specific diseases. Polymers such as chitosan, alginate, polyethylene glycol and poly (D, L-lactic acid), poly (D, L-glycolic acid) were approved by the USFDA. These polymers were utilized for the fabrication of polymerionic nanoparticles for clinical applications. The polymeric nanoparticles improved safety and efficacy compared with the free drug, which was observed to have an enhanced circulation half-life and prolonged drug release [52, 53]. The polymeric nanocarrier systems are used for both treatment and diagnostic purposes. Nanotheranostics hold significant promise in medicine and biomedical research due to their ability to simultaneously deliver therapy and provide imaging [54]. Polymers are particularly advantageous in this field because of their biocompatibility, customizable properties, and cost-effectiveness. They are effective in encapsulating poorly soluble compounds and safeguarding them from degradation until they reach the target site.

3.5.1 Biodegradable Polymeric Nanoparticles

The rapid advancement of nanoscience and nanotechnology has sparked great optimism for significant progress in medical diagnostics and treatment [55]. Applications include in vitro and in vivo diagnostics, the development of biodegradable materials and drug delivery systems. The growing interest in nanodrug delivery systems in biomedical applications is due to distinct characteristics, including a high surface-to-mass ratio, quantum effects and the capability to load active substances. Integrating polymer science with nanotechnology has revolutionized drug delivery systems, nano-implants, and biomedical engineering. Even though nanoparticles are typically defined as structures smaller than 100 nm. The drug delivery systems often range from 50 to 500 nm, as they generally consist of an active pharmaceutical compound and a carrier compound. Polymeric nanoparticles (PNPs) are particularly valuable for targeted drug delivery, enhancing the therapeutic efficacy of both hydrophobic and hydrophilic drug molecules and several biologically active substances. They offer improved bioavailability, safety, biocompatibility, reduced toxicity, enhanced permeability and customizable release times.

Nanomedicines are typically formulated as nano capsules-where the drug is enclosed within a polymer shell-or nanospheres, in which bioactives are dispersed throughout a polymer matrix. The transition from macro to nano drug delivery systems effectively addresses pharmacological challenges and offers significant advantages over conventional dosage forms. The success of nano-drug delivery systems relies on the target site of action, correct drug selection, and its selection of a compatible biodegradable polymer. The effectiveness of polymeric nanoparticles in reaching their target site depends on several key factors, including particle size, zeta potential, and hydrophobicity or hydrophilicity. The size of PNPs is particularly crucial as it influences their capability to navigate biological barriers and reach the desired site of action. Additionally, the surface charge of PNPs significantly impacts their behavior within the body, determining whether they continue to circulate in the bloodstream or interact with cell membranes.

3.5.2 Polymeric Supermagnetic Nanoparticles

A new class of magnetic nanocarriers, including superparamagnetic $La_{0.7}Sr_{0.3}$ MnO_3 nanoparticles (SPMNPs) and ferrite-centered superparamagnetic nano drug delivery systems show great promise in advanced cancer theranostics [56]. These combined systems of chemotherapy and visualization-guided diagnosis therapies, like magnetic hyperthermia. SPMNPs are particularly effective because they can function both as a nano-carrier for targeted drug delivery and as controlled heat sources, utilizing magnetic hysteresis and superparamagnetic properties [57]. The success of SPMNPs in therapeutic applications relies on their intrinsic magnetization and key biophysical characteristics, including biocompatibility, surface modification, structural stability and specific absorption rate under physiological and tumor environments. These properties can be enhanced through surface modification with biomolecules or chemical compounds. Magnetic fluid hyperthermia (MFH) offers a noninvasive with externally controllable treatment method. However, due to its limited efficacy in completely suppressing tumors, MFH is often combined with chemotherapy.

3.5.3 Polymer Gold Nanoparticles

Gold nanoparticles (GNPs) are minute particles made of gold, sized below 100 nm. Gold nanoparticles have characteristics such as ease of surface modification; due to these properties, gold nanoparticles conjugated with polymer are the most suitable carrier for sustained drug release. The gold nanoparticles were prepared in various shapes, such as nanorods, nano-shells, nanospheres, nano-stars, nanoclusters, and nanocubes [58]. Polymer-conjugated gold nanoparticles are excellent nanocarriers for drug delivery systems, and they have the characteristics of prolonged circulation time, higher drug loading, better stability, and lesser toxicity. Polymer gold nanoparticles have the ability of optical and photothermal properties, making them useful for diagnostic purposes due to their capability to engage NIR light and X-rays, as well as their surface-enhanced Raman scattering capabilities. Many of the anti-cancer and imaging agents are loaded on the gold nanoparticle through covalent and ionic bonds for better therapeutic effectiveness.

In a study, polymer-stabilized GNPs were synthesized by reducing Au III to Au0 using either chitosan or polyvinyl alcohol. These polymers served as simultaneously in situ reducing agents and stabilizers. The nanoparticles' size, surface morphology, and polydispersity index were analyzed using microscopic techniques such as transmission electron microscopy, scanning electron microscopy, and dynamic light scattering. The mucoadhesive properties of chitosan make this system particularly effective for antimicrobial activities, including antibacterial and antifungal effects. The study also explored the mechanism behind the antimicrobial action. Remarkably, chitosan-coated gold nanoparticles (GNPs) showed significant antimicrobial efficacy even at low doses. Biochemical assays, including malondialdehyde (MDA), DNA fragmentation and heat shock protein overexpression, demonstrated the antimicrobic properties of the nanoparticles. Furthermore, the chitosan-GNPs systems interacted with microbial DNA, inhibiting DNA repair enzymes. Importantly, in vitro (rat peritoneal macrophages) and in vivo (Wistar rat) analyses indicated minimal cytotoxicity [59].

3.5.4 Polymeric Micelles

Polymeric micelles are colloidal structures composed of self-assembling block copolymers. The hydrophobic drug is loaded in a hydrophobic core, and colloidal stabilization is ensured by the hydrophilic layer in aqueous media. As a result of these properties, polymeric micelles gain attention for developing nanotheranostics loaded with target moieties and contrasting agents [60]. The various types of polymeric micelles were developed for theranostics purposes for cancer treatment. The complexity of developing polymeric micelles opens opportunities for novel strategies for treating cancer. The real-time distribution monitoring of polymeric micelles in the blood circulation and target site drug release from the formulation would empower the efficiency of treatment and improve the treatment with respect to disease-specific individualities in various patients [61]. One of the studies using multifunctional polymeric micelles loaded with an anticancer moiety and having MRI contrasting properties was published by Kaida et al. In that study, multifunctional cancer-targeted micelles encapsulated with chemotherapeutics and a contrast agent for diagnosing and treating pancreatic tumor and small liver metastasis [62]. Polymeric micelles were prepared with poly (ethylene glycol)-b-poly (glutamic acid) self-assembling block copolymers, in which Gd^{3+} and chemotherapeutic agents, (1R,2R)-1,2-diaminocyclohexane platinum(II) (DACHPt), encapsulated in the core. The efficacy of the Gd-DTPA–DACHPt entrapped polymeric micelles as a theranostics system was assessed on an orthotopic human pancreatic (BxPC3) xenograft model. The Gd-DTPA–DACHPt encapsulated polymeric micelles were injected intravenously. The polymeric micelles were observed to accumulate vastly inside the tumor, leading to high results in MRI imaging signals. In contrast, no improvement was observed in the case of free Gd-DTPA. A high MRI signal at the tumor region monitors the efficacy of the therapy. The MRI signal monitored the accumulation of polymeric micelles and the change in the tumor size. The high resolution of the MRI signal can effortlessly distinguish between affected and healthy tissue. In the healthy tissue region, there is no accumulation of polymeric micelles, as confirmed by the MRI signal. The study results revealed that polymeric micelles have great potential for cancer treatment.

3.6 Liposome-Based Theranostic Agents

Liposomes are spherical nanovesicles made of natural and synthetic phospholipids. They are already placed in the market as a drug delivery carrier with the entrapment potential of small and large molecules. The hydrophilic molecule is entrapped inside the core, and hydrophobic molecules are intercalated in the lipid bilayer. Due to all these characteristics, the lipid-based carrier is considered for developing nanotheranostics by incorporating both therapeutic and contrasting agents [63, 64]. In a report study, PEGylated liposomes were encapsulated with both doxorubicin and gadolinium derivative (Gd-DOTA monoamide) as a therapeutic and MRI contrasting agent published by Grange et al. [65]. The liposomal formulation was studied in a mouse model of human Kaposi's sarcoma overexpressing neural cell adhesion molecules (NCAM). The liposomes are surface-modified with C3d peptide through an active targeting approach to target the NCAM. The surface-modified liposomes show better therapeutic efficiency and lower toxicity than uncoated liposomes and free doxorubicin solution. MRI studied the effectiveness of the theranostics liposomal therapy through the accumulation of liposomes inside the tumor and the decrease of the tumor volume after administration of C3d-coated liposomes. The tumor growth inhibition validates that a contrast agent does not significantly influence the release of drugs from the formulation. C3d-based theranostics liposomes execute the in vivo imaging of the liposome's biodistribution and concurrently evaluation of the therapeutic response using MRI. The theranostics liposomes have the ability to track the real-time treatment of diseases [66].

3.7 Dendrimers-Based Nanotheranostic Agents

Dendrimers are highly branched systems consisting of monodisperse polymers. It can be prepared with a definite size, shape, and surface modification with various ligands. Because of their exceptional characteristics, dendrimers have gained interest as versatile carrier systems for developing nanotheranostics agents. It can use as a theranostics agent in various diseases such as cancer. All the characteristics of dendrimers exhibit a stable and desired pharmacokinetic biodistribution that ensures effectiveness [67]. It can potentially carry multiple imaging and therapeutic agents within its internal cavities. Dendrimers can be developed as multifunctional theranostics agents, which can concurrently target tumor cells, imaging tumor sites, and deliver chemotherapeutic agents. The dendrimers inside the core can encapsulate therapeutics, whereas the surface functional groups can be used to build complexes [68]. It has been used in the field of biomedical, mainly for the diagnosis, the transfer of genes through cell membranes, and regenerative medicine [69].

3.8 Silica-Based Nanotheranostics Agents

Currently, silica-based nanoparticles have secured the attention of researchers due to the potential of theranostics properties. Mesoporous silica nanoparticles exhibit significant advantages such as high surface area and tailored mesoporous structure compared to conventional nanocarriers [70]. In addition to that, it also exhibits various morphologies, customized pore sizes, tailored surfaces, and biocompatibility [71]. The functionalized targeting moieties specifically bid to the cancer cells, not the healthy cells, and the contrasting agents help in the noninvasive detection and monitoring of disease. The multi-functional properties, such as the loading of various organic and inorganic molecules and attachment with various polymers to get the controlled release of the drugs, make the system a suitable candidate for theranostics system [72]. Mesoporous silica-based nanoparticle imaging agents show good response in diagnosing hepatocellular carcinoma, which has the potential of a multifunctional nanocarrier system with desired size and surface modification for the best use of theranostics purposes [73].

3.9 Quantum Dots

Quantum dots (QDs) are tiny particles of size less than 10 nm that can potentially be used in diagnosis and therapeutic delivery. It has properties of high fluorescence and a wide emission spectrum, predominantly in the near-infrared area ($\lambda > 700$ nm). It has the potential to ease surface modification to enhance their stability, biocompatibility, and solubility. The surface modification of the quantum dots delivers improved specificity to targets [74]. Due to the nano size, it can be used in broad applications such as fluid mixture textures and polymer frameworks. The surface functionalization of quantum dots can help site-specific targeting and accumulation, leading to high constracting imaging and effectiveness [75]. Various limitations should be considered, and a few preparation approaches are preferred in designing the preliminary quantum dots for tumor growth monitoring and disease therapy. In several studies reported, quantum dots were used for the in vitro fluorescent image of malignant cells from various cancers, such as melanoma, ovarian, breast, pancreatic, glioblastoma, epidermoid ovarian, lung, hepatocellular, and adenocarcinoma. Modern QDs are engineered for diverse applications by functionalizing them with biocompatible molecules and enhancing their hydrophilic properties through suitable modifications. They can be precisely designed with sizes ranging from 2 to 4 nm, displaying unique optical, fluorescent, and electronic properties. The manufacturing process influences the size of self-assembled QDs, where longer reaction times lead to larger particle sizes. This growth can be controlled by lowering the temperature or intermittently withdrawing the solution and allowing it to cool to room temperature.

QDs are highly valued for their exceptional features, including biocompatibility, photoluminescence (PL), morphology and size control. However, they also present concerns of toxicity attributed to the presence of heavy metals, even at low concentrations. Among various types, including graphene quantum dots, cadmium quantum dots, and carbon dots (CDs). CDs have gained prominence due to their cost-effectiveness, ease of large-scale synthesis, favorable chemical composition, straight forward.

4 Personalized Medicine and the Use of Nanotheranostics

The body of individuals is composed of variable genetic makeup, which leads to inter-patient variability [76]. Hence, the concept of "one fits all" with the use of single therapeutics is not conceivable [76]. For example, most of the nanomedicines passively accumulated at the tumor site due to the enhanced permeation and retention (EPR) effect caused by the leaky vasculature and lower lymphatic drainage at the tumor site. However, it has been observed that patients with the same tumor type possess variations in the structural integrity and density of tumor neovasculature. Hence, rigorous evaluation of the leakage of vasculature and site-specific drug accumulation allows the evaluation of treatment efficacy specific to a person [77].

Personalized medicine has recently been considered a suitable strategy to address unmet medical needs. The patient-tailored, pre-sized therapy using therapeutics is called personalized medicine [76]. Personalized medicines are designed to administer the right drug to the right patient at the right moment. When nanotheranostics are used as personalized medicine, it is called personalized nanomedicine [77]. Nanomedicines offer ultra-small size, higher surface-to-volume ratio, greater flexibility, surface functionalization, and targeting ability [78]. Multiple omic strategies, such as pharmacoproteomics and pharmacogenomics, are adopted to develop patient-specific personalized therapy. Such strategies allow the study of detailed molecular profiles, biomarkers, and genetic makeup of each patient that contributes to disease progressions, which can be utilized to develop personalized medicine [77]. Personalized medicine also includes the development of strategies for disease diagnosis and predicting the therapeutic effect [77]. In such cases, nanotheranostics, which contain therapeutic and imaging agents, could be a good personalized therapy for earlier diagnosis and treatment of diseases. Since each individual's body responds differently to the therapeutics, the noninvasive monitoring of drug accumulation at the target diseased site can be investigated. This will allow the screening of the patients towards specific therapy from others who might need a different therapeutic option. Early identification of the effectiveness of treatment permits better management of the disease. In addition, the accumulation of nanotheranostics in healthy tissue allows the prediction of possible side effects and risks to patients. Based on the results, a strategic, precise, personalized therapy could be optimized to obtain maximum therapeutic benefits with negligible toxicity and the best safety profile [77]. Multiple nanotheranostics, such as polymeric nanoparticles (chitosan, alginate, dextran, PEG, PCL, PLGA, EGCG), polymersomes, liposome, nanobialys, wormlike-polymer vesicles, micelles, metallic and metal oxide nanoparticles (iron nanoparticle, iron oxide nanoparticle, gold nanoparticle, zinc oxide nanoparticle), gold nanoshells, magnetic nanoparticles (nickel, gadolinium, manganese, magnetite or maghemite), upconversion nanoparticles, dendrimers, cyclodextrin nanoparticle, quantum dots, mesoporous silica nanoparticle (MSNs), silica lipid hybrid, gold nanorods, and micellar hybrid nanoparticles have been studied as potential theranostics against various therapeutic purposes (chemotherapy, photothermal therapy (PTT), photodynamic therapy (PDT), magnetic hyperthermia, immunotherapy, gene therapy, gene editing, biosensing, thrombolysis), bioimaging (in vivo molecular imaging, image-guided therapy), and sensing [77–81].

5 Applications

5.1 Monitoring Molecular Targets Associated with Disease Progression

Molecular imaging with the use of theranostics is widely used to image and monitor the molecular targets associated with disease progression [76]. Currently, noninvasive techniques are used to find the presence and levels of molecular targets associated with disease in real-time via a whole-body scan. Such techniques are useful in recognizing disease stage and planning the therapy, such as timing and dose [76]. For example, a widely used chemotherapeutic drug, doxorubicin, exerts anticancer activity by inducing apoptosis. Caspases are protease enzymes that play a vital role in apoptosis. They induce a caspase cascade and ultimately cause cell death. Among them, caspase-3 is mainly regarded as a vital and direct mediator in the process of apoptosis [76]. Hyaluronic acid-based polymeric nanotheranostic functionalized with caspase-3 substrate peptide-based probe was designed to investigate caspase-3 activity. The probe consists of Cy5.5 dye and black hole quencher-3 (BHQ-3), which are covalently linked with a short bond. The bond serve as a hydrolyzable substrate for caspase-3. Upon hydrolysis, the probe is stimulated and develops a strong fluorescence in apoptotic cells in real time. Thus, such nanotheranostics can be used to monitor the efficacy of apoptosis-inducing drugs [76]. Again, by changing suitable hydrolyzable substrates, such a nanoplatform can be used as imaging probes to detect other important proteases in cancer biology [76]. MSNs with various caspase-specific substrates have been studied to simultaneously detect three types of caspases (caspase-3, caspase-8, and caspase-9) via multiplexed imaging. This nanosystem demonstrated its in vivo potential to monitor the therapeutic efficacy of apoptosis-related therapy in individual patients with cancer in real time. Thus, such imaging nanoprobes are useful for monitoring specific molecular targets associated with disease progression. In combination with therapeutics, such imaging nanoprobes can be used to develop nanotheranostics for achieving diagnostic and therapeutic benefits [76].

5.2 To Monitor Downstream Molecular Targets Affected by Treatment

The treatment efficacy can be estimated by monitoring the level of some molecules in a downstream molecular process and targeted molecules that are affected by the treatment [76]. siRNA has recently fascinated considerable interest as a successful therapy for treating multiple diseases due to its specific therapeutic benefits. For example, in one of the studies, thiolated glycol chitosan nanoparticles loaded with tumor necrosis factor (TNF)-α targeted polymerized siRNA have been verified for treating rheumatoid arthritis. TNF-α plays a crucial role in the pathogenesis of rheumatoid arthritis due to the induction of chronic inflammation via stimulation of multiple inflammatory mediators and cytokines. The TNF-α-targeted siRNA-loaded nanoparticles demonstrated excellent gene silencing for TNF-α. The micro CT or immunoblot analysis showed significant inflammation and bone erosion inhibition in arthritis mice [76]. The progression of disease following the siRNA-loaded nanoparticles therapy was examined by monitoring the matrix metalloproteinase-3 (MMP-3) level in a downstream process that is affected by the therapy through in vivo fluorescence bioimaging. MMPs play a significant role in the inflammation of the synovial region and the degradation of articular cartilage. Among various MMPs, MMP-3 plays a vital role in the pathogenesis of rheumatoid arthritis. The MMP is known to be downregulated due to the inhibition of TNF-α in arthritis treatment. Glycol chitosan nanoparticles containing MMP-3-specific cleavable substrate peptide-based probes (Cy5.5 and quencher BHQ-3) at the surface were examined to monitor the level of MMP-3. A significant reduction of NIR fluorescence was observed from NP compared to the untreated control in the collagen-induced arthritic mouse model. Thus, such theranostics are useful for monitoring the molecular targets of a downstream process affected by the treatment. Thus, they can be used as potential personalized therapeutics to assess the clinical efficacy of the treatment [76].

6 Molecular Imaging Methods for Targeted Drug Delivery/Imaging Modalities for Theranostics

Molecular imaging is widely used for diagnosis and targeted therapeutic delivery [80]. Diagnostic imaging is required to identify the presence and level of specific molecular targets of certain diseases [76]. Such techniques are widely used to diagnose and treat cancer by visualizing the tumor microenvironment's location, size, and progression. Molecular imaging offers the development of personalized medicine for targeted delivery to the intended location, reducing the off-target effects and toxicity [80]. Nanotheranostics are comprised of diagnostics and therapeutics in a single platform that offers simultaneous visualization and site-specific delivery for personalized remedies [80]. They contain multiple contrast agents like radionuclides, fluorescent dyes, and gadolinium, which are visualized by molecular imaging methods to track the physiological and pathological processes in the body [80]. Various noninvasive imaging tools are applied to visualize in vivo molecular targets, such as ultrasound (US) imaging, optical imaging, CT, MR imaging, and nuclear imaging (positron emission tomography (PET) and single photon emission computed tomography (SPECT)). These techniques are broadly classified into (i) primarily molecular imaging techniques and (ii) anatomical/morphological imaging techniques. The molecular imaging techniques (PET, optical imaging, and SPECT) are characterized by high sensitivity. The anatomical imaging tools (US, CT, and MR) are characterized by high spatial resolution [76].

6.1 Magnetic Resonance Imaging (MRI)

MRI is a widely used noninvasive imaging tool to obtain cross-sectional images of soft and hard tissues. It utilizes magnetic fields and radio waves for imaging. MRI possesses good resolution; however, it has low sensitivity and takes a prolonged time for imaging. Materials having magnetic properties, such as ferromagnetic, paramagnetic, and superparamagnetic iron oxides, are used in MRI [80]. Superparamagnetic ultrasmall iron oxide has been utilized in MRI and evaluated for the inhibition of liver metastasis with targeted delivery [80]. Gold nanoparticles are also used in MRI as a contrast agent due to their low cytotoxicity. During exposure to near-infrared radiation, it generates heat to kill cancerous cells, hence acting as PTT [80].

Gadolinium (Gd) is the most commercialized contrast agent, but it possesses a short life in blood due to its low molecular size. In contrast, Gd-Gd-diethylenetriaminepentaacetic acid (Gd-DTPA) is a macromolecular MRI contrast agent with prolonged systemic life and limited clearance [80]. Composite nanoparticles containing bismuth and iron oxide have been evaluated as dual-mode imaging agents. Bismuth and iron oxide are used as X-ray attenuating agents and contrast agents for CT scans and MRI, respectively. The in vivo CT scans and MRI with the composite NPs demonstrated excellent contrast in both imaging modalities [80]. MRI-based contrast agents can be integrated with therapeutics (drugs, thermal therapy, proteins, genes, and combinatorial therapies) to produce hybrid systems to provide treatment and diagnosis of cancer in a personalized strategy [80]. MRI has been used to diagnose synovitis, tenosynovitis, bone marrow edema, and erosions through multiplanar observation of soft tissues and bone in three dimensions [82].

6.2 Computed Tomography (CT)

The CT scan utilizes X-ray to provide detailed anatomy of internal structures, tumor size, location, and spread. It offers the advantages of intense penetration ability and excellent resolution; however, it cannot distinguish tissues with similar physical attributes. CT is mostly combined with other imaging techniques to achieve improved diagnosis. Contrast agents, such as tungsten, iodine, gold nanoparticles, and elements with high atomic number, are used to improve the efficiency of the CT scan. Polyethylene glycol-based nanoparticles containing iodine have been investigated as a contrast agent for CT in cancer biology. A dual contrast agent comprising holmium and barium has been explored for examining breast cancer. 2-DG (2-deoxy-D-Glucose)-labeled gold nanoparticles have been exploited as a contrast agent in CT for in vitro cellular uptake study in lung cancer cells (A-549). Currently, contrast-enhanced computed tomography is also used to determine the progression of renal cell carcinomas [80]. CT is used for the diagnosis of fractures, brain, inflammatory intestinal problems, and spinal cord inflammation, soft tissues, and blood arteries [83].

6.3 Single Photon Emission Computed Tomography (SPECT)

SPECT is a noninvasive, highly responsive, well-tolerated imaging technique used for diagnosis and therapy. In SPECT, a radioactive tracer is injected into the patient's systemic circulation. The tracer reaches the targeted tissue, builds up, and produces gamma rays, which are visualized by 2D and 3D images [80]. SPECT is utilized to identify the size and location of the tumor for diagnosis, determine therapeutic efficacy by imaging during the treatment period, and guide biopsies and surgeries. It is especially useful for diagnosing the stages of various cancers (thyroid and bone cancer). Methylene diphosphonate (^{99m}Tc MDP) was used as a tracer for identifying bone metastasis. Bioluminescent proteins and fluorescent dyes specific to cancer cells are utilized to provide real-time optical imaging and early diagnosis [80]. Indium-labeled liposomes and microcolloids have been evaluated as contrast agents for SPECT. However, it possesses low resolution and may not be as accurate as a CT or MRI for the detection of small lesions and tumors. The radioactive tracers in SPECT may lead to radiation exposure to patients and medical personnel [80]. SPECT/PET has been used for the diagnosis of stroke, atherosclerosis, Alzheimer's disease, and multiple autoimmune diseases [82].

6.4 Positron Emission Tomography (PET)

PET scan utilizes radio-labeled tracers that selectively bind to specific markers overexpressed markers in cancer. It traces metabolic activity of viable cancer cells, which helps in the earlier diagnosis of cancer stage [80]. It possesses excellent infinite penetration, sensitivity, and quantification ability. In addition to other imaging tools, such as MRI, CT, the PET offers a dual imaging platform [80]. Fluorodeoxyglucose-18 (FDG/F-18) is a radioactive tracer widely used for PET/CT scans to detect cancer, such as hepatocellular carcinoma, head and neck cancer, and other disease conditions [80]. Tumor-infiltrating lymphocytes and prostate-specific membrane antigen are utilized as biomarkers for head and neck squamous cell carcinoma and prostate cancer using PET/CT. The technique showed greater sensitivity compared to traditional tomography for the identification of pelvic lymphadenopathy and lymph node metastases [80]. Radio-ligands, such as Ga (68 Ga-NOTA-AE105), have been studied to identify urokinase receptors in breast and urinary tract malignancies using PET/CT [80]. Cerium oxide nanoparticles containing Zr-89 radioisotope demonstrated comprehensive PET scanning outcomes with therapeutic efficacy [80].

6.5 Photoacoustic Tomography (PAT)

PAT is a noninvasive technique that integrates optical and ultrasound techniques to visualize in vivo tissue structures. When a laser beam is impacted on a tissue, it produces ultrasonic waves, which the ultrasound transducer detects to construct the tissue structure. This technique is widely used to diagnose cancer, monitor cancer spread, and track the therapeutic effects of its treatment. Cancer cells possess overexpression of melanin, lipids, and hemoglobin, which possess strong absorption of light, and the resulting photoacoustic waves can be detected to detect cancer tissues. Compared to CT and MRI, the PAT provides functional information about the target tissue (distribution of specific molecules, metabolic activity, and oxygenation of blood vessels). Gold nanoparticles and nanorods are explored as contrast agents in PAT. During excitation at the resonance wavelength, it undergoes photothermal conversion and produces photoacoustic waves detected by an ultrasound transducer. Reporter gene products (proteins encoded by reporter genes) are used for improved PAT imaging and cellular events of deeper tissues. These reporter genes are introduced into the living body to provide specific cellular activities. Besides gold, palladium, silver, and platinum nanoparticles have been explored to improve sensitivity and contrast in PAT for cancer detection owing to their strong photoacoustic signal generation and optical absorption capacity. Palladium nanomaterials demonstrated improved high photothermal stability and photothermal conversion efficiency compared to gold and silver nanoparticles. Silver nanoparticles have been investigated as nontoxic contrast agents in both in vivo and in vitro pancreatic cancer models. Such nanoparticles may be anchored with targeting ligands, such as chemicals or antibodies, for active cancer targeting overexpressed markers and improved imaging due to the increased concentration of contrast agents at the tumor site. Hyaluronic acid-modified gold-platinum nanoparticles have been studied as noninvasive contrast agents for skin cancer photothermal therapy and imaging [80].

7 Nanotheranostics in Cardiovascular Diseases

The new scientific area of theranostics uses diagnostic techniques and targeted therapy to give a personalized healthcare approach for each patient. PNPs have been employed in nanotheranostics for the past two decades to monitor medication release and evaluate therapy efficacy for diverse diseases in real time. This is because they make excellent candidates for the concurrent administration of imaging and therapeutic drugs [84].

Multifunctional nanomedicines significantly impact the treatment of cardiovascular diseases (CVDs). CVDs, which include various illnesses impacting the heart and blood arteries, are the primary cause of death in the European Union, resulting in approximately 2.0 million fatalities per year. More than 50% of all deaths are caused by CVDs (45% for women and 38% for men). Existing methodologies for the early detection and management of CVDs are constrained, and their efficacy in preventing CVDs remains contentious despite significant clinical advancements. Given these facts, a primary issue in nanomedicine is the development of novel techniques for improving CVDs imaging and treatment. With a rapidly rising growth rate, atherosclerosis is the main cause of coronary heart disease and requires special attention in both diagnosis and therapy. This is why the use of nanotheranostics, which enables noninvasive in vivo imaging along with the simultaneous evaluation of the therapeutic effect and the prediction of treatment outcomes, is so popular. Imaging in cardiovascular diseases can assess treatment efficacy and is crucial for optimizing future therapeutic protocols based on individual responses.

7.1 MRI and Anti-angiogenic Therapy

Various approaches have been established to standardize vascular growth and diminish the incidence of intraplaque hemorrhage, thus facilitating plaque stability, with angiogenesis playing a crucial part in advancing atherosclerotic lesions. The instability of high plaque and its tendency to rupture, resulting in myocardial infarction and stroke, are associated with the growth of neo-vasculature [85]. The extensive expression of αvβ3-integrin, a marker of adhesion molecules in vascular epithelial cells, verifies the significant development of new vessels within the atherosclerotic plaque [84]. The efficacy of fumagillin and its water-soluble analogue, TNP-460, with other antiangiogenic agents, has been established in both animal studies and human clinical trials [86]. Nonetheless, significant neurocognitive side effects are associated with the substantial dosage required for therapeutic benefit [87]. Integrin-targeted PFC-loaded nanoparticles and liposomes have been developed to examine the newly growing vasculature to improve the therapy of plaque progression. Furthermore, fumagillin-encapsulated αvβ3-targeted Gd^{3+}-based nanoparticles have been proposed as theranostic approaches to enhance atherosclerosis treatment [88]. These NPs have an antiangiogenic impact that decreases the extent of the neo-vessels, lowering the risk of rupture and plaque vulnerability by specifically targeting the neo-vasculature within the plaque. The presence of Gd^{3+} enabled MRI to monitor the dispersion of nanoparticles and detect integrin expression, indicating plaque angiogenesis. The extent of plaque angiogen esis might be quantified, enabling real-time, noninvasive assessment of therapy efficacy and patient prognosis. Hyperlipidemic New Zealand rabbits were utilized to evaluate the effectiveness of theranostic systems employing paramagnetic αvβ3-targeted Gd^{3+}-based nanoparticles loaded with fumagillin. The minor thickening of the aortic wall observed in the T1-weighted MRI indicated that the high cholesterol diet in this animal model resulted solely in initial-stage atherosclerosis. Nevertheless, at this preliminary phase, the αvβ3-targeted NPs injection produced a significant contrast increase ($16.7 \pm 1.1\%$), enabling visualization of the plaque's neovasculature. The signal was augmented by $10.8 \pm 1.1\%$ due to nontargeted nanoparticles. A week after the first injection, there was no hint of a signal rise on MR imaging, indicating that the NPs had been injected and could no longer be detected. A 60–80% reduction in signal strength was noted following the second injection of αvβ3-targeted nanoparticles, compared to the initial injection. Histological investigation verified that the therapy effectively diminished atherosclerosis-related angiogenesis and αvβ3 integrin expression, as indicated by the reduction in MRI signal. After NPs therapy, the robust T1-weighted signal augmentation of the aortic wall observed on the initial treatment day markedly decreased, but the control group administered unloaded NPs showed no alterations. The tissue distribution and signal decline post-treatment can be employed to identify carrier accumulation, assess disease status, predict therapeutic efficacy, and subsequently monitor the effectiveness of the treatment regimen. Consequently, these NPs offer the opportunity for comprehensive therapeutic monitoring. After various treatments, the efficacy of the identical NPs was scrutinized in greater detail [89]. Four weeks after a distinctive injection of fumagillin-loaded nanoparticles, no differences in the MR signal were seen between the treated and control groups, despite a steady signal enhancement drop of 50–75% compared to controls at week 2. A subsequent injection of NPs elicited the same reaction, demonstrating that the previous medicine administration did not influence the treatment. A solitary administration of this nanotheranostic effectively reduced the neovasculature in the aortic wall, despite fumagillin being employed in these studies at concentrations up to 50,000 times lower than the total oral dose of TNP6470 utilized in prior trials [87]. Consequently, this regimen is expected to mitigate the adverse effects of traditional TNP-470 therapy. The adoption of this technology in clinical settings may facilitate the early identification of patients, many of whom are asymptomatic, at elevated risk for cardiovascular issues due to plaques characterized by significant neovascularization. Moreover, noninvasive monitoring of disease development could facilitate the rapid identification of patients who react to treatment and the assessment of the long-term effects of antiangiogenic therapy.

7.2 MRI and Anti-restenotic Therapy

Restenosis, vascular remodeling mainly brought on by the production of extracellular matrix and the activation, migration, and proliferation of smooth muscle cells, is one of the adverse outcomes of angioplasty [90]. Since $\alpha v\beta 3$ integrins are highly expressed in stretch-activated smooth muscle cells, these cells could be candidates for multifunctional NPs delivery. For example, rapamycin-loaded $\alpha v\beta 3$-targeted paramagnetic nanoparticles were developed for vessel imaging after balloon angioplasty and concomitant anti-restenotic drug delivery, allowing for the monitoring of their effectiveness [91]. The effectiveness has been tested in vivo on New Zealand white rabbits. The animals were fed a high-fat diet for 4 months to induce atherogenesis. They then suffered from femoral artery balloon overstretch, which caused damage to the vessels. The injury, $\alpha v\beta 3$-targeted paramagnetic nanoparticles containing rapamycin, were administered into the compromised vascular lumen and retained for 5 mins. After the suction of unbound nanoparticles and the restoration of blood flow, the animals underwent MRI scanning once the incisions were sutured. A second MRI of the femoral artery was carried out two weeks later. After $\alpha v\beta 3$-targeted paramagnetic NPs were administered, blood flow was restored and a notable signal amplification was observed 30–40 mins later, confirming that NPs had accumulated and remained in the injured artery wall segments. However, no signal amplification was observed after injecting nontargeted paramagnetic NPs or saline solution. The MR scans of all treated groups, obtained within one h post-angioplasty and therapy, did not indicate any artery wall dissections or flow obstructions. Animals exposed to saline solution or targeted or untargeted empty nanoparticles had vascular problems two weeks later. Mice administered rapamycin-loaded αvβ3-targeted paramagnetic nanoparticles exhibited only negligible abnormalities. The data clearly illustrated the benefits of localized rapamycin delivery by targeted nanoparticles, which inhibited lumen occlusion and restenosis while not delaying endothelium repair after balloon injury. Furthermore, the noninvasive monitoring facilitated the assessment of the treatment's efficacy.

7.3 Optical Imaging and Photodynamic Therapy

Macrophages are a significant component of atherosclerotic lesions. They regulate the accumulation of oxidized lipoproteins, the release of inflammatory cytokines, and the activity of extracellular proteinases [92], all of which are crucial for tissue remodeling and facilitate the degradation of the fibrous cap of plaques, leading to plaque rupture. Macrophages provide a significant target for detecting and controlling vulnerable plaques due to their essential role in the genesis and advancement of atherosclerotic lesions. Cross-linked dextran-coated iron oxide (CLIO) nanoparticles exhibit a pronounced attraction for inflammatory macrophages, as demonstrated by multiple investigations [93]. CLIO nanoparticles may serve as a compelling platform for advancing nanotheranostics in atherosclerotic lesions. A potent chlorine-based photosensitizer mesotetra(hydroxyphenyl)chlorin (THPC) and a near-infrared fluorophore were incorporated to produce functionalized CLIO [94]. The unique spectral properties of these compounds enable the monitoring of nanoparticles distribution and the execution of photodynamic therapy. An Apo $E^{-/-}$ mouse model of atherosclerosis has been utilized to evaluate in vivo efficacy. The carotid arteries in the AF750 channel that were surgically exposed were imaged using intravitreal laser scanning fluorescence microscopy 24 h following the CLIO-THPC injection. The positioning of the nanoparticles within the carotid atheroma lesion, namely in regions rich in macrophages and foam cells, was confirmed using fluorescence imaging. Following the imaging session, mice were subjected to 650 nm laser light to activate the macrophage-targeted photosensitizer. Mice were sutured and allowed a period for recovery post-treatment. One day post-treatment, a cohort of mice was euthanized, and histological analysis of their carotid arteries revealed that the CLIO-THPC-treated mice had a greater proportion of apoptotic cells ($55.5 \pm 5.1\%$ of the plaque area) compared to the control group ($0.796 \pm 0.10\%$ of the plaque area). After the mice in the other group received a second dose of the nanotheranostic, they were scanned a week later. The treatment's efficacy was validated by the mice's minimal absorption of CLIO-THPC following PDT administration. These findings demonstrated the potential of using such nanotheranostic methods for the detection and treatment of atherosclerotic lesions.

7.4 Ultrasonography and Thrombolytic Therapy

Acute myocardial infarction and stroke are commonly managed with thrombolytic agents [95]. The predominant thrombolytic agent employed in clinical settings is recombinant tissue plasminogen activator (rtPA). It has been shown to assist patients with thromboembolic illnesses by stimulating plasmin production, which leads to the lysis of fibrin clots. Utilizing an external magnetic field, super paramagnetic iron oxide (SPIO) nanoparticles can facilitate the delivery of thrombolytic agents to the target location. Again, the NPs' biodistribution and the therapy's effectiveness may be monitored in real time without invasive procedures. Ma et al. [96] developed polyacrylic acid-coated magnetic nanoparticles (PAA-MNP) with covalently linked rtPA (PAA–MNP–rtPA). By injecting a clot at the aorta and iliac artery bifurcation to reduce the left hind limb's perfusion, these theranostic NPs' effectiveness was evaluated in vivo in a rat embolic model. A laser Doppler perfusion imager was employed to monitor skin tissue perfusion in the hind limbs for 120 mins post-injection of the clot, after administering PAA–MNP–rtPA, free rtPA, or PAA-MNP. When delivered by magnetic guidance, PAA-MNP-rtPA produced more substantial thrombolysis and restoration of tissue perfusion compared to controls. The optimal thrombolysis induced by PAA-MNP-rtPA occurred at an rtPA dosage below 20% of the usual dose. These results provide promising paths toward successful in vivo reperfusion and thrombolysis. Doppler perfusion monitoring facilitates the timely assessment of treatment efficacy, allowing for reevaluating of thrombolytic therapy in patients.

7.5 Nanotheranostics for Myocardial Infarction (MI)

MI, the most severe and prevalent type of CVD, is the leading cause of morbidity and mortality worldwide [97]. Annually, almost seven million individuals are affected by myocardial infarction, resulting in healthcare expenses of over $450 billion for the economy. Despite tremendous improvements in patient care and lifestyle changes, MI remains a threat to global health, contributing 15% of the world's yearly death rate [98]. The predominant cause of myocardial infarction is the sudden onset of ischemia due to partial or complete obstruction of blood flow in several coronary arteries, resulting from the corrosion, rupture, ulceration, or fissuring of atherosclerotic plaque. Myocyte necrosis occurs due to diminished blood flow and subsequent oxygen deficiency in the perfused myocardium. Type 1 myocardial infarction is the designation assigned to this ischemia-associated myocardial infarction [99]. Type 2 myocardial infarction, while spontaneous, mostly originates from an abrupt imbalance between oxygen supply and demand, leading to ischemia, myocardial damage, and necrosis. Sudden unexpected cardiac death, characterized by cardiac arrest, is designated as type 3 MI. It often presents with symptoms indicative of myocardial ischemia or the development of a new thrombus in a coronary artery, identified through angiography and/or autopsy [99]. MI types 4 and 5 are classified as periprocedural infarctions due to their occurrence following surgical procedures [100]. Type 4 myocardial infarction is subdivided into categories 4a and 4b. The 4a signifies a myocardial infarction associated with percutaneous coronary intervention, characterized by high cardiac troponin levels and symptoms indicative of myocardial ischemia, including new electrocardiogram abnormalities or a reduction in viable myocardium. Thrombotic blockage of a coronary stent, as revealed by angiography or autopsy, is linked to type 4b, which is further defined by fluctuations in cardiac biomarker levels. Finally, coronary artery bypass grafting is associated with type 5 MI, characterized by elevated cardiac biomarker levels, newly obstructed native coronary arteries, loss of viable myocardium, or new regional wall motion abnormalities [97].

Stem cell transplantation is a state-of-the-art technique for repairing the injured myocardium after MI and restoring normal cardiac function within the infarct. As previously mentioned, preserving the survival of stem cells during transplantation and enhancing their retention rate in the myocardium, currently ranging from 0.1% to 15%, remains challenging. Gomez-Mauricio et al. developed an innovative theranostic approach to facilitate cardiac recovery post-myocardial infarction by administering and monitoring adipose-derived stem cells encapsulated in magnetic resonance-labeled microcapsules. Magnetic alginate beads were synthesized by extruding a combination of alginate solution, Endorem (commercially available superparamagnetic iron oxide nanoparticles), and adipose-derived stem cells (ASCs) cell pellet into a calcium chloride solution. The ultimate particles of magnetic resonance-labeled alginate-PLL-alginate microcapsules, referred to as "magnetocapsules," were synthesized through cross-linking magnetic alginate beads with PLL solutions, followed by an additional alginate solution. These capsules maintained high cellular vitality of ASCs for 21 days in culture, and the magnetic labeling did not affect their survival. ASCs were successfully retained inside the myocardium, as demonstrated by the identification of ASCs-encapsulated magnetocapsules in the infarcted myocardium of a porcine myocardial infarction model for up to 30 days post-injection, with no decrease in the intensity of the iron oxide label, as per in vivo MRI imaging. Moreover, these stem cell-encapsulated magneto capsules improved cardiac function by diminishing the infarct size and increasing the left ventricular ejection fraction as evaluated by MRI [101]. In another work, polymeric nanoparticles were utilized to enhance cardiac regeneration by a theranostic approach. A lipid-based copolymer of DSPE-PEG-Maleimide was synthesized to encapsulate a semiconductor polymeric contrast agent (PCPDTBT). Cell-penetrating peptides were added to functionalize the NPs further and improve their labeling efficacy for cardiomyocytes derived from human embryonic stem cells. Ultrasound and photoacoustic imaging facilitated high spatial resolution real-time observation of the marked cells' transit, engraftment, and localization following their transplantation into living mouse hearts. This study [102] highlighted the advantages of employing polymeric nanoparticles for photoacoustic imaging to monitor the fate of heart-transplanted cells in regenerative therapies.

Theranostic nanosystems are additionally utilized for precise imaging and diagnostics of ischemia/reperfusion illnesses, along with the controlled delivery of therapeutic agents. Kang et al. developed micellar nanoparticles utilizing

H_2O_2-scavenging boronate polymers linked with near-infrared fluorescent dyes to visualize obstructive thrombus and inhibit further thrombus development. Boronate not only scavenges H_2O_2 but also releases hydroxybenzyl alcohol (HBA), an antioxidant, during H_2O_2-mediated oxidation. Fibrin-targeting lipopeptides were incorporated onto the surface of nanoparticles to improve site-specific delivery and minimize off-target effects, as fibrin is a critical and ubiquitous component of thrombi. At a concentration of 100 μg/mL, these nanoparticles, designated as fibrin-targeted imaging and antithrombotic nanomedicine (FTIAN) effectively inhibited cytotoxicity and exhibited anti-inflammatory and antioxidant properties on arterial endothelial cells and H_2O_2-stimulated macrophages. They displayed antiplatelet characteristics by significantly diminishing the soluble CD40 ligand (sCD40L) expression on activated platelets. Significantly, the utilization of FTIAN in lieu of HBA alone resulted in markedly enhanced results. Additionally, in vivo investigations utilizing a murine carotid thrombosis model administered with FTIAN nanoparticles demonstrated a significant reduction in platelet activation and inflammation, as seen by a decline in TNF-α and sCD40L levels, respectively. At a lower dosage of the anti-thrombotic drug tirofiban, these NPs also showed impressive ROS generation suppression and thrombus formation inhibition. Ultimately, the NPs enhanced the fluorescence and photoacoustic intensity of fibrin-rich clots obstructing arteries, so serving as dual imaging agents. Ultimately, these novel NPs exhibited potential as thrombus-specific theranostic agents for treating cardiovascular diseases primarily caused by thrombosis, including type 1 myocardial infarction [103]. Polymeric nanoparticles were utilized in a separate investigation to encapsulate therapeutic agents and employed imaging techniques to monitor dosage and delivery. A mouse model of myocardial ischemia/reperfusion was employed to examine chitosan hydrogel nanoparticles radiolabeled with technetium-99 m and encapsulating VEGF81–91 peptides. SPECT demonstrated that they augment vascular density and mitigate perfusion deficits in the ischemic region of the myocardium, while also promoting radionuclide retention of up to 9.2%. Consequently, these nanoparticles presented intriguing alternatives for quantitative radiological imaging and the controlled delivery of VEGF peptides, which may facilitate the customization of myocardial ischemia therapy [104].

8 Nanotheranostics for Neurological Disorders

The diseases enumerated below illustrate that various attributes of nanotheranostic systems can significantly beneficial the development of next-generation neuro-targeting nanotherapeutics, serving dual roles as therapeutic and diagnostic agents to ameliorate severe neurological disorders.

8.1 Glioma (Brain Tumors)

The incidence of cancer-related fatalities has significantly increased over the past many years. Currently, most therapies utilize chemotherapy and radiation therapy. These approaches, however, lack specificity and obliterate both malignant and benign tissues. Due to the blood-brain barrier (BBB), the majority of anticancer pharmaceuticals cannot penetrate the brain. Theranostic nanomedicines are being developed to improve current cancer therapies and circumvent their limitations [105]. Ruan et al. developed a hybrid peptide-based nanoparticulate theranostic system, incorporating gold and gelatin nanoparticles, to assess doxorubicin (DOX) administration and its anti-glioma efficacy. The system was surface anchored utilizing the glioma targeting sequence to facilitate active targeting. Following a 24-h incubation with matrix metalloproteinase-2, the dimensions of G-AuNPs-DC-RRGD diminished from 188.2 nm to 55.9 nm, with DOX and Cy5.5 exhibiting pH-dependent release. G-AuNPs-DC-RRGD demonstrated cellular penetration, exhibited superior glioma targeting and accumulation efficacy, and facilitated the transfer of DOX to tumors. These approaches facilitated simultaneous tracking of drug delivery potential and therapeutic effects in vivo by fluorescence imaging [106]. In another work, Ching et al. utilized DOX-encapsulated microbubbles (DOXSPIO-MBs) combined with SPIO for the diagnosis and treatment of a brain tumor. The BBB was accessed locally and noninvasively by a targeted ultrasound technique. Microbubbles responsive to focused ultrasound (FUS) operated as stable nanosystems designed to facilitate the opening of the BBB. The transport of therapeutic drugs was effective following FUS exposure in the rat C6 glioma model, and these microbubbles may serve as valuable instruments for future image-guided diagnosis and treatment of brain tumors [107]. Sun and colleagues developed highly selective and biocompatible nanoprobes by functionalizing the surface of PEG-coated iron oxide nanoparticles with the glioma-targeting peptide chlorotoxin (NP-PEG-CTX nanoprobe). This study emphasized the preferential accumulation of NP-PEG-CTX nanoprobes in glioma tissues, followed by in vitro evidence of contrast enhancement in MRI in both xenografts and 9 L cells. The internalization of NP-PEG-CTX nanoprobes into the cytoplasm of 9 L cells was further confirmed using TEM imaging. The developed nanoprobe offered a potential platform for the concurrent diagnosis and treatment of lethal gliomas and other neurological disorders, characterized by specific targeting and biocompatibility [108].

8.2 Alzheimer's Disease (AD)

Alzheimer's disease (AD), a highly widespread neurodegenerative disorder, is marked by the accumulation of intracellular neurofibrillary tangles formed by phosphorylated tau protein and plaques resulting from β-amyloid aggregation [109]. Notwithstanding the participation of a varied team of researchers, no effective treatment for this ailment has been developed to date. In clinical trials, drugs for AD primarily focus on the central nervous system as its site of action. Nonetheless, the BBB's role as a protective and highly selective barrier in the central nervous system inhibits the passage of medications at elevated concentrations [110]. A prime example of "image-guided therapy" for Alzheimer's disease is the "sense and treat system" developed by Zhi Du and colleagues, which simultaneously diagnoses and targets amyloid aggregates associated with the condition. They utilized magnetic nanoparticles (MNPs) featuring two moieties on its surface: a naphthalimide-based fluorescence probe (NFP) to target the exposed hydrophobic regions of Aβ oligomers, and the Aβ-targeting peptide KLVFF. The disaggregation of Aβ aggregates, induced by the localized heat generated by MNPs and detectable by the fluorescence probe, allowed the complex (MNP@NFP-pep) to monitor alterations in Aβ morphology. The results indicated its potential application as a real-time "sense and treat" system for the treatment of AD [111]. A different research team developed a tau-targeted multifunctional nanocomposite (CeNC/IONC/MSN-T807-MB) for the theranostics of AD. This comprised mesoporous silica nanoparticles (MSNs) functionalized with amino-T807 for targeted tau engagement, iron oxide nanocrystals (IONCs) serving as an MRI contrast agent, and cerium nanocrystals (CeNCs) that inhibit tau hyperphosphorylation. Methylene blue (MB) was administered to prevent tau aggregation. Research performed both in vitro and in vivo revealed a significant decrease in tau hyperphosphorylation and neuroprotection in AD models, offered an alternative to amyloid-β therapies [112]. Zhang et al. developed a targeted delivery method utilizing dual-functional PEGylated poly(lactic acid) nanoparticles (PEG-PLA NPs) for the early detection and treatment of AD. Dual targeting was achieved by surface functionalization of the nanoparticles with TGN peptide, a blood-brain barrier targeting ligand, and QSH peptide, an Aβ42 targeting ligand. Ex vivo imaging assays and cerebral distribution analyses of nanoparticles utilizing the near-infrared dye DiR as a probe confirmed the dual targeting efficacy of the established approach. In the brains of an AD-induced animal model, the dual-targeted polymeric nanoparticles exhibited precise and improved targeted delivery to amyloid plaques, indicated a promising therapeutic approach for Alzheimer's diagnosis and therapy [113].

8.3 Parkinson's Disease (PD)

Parkinson's disease (PD) is the most common age-related neurodegenerative disorder behind AD. Neuropathology indicates that PD is marked by a gradual degeneration of dopaminergic neurons in the substantia pars compacta, leading to diminished dopamine levels and notable cognitive and motor deficits [114]. L-3,4-dihydroxyphenylalanine (levodopa), a dopamine precursor that efficiently traverses the blood-brain barrier via the large neutral amino acid transporter (LAT1), has lately assumed the role of dopamine [115]. L-DOPA is presently an effective medicine for those with PD [116]. Birgitte and her colleagues developed a distinctive manganese oxide nanoparticles (MONPs) modified with L-DOPA, capable of simultaneously releasing Mn^{2+} ions and L-DOPA in water upon disintegration. L-DOPA was utilized as a stabilizer and therapeutic agent for PD, while MONPs served as an

MRI contrast agent for imaging and diagnosis via a time-dependent change in MR contrast [117]. Niu et al. developed a nano-gene delivery technique employing magnetic Fe_3O_4 nanoparticles coated with oleic acid and integrated with a short hairpin RNA (shRNA) plasmid to inhibit α-synuclein expression in neurons. N-isopropylacrylamide derivative (NIPAm-AA) was photo-immobilized onto oleic acid molecules to render the system responsive to pH and temperature for effective gene release. Nerve growth factor was similarly adsorbed onto NIPAm-AA to enhance neuronal absorption through receptor-mediated endocytosis. The findings from both in vitro and in vivo studies demonstrated that this method effectively prevented apoptosis and rehabilitates a PD model [94]. Hu et al. employed lactoferrin (Lf)-conjugated PEG-PLGA nanoparticles (Lf-NPs) to develop a biodegradable drug delivery method for the brain. The nanoparticles were infused with urocortin (UCN), a peptide that protects against PD, while lactoferrin served as a brain-targeting ligand. Both qualitative and quantitative investigations of Lf-NPs uptake demonstrated a substantial accumulation of Lf-NPs in bEnd.3 cells via clathrin-mediated endocytosis. Lf-NPs infused with UCN significantly mitigated striatal damage induced by 6-OHDA. The findings indicated that UCN-loaded Lf-NPs could serve as effective delivery vehicles for brain medicines in the treatment of PD [118].

8.4 Neurovascular Diseases

Atherosclerosis and stroke are prominent neurovascular disorders that are among the primary causes of mortality and disability globally. The existing treatments are constrained by toxicity, rapid clearance, and insufficient targeting of medications. By modifying nanoparticles with specific ligands or peptides for precise vascular targeting, nanotherapeutic systems offer a potential alternative and enhance treatment efficacy [119]. McCarthy and his team developed fluorescent nanoparticles targeting macrophages for the treatment of atherosclerosis. By integrating dextran-coated magnetic nanoparticles with near-infrared dye, magnetic fluorescent nanoparticles (MFNPs) were synthesized and later evaluated for their efficacy in photodynamic therapy. The MFNPs generated singlet oxygen species upon stimulation at 646 nm, destroying macrophages while preserving adjacent normal cells [120]. Myerson et al. explored the application of nanotheranostics for thrombosis treatment by synthesizing perfluorocarbon nanoparticles functionalized with bivalirudin. These nanoparticles offered a promising theranostic strategy for clot management by selectively targeting thrombin at active coagulation sites, adhering to the thrombus and inhibiting platelet deposition by diminishing local thrombin activity [121]. Sun and colleagues employed VEGF-loaded, IR800-conjugated graphene oxide nanoparticles to induce therapeutic angiogenesis in a murine hind limb model [122].

9 Challenges and Limitations

The synthesis of nanoparticles (1–200 nm) and the identification of unique physical and chemical features not present in their bulk counterparts are two factors contributing to the significant impact of nanotechnology on theranostics. Despite the numerous achievements in nanotechnology, significant limitations persist. The implementation of nanotheranostics remains constrained by design, fabrication, biomolecular interactions, regulatory compliance, risk factors, and safety issues. These restrictions must be considered when developing an effective nanotherapeutic platform.

9.1 Specificity and Selectivity

Due to their unique properties, nanoparticles are often utilized in cancer treatment; yet, challenges regarding selectivity and specificity persist. Their ability to selectively target cancer cells by modifying their surface with ligands, antibodies, or peptides is referred to as specificity. Utilizing tumor-specific properties, such as vascular permeability, selectivity ensures that nanoparticles accumulate in tumors while sparing healthy cells. To reduce harm to healthy tissues, they might be engineered to release medications within tumors in a controlled manner. To achieve successful targeting, biological barriers must be surmounted. The combination of specificity and selectivity reduces side effects and enhances therapeutic efficacy. Nanoparticle-based cancer therapies must evaluate issues such as tumor heterogeneity, microenvironmental changes, and potential off-target effects [123].

9.2 Biocompatibility

Biocompatibility denotes the capacity of nanoparticles to avoid causing harm to human tissue. The clinical application of specific nanoparticles may be limited due to their possible toxicity or immunological reactions. The primary issue with nanoparticles is their toxicity. The movement of nanoparticles through the bloodstream is also affected by their curvature and morphology. If nanoparticles lack adequate biocompatibility, they may induce inflammation, immune responses, or accumulation in unintended regions. Biocompatibility is crucial when utilizing immune cells to target cancer cells in immunotherapies to prevent complications such as cytokine storms or autoimmune responses. Scientists and medical experts continually strive to create biocompatible medicines that efficiently target cancer cells while reducing damage to healthy tissues [124].

9.3 Pharmacokinetics

Nanoparticle-based drug delivery techniques exhibit superior pharmacokinetics and can efficiently target tumors by traversing the circulatory system. They must ensure controlled and uniform medication delivery. The kinetics of drug adsorption and desorption on carbon nanomaterials, considering aggregation, surface chemistry, and purity, is a primary focus of study. Substantial challenges persist, including inadequate tumor penetration, difficulties in predicting the efficiency of targeted therapies, complex nanoformulations, inconsistent compatibility with combination treatments, and knowledge deficiencies regarding tumor biology and progression.

9.4 Manufacturing and Scale-Up

Producing homogeneous nanoparticles is challenging; nonetheless, reducing particle size alters surface energy and increases the surface-area-to-volume ratio. Batch-to-batch variations affect the safety, efficacy, and repeatability of nanotheranostics. Outcomes from diverse methodologies vary, requiring re-optimization. The production of nanoparticles is complex and labor-intensive, with stability issues emerging due to external factors that induce particle alterations over time.

9.5 Cost-Effectiveness

The cost of building nanotheranostic platforms limits patient access. Noble metal nanoparticles (MNPs) offer less invasive cancer therapies; considerable research focuses on the stability, biocompatibility, and functionalization of gold and silver. Palladium and platinum have been used together for biomedical purposes, nonetheless, metals such as ruthenium and iridium remain inadequately researched due to their elevated cost and scarcity [125].

9.6 Clinical Translation

The stringent regulatory approval processes render the clinical implementation of nanotheranostic systems challenging, despite promising preclinical results. Theranostic nanosystems have developed more during the past decade; yet, none have attained FDA approval for clinical application. Linking nanoparticles' chemical and physical properties to their biological interactions presents a significant challenge. Clinical studies are complex, time-intensive, and uncertain endeavors necessary for assessing efficacy, potential side effects, and nanoparticle behavior throughout the body. Substantial obstacles are posed by factors associated with chemistry, manufacturing, and control and good manufacturing practices. Complying with regulatory standards and transitioning these technologies from laboratory research to practical medical applications requires the optimization of existing procedures and the enhancement of production standards [126, 127].

10 Summary and Conclusion

Nanotheranostic platforms have significant potential for diagnosis and treatment, despite challenges related to safety, efficacy, accessibility, and regulatory approval. Nanoparticles have been explored as potential treatments in chemotherapy, radiation, phototherapy, gene therapy, and immunotherapy. The significance of functional nanomaterials in the early detection and targeted therapy of cancer is underscored by advancements in nanotheranostics. Researchers are focusing on multifunctional nanotheranostic agents for precise medication delivery and imaging. However, challenges arise due to complexities like as toxicity, synergistic effects, and ambiguous mechanisms. Efforts are underway to incorporate nanotheranostics into clinical investigations, emphasizing molecular imaging, targeted drug delivery, and interactions with biological systems. Collaboration between researchers and pharmaceutical firms is essential for the advancement and commercialization of innovative nanotheranostic medicines.

11 Future Prospectives

Nanotheranostics is an innovative approach that could transform medical treatment. It has been widely utilized in recent years for targeted drug administration, therapy, and imaging of central nervous system illnesses. Moreover, it facilitates the monitoring of drug accumulation, tissue distribution, and pharmacokinetics, thereby enhancing our understanding of neurological illnesses. Despite advancements in nanotheranostic use via in vitro studies, in vivo validation remains challenging. Increased innovation is essential to guarantee safety and efficacy before clinical application. Advanced biomaterials capable of subcellular interaction are being developed. Nanotheranostics holds prospective uses in prognostic assessment, therapeutic efficacy evaluation, and disease progression monitoring. Moreover, its integration with delivery systems for pharmaceuticals, growth factors, and stem cells may enhance tissue regeneration and offer improved treatments for neurological disorders.

References

1. Scannell, J. W., Blanckley, A., et al. (2012). Diagnosing the decline in pharmaceutical R&D efficiency. *Nature Reviews Drug Discovery, 11*(3), 191–200.
2. Hamburg, M. A., et al. (2010). The path to personalized medicine. *New England Journal of Medicine, 363*(4), 301–304.
3. Chen, X. S. (2011). Introducing theranostics journal from the editor-in-chief. *Theranostics, 1*, 1–2.

4. Chen, X. S. (2012). One year after a successful start of theranostics. *Theranostics, 2*(1), 1–2.
5. Swierczewska, M., Liu, G., et al. (2012). High-sensitivity nanosensors for biomarker detection. *Chemical Society Reviews, 41*(7), 2641–2655.
6. Chikkaveeraiah, B. V., Bhirde, A. A., et al. (2012). Electrochemical immunosensors for detection of cancer protein biomarkers. *ACS Nano, 6*(8), 6546–6561.
7. Shao, H., Min, C., et al. (2012). Magnetic nanoparticles and micro-NMR for diagnostic applications. *Theranostics, 2*(1), 55–65.
8. Farokhzad, O. C., et al. (2006). Nanomedicine: Developing smarter therapeutic and diagnostic modalities. *Advanced Drug Delivery Reviews, 58*(14), 1456–1459.
9. Swierczewska, M., Lee, S., et al. (2011). Moving theranostics from bench to bedside in an interdisciplinary research team. *Therapeutic Delivery, 2*(2), 165–170.
10. Xie, J., Lee, S., et al. (2010). Nanoparticle-based theranostics agents. *Advanced Drug Delivery Reviews, 62*(11), 1064–1079.
11. Sun, Z., et al. (2025). Preparation, characterization, and binding mechanism of quercetin-loaded composite nanoparticles based on zein-soybean protein isolate. *Food Chemistry, 463*, 141359.
12. Adul-Rasool, A. A., et al. (2024). 0, 1, 2, 3D nanostructures, types of bulk nanostructured materials, and drug nanocrystals: An overview. *Cancer Treatment and Research Communications, 40*, 100834.
13. Peng, S., & Li, P. (2024). *Nanomaterials and nanotechnology: Basic, preparation and applications*. Springer.
14. Campos, M. T., et al. (2025). Self-assembled inorganic nanomaterials for biomedical applications. *Nanoscale, 17*, 5526.
15. Singh, A., et al. (2025). Nanotheranostics: The fabrication of theranostics with nanoparticles and their application to treat the neurological disorders. *Recent Patents on Nanotechnology, 19*(1), 17–34.
16. Sanson, C., Diou, O., et al. (2011). Doxorubicin-loaded magnetic polymersomes: Theranostic nanocarriers for MR imaging and magneto-chemotherapy. *ACS Nano, 5*(1), 1122–1140.
17. Pan, D., Caruthers, S. D., et al. (2008). Ligand-directed nanobialys as theranostic agent for drug delivery and manganese-based magnetic resonance imaging of vascular targets. *Journal of the American Chemical Society, 130*(30), 9186–9187.
18. Smith, D. A., Vaidya, S. S., et al. (2010). Ultrasound-triggered release of recombinant tissue type plasminogen activator from echogenic liposomes. *Ultrasound in Medicine and Biology, 36*(1), 145–157.
19. Wang, A. Z., Bagalkot, V., et al. (2028). Superparamagnetic iron oxide nanoparticle–aptamer bioconjugates for combined prostate cancer imaging and therapy. *ChemMedChem, 3*(8), 1311–1315.
20. Singh, F., Dilnawaz, S., et al. (2011). Long circulating lectin conjugated paclitaxel loaded magnetic nanoparticles: A new theranostic avenue for leukemia therapy. *PLoS One, 6*(11), 26803.
21. Ling, Y., Wei, K., et al. (2011). Dual docetaxel/super paramagnetic iron oxide loaded nanoparticles for both targeting magnetic resonance imaging and cancer therapy. *Biomaterials, 32*(30), 7139–7150.
22. Yallapu, M. M., Othman, S. F., et al. (2011). Multifunctional magnetic nanoparticles for magnetic resonance imaging and cancer therapy. *Biomaterials, 32*(8), 1890–1905.
23. Lanza, G. M., Yu, X., et al. (2002). Targeted anti-proliferative drug delivery to vascular smooth muscle cells with a magnetic resonance imaging nanoparticle contrast agent: Implications for rational therapy of restenosis. *Circulation, 106*(22), 2842–2847.
24. Veiseh, O., Kievit, F. M., et al. (2010). Chlorotoxin-bound magnetic nanovector tailored for cancer cell targeting, imaging, and siRNA delivery. *Biomaterials, 31*(29), 8032–8042.
25. De Smet, M., Langereis, S., et al. (2010). Temperature-sensitive liposomes for doxorubicin delivery under MRI guidance. *Journal of Controlled Release, 143*(2), 120–127.

26. Guthi, J. S., Yang, S.-G., et al. (2009). MRI-visible micellar nanomedicine for targeted drug delivery to lung cancer cells. *Molecular Pharmaceutics, 7*(1), 32–40.
27. Tang, Y., Lei, T., et al. (2010). Simultaneous delivery of chemotherapeutic and thermal-optical agents to cancer cells by a polymeric (PLGA) nanocarrier: An in vitro study. *Pharmaceutics, 27*(10), 2242–2253.
28. Wang, C., Cheng, L., et al. (2011). Drug delivery with upconversion nanoparticles for multi-functional targeted cancer cell imaging and therapy. *Biomaterials, 32*(30), 1110–1120.
29. Quintana, A., Raczka, E., et al. (2002). Design and function of a dendrimer-based therapeutic nanodevice targeted to tumor cells through the folate receptor. *Pharmaceutical Research, 19*(10), 1310–1316.
30. Heo, D. N., Yang, D., et al. (2012). Gold nanoparticles surface-functionalized with paclitaxel drug and biotin receptor as theranostic agents for cancer therapy. *Biomaterials, 33*(4), 856–866.
31. Abdalla, M. O., Karna, P., et al. (2011). Enhanced noscapine delivery using uPAR-targeted optical-MR imaging trackable nanoparticles for prostate cancer therapy. *Journal of Controlled Release, 149*(3), 314–322.
32. Das, M., Mishra, D., et al. (2009). Biofunctionalized, phosphonate-grafted, ultrasmall iron oxide nanoparticles for combined targeted cancer therapy and multimodal imaging. *Small, 5*(3), 2883–2893.
33. Santra, S., Kaittanis, C., et al. (2009). Drug/dye-loaded, multifunctional iron oxide nanoparticles for combined targeted cancer therapy and dual optical/magnetic resonance imaging. *Small, 5*(2), 1862–1868.
34. Alexiou, C., Arnold, W., et al. (2000). Locoregional cancer treatment with magnetic drug targeting. *Cancer Research, 60*(23), 6641–6648.
35. Li, X., Qian, Y., et al. (2011). Amphiphilic multi-arm star block copolymer based multifunctional unimolecular micelles for cancer targeted drug delivery and MR imaging. *Biomaterials, 32*(27), 6595–6605.
36. Liu, T., Li, X., et al. (2012). Multifunctional pH-disintegrable micellar nanoparticles of asymmetrically functionalized β-cyclodextrin-based star co-polymer covalently conjugated with doxorubicin and DOTA-Gd moieties. *Biomaterials, 33*(10), 2521–2531.
37. De Smet, M., Heijman, E., et al. (2011). Magnetic resonance imaging of high intensity focused ultrasound mediated drug delivery from temperature-sensitive liposomes: An in vivo proof-of-concept study. *Journal of Controlled Release, 150*(1), 102–110.
38. Gao, Z., Kennedy, A. M., et al. (2008). Drug-loaded nano/microbubbles for combining ultrasonography and targeted chemotherapy. *Ultrasonics, 48*(3), 260–270.
39. Park, J. H., von Maltzahn, G., et al. (2008). Micellar hybrid nanoparticles for simultaneous magneto fluorescent imaging and drug delivery. *Angewandte Chemie International Edition, 47*(41), 7284–7288.
40. Nasongkla, N., Bey, E., et al. (2006). Multifunctional polymeric micelles as cancer-targeted, MRI-ultrasensitive drug delivery systems. *Nano Letters, 6*(11), 2427–2430.
41. Wang, W., et al. (2024). Utilization of nanomaterials in MRI contrast agents and their role in therapy guided by imaging. *Frontiers in Bioengineering and Biotechnology, 12*, 1484577.
42. Chen, L., et al. (2024). Silicon-containing nanomedicine and biomaterials: Materials chemistry, multi-dimensional design, and biomedical application. *Chemical Society Reviews, 53*, 1167.
43. Rahmanian, M., et al. (2024). Current trends in triblock copolymers-based multifunctional nanotheranostics for cancer treatment. *Journal of Drug Delivery Science and Technology, 99*, 105985.
44. Ullah, A., et al. (2025). Advancing therapeutic strategies with polymeric drug conjugates for nucleic acid delivery and treatment. *International Journal of Nanomedicine, 25*, 52.
45. Kumar, P., et al. (2024). Catalyzing innovation: Exploring iron oxide nanoparticles – Origins, advancements, and future application horizons. *Coordination Chemistry Reviews, 507*, 215750.

46. Butt, A., & Bach, H. (2025). Nanomedicine and clinical diagnostics part I: Applications in conventional imaging (MRI, X-ray/CT, and ultrasound). *Nanomedicine, 20*(2), 167–182.
47. Dastgheib, Z. S., et al. (2024). Gold nanostructures in melanoma: Advances in treatment, diagnosis, and theranostic applications. *Heliyon, 10*(15), e35655.
48. Hirulkar, R., et al. (2024). Nanotheranostics: Clinical status, toxicity, regulatory consideration, and future prospects. In *Nanotheranostics for diagnosis and therapy* (pp. 249–285). Springer.
49. Ko, H.-J., Suh, J.-S., et al. (2007). Multifunctional magneto-polymeric nanohybrids for targeted detection and synergistic therapeutic effects on breast cancer. *Angewandte Chemie International Edition, 46*(46), 8836–8839.
50. Kuna, K., et al. (2024). Emerging natural polymer-based architecture nanotherapeutics for the treatment of cancer. *International Journal of Biological Macromolecules, 262*, 129434.
51. Luk, B. T., & Zhang, L. (2014). Current advances in polymer-based nanotheranostics for cancer treatment and diagnosis. *ACS Applied Materials & Interfaces, 6*(24), 21859–21873.
52. Liu, S., Hou, X., et al. (2022). Lipid perfluorohexane nanoemulsion hybrid for MRI-guided high intensity focused ultrasound therapy of tumors. *Frontiers in Bioengineering and Biotechnology, 10*, 846446.
53. Peng, S.-L., Lai, C.-H., et al. (2020). Nanotheranostics with the combination of improved targeting, therapeutic effects, and molecular imaging. *Frontiers in Bioengineering and Biotechnology, 8*, 8.
54. Liu, B., et al. (2024). Exploring treatment options in cancer: Tumor treatment strategies. *Signal Transduction and Targeted Therapy, 9*(1), 175.
55. Tantray, J., et al. (2025). Nanotechnology-based biomedical devices in cancer diagnostics and therapy. *Medical Oncology, 42*(2), 50.
56. Graham, W., et al. (2025). Magnetic nanoparticles and drug delivery systems for anti-cancer applications: A review. *Nanomaterials (Basel), 15*(4), 285.
57. Singh, V. (2024). Theranostics: Integrated diagnostics and therapy using nanomedicine. In *Nanomedicine* (pp. 505–530). Springer.
58. Zhu, S., Wang, X., et al. (2020). Gold nanocluster grafted conjugated polymer nanoparticles for cancer cell imaging and photothermal killing. *Colloids and Surfaces A: Physicochemical and Engineering Aspects, 597*, 124764.
59. Liu, Y., et al. (2024). Targeted drug delivery strategies for the treatment of hepatocellular carcinoma. *Molecules, 29*(18), 4405.
60. Nishiyama, N., & Kataoka, K. (2006). Current state, achievements, and future prospects of polymeric micelles as nanocarriers for drug and gene delivery. *Pharmacology & Therapeutics, 112*(3), 630–648.
61. Pan, D., Caruthers, S. D., et al. (2008). Ligand-directed nanobialys as theranostic agents for drug delivery and manganese-based magnetic resonance imaging of vascular targets. *Journal of the American Chemical Society, 130*(30), 9186–9187.
62. Kaida, S., Cabral, H., et al. (2010). Visible drug delivery by supramolecular nanocarriers directing to single-plate formed diagnosis and therapy of pancreatic tumor model. *Cancer Research, 70*(17), 7031–7041.
63. Koning, G. A., & Krijger, G. C. (2007). Targeted multifunctional lipid-based nanocarriers for image-guided drug delivery. *Anticancer Agents in Medicinal Chemistry, 7*(4), 425–440.
64. Schwendener, R. A. (2007). Liposomes in biology and medicine. *Advances in Experimental Medicine and Biology, 620*, 117–128.
65. Grange, C., Geninatti-Crich, S., et al. (2010). Combined delivery and magnetic resonance imaging of neural cell adhesion molecule-targeted doxorubicin-containing liposomes in experimentally induced Kaposi's sarcoma. *Cancer Research, 70*(5), 2180–2190.
66. Cittadino, E., Ferraretto, M., et al. (2012). MRI evaluation of the antitumor activity of paramagnetic liposomes loaded with prednisolone phosphate. *European Journal of Pharmaceutical Sciences, 45*(4), 436–441.
67. Saluja, V., Mishra, Y., et al. (2021). Dendrimer-based cancer nanotheranostics: An overview. *International Journal of Pharmaceutics, 600*, 120485.

68. Sk, U. H., & Kojima, C. (2015). Dendrimers for theranostic applications. *Biomolecular Concepts, 6*(3), 205–217.
69. Ali, I., Alsehli, M., et al. (2020). Progress in polymeric nano-medicines for theranostic cancer treatment. *Polymers, 12*(3), 598.
70. Li, L., Liu, T., et al. (2014). Multifunctional silica-based nanocomposites for cancer nanotheranostics. *Journal of Biomedical Nanotechnology, 10*(9), 1784–1809.
71. Singh, R. K., Patel, K. D., et al. (2017). Progress in nanotheranostics based on mesoporous silica nanomaterial platforms. *ACS Applied Materials & Interfaces, 9*(12), 10309–10337.
72. Živojević, K., Mladenović, M., et al. (2021). Advanced mesoporous silica nanocarriers in cancer theranostics and gene editing applications. *Journal of Controlled Release, 337*, 193–211.
73. Peng, X., Lin, G., et al. (2021). Mesoporous silica nanoparticle-based imaging agents for hepatocellular carcinoma detection. *Frontiers in Bioengineering and Biotechnology, 9*, 9.
74. Mendes, L. P., Lima, E. M., et al. (2018). Targeted nano theranostics for selective drug delivery in cancer. In J. Conde (Ed.), *Handbook of nanomaterials for cancer theranostics* (pp. 245–277). Elsevier.
75. Drbohlavova, J., Adam, V., et al. (2009). Quantum dots—characterization, preparation, and usage in biological systems. *International Journal of Molecular Sciences, 10*(2), 656–673.
76. Ryu, J. H., et al. (2014). Theranostic nanoparticles for future personalized medicine. *Journal of Controlled Release, 190*, 477–484.
77. Mura, S., & Couvreur, P. (2012). Nanotheranostics for personalized medicine. *Advanced Drug Delivery Reviews, 64*(13), 1394–1416.
78. Sharma, S. (2014). Nanotheranostics in evidence based personalized medicine. *Current Drug Targets, 15*(10), 915–930.
79. Kim, T. H., Lee, S., et al. (2013). Nanotheranostics for personalized medicine. *Expert Review of Molecular Diagnostics, 13*(3), 257–269.
80. Gupta, D., et al. (2024). Recent nanotheranostic approaches in cancer research. *Clinical and Experimental Medicine, 24*(1), 8.
81. Bose, R., et al. (2023). Cyclodextrin nanoparticles in targeted cancer theranostics. *Frontiers in Pharmacology, 14*, 1218867.
82. Goyal, A. K., Ramchandani, M., et al. (2023). Recent advancements, challenges, and future prospects in usage of nanoformulation as theranostics in inflammatory diseases. *Journal of Nanotheranostics, 4*(1), 106–126.
83. MacRitchie, N., et al. (2020). Molecular imaging of inflammation-current and emerging technologies for diagnosis and treatment. *Pharmacology & Therapeutics, 211*, 107550.
84. Takada, Y., Ye, X., & Simon, S. (2007). The integrins. *Genome Biology, 8*(5), 215.
85. Moreno, P. R., Purushothaman, K. R., et al. (2004). Plaque neovascularization is increased in ruptured atherosclerotic lesions of human aorta: Implications for plaque vulnerability. *Circulation, 110*(14), 2032–2038.
86. Castronovo, V., & Belotti, D. (1996). TNP-470 (AGM-1470): Mechanisms of action and early clinical development. *European Journal of Cancer, 32A*(14), 2520–2527.
87. Moulton, K. S., Heller, E., et al. (1999). Angiogenesis inhibitors endostatin or TNP-470 reduce intimal neovascularization and plaque growth in apolipoprotein E-deficient mice. *Circulation, 99*(13), 1726–1732.
88. Winter, P. M., Neubauer, A. M., et al. (2006). Endothelial ανβ3 integrin-targeted fumagillin nanoparticles inhibit angiogenesis in atherosclerosis. *Arteriosclerosis, Thrombosis, and Vascular Biology, 26*(9), 2103–2109.
89. Winter, P. M., Caruthers, S. D., et al. (2008). Antiangiogenic synergism of integrin-targeted fumagillin nanoparticles and atorvastatin in atherosclerosis. *JACC. Cardiovascular Imaging, 1*(5), 624–634.
90. Panyam, J., & Labhasetwar, V. (2003). Biodegradable nanoparticles for drug and gene delivery to cells and tissue. *Advanced Drug Delivery Reviews, 55*(3), 329–347.

91. Cyrus, T., Zhang, H., et al. (2008). Intramural delivery of rapamycin with alphavbeta3-targeted paramagnetic nanoparticles inhibits stenosis after balloon injury. *Arteriosclerosis, Thrombosis, and Vascular Biology, 28*(5), 820–826.
92. Buja, L. M., McAllister, H. A., et al. (2007). *Atherosclerosis: Pathologic anatomy and pathogenesis cardiovascular medicine* (Vol. 25, pp. 1581–1591). Springer.
93. Jaffer, F. A., Nahrendorf, M., et al. (2006). Cellular imaging of inflammation in atherosclerosis using magnetofluorescent nanomaterials. *Molecular Imaging, 5*(2), 85–92.
94. McCarthy, J. R., Korngold, E., et al. (2010). A light-activated theranostic nanoagent for targeted macrophage ablation in inflammatory atherosclerosis. *Small, 6*(18), 2041–2049.
95. White, H. D., & Van de Werf, F. J. J. (1998). Thrombolysis for acute myocardial infarction. *Circulation, 97*(16), 1632–1646.
96. Ma, Y. H., Wu, S.-Y., et al. (2009). Magnetically targeted thrombolysis with recombinant tissue plasminogen activator bound to polyacrylic acid-coated nanoparticles. *Biomaterials, 30*(19), 3343–3351.
97. Thygesen, K., et al. (2012). Third universal definition of myocardial infarction. *Journal of the American College of Cardiology, 60*(16), 1581–1598.
98. Joshua, C. J., et al. (2018). *Epidemiology of myocardial infarction* (p. e707102). Intech. https://doi.org/10.5772/intechopen.74768
99. Thygesen, K., et al. (2007). Universal definition of myocardial infarction. *European Heart Journal, 28*(20), 2525.
100. Mohan, J. C., & Narula, J. (2012). New universal definition of myocardial infarction: Global implications, applicability, and need for flexibility. *Global Heart, 7*(4), 377–380.
101. Gomez-Mauricio, R. G., et al. (2013). A preliminary approach to the repair of myocardial infarction using adipose tissue-derived stem cells encapsulated in magnetic resonance-labeled alginate microspheres in a porcine model. *European Journal of Pharmaceutics and Biopharmaceutics, 84*(1), 29–39.
102. Qin, X., et al. (2018). Photoacoustic imaging of embryonic stem cell-derived cardiomyocytes in living hearts with ultrasensitive semiconducting polymer nanoparticles. *Advanced Functional Materials, 28*(1), 1704939.
103. Kang, C., et al. (2017). Fibrin-targeted and H2O2-responsive nanoparticles as a theranostics for thrombosed vessels. *ACS Nano, 11*(6), 6194–6203.
104. Hwang, H., et al. (2014). Peptide-loaded nanoparticles and radionuclide imaging for individualized treatment of myocardial ischemia. *Radiology, 273*(1), 160–167.
105. Alshehri, S., et al. (2021). Progress of cancer nanotechnology as diagnostics, therapeutics, and theranostics nanomedicine: Preclinical promise and translational challenges. *Pharmaceutics, 13*(1), 24.
106. Chen, Q., et al. (2018). Tau-targeted multifunctional nanocomposite for combinational therapy of Alzheimer's disease. *ACS Nano, 12*(2), 1321–1338.
107. Connolly, B. S., & Lang, A. E. (2014). Pharmacological treatment of Parkinson disease: A review. *JAMA, 311*(16), 1670–1683.
108. Du, Z., et al. (2018). Rational design of a "sense and treat" system to target amyloid aggregates related to Alzheimer's disease. *Nano Research, 11*(4), 1987–1997.
109. Dubois, B., et al. (2014). Advancing research diagnostic criteria for Alzheimer's disease: The IWG-2 criteria. *The Lancet Neurology, 13*(6), 614–629.
110. Fan, C. H., et al. (2013). SPIO-conjugated, doxorubicin-loaded microbubbles for concurrent MRI and focused-ultrasound enhanced brain-tumor drug delivery. *Biomaterials, 34*(14), 3706–3715.
111. Hu, K., et al. (2011). Lactoferrin conjugated PEG-PLGA nanoparticles for brain delivery: Preparation, characterization and efficacy in Parkinson's disease. *International Journal of Pharmaceutics, 415*(1–2), 273–283.
112. Johnson, K. K., et al. (2021). Preclinical cancer theranostics—From nanomaterials to clinic: The missing link. *Advanced Functional Materials, 31*(43), 2104199.

113. Kievit, F. M., et al. (2011). Cancer nanotheranostics: Improving imaging and therapy by targeted delivery across biological barriers. *Advanced Materials, 23*(36), H217–H247.
114. Kubrusly, R. C., et al. (2003). l-DOPA supply to the neuro retina activates dopaminergic communication at the early stages of embryonic development. *Journal of Neurochemistry, 86*(1), 45–54.
115. Kumar, D., et al. (2023). Recent advances in noble metal nanoparticles for cancer nanotheranostics. *Journal of Nanotheranostics, 4*(2), 150–170.
116. Lee, Y. T., et al. (2018). Molecular targeted therapy: Treating cancer with specificity. *European Journal of Pharmacology, 834*, 188–196.
117. Li, X., et al. (2012). Biocompatibility and toxicity of nanoparticles and nanotubes. *Journal of Nanomaterials, 2012*(1), 548389.
118. McDonagh, B. H., et al. (2016). L-DOPA-coated manganese oxide nanoparticles as dual MRI contrast agents and drug-delivery vehicles. *Small, 12*(3), 301–306.
119. Myerson, J. W., et al. (2014). Thrombin-inhibiting nanoparticles rapidly constitute versatile and detectable anticlotting surfaces. *Nanotechnology, 25*(39), 395101.
120. Niu, S., et al. (2017). Inhibition by multifunctional magnetic nanoparticles loaded with alpha-Synuclein RNAi plasmid in a Parkinson's disease model. *Theranostics, 7*(2), 344.
121. Pardridge, W. M., et al. (2005). The blood-brain barrier: Bottleneck in brain drug development. *NeuroRx, 2*(1), 3–14.
122. Potschka, H., et al. (2010). *Targeting the brain – Surmounting or bypassing the blood–brain barrier* (Handbook of experimental pharmacology) (Vol. 197, pp. 411–431). Springer.
123. Ruan, S., et al. (2015). Matrix metalloproteinase triggered size-shrinkable gelatin-gold fabricated nanoparticles for tumor microenvironment sensitive penetration and diagnosis of glioma. *Nanoscale, 7*(21), 9487–9496.
124. Sun, C., et al. (2008). In vivo MRI detection of gliomas by chlorotoxin-conjugated superparamagnetic nanoprobes. *Small, 4*(3), 372–379.
125. Sun, Z., Huang, P., Tong, G., Lin, J., Jin, A., Rong, P., Zhu, L., Nie, L., Niu, G., Cao, F., & Chen, X. (2013). VEGF-loaded graphene oxide as theranostics for multi-modality imaging-monitored targeting therapeutic angiogenesis of ischemic muscle. *Nanoscale, 5*(15), 6857–6866.
126. Wen, Z., et al. (2011). Odorranalectin-conjugated nanoparticles: Preparation, brain delivery and pharmacodynamic study on Parkinson's disease following intranasal administration. *Journal of Controlled Release, 151*(2), 131–138.
127. Zhang, C., et al. (2014). Dual-functional nanoparticles targeting amyloid plaques in the brains of Alzheimer's disease mice. *Biomaterials, 35*(1), 456–465.

MIX
Papier aus verantwortungsvollen Quellen
Paper from responsible sources
FSC® C105338

If you have any concerns about our products,
you can contact us on
ProductSafety@springernature.com

In case Publisher is established outside the EU,
the EU authorized representative is:
Springer Nature Customer Service Center GmbH
Europaplatz 3, 69115 Heidelberg, Germany

Printed by Libri Plureos GmbH
in Hamburg, Germany